Natural P

Natural Products

Drug Discovery and Therapeutic Medicine

Edited by

Lixin Zhang, PhD

*Guangzhou Institute of Biomedicine and Health,
Chinese Academy of Sciences,
Guangzhou, China, and
SynerZ Pharmaceuticals Inc.,
Lexington, MA*

Edited by

Arnold L. Demain, PhD

*Charles A. Dana Research Institute (R.I.S.E.),
Drew University, Madison, NJ*

HUMANA PRESS ✸ TOTOWA, NEW JERSEY

© 2005 Humana Press Inc.
999 Riverview Drive, Suite 208
Totowa, NJ 07512

www.humanapress.com

All rights reserved. No part of this book may be reproduced, stored in a retrieval system, or transmitted in any form or by any means, electronic, mechanical, photocopying, microfilming, recording, or otherwise without written permission from the Publisher.

The content and opinions expressed in this book are the sole work of the authors and editors, who have warranted due diligence in the creation and issuance of their work. The publisher, editors, and authors are not responsible for errors or omissions or for any consequences arising from the information or opinions presented in this book and make no warranty, express or implied, with respect to its contents.

Production Editor: Tracy Catanese

Cover design by Patricia F. Cleary

For additional copies, pricing for bulk purchases, and/or information about other Humana titles, contact Humana at the above address or at any of the following numbers: Tel.: 973-256-1699; Fax: 973-256-8341; E-mail: orders@humanapr.com or visit our website: www.humanapress.com

This publication is printed on acid-free paper. ∞
ANSI Z39.48-1984 (American National Standards Institute) Permanence of Paper for Printed Library Materials.

Photocopy Authorization Policy:
Authorization to photocopy items for internal or personal use, or the internal or personal use of specific clients, is granted by Humana Press Inc., provided that the base fee of US $30.00 per copy is paid directly to the Copyright Clearance Center at 222 Rosewood Drive, Danvers, MA 01923. For those organizations that have been granted a photocopy license from the CCC, a separate system of payment has been arranged and is acceptable to Humana Press Inc. The fee code for users of the Transactional Reporting Service is: [1-58829-383-1/05 $30.00].

Printed in the United States of America.   10 9 8 7 6 5 4 3 2 1

eISBN: 1-59259-920-6

Library of Congress Cataloging-in-Publication Data

Natural products : drug discovery and therapeutic medicine / edited by Lixin Zhang, Arnold L. Demain.
   p. ; cm.
 Includes bibliographical references and index.
 ISBN 1-58829-383-1 (alk. paper)
 1. Pharmacognosy.
 [DNLM: 1. Drug Design. 2. Pharmacognosy. 3. Biological Factors--therapeutic use. 4. Biological Products--therapeutic use. 5. Pharmacogenetics. QV 752 N285 2005] I. Zhang, Lixin. II. Demain, A. L. (Arnold L.), 1927-
  RS160.N38 2005
  615'.321--dc22
                                                                                             2004026635

Dedication

I wish to dedicate this book to honor Professor Arnold L. Demain's 60 years experience as a pioneer and a mentor in the field of natural product-based drug discovery. In 1954, he received his PhD from the University of California, Davis and Berkeley, in Microbiology, and joined Merck and Co. as a research microbiologist. By 1965, he had become the Founder and Head of the Department of Fermentation Microbiology at Merck. In 1969, he became a full Professor at MIT. He was elected to the National Academy of Sciences in 1994. Arny is one of the world's leading industrial microbiologists and a pioneer in research on the elucidation and regulation of the biosynthetic pathways leading to penicillins and cephalosporins. He has led the way to the development of the β-lactam industry. His current interests are in the area of industrial microbiology and biotechnology, including industrial fermentation, antibiotics, enzymes, secondary metabolism, biofuels, and bioconversions. During his tenure, Arny trained a group of visiting scholars, postdocs, and students from all over the world, which is now internationally renowned as "Arny's Army." Approximately every 2 years, there is a unique scientific symposium, bringing together key academic and industrial professionals in industrial microbiology and biotechnology, called "A Celebration of Arny's Army & Friends." Continuing the success of the four previous meetings (in 1995 in Cambridge, Massachusetts; in 1997 in Nara, Japan; in 1999 in Gent, Belgium; and in 2001 in Merida, Mexico), the fifth symposium will be held in Shanghai, China on June 27–29, 2005.

Arny is a tireless advocate who would use every possible opportunity to promote natural product-based drug discovery. His vision, inspiration, and leadership contributed significantly to the soon-to-come renaissance of natural products. As we reflect on the history, it is abundantly clear that we benefit from his wisdom to this day.

Lixin Zhang, PhD

Acknowledgments

We thank all the contributors for their support of this project. In addition, we thank Production Editor Tracy Catanese from Humana Press, Inc. for her assistance, guidance, many helpful discussions, as well as her encouragement to finish this book on time. Special thanks to Professors Marcia S. Osburne, Guangyi Wang, Richard Roberts, John Collier, John M. Barberich, and G. Alexander Fleming for their help in editing the book and supporting the work. We are indebted to our wives Jun Kuai and Jody and Lixin's children Peijin and Powell, as well as Lixin's parents-in-law Jingyuan Kuai and Lanying Yu for their encouragement and moral support. Finally, we would like to express our gratitude and appreciation to the staff at Humana Press for their fine work in turning the manuscript into a finished book.

Lixin Zhang, PhD
Arnold L. Demain, PhD

Preface

It seems appropriate to emphasize the topic of natural products at a time when new compounds are desperately needed to combat the current problems of antibiotic resistance, emergence of new diseases, continued presence of old, unconquered diseases, and the toxicity of certain present-day medical products. Despite such needs, today's output from the pharmaceutical industry has decreased markedly as a result of mega-mergers among the large pharmaceutical companies, and the downgrading of natural-product discovery efforts in favor of high throughput screening of synthetic compounds made by combinatorial chemistry. The latter may appear surprising because at least half of the antibiotics and antitumor agents approved by the FDA have been natural products, derivatives of natural products, or synthetic compounds inspired by natural product chemistry. However, it is a matter of economics. The extremely high costs to the large companies of purchasing or developing genomics, proteomics, and bioinformatics have left little funding available for the more tedious screening of natural products. Even so, there is some hope. The continuing success of biopharmaceutical products from the biotechnology industry points to the ever-increasing success of natural compounds, albeit that of large molecules. Some of these smaller companies are directing part of their efforts toward small-molecule natural-product screening. A few are emphasizing biodiversity by either harnessing environmental DNA in the metagenomic effort or discovering means of growing the uncultured microbes of the past and learning how to induce secondary metabolism in these organisms. Other companies are emphasizing combinatorial biosynthesis to yield new derivatives or DNA shuffling to rapidly increase the levels of production. Future success is not a matter of the old vs the new; it is dependent on learning how to apply the exciting methodologies of genomics, proteomics, combinatorial chemistry, DNA shuffling, combinatorial biosynthesis, biodiversity, bioinformatics, and high-throughput screening to rapidly evaluate the activities in extracts as well as purified components derived from microbes, plants, and marine organisms.

There have been concomitant advances and an explosion of information in the field of natural products and it is therefore timely to review both basic and applied aspects. *Natural Products: Drug Discovery and Therapeutic Medicine* addresses historical aspects of natural products and the integration of approaches to their discovery, microbial diversity, specific groups of products (Chinese herbal drugs, antitumor drugs from microbes and plants, terpenoids, and arsenic compounds), specific sources (the sea, rainforest endophytes, and Ecuadorian biodiversity), and methodology (high-performance liquid chromatography profiling, combinatorial biosynthesis, genomics, bioinformatics, and strain improvement by modern genetic manipulations). We consider past successes, the excitement of the present, and our thoughts on the future. We hope that this book will inspire industrial and academic researchers, practitioners, and developers, as well as administrators, to look again at Nature for the future gifts that will solve unmet medical needs and make the world a safer place in which to live.

Lixin Zhang, PhD
Arnold L. Demain, PhD

Contents

DEDICATION V
ACKNOWLEDGMENT VII
PREFACE IX
CONTRIBUTORS XIII

PART I FUNDAMENTAL ISSUES RELATED TO NATURAL PRODUCT-BASED DRUG DELIVERY

1 Natural Products and Drug Discovery
   Arnold L. Demain and Lixin Zhang 3

PART II STRATEGIES

2 Integrated Approaches for Discovering Novel Drugs
   From Microbial Natural Products
   Lixin Zhang 33

3 Automated Analyses of HPLC Profiles of Microbial Extracts:
   A New Tool for Drug Discovery Screening
   José R. Tormo and Juan B. García 57

4 Manipulating Microbial Metabolites for Drug Discovery
   and Production
   C. Richard Hutchinson 77

5 Improving Drug Discovery From Microorganisms
   Chris M. Farnet and Emmanuel Zazopoulos 95

6 Developments in Strain Improvement Technology:
   *Evolutionary Engineering of Industrial Microorganisms
   Through Gene, Pathway, and Genome Shuffling*
   Stephen B. del Cardayré 107

PART III SPECIFIC GROUPS OF DRUGS

7 The Discovery of Anticancer Drugs From Natural Sources
   David J. Newman and Gordon M. Cragg 129

8 Case Studies in Natural-Product Optimization: *Novel Antitumor Agents
   Derived From* Taxus brevifolia *and* Catharanthus roseus
   Jian Hong and Shu-Hui Chen 169

9 Terpenoids As Therapeutic Drugs and Pharmaceutical Agents
   Guangyi Wang, Weiping Tang, and Robert R. Bidigare 197

10 Challenges and Opportunities in the Chinese Herbal Drug Industry
   Wei Jia and Lixin Zhang 229

11   Arsenic Trioxide and Leukemia: *From Bedside to Bench*
     Guo-Qiang Chen, Qiong Wang, Hua Yan, and Zhu Chen 251

PART IV   MICROBIAL DIVERSITY

12   New Methods to Access Microbial Diversity
     for Small Molecule Discovery
     Karsten Zengler, Ashish Paradkar, and Martin Keller 275

13   Accessing the Genomes of Uncultivated Microbes
     for Novel Natural Products
     **Asuncion Martinez, Joern Hopke, Ian A. MacNeil,
     and Marcia S. Osburne** 295

PART V   SPECIFIC SOURCES

14   New Natural-Product Diversity From Marine Actinomycetes
     Paul R. Jensen and William Fenical 315

15   Novel Natural Products From Rainforest Endophytes
     Gary Strobel, Bryn Daisy, and Uvidelio Castillo 329

16   Biological, Economic, Ecological, and Legal Aspects
     of Harvesting Traditional Medicine in Ecuador
     Alexandra Guevara-Aguirre and Ximena Chiriboga 353

INDEX 371

Contributors

ROBERT R. BIDIGARE, PhD • *Department of Oceanography, School of Ocean and Earth Sciences and Technology, University of Hawaii at Manoa, Honolulu, HI*

GUO-QIANG CHEN, PhD • *Dept. of Pathophysiology, Shanghai Second Medical University, and Health Science Center, Shanghai Institutes for Biological Sciences, Chinese Academy of Sciences, Shanghai, People's Republic of China*

SHU-HUI CHEN, PhD • *Discovery Chemistry Research & Technology, Lilly Research Laboratories, Eli Lilly and Company, Indianapolis, IN*

ZHU CHEN, PhD • *Shanghai Institute of Hematology, Rui-Jin Hospital, Shanghai Second Medical University, Shanghai, People's Republic of China*

XIMENA CHIRIBOGA, PhD • *Fundación GEA, Proyectos Ambientales; and Extracta Ecuador S.A., Quito, Ecuador*

STEPHEN B. DEL CARDAYRÉ, PhD • *Codexis, Redwood City, CA*

UVIDELIO CASTILLO, PhD • *Department of Plant Sciences, Montana State University, Bozeman, MT*

GORDON M. CRAGG, DPhil • *Natural Products Branch, Developmental Therapeutics Program, Division of Cancer Treatment and Diagnosis, NCI-Frederick, MD*

BRYN DAISY, BS • *Department of Plant Sciences, Montana State University, Bozeman, MT*

ARNOLD L. DEMAIN, PhD • *Charles A. Dana Research Institute (R.I.S.E.), Drew University, Madison, NJ*

CHRIS M. FARNET, PhD • *Ecopia BioSciences Inc., Montreal, Quebec, Canada*

WILLIAM FENICAL, PhD • *Center for Marine Biotechnology and Biomedicine, Scripps Institution of Oceanography, University of California, San Diego, CA*

JUAN B. GARCÍA • *Centro de Investigación Básica (CIBE), Merck Research Laboratories (MRL), Merck, Sharp & Dohme de España S.A., Madrid, Spain*

ALEXANDRA GUEVARA-AGUIRRE • *Fundación GEA, Proyectos Ambientales; and Extracta Ecuador S.A., Quito, Ecuador*

JIAN HONG, PhD • *Discovery Chemistry Research & Technology, Lilly Research Laboratories, Eli Lilly and Company, Indianapolis, IN*

JOERN HOPKE, PhD • *Cambridge Genomics Center, Aventis Pharmaceuticals Inc., Cambridge, MA*

C. RICHARD HUTCHINSON, PhD • *Kosan Biosciences, Hayward, CA*

PAUL R. JENSEN, MS • *Center for Marine Biotechnology and Biomedicine, Scripps Institution of Oceanography, University of California, San Diego, CA*

WEI JIA, PhD • *School of Pharmacy, Shanghai Jiao Tong University, Shanghai, People's Republic of China*

MARTIN KELLER, PhD • *Diversa Corporation, San Diego, CA*

IAN A. MACNEIL, PhD • *ActivBiotics, Inc., Lexington, MA*

ASUNCION MARTINEZ, PhD • *Massachusetts Institute of Technology, Cambridge, MA*

DAVID J. NEWMAN, DPhil • *Natural Products Branch, Developmental Therapeutics Program, Division of Cancer Treatment and Diagnosis, NCI-Frederick, MD*

MARCIA S. OSBURNE, PhD • *ActivBiotics, Inc., Lexington, MA*

ASHISH PARADKAR, PhD • *Diversa Corporation, San Diego, CA*

GARY STROBEL, PhD • *Department of Plant Sciences, Montana State University, Bozeman, MT*

WEIPING TANG, PhD • *Department of Chemistry, Stanford University, Stanford, CA*

JOSÉ R. TORMO, PhD • *Centro de Investigación Básica (CIBE), Merck Research Laboratories (MRL), Merck, Sharp & Dohme de España S.A., Madrid, Spain*

GUANGYI WANG, PhD • *Hawaii Natural Energy Institute, Department of Oceanography, School of Ocean and Earth Sciences and Technology, University of Hawaii at Manoa, Honolulu, HI*

QIONG WANG, PhD • *Shanghai Institute of Hematology, Rui-Jin Hospital, Shanghai Second Medical University, Shanghai, People's Republic of China*

HUA YAN, MD, PhD • *Shanghai Institute of Hematology, Rui-Jin Hospital, Shanghai Second Medical University, Shanghai, People's Republic of China*

EMMANUEL ZAZOPOULOS, PhD • *Ecopia BioSciences, Inc., Montreal, Quebec, Canada*

KARSTEN ZENGLER, PhD • *Diversa Corporation, San Diego, CA*

LIXIN ZHANG, PhD • *Guangzhou Institute of Biomedicine and Health, Chinese Academy of Sciences, Guangzhou, China, and SynerZ Pharmaceuticals Inc., Lexington, MA*

PART I
FUNDAMENTAL ISSUES RELATED TO NATURAL PRODUCT-BASED DRUG DELIVERY

1
Natural Products and Drug Discovery

Arnold L. Demain and Lixin Zhang

Summary

For more than 50 yr, natural products have served us well in combating infectious bacteria and fungi. During the 20th century, microbial and plant secondary metabolites helped to double our life span, reduced pain and suffering, and revolutionized medicine. The increased development of resistance to older antibacterial, antifungal, and antitumor drugs has been challenged by (1) newly discovered antibiotics (e.g., candins, epothilones); (2) new semisynthetic versions of old antibiotics (e.g., ketolides, glycylcyclines); (3) older underutilized antibiotics (e.g., teicoplainin); and (4) new derivatives of previously undeveloped narrow-spectrum antibiotics (e.g., streptogramins). In addition, many antibiotics are used commercially, or are potentially useful in medicine for purposes other than their antimicrobial action. They are used as antitumor agents, enzyme inhibitors including powerful hypocholesterolemic agents, immunosuppressive agents, antimigraine agents, and so on. A number of these products were first discovered as antibiotics that failed in their development as such, or as mycotoxins.

It is unfortunate that the pharmaceutical industry has downgraded natural products just at the time that new assays are available and major improvements have been made in detection, characterization, and purification of small molecules. With the advent of combinatorial biosynthesis, thousands of new des4œatives can now be made by a biological technique complementary to combinatorial chemistry. Furthermore, only a minor proportion of bacteria and fungi, i.e., 0.1–5%, have thus far been examined for secondary metabolite production. New methods are being developed to cultivate the so-called unculturable microbes from the soil and the sea. High-throughput screening (HTS) of combinatorial chemicals has not provided the numbers of high-quality leads that were anticipated. It has virtually eliminated the most unique source of chemical diversity, i.e., natural products, from the playing field, in favor of combinatorial chemistry. Combinatorial chemistry mainly yields minor modifications of present-day drugs and absolutely requires new scaffolds on which to build. Although comparative genomics is capable of disclosing new targets for drugs, the number of targets is so large that it requires tremendous investments of time and money to set up all the screens necessary to exploit this resource. This can be handled only by HTS methodology, which demands libraries of millions of chemical entities. Although such targets would be excellent for screening natural products, the industry has failed to exploit this unique opportunity and has opted to save funds by eliminating natural-product departments or decreasing their relevance in the hunt for new drugs. It is clear that the future success of the pharmaceutical industry depends on the combining of complementary technologies such as natural product discovery, HTS, integrative and systems biology, combinatorial biosynthesis, and combinatorial chemistry.

Key Words: Antibiotics; antitumor; immunosuppressants; hypocholesterolemics; enzyme inhibitors; drug discovery; natural products; combinatorial biosynthesis.

1. Introduction

Natural products have been an overwhelming success in our society (**Fig. 1**). They have reduced pain and suffering, and revolutionized medicine by facilitating the transplantation of organs. Natural products are the most important anticancer and anti-infective agents. More than 60% of approved and pre-new drug application (NDA) candidates are either natural products or related to them, not including biologicals such as vaccines and monoclonal antibodies *(1)*.

Many natural products have reached the market without chemical modification, a testimony to the remarkable ability of microorganisms to produce small, drug-like molecules. Indeed, the potential to commercialize a compound without chemical modification distinguishes natural products from all other sources of chemical diversity and fuels efforts to discover new compounds. Nature apparently optimizes certain compounds through many centuries of evolution. In these cases, production of the product directly by microbial fermentation is much more economical than using synthetic chemistry, e.g., steroids, β-lactams, erythromycin. In other cases, the natural molecule was not used itself but served as a lead molecule for manipulation by chemical or genetic means, e.g., cephalosporins, rifampicin. In these instances, the natural product presented important structural motifs and pharmacophores, which were then optimized via "semi-synthesis" to yield drugs with improved properties.

Secondary metabolism has evolved in nature in response to needs and challenges of the natural environment. Nature has been continually carrying out its own version of combinatorial chemistry *(2)* for the over 3 billion years during which bacteria have inhabited the earth *(3)*. During that time, there has been an evolutionary process going on in which producers of secondary metabolites evolved according to their local environments. If the metabolites were useful to the organism, the biosynthetic genes were retained, and genetic modifications further improved the process. Combinatorial chemistry practiced by nature is much more sophisticated than that in the laboratory, yielding exotic structures rich in stereochemistry, concatenated rings, and reactive functional groups *(2)*. As a result, an amazing variety and number of products have been found in nature. The total number of natural products produced by plants has been estimated to be over 500,000 *(4)*. One-hundred sixty-thousand natural products have been identified *(5)*, a value growing by 10,000 per year *(6)*. About 100,000 secondary metabolites of molecular weight less than 2500 have been characterized, half from microbes and the other half from plants *(7–9)*.

It is not generally appreciated that a number of synthetic products of wide medical use have a natural origin from microbial, plant, and even animal systems. The predecessor of aspirin has been known since the fifth century BC, at which time it was extracted from willow tree bark by Hippocrates. It probably was used even earlier in Egypt and Babylonia for fever, pain, and childbirth *(10)*. Salicylic acid derivatives have been found in plants such as white willow, wintergreen, and meadowsweet. Synthetic salicylates were produced on a large scale in 1874 by the Bayer Company in Germany. In 1897, Arthur Eichengrun at Bayer discovered that its acetyl derivative was able to reduce its acidity, bad taste, and stomach irritation *(11)*; thus was born aspirin, of which 50 billion tablets are consumed each year. The drugs Acyclovir (ZoviraxR) used against herpes virus and Cytarabine (Cytostar®) for non-Hodgkin's lymphoma were originally isolated from a sponge *(12)*. Drugs inhibiting human immu-

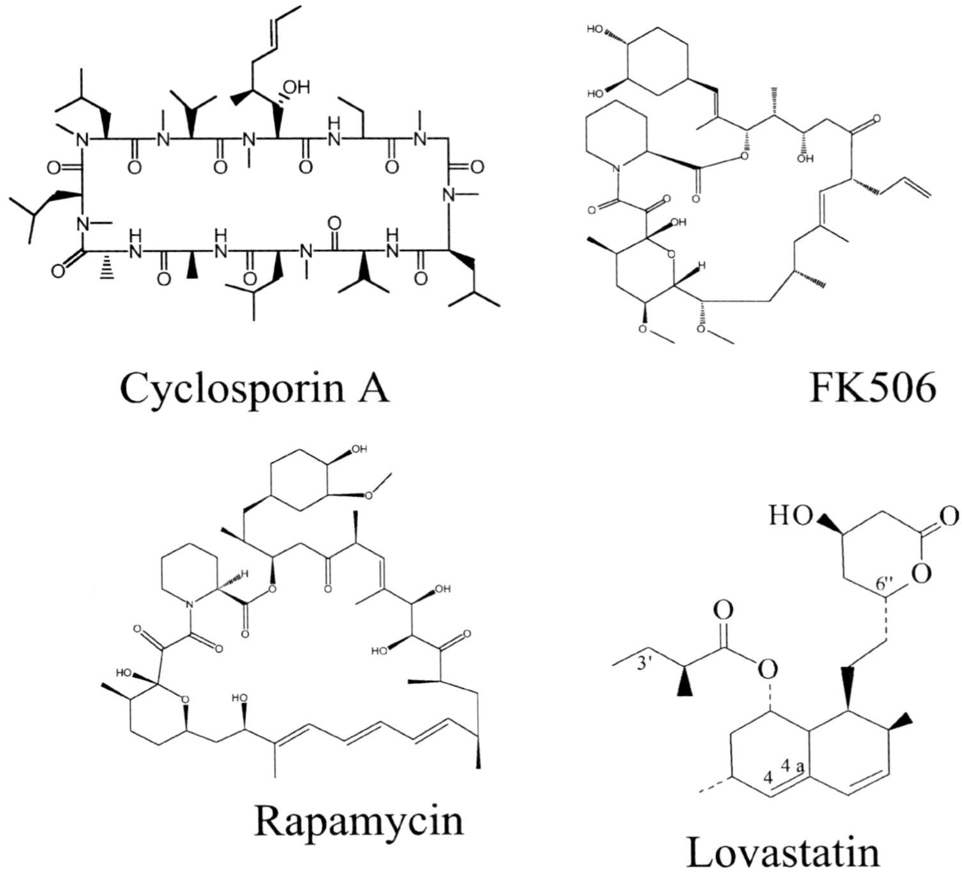

Cyclosporin A

FK506

Rapamycin

Lovastatin

Fig. 1. Structures of some important natural products not used as antibiotics.

nodeficiency virus (HIV) reverse-transcriptase and protease were derived from natural product leads screened at the National Cancer Institute *(13)*. Angiotensin-converting enzyme (ACE) inhibitors, widely used for hypertension and congestive heart failure, are chemicals based on peptides isolated from snake venom *(14,15)*.

In the last decade, some large pharmaceutical companies, emphasizing combinatorial chemistry, left natural products and attempted to fill the void with large numbers of synthetic molecules. Unfortunately, the chemistry employed did not create sufficiently diverse or pharmacologically active molecules. Fortunately, some small biotechnology companies have revitalized the interest in natural products. Approaches such as diversity-oriented synthesis, which mimics the structures of natural products, are emerging for drug discovery *(15a)*.

Highly diverse and selective synthetic compounds resulting from these efforts could be useful for chemical genomics and chemical genetics, to uncover disease-relevant protein targets and to understand critical biological pathways *(15b)*. Such information could push biology forward toward an understanding of the initiation and progression of diseases, and shed light on new therapeutic intervention.

More natural product research is needed due to: unmet medical needs; remarkable diversity of structures and activities; utility as biochemical probes; novel and sensitive assay methods; improvements in isolation, purification, and characterization; and new production methods *(16)*.

The enormous diversity of microorganisms is a factor that must be kept in mind for future drug development. Bacteria have existed on earth for over 3-billion years, and eukaryotes have been around for 1 billion years. Because 95–99.9% of organisms existing in nature have not yet been cultured, only a minor proportion of bacteria and fungi have thus far been examined for secondary metabolite production. It has been estimated that 1 g of soil contains 1000 to 10,000 species of undiscovered prokaryotes *(17)*.

Estimates of the number of described fungal species vary from about 65,000 to 250,000, but as many as 10 million might exist in nature *(18)*. Their total weight is thought to be higher than that of humans. Of the fungal species that have been described, only about 16% have been cultured. The use of fungal ecology in the search for new drugs is extremely important. The estimated number of fungal species is more than five times the predicted number of plant species and fifty times the estimated number of bacterial species. Some previously unrecognized and uncultivated microbes can be isolated from the environment by encapsulating cells in gel microdroplets under low nutrient flux conditions and detecting microcolonies by flow cytometry *(19)*. In addition to new ways of culturing microbes, accessing the diversity of "environmental DNA" (also called metagenomic DNA) is an exciting area of research *(20)*.

The concept that microbial strains must be isolated from different geographical and climatic locations around the world in order to insure diversity in collections still gathers support *(21,22)*.

2. Antibiotics for Human Therapy

The selective action exerted on pathogenic bacteria and fungi by microbial secondary metabolites ushered in the antibiotic era. For more than 50 yr, we have benefited from this remarkable property of wonder drugs such as penicillins, cephalosporins, tetracyclines, aminoglycosides, chloramphenicol, and macrolides. They have been crucial in the increase in average life expectancy in the United States from 47 yr in 1900 to 74 yr for males and 80 yr for women in 2000 *(2,23)*. Antibiotics have been virtually the only drugs utilized for chemotherapy against pathogenic microorganisms. They are defined as low-molecular-weight organic natural products (secondary metabolites, or idiolites) made by microorganisms, which are active at a low concentration against other microorganisms. Of the 12,000 antibiotics known in 1995, 55% were produced by filamentous bacteria (actinomycetes) of the genus *Streptomyces*, 11% from other actinomycetes, 12% from nonfilamentous bacteria, and 22% from filamentous fungi *(8,24)*. New bioactive products from microbes have been discovered at an amazing pace: 200 to 300 per year in the late 1970s and increasing to 500 per year by 1997. However, recently the number has dropped off as a result of the misguided loss of interest by some large pharmaceutical companies in the discovery of new antibiotics and even in natural products (discussed later).

More than 350 agents have reached the world market as antimicrobials (anti-infectives [defined as antibiotics, both natural and semi-synthetic] plus strictly synthetic chemicals) *(25)*. The antibiotics include cephalosporins (45%), penicillins (15%), tetracyclines (6%), macrolides (5%), aminoglycosides, ansamycins, glycopeptides, and polyenes *(24)*. The synthetics include quinolones (11%) and the azoles. Of the 25 top-selling drugs in 1997, 42% were natural products or derived from natural products *(22)*; of these, antibiotics contributed 67% of sales.

The worldwide market for antibiotics in 1996 was $24 billion *(26)* and today is about $35 billion. Antimicrobials, including antibiotics, synthetics, and antiviral agents, had sales of $55 billion in 2000. The market for cephalosporins was $9.9 billion *(27)*; sales of penicillins amounted to $8.2 billion and that of other β-lactams was $1.5 billion, making a total of $19.6 billion for β-lactam antibiotics. Markets for other groups were $6.4 billion for quinolones (synthetic antibacterials), $5.2 billion for macrolides (including $3.5 billion for erythromycins) *(28)*, $4.2 billion for antifungals (including $2 billion for the synthetic azoles) *(29)* and antiparasitics, $1.8 billion for aminoglycosides, and $1.4 billion for tetracyclines. The antiviral market was $10.2 billion (vaccines excluded). All other antimicrobials had a market of $ 6.1 billion. It is expected that the antimicrobial market will continue to grow as a result of (1) a worldwide aging population; (2) increasing numbers of immunocompromised patients, mainly infected with HIV, who often require longer courses of anti-infective treatment; and (3) increasing microbial resistance worldwide.

Antimicrobials with markets over $1 billion include Augmentin ($2 billion); the quinolones ciprofloxacin (Cipro; $1.8 billion) and levofloxacin/ofloxacin (Levaquin, Floxin; $1.1 billion); the semi-synthetic erythromycins azithromycin (Zithromax; $1.5 billion) and clarithromycin (Biaxin; $1.2 billion); and the semi-synthetic cephalosporin ceftriazone (Rocephin; $1.1 billion) *(30)*. Augmentin is a combination of a semi-synthetic penicillin and an inhibitor of penicillinase (clavulanic acid). Combined sales of the glycopeptides vancomycin and teicoplanin were $1 billion per year *(31)*.

Unfortunately, in 1969, US Surgeon General William H. Stewart stated to Congress: "The time has come to close the book on infectious disease" *(30)*. Companies began to exit from the antibiotic area. There was great difficulty and a high cost of isolating novel antibiotic structures and agents with new modes of action because the chance of finding useful antibiotics from microbes was very low. The following are experiences of workers in the field:

1. Only one in 10,000 to 150,000 compounds made it into medical practice *(32–34)*.
2. Only 3 usable antibiotics were isolated from a 10-yr screen of 400,000 cultures of microorganisms *(32)*.
3. Of 5000 compounds evaluated, 5 entered human clinical trials and of these, only 1 was approved by the US Food and Drug Administration (FDA) *(35)*.
4. Of 5000 to 10,000 compounds screened, 250 leads entered preclinical testing, 5 drug candidates entered phase I, 4 passed into phase IIa, 1.5 passed phase IIb, 1.2 passed phase III, and 1 drug was approved by FDA *(36)*.

Similar experiences were encountered with products from plants *(37)*. Although chemists had been successfully improving natural antibiotics by the technique of semi-synthesis for many years, new screening techniques were sorely needed in 1970 to isolate new bioactive molecules from nature. A few companies downgraded their efforts in natural product discovery in favor of large investments in recombinant DNA technology. Indeed, the number of anti-infective investigational new drugs (INDs) declined by 50% from the 1960s to the late 1980s *(38)*. Many felt that the golden era of antibiotic discovery was over, but this was far from the truth. Owing mainly to the development of novel target-directed screening procedures by some forward-thinking companies, important new antibiotics appeared on the scene and became commercial successes in the 1970s and 1980s. These included cephamycins (e.g., cefoxitin) *(39)*, fosfomycin *(40)*, carbapenems (e.g., thienamycin) *(41)*, monobactams (e.g., aztreonam), glycopeptides (e.g., vancomycin, teicoplanin), aminoglycosides (e.g., amikacin, sisomicin), as well as semisynthetic versions of cephalosporins and macrolides.

Another factor limiting new antibiotic discovery is the shift by major pharmaceutical companies to pursue larger drug markets such as depression, heartburn, and erectile dysfunction. This shift to lifestyle problems and chronic complaints is of major concern to infectious-disease physicians. The hard fact is that there is not much economic incentive to create new antibiotics. The market potential for a new antibiotic is estimated to be $200 million to $400 million in sales per year. In contrast, drugs to treat chronic conditions are often used by patients daily for the rest of their lives and are much more likely to reach blockbuster status (over $1 billion/yr). The FDA approved only two new antibiotics in 2003, and none was approved in 2002. Such numbers are only a fraction of the numbers of antibiotics approved in the 1980s and early 1990s. From 1983 to 1992, thirty new antibiotics won FDA approval. Over the next 10 years, 17 were approved. Serious medical consequences could result from this situation. Many types of bacteria are developing resistance to existing classes of antibiotics (discussed later). Nosocomial infections (infections acquired in hospitals) pose a particular threat to ill patients. Governments must become involved in reversing the trend away from antibiotic research and development.

Microbiologists have known for years that technology had not won the war against infectious microorganisms due to resistance development in pathogenic microbes. Indeed, technology will never win this war, and we must be satisfied to merely stay one step ahead of the pathogens for a long time to come *(42)*. Thus, the search for new drugs must not be stopped. New antibiotics are continually needed because of the following:

1. Resistant pathogens are developing, e.g., enterococci resistant to all antibiotics *(43)*.
2. New diseases are evolving, e.g., acquired immunodeficiency syndrome (AIDS), Hanta virus, Ebola virus, *Cryptospiridium*, Legionnaire's disease, Lyme disease, *Escherichia coli* 0157:H7. The World Health Organization concluded that at least thirty new diseases emerged in the 1980s and 1990s *(44)*.
3. Naturally resistant bacteria exist, e.g., *Pseudomonas aeruginosa*, causing fatal wound infections, burn infections, and chronic and fatal infections of lungs in cystic fibrosis patients; also *Stenotrophomonas maltophilia, Enterococcus faecium, Burkholderia cepacia*, and *Acinetobacter baumanni (45)*. Furthermore, tuberculosis had never been defeated.
4. Some of the compounds in use are relatively toxic *(24)*.

The resistant bacteria that have developed over the years are generally uninhibited by most commercial antibiotics *(46)*. Enterococci resistant to all antibiotics have arisen *(43)*. Other organisms exist that are not normally virulent but do infect immunocompromised patients *(47)*. In recent years, there has been great concern about resistance development among Gram-positive pathogens, the so-called methicillin-resistant bacteria. Clinical isolates of penicillin-resistant *Streptococcus pneumoniae*, the most common cause of bacterial pneumonia, increased 60-fold in the United States from 1987 to 1992 *(48)*. Methicillin-resistant *Staphylococcus aureus* (MRSA) infections increased to an alarming extent *(49)*. Recently, the glycopeptide vancomycin has been the molecule of choice to treat infections caused by such organisms; however, vancomycin resistance is developing, especially in the case of nosocomial *Enterococcus* infections. Fortunately, some vancomycin-resistant enterococci (VRE) are treatable by the related glycopeptide antibiotic teicoplanin. Of the three different resistance mechanisms in different strains of VRE, two of these, Van B and Van D, are induced by vancomycin but not by teicoplanin *(50,51)*; Van A is induced by either antibiotic. Teicoplanin also has fewer side effects than vancomycin, and a longer half-life in the body *(52,53)*. A major problem today is tuberculosis, which is infecting 2 billion people worldwide. Each year, 9 million new cases are diagnosed and 2.6 million people die. Resistance is developing to the combination treatment of isoniaizid and rifamycin *(53a)*.

Exploitation of old and underutilized antibiotics is occurring via semi-synthesis of drugs active against resistant bacteria. One group of useful narrow-spectrum compounds are the streptogramins, which are synergistic pairs of antibiotics made by single microbial strains. The pairs are constituted by a (group A) polyunsaturated macrolactone containing an unusual oxazole ring and a dienylamide fragment, and a (group B) cyclic hexadepsipeptide possessing a 3-hydroxypicolinoyl exocyclic fragment. Such streptogramins include virginiamycin and pristinamycin *(54)*. Pristinamycin, made by *Streptomyces pristinaespiralis*, is a mixture of a cyclodepsipeptide (pristinamycin I) and a polyunsaturated macrolactone (pristinamycin II). Although the natural streptogramins are poorly water-soluble and cannot be used intravenously, new derivatives have been made by semi-synthesis and mutational biosynthesis. Synercid (RP59500) is a mixture of two water-soluble semisynthetic streptogramins, quinupristin (RP57669) and dalfopristin (RP54476), developed by Rhone Poulenc-Rorer (now Aventis) *(55)*, and approved in 1999 for resistant bacterial infections *(56)*. Synercid is useful against β-lactam-resistant *S. pneumoniae (57)*. The two Synercid components synergistically (100-fold) inhibit protein synthesis and are active against VRE and MRSA *(58,59)*. Synergistic action of the streptogramins is due to the fact that the B component blocks binding of aminoacyl-tRNA complexes to the ribosome while the A component inhibits peptide bond formation and distorts the ribosome, promoting the binding of the B component *(60)*.

Semisynthetic tetracyclines, e.g., glycylcyclines, are being developed by Wyeth for use against tetracycline-resistant bacteria *(61)*. The 9-*t*-butylglycylamido derivative of minocycline called GAR-936 is a novel glycylcycline *(62)*. It is active in vitro and in vivo (with mice) against resistant Gram-positive, Gram-negative, and anaerobic bacteria possessing the ribosomal protection resistance mechanism or the active efflux mechanism.

New vancomycin derivatives have also been made, which are active against vancomycin-resistant Gram-negative bacteria *(63)*. These have modified carbohydrate moieties and are inhibitors of the transglycosidation reactions of peptidoglycan biosynthesis without binding to D-ala-D-ala, the traditional mode of action of vancomycin. They not only act on resistant bacteria but are more active than vancomycin on sensitive bacteria

(64). They inhibit the earlier transglycosylation step of cell-wall synthesis rather than the main vancomycin target, transpeptidation. Even the sugars themselves have antibacterial activity. An exciting molecule in the new antibiotic arena is the lipoglycodepsipeptide ramoplanin, produced by *Actinoplanes* ATCC 33076 *(65)*. Ramoplanin inhibits cell-wall synthesis in Gram-positive bacteria by a new mechanism: binding to lipid I and II intermediates at a site different from vancomycin's target N-acyl-D-ala-D-ala dipeptide. By binding to the lipid intermediates, these substrates are physically prevented from being acted upon by the late peptidoglycan enzymes MurG and the transglycosidases *(66,67)*. Ramoplanin is 2- to 10-fold more active than vancomycin and is active against MRSA, VRE, and pathogens resistant to ampicillin and erythromycin *(68)*.

Semisynthetic erythromycins have been very successful *(69)*. Modified macrolides include clarithromycin (Biaxin of Taisho), azithromycin (Zithromax of Pliva), and the ketolide telithromycin (Ketek of Aventis). Whereas the first two showed improved acid stability and bioavailability over erythromycin A, they showed no improvement against resistant strains *(70)*. On the other hand, the ketolides (third-generation semisynthetic erythromycins, 6-methyl-3-oxoerythromycin derivatives, i.e., polyketides containing a 14-membered macrolide ring and a C-3 keto group in place of the C-3 cladinose in erythromycin A) act against macrolide-sensitive and macrolide-resistant bacteria *(71–73)*. All of the above semisynthetic erythromycins are effective agents for upper respiratory tract infections and can be given parenterally or orally. The ketolides include telithromycin (Ketek of Aventis; formerly called HMR-3647) *(74,75)*, which has been approved, and ABT-773 (Abbott), which is in Phase II clinical development. ABT-773 has broad-spectrum activity, including anaerobes and intracellular pathogens. It binds to its ribosomal target with 10- to 100-fold greater affinity than erythromycin A, has increased uptake and/or reduced efflux, and is bactericidal. Ketolides are also active against penicillin- and erythromycin-resistant *S. pneumoniae* and many other bacteria, such as *Hemophilus influenzae*, group A streptococci, *Legionella* spp., *Chlamydia* spp., and *Mycoplasma pneumoniae (76)*. They are produced semi-synthetically from erythromycin *(77)*. Telithromycin is bacteriostatic, active orally, and of great importance for community-acquired respiratory infections. Of great interest is its low ability to select for resistance mutations and to induce cross-resistance. It does not induce MLS_B resistance, a problem with other macrolides.

A number of previously discovered compounds are now available for development that were not previously exploited because of their narrow antibacterial spectrum, which was restricted to Gram-positive bacteria. At that time (in the 1970s and 1980s), breadth of spectrum was the commercial goal, but today, an important aim is to inhibit resistant Gram-positive pathogens. An example is the lipopeptide daptomycin, produced by *Streptomyces roseosporus*, which acts on Gram-positive bacteria, including VRE, MRSA, and penicillin-resistant *S. pneumoniae (78)*. It kills by disrupting plasma membrane function without penetrating into the cytoplasm. Daptomycin was discovered by Eli Lilly and Co. in the early 1980s and licensed to Cubist Pharmaceuticals, Inc., in 1997. It was approved in 2003.

Efforts are proceeding to screen for or design compounds that interfere with resistance mechanisms *(79)*. Such a strategy has long been successful with clavulanic acid, a natural β-lactamase inhibitor. The plant natural product 5'-methoxyhydnocarpin has been found to inhibit the NorA multidrug-resistance pump and to potentiate norfloxacin activity *(80)*. A successful inhibitor of the multidrug-resistance pump could be of tremendous help in combating resistant bacteria.

About 200 species of fungi are pathogenic to mammals. Most such infections are self-limiting, but they can be deadly to immunocompromised patients; e.g., systemic fungal infections are responsible for the deaths of 50% of leukemia patients. Fungal infections are a real problem today, having doubled from the 1980s to the 1990s, with bloodstream infections increasing fivefold, with an observed mortality of 55% *(81)*. There is an increasing incidence of candidiasis, cryptococcosis, and aspergillosis, especially in AIDS patients *(82)*. Aspergillosis failure rates exceed 60%. Fungal infections occur often after transplant operations: 5% for kidney, 15–35% for heart and lung, up to 40% for liver transplants, usually (80%) by *Candida* and *Aspergillus* spp. *(83)*. Pulmonary aspergillosis is the main factor involved in deaths of recipients of bone-marrow transplants, and *Pneumocystis carinii* is the number one cause of death in patients with AIDS from Europe and North America *(84)*.

The four classes of antifungal antibiotics in use today are the natural polyenes and the synthetic azoles, allylamines, and fluoropyrimidines *(85)*. The first three classes target ergosterol, the major fungal sterol in the cell membrane. Polyene macrolides selectively bind ergosterol and destabilize fungal membranes, leading to leakage of cell components and subsequent cell death. Amphotericin B is a polyene macrolide that was introduced in 1956 and has been the standard for antifungal therapy, since it was the only effective agent against systemic fungal infection. Although highly active against a wide range of fungi, it must be administered intravenously and is somewhat toxic, producing a range of side effects. The azoles (e.g., fluconazole, itraconazole, and ketoconazole) interfere with the biosynthesis of sterols and other membrane lipids that comprise the fungal cell membrane. They do this by inhibiting cytochrome P450-dependent lanosterol 14-α-demethylase, which is responsible for converting lanosterol to ergosterol. The lack of ergosterol in the cell membrane leads to cell permeability and death. Each of the azoles has a different spectrum of effectiveness and defined limitations. For example, fluconazole is highly effective against *Cryptococcus*, a serious infection common in AIDS patients, but is ineffective against *Aspergillus* and has limited effectiveness against certain *Candida* species. Itraconazole has the broadest range of activity and the fewest side effects among the azoles, but suffers from unpredictable bioavailability, varying between patients, and frequent drug interactions. Ketoconazole is the most effective azole against chronic, indolent forms of endemic fungal infections, but is associated with clinically important toxic effects, including hepatitis. Allylamines (e.g., terbinafine) inhibit squalene epoxidase, another enzyme leading to ergosterol. Fluoropyrimidines (e.g., 5-fluorocytosine) are pyrimidine analogs that are selectively converted by a fungal enzyme to its active nucleoside, which interferes with DNA synthesis in fungi. Their spectrum of activity is fairly limited, and drug resistance develops if they are used alone. For that reason, 5-fluorocytosine is utilized in combination with amphotericin B or fluconazole. Toxicity is frequent, however, which includes mucositis and myelosuppression *(85a)*. Usage of these antifungal compounds is becoming limited especially due to development of resistance to the azoles and toxicity of the polyenes.

The cyclic lipopeptides, known as candins (or echinocandins or pneumocandins), inhibit (1,3)-β-glucan synthase and thus the biosynthesis of the 1,3 glucan layer of the *Candida albicans* cell wall; they are relatively nontoxic. Although the earlier discovered papulocandins failed due to a spectrum restricted to *Candida* and lack of in vivo activity, the parenteral semisynthetic caspofungin of Merck (also known as pneumocandin, L-743,872, MIC 991, or Cancidas) produced by *Zalerion arboricola* inhibits the same enzyme and was approved in 2000 *(86–88)*. It is active against many species of *Candida*, including *C. albicans*, many species of *Aspergillus*, and *Histoplasma*, and can be administered as an aerosol for prophylaxis against *P. carinii*, a major cause of death in HIV patients *(29)*. It is more active and less toxic than amphotericin B. Another semisynthetic candin is micafungin (FK-463), which was recently approved in Japan. Lilly's anidulafungin (V-echinocandin, LY-303366A) produced by *Aspergillus nidulans* var. *echinulatus (89)*, was licensed to Versicor (now Vicuron Pharmaceuticals) and is awaiting FDA approval. In contrast to the currently used azoles, which are fungistatic, candins are fungicidal.

3. Novel Applications of Antibiotics

An extremely important concept for the further development of natural products is that compounds that possess antibiotic activity also possess other activities. Some of these activities had been quietly exploited in the past, and it became clear in the 1980s that such a scope should be expanded. Thus, a broad screening of antibiotically active molecules for antagonistic activity against organisms other than microorganisms, as well as for activities useful for pharmacological or agricultural applications, was proposed by Umezawa *(90,91)* in order to yield new and useful lives for "failed antibiotics" (for a review, see ref. 92). This resulted in the development of a large number of simple in vitro laboratory tests (e.g., enzyme inhibition screens) to detect, isolate, and purify useful compounds. As a result, we entered a new era in which microbial metabolites were applied to diseases heretofore only treated with synthetic compounds, i.e., diseases not caused by bacteria and fungi, and huge successes were achieved.

3.1. Anticancer Drugs

In 2000, 57% of all drugs in clinical trails for cancer were either natural products or their derivatives *(93)*. The drug cytarabine (Cytostar) for non-Hodgkin's lymphoma was originally isolated from a sponge *(12)*. Most of the important antitumor compounds used for chemotherapy of tumors are microbially produced antibiotics *(94,95)*. These include actinomycin D, mitomycin, bleomycins, and the anthracyclines daunorubicin and doxorubicin.

Metastatic testicular cancer, although rather uncommon (1% of male malignancies in the United States; 80,000 in the year 2000 as compared to 190,000 cases of prostate cancer), is the most common carcinoma in men aged 15 to 35 yr. It is of significance that the cure rate for disseminated testicular cancer was 5% in 1974, whereas today it is 90%, mainly as a result of combination chemotherapy with the natural products bleomycin and etoposide and the synthetic cisplatin *(96)*.

The recent successful molecule Taxol (paclitaxel) was originally discovered in plants *(97)* and later reported to be a fungal metabolite *(98)*. It was approved for breast and ovarian cancer, and is the only antitumor drug on the market that acts by blocking depolymerization of microtubules. In addition, taxol promotes tubulin polymerization and inhibits rapidly dividing mammalian cancer cells *(99)*. In 2000, taxol sales amounted to over $1 billion and was Bristol Myers-Squibb's third largest selling product *(100)*. Today, the sales have escalated to $9 billion *(101)*.Taxol has antifungal activity by the same mechanism as above, especially against oomycetes *(102,103)*. These are water molds exemplified by plant pathogens such as *Phytophthora*, *Pythium*, and *Aphanomyces*.

Epothilone was originally discovered as a myxobacterial product with weak antifungal activity against rust fungi *(104)*. Later, it was shown to stabilize microtubules, the same mechanism as possessed by taxol *(105)*, and to be active against taxol-resistant tumor cells *(106)*. Epothilone polyketides are more water soluble than taxol. They are produced by *Sorangium cellulosum*, which is a very slow growing bacterium (16 h doubling time) and a low producer (20 µg/mL). The epithilone gene cluster has been cloned, sequenced, characterized, and expressed in the faster growing *Streptomyces coelicolor (107,108)*. A number of derivatives are in clinical trials.

A very exciting development has been the attachment of the extremely potent but extremely toxic enediyne antitumor drug calicheamicin to a humanized monoclonal antibody. The conjugated product Mylotarg™ (orgemtuzumab, ozogamicin) of Wyeth was approved for use against acute myeloid leukemia (AML) *(109)*. The monoclonal antibody was designed to direct the antitumor agent to the CD-33 antigen, which is a protein commonly expressed by myeloid leukemic cells.

3.2. Immunosuppressive Agents

Cyclosporin A was originally discovered as a narrow-spectrum antifungal peptide produced by the mold *Tolypocladium nivenum* (previously *Tolypocladium inflatum*) *(110)*. Discovery of its immunosuppressive activity led to its use in heart, liver, and kidney transplants, and to the overwhelming success of the organ transplant field. Sales of cyclosporin A reached $1 billion in 1994 *(111)*.

Although cyclosporin A was the only product on the market for many years, two other products, produced by actinomycetes, have been very successful. These are FK-506 (tacrolimus) *(112)* and rapamycin (sirolimus) *(113)*, both narrow-spectrum polyketide antifungal agents, which are 100-fold more potent than cyclosporin as immunosuppressants, and less toxic. Tacrolimus and rapamycin have both received FDA approval and are on the market. Tacrolimus was almost abandoned by the Fujisawa Pharmaceutical Co. after initial animal studies showed dose-associated toxicity. However, Thomas Starzl of the University of Pittsburgh, realizing that the immunosuppressant was 30- to 100-fold more active than cyclosporin, tried lower doses, which were very effective and nontoxic, thus saving the drug and many patients after that, especially those that were not responding to cyclosporin *(114)*. Since its introduction (1993 in Japan; 1994 in the United States), tacrolimus has been used for transplants of liver, kidney, heart, pancreas, lung, and intestines, and for prevention of graft-vs-host disease. Wyeth's sirolimus does not exhibit the nephrotoxicity of cyclosporin A and tacrolimus, and is synergistic with both compounds in immunosuppressive action *(115)*. By combining sirolimus with either, kidney toxicity is markedly reduced. Owing to a different mode of action, sirolimus has advantages over cyclosporin and tacrolimus *(114)*. Although it has been reported that cyclosporin A promotes tumor growth and many transplant patients are killed by tumors, sirolimus has the advantage of inhibiting tumor growth by interfering with angiogenesis *(116)*. Studies on the mode of action of these immunosuppressive agents have markedly expanded current knowledge of T-cell activation and proliferation *(117)*. The sirolimus analog everlimus also has immunosuppressive activity, and is in phase III clinical trials. Another analog, CCI-779, is in phase II against tumors of the breast, renal cell carcinoma, and non-Hodgkin's lymphoma. Sirolimus has found a new use in cardiology because sirolimus-impregnated stents are less prone to proliferation and restenosis, which usually occur after treatment of coronary artery disease.

Yeasts and filamentous fungi are inhibited by cyclosporin A, tacrolimus, and sirolimus. Susceptible fungi include *Candida albicans, Cryptococcus neoformans, Coccidiodes immitis, Aspergillus niger, A. fumigatus,* and *Neurospora crassa (118,119)*. Two nonimmunosuppressive analogs of cyclosporin A are active against *C. neoformans (120)*. A nonimmunosuppressive tacrolimus derivative, a C18-hydroxy C21-ethyl analog called L-685, 818, inhibits *C. neoformans (121)*. Recently, a topical preparation of tacrolimus was shown to be very active against atopic dermatitis, a widespread skin disease. The ascomycins, structurally related to tacrolimus, have anti-inflammatory action and are being clinically examined for topical treatment of atopic dermatitis, allergic contact dermatitis, and psoriasis *(122)*. One ascomycin macrolactam derivative, SDZ ASM 981, is in clinical studies. Sirolimus, cyclosporin A, and tacrolimus are also able to reverse multidrug resistance to antitumor agents in mammalian cells *(123)*.

A very old broad-spectrum antibiotic compound, mycophenolic acid, has an amazing history. The unsung hero of the story is Bartolomeo Gosio (1863–1944), the Italian physician who discovered the compound in 1893 *(124)*. Gosio isolated a fungus from spoiled corn, which he named *Penicillium glaucum*, which was later reclassified as *Penicillium brevicompactum*. He isolated crystals of the compound from culture filtrates in 1896 and found it to inhibit growth of *Bacillus anthracis*. This was the first time an antibiotic had been crystallized and the first time that a pure compound had ever been shown to have antibiotic activity. The work was forgotten, but fortunately the compound was rediscovered by Alsberg and Black *(125)* of the US Department of Agriculture, and given the name mycophenolic acid. They used a strain originally isolated from spoiled corn in Italy called *Penicillium stoloniferum*, a synonym of *P. brevicompactum*. The chemical structure was elucidated many years later by Raistrick and coworkers in England *(125a)*. Mycophenolic acid has antibacterial, antifungal, antiviral, antitumor, antipsoriasis, and immunosuppressive activities. It was never commercialized as an antibiotic because of its toxicity, but its 2-morpholinoethylester was approved as a new immunosuppressant for kidney transplantation in 1995 and for heart transplants in 1998 *(126)*. The ester derivative is called mycophenolate mofetil (CellCept) and is a prodrug that is hydrolyzed to mycophenolic acid in the body.

3.3. Hypocholesterolemic Agents

High blood cholesterol leads to atherosclerosis, which is a causal factor in many types of coronary heart disease and a leading cause of human death. Only 30% of the cholesterol in the human body comes from the diet. The remaining 70% is synthesized by the body, mainly in the liver *(127)*. Many people cannot control their cholesterol at a healthy level by diet alone, but must depend on hypocholesterolemic drugs. The statins inhibit *de novo* production of cholesterol in the liver. These extremely successful cholesterol-lowering agents are substituted hexahydronaphthalene lactones. They also have antifungal activities, especially against yeasts. Brown et al. *(128)* discovered the first member of this group, compactin (ML-236B), as an antifungal product of *P. brevicompactum*. Independently, Endo et al. *(129)* discovered compactin in broths of *Penicillium citrinum* as an inhibitor of 3-hydroxy-3-methylglutaryl coenzyme A reductase, the regulatory and rate-limiting enzyme of cholesterol biosynthesis. Later, Endo *(130)* and Alberts et al. *(131)* reported on the independently discovered more active methylated form of compactin known as lovastatin (monacolin K; mevinolin) in broths of *Monascus ruber* and *Aspergillus terreus* respectively. Statins were a success because they reduced total plasma cholesterol by 20–40%, whereas the previously used fibrates reduced it by only 10–15% *(132)*.

Lovastatin was approved by the FDA in 1987, when clinical tests in humans showed a lowering of total blood cholesterol of 18 to 34%, a 19 to 39% decrease in low-density lipoprotein cholesterol ("bad cholesterol"), and a slight increase in high-density lipoprotein cholesterol ("good cholesterol"). Another successful derivative is pravastatin, which is produced by hydroxylation of compactin using actinomycetes *(133,134)*. The market for the statins amounted to $19 billion in 2002 *(135)*. Merck's Zocor, a semisynthetic derivative of lovastatin, had sales of $7.2 billion in 2002, and Pfizer's synthetic statin Lipitor had a market of $8 billion, making it the world's leading drug in 2002.

Natural statins are produced by many fungi: *Aspergillus terreus* and species of *Monascus, Penicillium, Doratomyces, Eupenicillium, Gymnoascus, Hypomyces, Paecilomyces, Phoma, Trichoderma*, and *Pleurotis (136)*. Although pravastatin is commercially made by bioconversion of compactin, certain strains of *Aspergillus* and *Monascus* can produce pravastatin directly *(137,138)*. Statins are not only useful for reduction in the risk of cardiovascular disease; they can also prevent stroke, reduce development of peripheral vascular disease, and they have antithrombotic and anti-inflammatory activities. Statins inhibit production of pro-inflammatory molecules and are in clinical trials for multiple sclerosis *(139)*. They also may be of use in other autoimmune diseases *(140)*.

3.4. Enzyme Inhibitors

In addition to the enzyme inhibitors used to lower cholesterol, another antibiotic has succeeded in the world of medicine. Clavulanic acid is a β-lactam with poor antibiotic activity, produced by *Streptomyces clavuligerus*. It is an inhibitor of penicillinase, and is thus included with penicillins in combination therapy of penicillin-resistant bacterial infections *(141)*. Clavulanic acid is used as a combination product with amoxycillin (Augmentin™) or ticarcillin (Timentin™), and has a market of over $1 billion *(142)*. Additional enzyme inhibitors with antibiotic activity are the candins (mentioned previously).

3.5. Other Applications

A lantibiotic, epidermin, is used for treatment of acne *(143)*. Clindamycin, a semisynthetic derivative of the antibiotic lincomycin, is an effective antimalarial drug, especially when used with quinine *(144)*. Tetracyclines may be useful against prion diseases by rendering prion aggregates susceptible to proteolytic attack *(145)*. Prion diseases include scrapie of sheep, spongiform encephalopathy of cattle, Creutzfeldt–Jakob disease, fatal insomnia, and Gerstmann–Sträussler–Scheinker disease in humans *(146)*. Both tetracycline and doxycycline reduce infectivity of prions. Although a number of other agents are known (e.g., quinacrine, polyanions, polyene antibiotics, anthracyclines [iododoxorubicin], chlorpromazine, Congo Red, tetrapyrroles, polyamines, antibodies, and certain peptides), these are unable to pass the blood–brain barrier and/or are toxic. The tetracyclines can pass through and are nontoxic. Clinical trials are planned. Violacein, a toxic antibiotic known for many years as a product of *Chromobacterium violaceum*, is being studied as an agent preventing gastric ulcers when complexed with β-cyclodextrin, which decreases its toxicity *(147)*.

4. Additional Enzyme Combounds

Desferal is a siderophore produced by *Streptomyces pilosus (148)*. Its high level of metal-binding activity has led to its use in iron-overload diseases (hemochromatosis), and aluminum overload in kidney dialysis patients.

Acarbose is a pseudotetrasaccharide that is used as an inhibitor of intestinal α-glucosidase in type I and type II diabetes and hyperlipoproteinemia. It is produced by *Actinoplanes* sp. SE50 *(149)*. Acarbose contains an aminocyclitol moiety, valienamine, which is responsible for the inhibition of intestinal α-glucosidase and sucrase. The resulting decrease in starch breakdown in the intestine is the basis for its medical use against diabetes in humans.

Another enzyme inhibitor is lipstatin, which is used to combat obesity and diabetes by interfering with gastrointestinal absorption of fat. Lipstatin is produced by *Streptomyces toxytricini* as a pancreatic lipase inhibitor. This lipase is involved in digestion of fat *(150)*. Orlistat, the commercial product, is tetrahydrolipstatin.

5. Mycotoxins As Sources of Useful Agents

It is difficult to accept the fact that even poisons can be harnessed as medically useful drugs, yet this is the case with the ergot alkaloids. However, there is a philosophy in Traditional Chinese Medicine (TCM) of "using poison against poison." In the 18th century, Dr. William Withering found that one of his patients with dropsy (congestive heart failure) improved remarkably after using a traditional herbal remedy. He discovered the active ingredient digitoxin (digitalis) in the leaves of foxglove. In contemporary medicine, digitalis is used to strengthen cardiac diffusion and regulate heart rhythm. Dr. Withering commented, "Poisons in small doses are the best medicine; and the best medicines in too large doses are poisons" *(150a)*.

The mycotoxins (toxins produced by molds) were responsible for fatal poisoning of humans and animals (ergotism) throughout the ages after consumption of bread made from grain contaminated with species of *Claviceps*. In the Middle Ages, the ergot alkaloids caused the disease known in Europe as "Holy Fire" or "St. Antony's Fire." This widespread epidemic disease produced gangrene, cramps, convulsions, and hallucinations. These early names of the disease relate to the care of patients by the monks of the Antoniter Brotherhood. A major epidemic occurred in the USSR during the famine of

1926–1927. It is amazing but true that these "poisons" are now used for angina pectoris, hypertonia, migraine headache, cerebral circulatory disorder, uterine contraction, hypertension, serotonin-related disturbances, inhibition of prolactin release in agaloactorrhea, reduction in bleeding after childbirth, and for prevention of implantation in early pregnancy *(151,152)*. Among their physiological activities are the inhibition of action of adrenalin, noradrenalin, and serotonin, and the contraction of smooth muscles of the uterus. Some of the ergot alkaloids possess antibiotic activity.

6. Combinatorial Biosynthesis

Many new products have been made by genetic methods involving modification or exchange of genes between organisms to create hybrid molecules; the technique is known as combinatorial biosynthesis *(153–155)*. Recombinant DNA (rDNA) methods are used to introduce genes coding for natural product synthetases into producers of other natural products or into nonproducing strains to obtain modified or hybrid antibiotics. The first demonstration of this involved gene transfer from a streptomycete strain producing the isochromanequinone antibiotic actinorhodin into strains producing granaticin, dihydrogranaticin, and mederomycin (which are also isochromanequinones). This led to the discovery of two new antibiotic derivatives, mederrhodin A and dihydrogranatirhodin. Since this breakthrough paper by Hopwood et al. *(156)*, many hybrid antibiotics have been produced by rDNA technology. New antibiotics can also be created by changing the order of the genes of an individual pathway in its native host *(157)*. Many clusters of natural product genes are modular and produce multifunctional enzymes with a high degree of plasticity. By interchanging genes within these clusters, hybrid enzymes can be produced that are capable of synthesizing "unnatural natural products" *(157a)*.

Thousands of new antibiotics have been made, including erythromycins *(28,158–164)*, spiramycins *(165,166)*, tetracenomycins *(167,168)*, anthracyclines *(169–172)*, and nonribosomal peptides such as modified surfactins *(173)*.

7. Closing Remarks

Going from the three-dimensional crystal structure of a protein to designing of a drug is very difficult, because many proteins with similar structures have entirely different functions *(174)*. For example, the three-dimensional structure of triosephosphate isomerase, resembling a barrel or a bagel, is found in 1 out of every 10 enzymes that catalyze very different reactions. Also many proteins having similar functions, e.g., L-aspartate aminotransaminase and D-amino acid aminotransferase, have no identity of sequence and different folds of their peptide chains, yet they both use the cofactor pyridoxal phosphate and catalyze transamination. Furthermore, one protein often catalyzes several different reactions. These facts suggest that screening for new drugs is a process that must continue into the foreseeable future.

Thirty-five years ago, there was a drop in interest in screening natural products as a result of the difficulty in finding new antibiotics, lack of interest on the part of the government to fund such work, and huge investments by the pharmaceutical industry in the emerging area of rDNA. Fortunately, a few companies remained in the game, wisely incorporating the newer knowledge of genetics with natural products screening.

Not only were new antibiotics discovered and developed, but there was a broadening of the search to include nonantibiotic applications of natural products. No longer were microbial sources looked upon solely as potential solutions for infectious diseases, but for other applications, such as cholesterol-lowering, immunosuppression, enzyme inhibition, and so on. The change in screening philosophy was accompanied by ingenious applications of molecular biology to detect receptor antagonists and agonists, and other agents inhibiting or enhancing cellular activities on a molecular level (175).

Natural products are unsurpassed in their ability to provide novelty and complexity (22). With respect to the number of chirality centers, rings, bridges, and functional groups in the molecule, natural products are spatially more complex than synthetic compounds (6). Synthetic compounds highlighted via combinatorial chemistry and in vitro high-throughput assays are based on small chemical changes to existing drugs, and of the thousands, perhaps millions, of chemical "shapes" available to pharmaceutical researchers, only a few hundred are being explored (176).

About $48 billion is spent annually on new drug development (177). Of this amount, $14 billion is spent on the drug discovery phase. The total research and development (R&D) annual spending of the top 50 pharmaceutical companies was $1.9 billion in 1978–80 and rose to $29 billion by 2000 (178)! The industry as a whole was reported to have spent $21 billion for R&D in 1998 and $32 billion in 2002 (179). However, another report put the R&D spending of the top 20 pharmaceutical companies at $37 billion in 2001 (180).

Clinical development time has doubled since 1982 to 5-2/3 years, and the total time to get a drug on the market is 12–15 yr (181,182). It may take even longer, as judged from the following estimate: 2–10 yr for discovery, 4 yr for preclinical testing, 1 yr of phase I (involving 20–30 healthy volunteers for safety and dosage), 1.5 yr for phase II (100–300 patient volunteers for efficacy and side effects), 3.5 yr for phase III (1000–5000 patient volunteers monitoring effects of long-term use), 1 yr of FDA review and approval, and 1 yr of postmarketing testing (183). The total time can thus be 14 to as long as 22 yr. The time commitment has not changed since 1999, but the estimated cost of doing this rose from $500–600 million to $900 million (179). Two-thirds of the cost is spent on leads that fail in the clinic (36). One-half of all potential drugs fail because of adsorption, distribution, metabolism, excretion, or toxicity (ADME/TOX) problems.

A few years ago, it was thought that combinatorial chemistry and high-throughput screening (HTS) would yield many new hits and leads, but the result has been disappointing, despite the extraordinary amount of money spent (184,185). After it was developed in the early 1990s, HTS methods achieved speed and miniaturization but discovery of new leads did not accelerate. HTS methods allowed 100,000 chemicals to be assayed per day, and combinatorial and other chemical libraries of 1 million compounds were available commercially. Despite this, no drugs had been approved that resulted from HTS by 1999 (186). Since 1998, R&D spending by the top 20 pharmaceutical companies increased from $26 billion in 1998 to $37 billion in 2001, but the number of NDAs decreased from 34 to 16. Whereas FDA drug applications (total of NDAs, INDs, orphan drug applications, and so on) peaked at 131 in 1996, the number dropped steadily to 78 in 2002 (187). In 2001, there was a 20-yr low in the number of new active substances approved by the FDA (188). The number was 37, and was part of a continuous drop since 1997. During 1978–1980, the average number of new chemical entities launched by the pharmaceutical industry was 43; in 1998–2000, the number had dropped to 33. Only 17 drugs were approved in 2002, compared to 32 or more per year in the late 1990s *(189)*.

The advent of combinatorial chemistry, HTS, genomics, and proteomics has "not yet delivered the promised benefits" *(183)*. Investment in genomics and HTS has had no effect on the number of products in preclinical development or phase I clinical trials. The problems are that HTS has not been applied to natural product libraries, and combinatorial chemistry has not been applied to natural product scaffolds *(190–192)*. Natural product collections have a much higher hit rate in high-throughput screens than do combinational libraries *(193)*. Breinbauer et al. *(194)* pointed out that the numbers of compounds in a chemical library are not the important point; it is the biological relevance, design, and diversity of the library, and that a scaffold from nature provides viable, biologically validated starting points for the design of chemical libraries. According to Sam Danishefsky, prominent synthetic chemist at Memorial Sloan-Kettering Cancer Center in New York, it is appropriate "to critically examine the prevailing supposition that synthesizing zillions of compounds at a time is necessarily going to cut the costs of drug discovery or fill pharma pipelines with new drugs any time soon" *(195)*.

Some companies have dropped the screening of their natural product libraries because they considered that such extracts were not amenable to HTS *(186)*. Even worse, we hear that combinatorial chemistry is replacing natural product efforts for discovery of new drugs, and that most companies have even dropped their natural product programs to support combinatorial chemistry efforts. This makes no sense, since the role of combinatorial chemistry, like those of structure–function drug design and recombinant DNA technology two and three decades ago, is that of complementing and assisting natural product discovery and development, not replacing them *(196)*. Instead of downgrading natural product screening, there is real opportunity in combining it with HTS, combinatorial chemistry, genomics, proteomics, and new discoveries being made in biodiversity.

Genomics will provide a huge group of new targets against which natural products can be screened *(197)*. By 2003, there were 79 sequenced genomes of bacteria *(198)*. Useful targets for antibacterial therapy revealed by genomics are (1) two-component signal transduction systems, (2) FtsA-FtsZ interaction for cell-division inhibition, (3) MurA for cell-wall synthesis inhibition, (4) chorismate biosynthesis, (5) isoprenoid biosynthesis, and (6) fatty acid biosynthesis *(199)*. Additional new targets for further discovery efforts include bacterial signal peptidases *(200)*, non-β-lactam inhibitors of β-lactamase, lipid A biosynthesis, tRNA synthetases *(201)*, DNA replication (helicase, encoded by *dnaB*; DNA gyrase subunit B, encoded by *gyrB*), enoyl-ACP reductase, protein secretion, intermediary metabolism (dihydrofolate reductase encoded by *folA*; UMP kinase encoded by *pyrH*), translation (methionyl-tRNA synthetase encoded by *metG*; elongation factor Tu encoded by *tufA[B]*). *E. coli* strains with low levels of such enzymes have been constructed; they are hypersusceptible to specific inhibitors of each target and useful for whole-cell assays of new antibacterials *(202)*.

A genomic comparison of the pathogenic *Haemophilus influenzae* with a nonpathogenic *E. coli* revealed 40 potential drug targets in the former *(203)*. Similarly, a comparison of genomes of *Helicobacter pylori* with *E. coli* and *H. influenzae* revealed 594 *H. pylori*-specific genes, of which 196 were known, 123 of the known genes being involved in known host-pathogen interactions and 73 targets of novel potential *(204)*. Bacterial pathogens contain about 2700 genes, of which current antibiotics target less than 25 *(205)*. When the *S. cerevisiae* genome sequence was announced in 1996, the functions of only one-third of its 6200 predicted genes were known *(206)*. By 2002, some 4400 yeast genes were characterized *(207)*. Of the genes of the *S. cerevisiae* genome, about a dozen are potential targets: chitin synthetase, β-1,4-glucan synthetase, tubulin, elongation factor 2, *N*-myristoyl transferase, acetyl-CoA carboxylase, inositol phosphoryl ceramide synthase, membrane ATPase, mannosyl transferase, tRNA synthetases, lanosterol dehydrogenase, and squalene epoxidase *(208)*.

Although the performance of the pharmaceutical industry has been dismal recently because of poor decisions, the biotechnology industry is doing very well. Between 1997 and 2002, 40% of the drugs introduced came from biotechnology companies. The five largest pharmaceutical companies have in-licensed from 6 to 10 products from biotechnology or specialty pharmaceutical companies, yielding 28–80% of their revenue. The biotechnology industry had two drug/vaccine approvals in 1982, none in 1983–1984, one in 1985, rising to 32 in 2000! The number of patents granted to biotechnology companies rose from 1500 in 1985 to 9000 in 1999. Some biotechnology companies are entering the area of natural product screening and, in the end, may save this valuable resource from falling into obscurity.

Acknowledgments

We thank the following colleagues, who provided helpful answers to our questions: Clifford Siporin, Richard J. White, and Michael Greenstein.

References

1. Cragg GM, Newman DJ, Snader KM. Natural products in drug discovery and development. J Nat Prods 1997;60:52–60.
2. Verdine GL. The combinatorial chemistry of nature. Nature 1996;384 Suppl:11–13.
3. Holland HD. Evidence for life on earth more than 3850 million years ago. Science 1998;275: 38–39.
4. Mendelson R, Balick M.J. The value of undiscovered pharmaceuticals in tropical forests. Econ Bot 1995;49:223–228.
5. Dictionary of Natural Products; London: Chapman and Hall/CRC Press, 2001.
6. Henkel T, Brunne RM, Müller H, Reichel F. Statistical investigation into the structural complementarity of natural products and synthetic compounds. Angew Chem Int Ed Engl 1999;38:643–647.
7. Fenical W, Jensen PR. Marine microorganisms: a new biomedical resource. In: Attaway DH, Zaborsky OR, eds. Marine Biotechnology I: Pharmaceutical and Bioactive Natural Products, Plenum, New York: 1993;419–475.
8. Berdy J. Are actinomycetes exhausted as a source of secondary metabolites? Proc. 9th Internat Symp Biol Actinomycetes; Part 1, Allerton, New York: 1995:3–23.
9. Roessner CA, Scott AI. Genetically engineered synthesis of natural products: from alkaloids to corrins. Ann Rev Microbiol 1996;50:467–490.
10. Kiefer DM. A century of pain relief. Todays chem at work 1997;6(12):38–42.
11. Shapiro S. Unsung aspirin hero. Mod Drug Disc 2003;December:9.

12. Rayl AJS. Oceans: medicine chests of the future? Scientist 1999;13(19):1,4.
13. Yang SS, Cragg GM, Newman DJ, Bader JP. Natural product-based anti-HIV drug discovery and development facilitated by the NCI development therapeutics program. J Nat Prods 2001;64:265–277.
14. Ondetti MA, Williams NJ, Sabo EF, Pluscec J, Weaver ER, Kocy O. Angiotensin-converting enzyme inhibitors from the venom of *Bothrops jaraca*. Isolation, elucidation of structure, and synthesis. Biochemistry 1971;10:4033–4039.
15. Patchett AA. Alfred Burger award address in medicinal chemistry. Natural products and design: interrelated approaches in drug discovery. J Med Chem 2002;45:5609–5616.
15a. Schreiber SL. Target-oriented and diversity-oriented organic synthesis in drug discovery. Science 2000;287:1964–1969.
15b. Mayer TU, Kapoor TM, Haggarty SJ, King RW, Schreiber SL, Mitchison TJ. Small molecule inhibitor of mitotic spindle bipolarity identified in a phenotype-based screen. Science 1999;286:971–974.
16. Clark AM. Natural products as a resource for new drugs. Pharmaceut Res 1996;13:1133–1141.
17. Torsvik V, Sorheim R, Goksoyr J. Total bacterial diversity in soil and sediment communities—a review. J Indust Microbiol 1996;17:170–178.
18. Hawksworth DL. The fungal dimension of biodiversity: magnitude, significance, and conservation. Mycol Res 1991;95:641–655.
19. Zengler K, Toledo G, Rappé M, et al. Cultivating the uncultured. Proc Natl Acad Sci USA 2002;99:15,681–15,686.
20. Handelsman J, Rondon MR, Brady SF, Clardy J, Goodman RM. Molecular biological access to the chemistry of unknown soil microbes: a new frontier for natural products. Chem Biol 1998;5:R245–R249.
21. Möller C, Weber G, Dreyfuss MM. Intraspecific diversity in the fungal species *Chaunopycnis alba*: implications for microbial screening programs. J Ind Microbiol 1996;17:359–372.
22. Bull AT, Ward AC, Goodfellow M. Search and discovery strategies for biotechnology: the paradigm shift. Microbiol Mol Biol Rev 2000;64:573–606.
23. Lederberg J. Pathways of discovery: infectious history. Science 2000;288:287–293.
24. Strohl W. Industrial antibiotics: today and the future. In: Strohl W. ed. Biotechnology of Antibiotics, 2nd ed. Marcel Dekker, New York: 1997:1–47.
25. Bronson JJ, Barrett JF. Quinolone, everninomycin, glycylcycline, carbapenem, lipopeptide and cephem antibiotics in clinical development. Curr Med Chem 2001;8:1775–1793.
26. Erdmann J. Bacteria resistant to drugs draw scrutiny of biofilms. Gen Eng News 1999;19(10):1,19,50,58.
27. Barber MS. The future of cephalosporins business. Chimica Oggi 2001;19(12):9–12.
28. McDaniel R, Thamchaipenet A, Gustafesson C, et al. Multiple genetic modifications of the erythromycin polyketide synthase to produce a library of novel "unnatural natural products." Proc Natl Acad Sci USA 1999;96:1846–1851.
29. Georgopapadakou NH. Update on antifungals targeted to the cell wall: focus on β-1,3-glucan synthase inhibitors. Exp Opin Invest Drugs 2001;10:269–280.
30. Wilson JF. Renewing the fight against bacteria. Scientist 2002;16(5):22–23.
31. Williams D H, Bardsley B. The vancomycin group of antibiotics and the fight against resistant bacteria. Angew Chem. Int Ed 1999;38:1172–1193.
32. Fleming ID, Nisbet LJ, Brewer SJ. Target directed antimicrobial screens. In: Bu'lock JD, Nisbet LJ, Winstanley DJ, eds. Bioactive Microbial Products: Search and Discovery, Academic, London: 1982:107–130.
33. Woodruff HB, Hernandez S, Stapley ED. Evolution of antibiotic screening programme. Hindustan Antibiot Bull 1979;21:71–84.

34. Woodruff HB, McDaniel LE. Antibiotic approach in strategy of chemotherapy. Soc Gen Microbiol Symp 1958;8:29–48.
35. Brennan M B. Drug discovery: filtering out failures early in the game. Chem Eng News 2000:78(23);63–74.
36. Wilson EK. Picking the winners. Chem Eng News 2002;80(17):35–39.
37. Jones CG, Firn RD. On the evolution of plant secondary chemical diversity. Phil Trans R Soc 1991;333:273–280.
38. DiMasi J, Seibring M, Lasagna L. New drug development in the United States from 1963 to 1992. Clin Pharmacol Ther 1994;55:609–622.
39. Stapley EO., Jackson M, Hernandez S, et al. Cephamycins, a new family of β-lactam antibiotics. I. Production by actinomycetes, including *Streptomyces lactamdurans* sp. n. Antimicrob Agents Chemother 1972;2:122–131.
40. Hendlin D, Stapley EO, Jackson M, et al. Phosphonomycin. A new antibiotic produced by strains of *Streptomyces*. Science 1969;166:122–123.
41. Birnbaum J, Kahan FM, Kropp, H, MacDonald JS. Carbapenems, a new class of beta-lactam antibiotics. Discovery and development of imipenem/cilastatin. Amer J Med 1985;78 Suppl 6A:3–21.
42. Lederberg J. Getting in tune with the enemy—microbes. Scientist 2003;17(11):20–21.
43. Chu DT, Plattner JJ, Katz L. New directions in antibacterial research. J Med Chem 1996;39: 3853–3874.
44. DaSilva E, Iaccarino M. Emerging diseases: a global threat. Biotech Adv 1999;17:363–384.
45. Tenover FC, Hughes JM. The challenges of emerging infectious diseases. JAMA 1996;275: 300–304.
46. Stephens C, Shapiro L. Bacterial protein secretion—a target for new antibiotics? Chem Biol 1997;4:637–641.
47. Morris A, Kellner JD, Low DE. The superbugs: evolution, dissemination and fitness. Curr Opin Microbiol 1998;1:524–529.
48. Breiman RF, Butler JC, Tenover FC, Elliot JA, Facklam RR. Emergence of drug-resistant pneumococcal infections in the United States. JAMA 1994;271:1831–1835.
49. Goldman DA, Weinstein RA, Wenzel RP, et al. Strategies to prevent and control the emergence and spread of antimicrobial-resistant microorganisms in hospitals. JAMA 1996;275: 234–240.
50. Leclercq R, Courvalin P. Resistance to glycopeptides in enererococci. Clin Inf Dis 1997;24: 545–555.
51. Perichon B, Reynolds P, Courvalin P. VanD-type glycopeptide-resistant *Entercoccus faecium* BM4339. Antimicrob Agents Chemother 1997;41:2016–2018.
52. Thompson GA, Smith JA, Kenny MT, Dulwoeth JK, Kulmala HK, Yuh L. Pharmakinetics of teicoplanin upon multiple dose administration to normal healthy male volunteers. Biopharm Drug Dispos 1992;33:213–220.
53. Wood MJ. The comparative efficacy and safety of teicoplanin and vancomycin. J Antimicrob Chemother 1996;37:209–222.
53a. Howden BP, Ward PB, Charles PG, et al. Treatment outcomes for serious infections caused by methicillin-resistant *Staphylococcus aureus* with reduced vancomycin susceptibility. Clin Infect Dis. 2004;38:521–528.
54. Barriere JC, Berthaud N, Beyer D, Dutka-Malen S, Paris JM, Desnottes JF. Recent developments in streptogramin research. Curr Pharm Des 1998;4:155–180.
55. Nichterlein T, Kretschmar M, Hof H. RP 59500, a streptogramin derivative, is effective in murine listerosis. J Chemother 1996;8:107–112.
56. Cimons M. FDA approves the antibiotic synercid for limited clinical uses. ASM News 1999;65:800–801.

57. Moellering RC. Quinupristin/dalfopristin: therapeutic potential for vancomycin-resistant enterococcal infections. J Antimicrob Chemother 1999;44 Topic A:25–30.
58. Stinson SC. Drug firms restock antibacterial arsenal. Chem Eng News 1996;74(39):75–100.
59. Moellering RC, Linden PK, Reinhardt J, Blumberg EA, Bompart F, Talbot GH. The efficacy and safety of quinupristin/dalfopristin for the treatment of infections caused by vancomycin-resistant *Enterococcus faecium*. J Antimicrob Chemother 1999;44:251–261.
60. Livermore DM. Quinopristin/dalfopristin and linezolid: where, when, which and whether to use? J Antimicrob Chemother 2000;46:347–350.
61. Sum P-E, Sum F-W, Projan SJ. Recent developments in tetracycline antibiotics. Curr Pharm Des 1998;4:119–132.
62. Petersen PJ, Jacobus NV, Weiss WJ, Sum PE, Testa RT. In vitro and in vivo antibacterial activities of a novel glycylcycline, the 9-*t*-butylglycylamido derivative of minocycline (GAR-936). Antimicrob Agents Chemother 1999;43:738–744.
63. Ge M, Chen Z, Russell HO, et al. Vancomycin derivatives that inhibit peptidoglycan biosynthesis without binding D-Ala-D-Ala. Science 1999;284:507–511.
64. Walsh C. Deconstructing vancomycin. Science 1999;284:442–443.
65. Ciabatti R, Kettenring JK, Winters G, Tuan G, Zerilli L, Cavalleri B. Ramoplanin (A-16686), a new glycolipodepsipeptide antibiotic. III. Structure elucidation. J Antibiot 1989;41:254–267.
66. Cudic P, Behenna DC, Kranz JK, et al. Functional analysis of lipoglycodepsipeptide antibiotic ramoplanin. Chem Biol 2002;9:897–906.
67. Jiang W, Wanner J, Lee RJ, Bounaud P-Y, Boger DL. Total synthesis of the ramoplanin A2 and ramoplanose aglycon. J Amer Chem Soc 2002;124:5288–5290.
68. Cavalleri B, Pagani H, Volpe G, Selva E, Parenti F. A-16686, a new antibiotic from *Actinoplanes*. I. Fermentation, isolation and preliminary physicohemical characteristics. J Antibiot 1989;37:309–317.
69. DoughertyTJ, Barrett JF. ABT-773: a new ketolide antibiotic. Exp Opin Invest Drugs 2001;10:343–351.
70. Henninger TC. Recent progress in the field of macrolide antibiotics. Exp Opin Ther Patents 2003;13:787–805.
71. Agouridas C, Denis A, Auger J-M, et al. Synthesis and antibacterial activity of ketolides (6-O-methyl-3-oxoerythromycin derivatives): a new class of antibacterials highly potent against macrolide-resistant and -susceptible respiratory pathogens. J Med Chem 1998;41:4080–4100.
72. Kaneko T, McArthur H, Sutcliffe J. Recent developments in the area of macrolide antibiotics. Exp Opin Ther Patents 2000;10:403–425.
73. Yassin HM, Dever LL. Telithromycin: a new ketolide antimicrobial for treatment of respiratory tract infections. Exp Opin Invest Drugs 2001;10:353–367.
74. Leclercq R. Overcoming antimicrobial resistance: profile of a new ketolide antibacterial, telithromycin. J Antimicrob Chemother 2001;48 Topic T1:9–23.
75. Bax R, Mullan N, Verhoef J. The millennium bugs—the need for and development of new antibacterials. J Antimicrob Agents 2000;16:51–59.
76. Denis A, Agouridas C, Auger J-M, et al. Synthesis and antibacterial activity of HMR 3647, a new ketolide highly potent against erythromycin-resistant and susceptible pathogens. Bioorg Med Chem Lett 1999; 9:3075–3080.
77. Borman S. Polyketide system cloned, characterized. Chem Eng News 1998;76(44):27–28.
78. Tally FP, Zeckel M, Wasilewski MM, et al. Daptomycin: a novel agent for Gram-positive infections. Exp Opin Invest Drugs 1999;8:1223–1238.
79. Wright GD. Resisting resistance: new chemical strategies for battling superbugs. Chem Biol. 2000;7:R127–R132.

80. Stermitz FR, Lorenz P, Tawara JN, Zenewicz LA, Lewis K. Synergy in a medicinal plant: Antimicrobial action of berberine potentiated by 5'-methoxyhydrocarpin, a multidrug pump inhibitor. Proc Natl Acad Sci USA 2000;97:1433–1437.
81. Tally FA, Wendler PA, Houman F. Translation targets for antifungal drug development. Abstr. S5, 5th Internat Conf Biotechnol Microb Prods: Novel Pharmacol Agrobiol Act, Williamsburg, 1997:19.
82. White TC. Antifungal drug resistance in *Candida albicans*. ASM News 1997;63:427–433.
83. Alexander BP, Perfect JR. Antifungal resistance trends towards the year 2000. Implications for therapy and new approaches. Drugs 1997;54:657–678.
84. Georgopapadakou NH. Antifungals: mechanism of action and resistance, established and novel drugs. Curr Opin Microbiol 1998;1:547–557.
85. DiDomenico B. Novel antifungal drugs. Curr Opin Microbiol 1999;2:509–515.
85a. Theis T, Stahl U. Antifungal proteins: targets, mechanisms and prospective applications. Cell Mol Life Sci 2004;61:437–455.
86. Schwartz RE, Sesin DF, Joshua H, et al. Pneumocandins from *Zalerion arboricola* I. Discovery and isolation. J Antibiot 1992;45:1853–1866.
87. Kurtz MB. New antifungal drug targets: a vision for the future. ASM News 1998;64:31–39.
88. Hoang AT. Caspofungin acetate: an antifungal agent. Amer J Health-Syst Pharm 2001;58:1206–1214.
89. Debono M. The echinocandins: antifungals targeted to the fungal cell wall. Exp Opin Invest Drugs 1994;3:821–829.
90. Umezawa H. Enzyme Inhibitors of Microbial Origin. University Park, Baltimore: 1972.
91. Umezawa H. Low-molecular-weight inhibitors of microbial origin. Annu Rev Microbiol 1982;36:75–99.
92. Demain AL. New applications of microbial products. Science 1983;219:709–714.
93. Cragg GM, Newman D. Antineoplastic agents from natural sources: achievements and future directions. Exp Opin Invest Drugs 2000;9:2783–2797.
94. Oki T, Yoshimoto A. Antitumor antibiotics. In: Perlman D, ed. Annual Reports on Fermentation Processes ,vol. 3. Academic, New York: 1979:215–251.
95. Tomasz M. Mitomycin C: small, fast and deadly (but very selective). Curr Biol 1995;2:575–579.
96. Einhorn LH. Curing metastic testicular cancer. Proc Natl Acad Sci USA 2002;99:4592–4595.
97. Wall ME, Wani MC. Campothecin and taxol: discovery to clinic. Cancer Res 1995;55:753–760.
98. Stierle A, Strobel G, Stierle D. Taxol and taxane production by *Taxomyces andreanae*, an endophytic fungus of Pacific yew. Science 1993;260:214–216.
99. Manfredi JJ, Horowitz SB. Taxol: an antimitotic agent with a new mechanism of action. Pharmacol Ther 1984;25:83–125.
100. Thayer AM. Busting down a blockbuster drug. Chem Eng News 2000;78(45):20–21.
101. Morissey SR. Maximizing returns. Chem Eng News 2003;81(37):17–20.
102. Strobel GA. Useful products from rainforest microorganisms. Part 1. Endophytes and taxol. Agro-Food Ind Hi-Tech 2002;13(2):30–32.
103. Strobel GA, Long DM. Endophytic microbes embody pharmaceutical potential. ASM News 1998;64:263–268.
104. Gerth K, Beodorf N, Hofle G, Irschik H, Reichenbach H. Epothilones A and B: antifungal and cytotoxic compounds from *Sorangium cellulosum* (Myxobacteria). Production, physicochemical and biological properties. J Antibiot 1996;49:560–563.
105. Bollag DM, McQueney PA, Zhu J, et al. Epothilones, a new class of microtubule-stabilizing agents with taxol-like mechanism of action. Cancer Res 1995;55:2325–2333.
106. Kowalski R J, Giannakakou P, Hamel E. Activities of the microtubule-stabalizing agents epothilones A and B with purified tubulin and in cells resistant to paclitaxel (Taxol®). J Biol Chem 1997;272:2534–2541.

107. Tang L, Shah S, Chung L, et al. Cloning and heterologous expression of the epothilone gene cluster. Science 2000;287:640–642.
108. Julien B, Shah S, Ziermann R, Goldman R, Katz L, Khosla C. Isolation and characterization of the epothilone biosynthetic gene cluster from *Sorangium cellulosum*. Gene 2000;249:153–160.
109. Borman S. Enediyne research continues apace. Chem Eng News 2000;78(11):47–49.
110. Borel JF, Feurer C, Gabler HU, Stahelin H. Biological effects of cyclosporin A: a new anti-lymphocytic agent. Agents & Actions 1976;6:468–475.
111. Stähelin HF The history of cyclosporin A (sandimmune) revisited: another point of view. Experientia 1996;52:5–13.
112. Kino T, Hatanaka H, Hashimoto M, et al. FK-506, a novel immunosuppressant isolated from *Streptomyces*. 1. Fermentation, isolation and physico-chemical and biological characteristics. J Antibiot 1987;40:1249–1255.
113. Vezina D, Kudelski A, Sehgal SN. Rapamycin (AY 22,989), a new antifungal antibiotic. 1. Taxonomy of the producing streptomycete and isolation of the active principle. J Antibiot 1975;28:721–726.
114. Amaya T, Hiroi J, Lawrence ID. Tacrolimus and other immunosuppressive macrolides in clinical practice. In: Omura S, ed. Macrolide Antibiotics: Chemistry, Biology and Practice, 2nd ed., Academic/Elsevier, San Diego: 2002:421–452.
115. Rohde J, Heitman J, Cardenas ME. The TOR kinases link nutrient sensing to cell growth. J Biol Chem 2001;276:9583–9586.
116. Pray L. Strange bedfellows in transplant drug therapy. Scientist 2002;16(7):36 only.
117. Liu J. FK506 and ciclosporin: molecular probes for studying intracellular signal transduction. Trends Pharm Sci 1993;14:182–188.
118. Cardenas ME, Sanfridson A, Cutler NS, Heitman J. Signal-transduction cascades as target for therapeutic intervention by natural products. Trends Biotech 1998;16:427–433.
119. Cruz MC, Goldstein AL, Blankenship J, et al. Rapamycin and less immunosuppressive analogs are toxic to *Candida albicans* and *Cryptococcus neoformans* via FKBP12-dependent inhibition of TOR. Antimicrob Agents Chemother 2001;45:3162–3170.
120. Cruz MC, Del Poeta M, Wang P, et al. Immunosuppressive and nonimmunosuppressive cyclosporine analogs are toxic to the opportunistic fungal pathogen *Cryptococcus neoformans* via cyclophilin-dependent inhibition of calcineurin. Antimicrob Agents Chemother 2000;44:143–149.
121. Odom A, Poeta MD, Perfect J, Heitman J. The immunosuppressant FK506 and its nonimmuno-suppressive analog L-685,818 are toxic to *Cryptococcus neoformans* by inhibition of a common target protein. Antimicrob Agents Chemother 1997;41:156–161.
122. Paul C, Graeber M, Stuetz A. Ascomycins: promising agents for the treatment of inflammatory skin diseases. Exp Opin Invest Drugs 2000;9:69–77.
123. Arceci RJ, Stieglitz K, Bierer BE. Immunosuppressants FK506 and rapamycin function as reversal agents of the multidrug resistance phenotype. Blood 1992;80:1528–1536.
124. Bentley R. Bartolomeo Gosio, 1863–1944: an appreciation. Adv Appl Microbiol 2001;48:229–250.
125. Alsberg CL, Black OF. Contributions to the study of maize deterioration. Biochemical and toxological investigations of *Penicillium puberulum* and *Penicillium stoloniferum*. USDA Bur. Plant Ind., Bull. No. 270, Govt. Printing Ofc., Washington, DC, 1913.
125a. Clutterbuck PW, Oxford AE, Raistrick H, Smith G. Studies in the biochemistry of microorganisms. XXIV. The metabolic products of the *Penicillium brevi-compactum* series. Biochem J 1932;26:1441–1458.
126. Lee WA, Gu L, Kikszal AR, Chu N, Leung K, Nelson PH. Bioavailability improvement of mycophenolic acid through amino ester derivitization. Pharmaceut Res 1990;7:161–166.

127. Bowden ME. Combating heart disease: rational drug design and cholesterol. Chem Heritage 2000;18(3):6–7,40–42.
128. Brown AG, Smale TC, King TJ, Hasenkamp R, Thompson RH. Crystal and molecular structure of compactin, a new antifungal metabolite from *Penicillium brevicompactum*. J Chem Soc Perkin Trans I 1976:1165–1170.
129. Endo A, Kuroda M, Tsujita Y. ML-236B and ML-236C, new inhibitors of cholesterogenesis produced by *Penicillium citrinin*. J Antibiot 1976;29:1346–1348.
130. Endo A. Monacolin K, a new hypocholsterolemic agent produced by a *Monascus* species. J Antibiot 1979;32:852–854.
131. Alberts AW, Chen J, Kuron G, et al. Mevinolin. A highly potent competitive inhibitor of hydroxymethylglutaryl-coenzyme A reductase and a cholesterol-lowering agent. Proc Natl Acad Sci USA 1980;77:3957–3961.
132. Knowles J, Gromo G. Target selection in drug discovery. Nature Revs/Drug Disc 2003;2:63–69.
133. Serizawa N, Matsuoka T. A two component-type cytochrome P-450 monooxygenase system in a prokaryote that catalyzes hydroxylation of ML-236B to pravastatin, a tissue-selective inhibitor of 3-hydroxy-3-methylglutaryl coenzyme A reductase. Biochim Biophys Acta 1991;1084: 35–40.
134. Peng Y, Demain AL. A new hydroxylase system in *Actinomadura* sp. cells converting compactin to pravastatin. J Indust Microbiol Biotechnol 1998;20:373–375.
135. Downton C, Clark I. Statins—the heart of the matter. Nature Revs/Drug Disc 2003;2:343–344.
136. Manzoni M, Rollini M. Biosynthesis and biotechnological production of statins by filamentous fungi and application of these cholesterol-lowering drugs. Appl Microbiol Biotechnol 2002;58:555–564.
137. Manzoni M, Rollini M, Bergomi S, Cavazzoni V. Production and purification of statins from *Aspergillus terreus* strains. Biotechnol Tech 1998;12:529–532.
138. Manzoni M, Bergomi S, Rollini M, Cavazzoni V. Production of statins by filamentous fungi. Biotechnol Lett 1999;21:253–257.
139. Brazil M. A new use for statins? Nature Revs/Drug Disc 2002;1:934.
140. Youssef S, Stüve O, Patarroyo JC, et al. The HMG-CoA reductase inhibitor, atorvastatin, promotes a Th2 bias and reverses paralysis in central nervous system autoimmune disease. Nature 2002;420(7):78–84.
141. Brown AG. Clavulanic acid, a novel β-lactamase inhibitor—a case study in drug discovery and development. Drug Design Devel 1986;1:1–21.
142. Jensen SE, Paradkar AS. Biosynthesis and molecular genetics of clavulanic acid. Ant v Leeuwenhoek 1999;75:125–133.
143. Jung G. Lantibiotics—ribosomally synthesized biologically active polypeptides containing sulfide bridges and α,β-didehydroamino acids. Angew Chem Int Ed Engl 1991;30:1051–1068.
144. Lell B, Kremser PG. Clindamycin as an antimalarial drug: review of clinical trials. Antimicrob Agents Chemother 2002;46:2315–2320.
145. Borman S. Potential treatment for prion diseases. Chem Eng News 2002;80(33):42 only.
146. Forloni G, Lussich S, Awan T, et al. Tetracyclines affect prion activity. Proc Natl Acad Sci USA 2002;99:10,849–10,854.
147. Duran N, Justo GZ, Melo PS, et al. Evaluation of the antiulcerogenic activity of violacein and its modulation by the inclusion complexation with β-cyclodextrin. Can J Physiol Pharmacol 2003;81:387–396.
148. Winkelmann G. Iron complex products (siderophores). In: Rehm HJ, Reed G, eds. Biotechnology, vol. 4. Weinheim: VCH, 1986:215–243.
149. Truscheit E, Frommer W, Junge B, Müller L, Schmidt DD, Wingender W. Chemistry and biochemistry of microbial α-glucosidase inhibitors. Angew Chem Intl Ed Engl 1981;20:744–761.

150. Weibel E K, Hadvary P, Hochuli E, Kupfer E, Lengsfeld H. Lipstatin, an inhibitor of pancreatic lipase, produced by *Streptomyces toxytricini*. J Antibiot 1987;40:1081–1085.
150a. Dec GW. Digoxin remains useful in the management of chronic heart failure. Med Clin North Am 2003;87:317–337.
151. Vining LC, Taber WA. Ergot alkaloids. In: Rose AH, ed. Economic Microbiology. Secondary Products of Metabolism, vol.3. Academic, London: 1979:389–420.
152. Bentley R. Microbial secondary metabolites play important roles in medicine; prospects for discovery of new drugs. Persp Biol Med 1997;40:364–394.
153. Hutchinson, CR. Antibiotics from genetically engineered microorganisms. In: Strohl W, ed. Biotechnology of Antibiotics, 2nd ed. Marcel Dekker, New York: 1997:683–702.
154. Hutchinson CR. Combinatorial biosynthesis for new drug discovery. Curr Opin Microbiol 1998;1:319–329.
155. McAlpine J. Unnatural natural products by genetic manipulation. In: Sapienza DM, Savage LM. eds. Natural Products II: New Technologies to Increase Efficiency and Speed, Internat Bus Comm, Southborough MA: 1998:251–278.
156. Hopwood DA, Malpartida F, Kieser HM, et al. Production of hybrid antibiotics by genetic engineering. Nature 1985;314:642–644.
157. Hershberger CL. Metabolic engineering of polyketide biosynthesis. Curr Opin Biotechnol 1996;7:560–562.
157a. Kennedy J, Auclair K, Kendrew SG, Park C, Vederas JC, Hutchinson CR. Modulation of polyketide synthase activity by accessory proteins during lovastatin biosynthesis. Science 1999;284:1368–1372.
158. Weber JM, Leung JO, Swanson SJ, Idler KB, McAlpine JB. An erythromycin derivative produced by targeted gene disruption in *Saccharopolyspora erythrea*. Science 1991;252:114–117.
159. Donadio S, McAlpine JB, Sheldon PJ, Jackson M, Katz L. An erythromycin analog produced by reprogramming of polyketide synthesis. Proc Natl Acad Sci USA 1993; 90:7119–7123.
160. Donadio S, Staver MJ, McAlpine JB, Swanson SJ, Katz L. Modular organization of genes required for complex polyketide biosynthesis. Science 1991;252:675–679.
161. Staunton J. Combinitorial biosynthesis of erythromycin and complex polyketides. Curr Opin Chem Biol 1998;2:339–345.
162. Pacey MS, Dirlam JP, Geldart RW, et al. Novel erythromycins from a recombinant *Saccharopolyspora erythraea* strain NRRL 23338 pIGI I. Fermentation, isolation and biological activity. J Antibiot 1998;51:1029–1034.
163. Xue Q, Ashley G, Hutchinson CR, Santi DV. A multi-plasmid approach to preparing large libraries of polyketides. Proc Natl Acad Sci USA 1999;96:11,740–11,745.
164. Rodriguez E, McDaniel R. Combinatorial biosynthesis of antimicrobials and other natural products. Curr Opin Microbiol 2001;4:526–534.
165. Epp JK, Huber MLB, Turner JR, Goodson T, Schoner BE. Production of a hybrid macrolide in *Streptomyces ambofaciens* and *Streptomyces lividans* by introduction of a cloned carbomycin biosynthetic gene from *Streptomyces thermotolerans*. Gene 1989;85:293–301.
166. Hara O, Hutchinson CR. A macrolide 3-*O*-acyltransferase gene from the midecamycin-producing species *Streptomyces mycarofaciens*. J Bacteriol 1992;174:5141–5144.
167. Decker H, Hutchinson CR. Transcriptional analysis of the *Streptomyces glaucescens* tetracenomycin C biosynthesis gene cluster. J Bacteriol 1993;175:3887–3892.
168. Wohlert S-E, Blanco G, Lombó F, et al. Novel hybrid tetracemomycins through combinatorial biosynthesis using a glycosyltransferase encoded by the *elm* genes in cosmid 16F4 and which shows a broad sugar substrate specificity. J Amer Chem Soc 1998;120:10,596–10,601.
169. Hwang CK, Kim HS, Hong YS, et al. Expression of *Streptomyces peucetius* genes for doxorubicin resistance and aklavinone 11-hydroxylase in *Streptomyces galilaeus* ATCC 31133 and production of a hybrid aclacinomycin. Antimicrob Agents Chemother 1995;39:1616–1620.

170. Kim H-S, Hong Y-S, Kim Y-H, Yoo O-K, Lee J-J. New anthracycline metabolites produced by the aklavinone 11-hydroxylase gene in *Streptomyces galilaeus* ATCC 31133. J Antibiot 1996;49:355–360.
171. Niemi J, Mäntsälä P. Nucleotide sequences and expression of genes from *Streptomyces purpurascens* that cause the production of new anthracyclines. J Bacteriol 1995;177:2942–2945.
172. Ylihonko K, Hakala J, Kunnari T, Mäntsälä P. Production of hybrid anthracycline antibiotics by heterologous expression of *Streptomyces nogalater* nogalamycin biosynthesis genes. Microbiology 1996;142:1965–1972.
173. Stachelhaus T, Schneider A, Marahiel M. Rational design of peptide antibiotics by targeted replacement of bacterial and fungal domains. Science 1995;269:69–72.
174. Borman S. From sequence to consequence. Chem Eng News 2001;79(48):31–33.
175. Tanaka H, Omura S. Screening of novel receptor-active compounds of microbial origin. In: Rehm HJ, Reed G, Kleinkauf H, von Doehren H. eds. Biotechnology, 2nd ed, vol.7. Weinheim: VCH, 1997:107–132.
176. Willis RC. Nature's pharma sea. Modern Drug Disc January 2002;32–38.
177. Anonymous. Drug discovery in the market place. Screening 2002;3(1):6.
178. Cunningham BC. Biotech and pharma: state of the relationship in the new millennium. Drug Devel Res 2002;57:97–102.
179. Agres T. Alliances eye early-stage drugs. Drug Disc Devel 2003;6(8):17–18.
180. Handen JS. The industrialization of drug discovery. Drug Disc Today 2002;7:83–85.
181. Watkins KJ. Fighting the clock. Chem Eng News 2002;80(4):27–34.
182. Burrill GS. Personalized medicine or blockbusterology. BioPharm 2002;15(4):46–50.
183. Ernst and Young. The Ernst & Young Fifteenth Annual Report on the Biotechnology Industry, Ernst & Young LLP 2000.
184. Ausman D.J. Screening's age of insecurity. Mod Drug Disc 2001;4(5):32–39.
185. Horrobin DF. Realism in drug discovery—could Cassandra be right? Nature Biotech 2001;19:1099–1100.
186. Fox S, Farr-Jones S, Yund MA. New directions in drug discovery. Gen Eng News 1999;19(21):10,36,56,66,80.
187. Warner S. Pipeline anxiety: scientists pumped into new roles. Scientist 2003;17(10),46–48.
188. Jacobs M. Pharmaceutical balancing act. Chem Eng News 2002;80(48):5.
189. Agres T. Biotech: leaner, but stronger. Drug Disc Devel 2003;6(7):17–19.
190. Demain AL. Prescription for an ailing pharmaceutical industry. Nature Biotech 2002;20:331.
191. Kingston DGI, Newman DJ. Mother nature's combinatorial libraries; their influence on the synthesis of drugs. Curr Opin Drug Disc Devel 2002;5:304–316.
192. Waldmann H, Breinbauer R. Nature provides the answer. Screening 2002;3(6):46–48.
193. Breinbauer R, Manger M, Scheck M, Waldman H. Natural product guided compound library development. Curr Med Chem 2002;9:2129–2145.
194. Breinbauer R, Vetter IR, Waldman H. From protein domains to drug candidates—natural products as guiding principles in the design and synthesis of compound libraries. Angew Chem Int Ed 2002;41:2879–2890.
195. Borman S. Organic lab sparks drug discovery. Chem Eng News 2002;80(2):23–24.
196. Paululat T, Tang Y-Q, Grabley S, Thiericke R. Combinatorial chemistry: the impact of natural products. Chim Oggi 1999;17:52–56.
197. Moir DT, Shaw KJ, Hare RS, Vovis GF. Genomics and antimicrobial drug discovery. Antimicrobial Agents Chemother 1999;43:439–446.
198. Mills SD. The role of genomics in antimicrobial discovery. J Antimicrob Chemother 2003;51:749–752.
199. McDevitt C, Payne DJ, Holmes DJ, Rosenberg M. Novel targets for the future development of antibacterial agents. J Appl Microbiol Symp Suppl 2002;92:28S–34S.

200. Black MT, Bruton G. Inhibitors of bacterial signal peptidases. Curr Pharmaceut Design 1998;4:133–154.
201. Bush K. Antimicrobial agents. Curr Opin Chem Biol 1997;1:169–175.
202. DeVito JA, Mills JA, Liu VG, et al. An array of target-specific screening strains for antibacterial discovery. Nature Biotech 2002;20:478–483.
203. Huynen MA, Diaz L, Bork P. Differential genome display. Trends Genet 1997;13:389–390.
204. Huynen M, Dandekar T, Bork P. Differential genome analysis applied to the species-specific features of *Helicobacter pylori*. FEBS Lett 1998;426:1–5.
205. Glass JI, Belanger AE, Robertson GT. *Streptococcus pneumonia* as a genomics platform for broad spectrum antibiotic discovery. Curr Opin Microbiol 2002;5:338–342.
206. Goffeau A, Barrell BG, Bussey H, et al. Life with 6000 genes. Science 1996;274:546, 563–567.
207. Auerbach D, Thaminy S, Hottiger MO, Stagljar I. The post-genomic era of interactive proteomics; facts and perspectives. Proteomics 2002;2:611–623.
208. Koltin Y. Targets for antifungal drug discovery. Annu Rep Med Chem 1990;25:141–148.

PART II
STRATEGIES

2
Integrated Approaches for Discovering Novel Drugs From Microbial Natural Products

Lixin Zhang

Summary

Historically, nature has provided the source for the majority of the drugs in use today. This owes in large part to their structural complexity and clinical specificity. However, only a small percentage of known microbial secondary metabolites have been tested as natural-product drugs. Natural-product programs need to become more efficient, starting with the collection of environmental samples, selection of strains, metabolic expression, genetic exploitation, sample preparation and chemical dereplication. A renaissance of natural products-based drug discovery is coming because of the trend of combining the power of diversified but low-redundancy natural products with systems biology and novel assays. This review will focus on integrated approaches for diversifying microbial natural-product strains and extract libraries, while decreasing genetic and chemical redundancy. Increasing the quality and quantity of different chemical compounds tested in diverse biological systems should increase the chances of finding new leads for therapeutic agents.

Key Words: Diversity; microbial natural products; drug discovery; redundancy; dereplication; synergy.

1. Introduction

The most well-known examples of natural product are antibiotics *(1)*. The "Golden Age of Antibiotics," from the 1940s to the 1970s, was sparked by the serendipitous discovery of penicillin by Alexander Fleming in 1928 and its development by Chain and Florey in the 1940s. Another remarkable milestone in the medicinal use of microbial metabolites and their derivatives was the introduction of the immunosuppressants cyclosporin A, FK-506 *(2)*, and rapamycin *(2,3)*. Other examples are the commercialization of the antihyperlipidemic lovastatin and the recent discovery of guggulsterone *(4)*. Microbial natural products have also been developed as antidiabetic drugs, hormone (ion-channel or receptor) antagonists, anticancer drugs, and agricultural and pharmaceutical agents *(5)*. Microorganisms not only produce secondary metabolites that affect cell growth, but also accumulate bioactive compounds that interact with valuable targets of cell metabolism and signaling that are not directly correlated with cell death *(6)*.

Drug discovery strategies for pharmaceutical and agrochemical applications are in a revolutionary period *(7)*. The completion of the Human Genome Project and the elucidation of dozens of microbial pathogens genomes have provided thousands of disease-related targets for screening. Automated instrument systems, robots, high-throughput screening (HTS) platforms, and high-throughput chemistry have provided powerful tools for screening large compound libraries in a cost-effective manner. Combinatorial chemical, synthetic chemical, and natural-product libraries provide abundant resources for target-based screening. The success and failure of drug discovery is coupled to the novelty and meaningfulness of the applied biological test systems as well as the amount and structural diversity of the test compounds available *(8)*.

In the early to mid-1990s, combi-chem companies attempted to fill the void with large numbers of new molecules. Unfortunately, it appears that the chemistry employed did not create sufficiently diverse or pharmacologically active molecules. It is clear that the future success of the pharmaceutical industry depends on the combining of complementary technologies such as natural-product discovery, HTS, genomics and proteomics, combinatorial biosynthesis, and combinatorial chemistry. The process of drug discovery for therapeutic and preventive medicines is facilitated by increasing knowledge of biological mechanisms, such as treatment efficiency, potential side effects, and the growing threat of drug resistance. A large amount of disease-relevant protein targets have been identified and validated from genomics, proteomics, and systems/ computational biology approaches. Novel targets and novel HTS assays and measurement systems are emerging to allow more sensitive, reliable, and low-background searches for new potential drugs among natural products.

The advantages and challenges of natural product-based drug discovery as compared to its synthetic chemistry counterpart are summarized as follows:

Advantages:
- Natural products offer unmatched chemical diversity with structural complexity and biological potency *(9)*.
- Natural products have been selected by nature for specific biological interactions. They have evolved to bind to proteins and have drug-like properties.
- Natural products are a main source of pharmacophores. Drugs such as cyclosporin A and FK-506 are not only active as immunosuppressants but also as antiviral, antifungal, and antiparasitic agents.
- Natural-product resources are largely unexplored. Novel discovery strategies will lead to novel bioactive compounds. Natural-product extracts are complementary to synthetic and combinatorial libraries. About 40% of natural-product diversity is not represented in synthetic compounds libraries.
- Natural products can guide the design of synthetic compounds *(10)*.
- Research on natural products has led to the discovery of novel mechanisms of action—for example, those of immunosuppressants and guggulsterone *(4)*.
- Natural products are powerful biochemical tools; they serve as "pathfinders" for molecular biology and chemistry, and in the investigations of cellular functions.

Main challenges:
- The lack of systematic exploitation of ecosystems for the discovery of novel microbial compounds had resulted in random sampling and has missed the true potential of many regions *(11)*.
- Little effort has focused on the isolation and cultivation of less culturable microorganisms. The discrepancy between the number of microbes detected by molecular methods

and the number of strains in culture, demonstrates that there remains a relatively untapped source of novel strains in all ecosystems *(12)*.
- The selection of strains has traditionally been based on morphology, rather than on the more powerful approaches of chemical diversity and genotype *(13)*.
- Lack of dereplication has resulted in a redundancy of strains and compounds within many natural-product extract libraries *(14)*.
- The characterization and isolation of active compounds from natural-product extracts are extremely labor intensive and time consuming *(15)*.
- The production of adequate quantities of the active compound needed for drug profiling may require extensive media optimization and scale-up *(16)*.

This review concentrates on the challenges of efficiently diversifying a library of microbial compounds, and does not deal with other problems such as preparation of samples, scale-up, chemical identification, and so on. Systematic approaches to maximize the biodiversity of microorganisms within a natural-product library are discussed from the following three perspectives: (1) isolation and selection of samples from diverse ecosystems; (2) manipulating microbial physiology to activate microbial natural-product biosynthetic machinery; and (3) genetically modifying strains for production of unnatural microbial natural products. By manipulating all three of these approaches, the diversity of an extract collection can be maximized, and in doing so, the chance of finding a "hit" can be increased. The quality of a microbial natural-product library is built on the dynamic equilibrium between diversification and reducing redundancy of microbial natural products. Therefore, strategies for obtaining high quality of a microbial natural-product library are discussed here.

2. Sample Collection and Selection From Diversified Ecosystems

Existing microbial natural-product drugs were originally isolated from all over the Earth. Microbes can sense, adapt, and respond to their environment quickly and help compete for defense and survival by generating unique secondary metabolites. These compounds are produced in response to stress. In diversifying microbial natural-product extract libraries, the greatest influence will undoubtedly be the genetic diversity of strains. By maximizing the types of samples collected and diversifying the isolation strategies, a highly diverse microbial collection can be generated.

2.1. Ecosystem Rationale

Collecting environmental samples for isolation of interesting microorganisms has often been conducted without defined strategies *(17)*. Such programs need to take into consideration the biogeography of ecosystems, number of samples collected, and isolation procedures. It is important to increase the number and diversity of sampling sites *(18)*, and it is especially important to look at underrepresented sites. Diverse regions such as the deep subsurface, the deep sea, and sites that have extreme temperature, salinity, or pH often generate novel microorganisms and therefore provide the potential for novel compounds *(19)*. Temperate ecosystems should not be excluded, because they also have the potential to provide many novel species, especially when novel isolation strategies are used. Cyclosporins, rapamycin, penicillin, and rifamycin, among others, were isolated from microorganisms collected in temperate regions.

It is still debated whether most microorganisms are cosmopolitan or endemic to specific geographic areas. There is a lack of detailed information in the field of geographical distribution of microorganisms *(19)*. In some cases, the presence of an endemic species can be detected; for example, several groups of bacteria appeared to be endemic to an ice microbial community *(20)*. However, the definition of a microbial species is difficult, especially for prokaryotes, which exchange parts of their genomes with sufficient ease to make difficult the biologically meaningful definition of a species *(21)*. In order to increase the chance of constructing a library of microorganisms with high diversity, the first step is to consider different geographic areas, including biodiversity hot spots *(22)*. There are more compounds in nature than possible molecular targets. One should concentrate on sampling in regions with different climates, fauna, and flora.

It is important to analyze properly the various ecosystems of a region. For example in Massachusetts, there are 13 eco-regions, from the Berkshire Highlands to the coastal regions of Cape Cod (http://www.state.ma.us/mgis/eco-reg.html), each of which has various subecosystems. Microbiologists should work closely with botanists and ecologists to obtain as many different samples and microorganisms as possible from one ecosystem to maximize the likelihood of finding novel strains and in turn novel chemicals. In almost all ecosystems, no matter how harsh, a group of organisms will grow and thrive. In these unique sites, we can expect to find unique metabolic pathways that have evolved to allow microorganisms to adapt and survive.

With the discovery that microbial symbionts were driving the metabolism of tubeworms in deep-sea hydrothermal vents, it was realized that an oasis of rich diversity could be found even in areas that were thought to be devoid of life. The deep sea, one such ecosystem, is actually a rainforest with a diversity of more than 10 million species, more than 60% of which are unknown *(23)*. In addition to the open ocean, there are diverse and dynamic areas such as mangroves, coral reefs, hydrothermal vents, and deep-sea sediments in which to search for microbes. Natural products have been isolated from marine invertebrates such as sponges, tunicates, mollusks, and bryozoans *(24)*. This not only demonstrates the numerous opportunities the oceans provide for discovering new compounds, but also validates the pharmacological value of exploring the oceans for novel compounds. There are some concerns about the isolation of marine microbes. Some researchers claim that this resource is not thoroughly explored because these organisms are hard to maintain in the laboratory environment. However, one successful case was the recent discovery of a new genus of actinomycetes found only in the marine environment, i.e., *Salinospora (25)*. One isolate produces salinosporamide A, a potent anticancer agent. Thus, the oceans can no longer be ignored *(26)*. Other regions that warrant further study are locations with extremes of pH, temperature, and salinity.

A defined sampling strategy must be adopted. To comprehensively explore a particular site, multiple discrete locations within the site must be sampled. Many types of samples should be selected in one ecosystem—e.g., soils, sediments, organic material, dung, dead animals, plants, and lichens. Soil still remains an important source because it carries higher populations of microbes than any other habitat *(27)*. DNA community analysis has proven that the number of types of organisms found within a microbial community is much higher than previously thought. One analysis of the reassociation kinetics of total bacterial DNA in a 30-g soil sample suggested that it contained more than 500,000 species *(21)*.

Plants and lichens offer niches for interaction between microorganisms and eukaryotic cells. Many natural products initially isolated from plants and animals were actually produced by microbial symbionts found within the tissue of the host *(23)*. In some incidences where the microorganism or symbiont could not be cultured axenically, the genes responsible for the production of the active compounds were attributed to the microorganism *(28)*.

Endophytes are microorganisms including unicellular bacteria *(29)*, actinomycetes *(30)*, and fungi *(31)* that spend part or all of their life cycle colonizing, either inter- or intracellularly, the healthy tissue of a plant *(32)*. Almost all vascular plants and mosses examined so far have endophytic bacteria or fungi within their tissues *(29)*. The number of strains found within the plant tissue can vary from a few to several hundred per plant. This relationship between the microorganism and the plant can range from mildly phytopathogenic to symbiotic. Endophytes produce a range of compounds *(16)*, some of which help the plant to survive and thrive in its ecosystem and some that help it fight off infection *(33)*. Plants therefore provide an obvious source for isolation of microorganisms that could potentially produce novel natural products. Of special interest are the large number of alkaloids and taxol produced by endophytic fungi *(16)*.

Lichens, symbiosis between fungi and cyanobacteria, are another source of microorganisms living in a unique and competitive environment. In each lichen sample, the fungus forms a thallus or lichenized stoma *(34,35)*. Furthermore, in addition to the symbiotic fungal strains, other fungi and bacteria live as endophytes inside the lichens, or as epiphytes on the lichens. The fungi within the lichen often produce unique secondary metabolites. Over 800 lichen secondary metabolites have been collected so far.

2.2. Isolation Strategies

Microbial diversity in the environment is far greater than reflected in most strain collections, due to the number of organisms that cannot be cultured using standard culture conditions *(36)*. Therefore, a vast majority of microorganisms in many samples remain unexplored. Molecular techniques allow detection of organisms that were missed using culture-dependent methods. Culture-independent methods, such as DNA clone libraries, have allowed identification of vast numbers of new organisms that are different from anything previously cultured. It is estimated that as few as 0.1–1% of the organisms living in the biosphere have been cultured and characterized in the laboratory setting. In one study, approx 10^7 bacteria were counted in 1 g of soil *(37)*, but as few as 0.1% were culturable using standard culture techniques. The other 99.9 % of the population may represent novel genetic diversity *(14)*, and may produce novel natural compounds. In 1987, when Dr. Carl Woese *(116)* proposed the five-kingdom phylogenetic tree, the bacteria were divided into twelve groups. The initial evaluation was done primarily with bacteria in culture. By 2000, the number of groups had expanded to 36, of which 13 do not have a representative in culture *(38)*. Approximately 6000 bacterial species have been described, but the number of bacteria that exist in nature is predicted to be as high as 600,000 *(39)*. The situation may even be more extreme for fungi. The currently accepted number of described species of fungi is 72,000, but the estimated figure of fungi that exist in nature is 1.5 million. This suggests that there are diverse novel microorganisms in the natural environment that could be used as sources for drug discovery. The argument against putting effort into culturing less-culturable organisms is that it is very time consuming and the techniques used for one organism may not be applicable to others. With this in mind, several biotech organizations are now seeking means to harness the potential of these less-culturable strains.

Microbial community analysis has revealed that the microorganisms in culture not only represent a small part of the population, but may not be the most prevalent in the natural environment *(12)*. It is expected that one of the largest efforts in the next decade will be exploring means to culture less-culturable organisms. It is thought that the reason for the enormous discrepancy between the total viable cell counts and those of culturable cells may be due to the following: (a) cell damage by oxidative stress, (b) formation of viable but nonculturable cells, (c) inhibition by high substrate concentrations, (d) induction of lysogenic phages upon starvation, and (e) lack of cell-to-cell communication in laboratory media *(41)*. Two main approaches have been used to enhance the resuscitation of less-culturable strains. The first is the addition of cell-signaling molecules and the second is the use of oligotrophic isolation media.

Microorganisms use pheromones to communicate with each other, both within and across species *(42)*. Microorganisms may require signaling from other organisms in order to grow, even if provided with the appropriate nutrients. The addition of growth factors to culture medium has been used successfully to increase the resuscitation of greater numbers of microorganisms and thus higher microbial diversity. The addition of pyruvate or catalase to reduce oxidative stress during isolation can increase the numbers and diversity of strains isolated *(43)*. The addition of cyclic AMP *(44)* and *N*-acyl-homoserine lactones have both been shown to increase the resuscitation of starved cultures under laboratory conditions *(41)*. In enterobacteria, cAMP is involved in the regulation of the majority of the genes expressed under starvation, including those coding for high-affinity sugar-transport systems *(45)*.

A second approach for increasing the resuscitation of less-culturable strains is the use of oligotrophic isolation media. It has been well documented that conventional media have extremely high concentrations of complex organic compounds compared with those present in the natural environment. Most isolation media allow for growth of only a selected group of strains and inhibit the majority of the natural population. Oligotrophic media not only allow the growth of less-culturable microorganisms but also prevent the overgrowth of fast-growing "microbial weeds" *(46)*. Using unamended site water as a growth medium, unique populations of microorganisms have been cultured *(47)*. A variation of this method is encapsulation of single cells within gel microdrops that contain low-nutrient media *(46)* or within specialized growth chambers incubated in site water *(42)*.

One argument against applying less-culturable strains to a drug-discovery program has been that they could not be cultured at a high enough cell density. The argument for including them is that although they may initially require the addition of growth factors or oligotrophic growth conditions, there is evidence that once cultured, the organisms can be grown out in nutrient-rich media. Using 960 cells cultured in microdrops, 67% of the cultures were able to grow to densities of $>10^7$ cells/mL *(46)*. This allows the cells to be cultured in a manner that could be easily applied to drug-discovery platforms.

An alternate approach to access unculturable microorganisms is to clone the DNA directly from uncultured microorganisms (*see* **Subheading 3.1.**).

3. Manipulating Microbial Physiology to Activate Microbial Natural-Product Machinery

In order to exploit the true potential of microorganisms, the physiological growth conditions used for generating extracts need to be diversified. And microbial metabolism can be influenced to produce qualitatively and quantitatively different chemical compounds. The physiology of secondary metabolism has often been neglected. Very few of the regulatory features of secondary metabolism have been elucidated *(48)*. The global situation in physiological regulation is very complex, as a result of the variety of microbes, the variety of biosynthetic pathways, and the variety of controls. Environmental conditions remain, however, a key element in the discovery and production of secondary metabolites. Strategies have to be developed in order to exploit the full metabolic potential of each microorganism in order to maximize chemical diversity. Biochemical pathways, induction, and regulation of secondary metabolism by internal molecules (such as the autoregulators) have been reviewed previously *(49,50)*.

3.1. Various Optima

The optimal conditions for biosynthesis of secondary metabolites are usually not identical to the ones for growth. In general, the optimal zones are narrower for secondary metabolite production. Physiological regulations vary with different microorganisms and different metabolic pathways. The qualitative and quantitative aspects for secondary metabolite production must be taken into consideration.

There are usually differences between the optimal carbon sources for growth and those that are good for secondary metabolism *(51)*. For example, glucose is an excellent carbon source for growth in most cases, but depresses the production of a series of secondary metabolites such as actinomycin, cephalosporin, ergot alkaloids, and tylosin. However, glucose does not interfere with the production of aflatoxin, aminoglycosides, or chloramphenicol *(52)*, and the production of anticapsin by *Streptomyces griseoplanus* is maximal at a concentration of glucose as high as 100 g/L *(53)*.

Secondary metabolic pathways are often negatively affected by nitrogen sources favorable to rapid growth. For example, ammonium salts inhibit the production of cephamycin, fusidin, and rifamycin *(54)*; however, some biosynthetic pathways are not affected, such as that for pyrroindomycins in *Streptomyces rugosporus (55)*. Optimal production of gibberellic acid by *Gibberella fujikuroi* in a defined medium requires a concentration of 22.5 mM ammonium sulfate. Complex natural sources of nitrogen such as soybeans and casamino acids are also good. The influence of amino acids on secondary metabolite production is very variable and can depend on the precursor or the natural inducer. Inorganic phosphate suppresses the synthesis of many secondary metabolites. Thus, the optimal phosphate concentration needed for production of secondary metabolites is generally lower than that required for growth. However, the optimal concentration can vary drastically between strains. The concentration can be as low as 0.08 mM for the synthesis of bacitracin by *Bacillus licheniformis*, or as high as 8 mM for the production of novobiocin by *Streptomyces griseus (56)*. In some incidences, high phosphate concentrations can even induce the biosynthesis of some metabolites *(57)*. Secondary metabolism often requires trace elements such as iron, zinc, and manganese. Once again, the optimal concentrations vary from process to process, but often range from less than 0.1 to 1×10^{-3} M *(58)*.

Optimal temperatures for the production of secondary metabolites are, in general, lower than for growth, but can vary considerably. For example, a temperature of 21°C is optimal for biosynthesis of cyclosporin by *Tolypocladium inflatum*, 25°C is optimal for the synthesis of streptomycin by *Streptomyces griseus*, and 28°C is best for nebramycin formation by *Streptomyces tenebrarius*. Most of the known secondary metabolites are produced under standard aeration conditions, but some require lower and some higher dissolved oxygen concentrations *(59)*. Extremely high aeration is required for optimal production of secondary metabolites by *Streptosporangium (60)*.

Incubation time is another key point and is dependent on the growth characteristics of the microorganism and the culture conditions. For example, actinomycetes can vary from 3 d for the maximum production of arylomycins by a strain of *Streptomyces (61)* to 12 d for maximum production of pramicidin S by a strain of *Actinomadura (62)*. The addition of adsorbents such as XAD-16 resin to liquid cultures can also enhance the concentration of secondary metabolites produced *(63)*. Most programs for the discovery of novel metabolites from microorganisms use liquid shaken cultures for cultivation of microorganisms. This provides an easy and well-controlled system. Solid-phase fermentation allows the biosynthesis of other metabolites, mainly related to the sporulation process *(64)*. Both types can be scaled up effectively *(65)*.

3.2. Selection of Culture Conditions

The optimal conditions for secondary metabolite production vary from microbe to microbe. The composition of media and the culture conditions have a great impact on the production of secondary metabolites. In a discovery program, one is working with a large series of unique and ubiquitous microorganisms. Multiple conditions are necessary in order to allow the expression of secondary metabolites. Both static and shaken liquid cultures should be incorporated. Different incubation temperatures must be chosen. Addition of elicitor compounds such as heavy metals, oils *(66)*, microbial or fungal cell-wall components, and dimethyl sulfoxide (DMSO) *(67)* can increase the biosynthesis of certain secondary metabolites. The media can include carbon and nitrogen sources at different concentrations, as well as other nutrients such as phosphate and trace elements or elicitors at various levels, and using a Greco-Latin square format. The goal is to have a good ratio between the number of strains (genetic diversity) and the number of culture conditions (metabolic expression) for each microbe in one collection.

Another powerful tool is the preselection of strains based upon growth in a series of media in microcultures or in small vials *(68)*. A standard format that is amenable to automation allows more than 20 media to be easily tested with each microorganism, at different temperatures, and in liquid as well as on solid media. This allows the selection of the best conditions for each strain, which can then be used to scale up to get larger volumes required for initial structure determination. This system is also adequate for a quick optimization program.

For each group of strains, a series of conditions can be chosen incorporating both shaken liquid cultures and stationary solid cultures. Typically, five media types are used for each batch of cultures, and three to five incubation conditions are chosen to include various temperatures. In order to enhance the chance of success in such a random process, the conditions used for one group must be rotated. The results of chemical profiling and scores in screening should be constantly analyzed in order to improve the system.

3.3. Physiological Exploitation of Talented Strains

Some microorganisms are able to produce a variety of compounds from different chemical families and are termed *metabolically talented (69)*. Most of the recently described microbial compounds are produced by actinomycetes, mostly *Streptomyces* strains, and by saprophytic filamentous fungi. One metabolically talented microorganism, *Streptomyces* sp. strain Go.40/10, synthesizes at least 30 different secondary metabolites, many of which are new compounds *(70)*. Some strains can synthesize more than 50 compounds, which can be detected only by classical chemical methods. Myxobacteria and fungi are also considered to be talented microbes *(71)*.

The genomes of actinomycetes (8 Mb) *(72)*, fungi (13–42 Mb) *(73)*, and myxobacteria (12 Mb) *(74)* are much larger than needed for all basic functions. Therefore, it is widely thought that part of the genome may contain genes for alternative metabolic pathways. For example, *Streptomyces coelicolor* A(3)2 is designated as a potent producer of secondary metabolites. It produces methylenomycin, prodigiosin, actinorhodin, and a calcium-dependent antibiotic. In addition, several formerly unknown gene clusters (polyketide syntheses type I and II, nonribosomal peptide synthases) have been found in its genome *(75)*. In *Streptomyces avermitilis* ATCC 31267, the producer of avermectin, 24 additional gene clusters have been sequenced *(76)*. Genomic data have suggested that the myxobacterium *Stigmatella aurantiaca* DW4/3-1 has a much broader capacity to produce a much broader group of natural products than those isolated to date from this organism *(77)*. Genes responsible for production of many compounds can be found in the genomes of nonproducing strains. The questions arise: Are these genes nonfunctional? Are the detection methods not powerful enough? Are these genes not expressed under standard growth conditions? Do the genes require external signaling to turn them on? More efficient detection methods, biochemical assays such as capillary electrophoresis *(78)*, or chemical methods will allow the discovery of large numbers of novel compounds *(79)*. As briefly described in **Subheading 3.1.** small changes in the culture conditions can have a major influence in the spectrum of secondary metabolites synthesized. For example, the fungal strain Sphaeropsidales strain F-24'707 is a producer of the antifungal compound cladospirone bisepoxide. When this strain was grown in a combination of different media and cultivation types, eight new spiro naphthalenes were isolated. There were previously only six known members of this class of compounds. The addition of inhibitors, such as tricyclazole, inhibited some pathways and therefore allowed the production of other compounds, such as two new spirobisnaphthalenes and a rare macrolide, mutolide *(79)*.

Many microorganisms do not readily express natural-product gene clusters when grown in the laboratory *(75)*. We have to find the right physiological signal to stimulate the molecular machinery. A systematic fermentation program should be conducted with "talented" strains and with representative strains of poorly known genera in order to maximize the number of compounds produced by each strain. Such an approach should include:

- Cultures grown in shaken liquid vessels and on solid media.
- Cultures grown with media of different composition

- Incubation at two or more temperatures.
- Incubation at two or more shaker speeds.
- Incubation for at least two different time periods.
- Media with at least two pH levels.
- Absorbents, enzyme inhibitors, elicitors, precursors, precursor analogs, and high concentration of salts should be added to the most productive fermentation media.

After selection of one or two potent media, the influence of the other factors can be analyzed using an experimental design, such as fractionated factorial or Plackett–Burman design *(80)*.

3.4. Co-Cultivation

Microbial communities also hold potential for the production of novel compounds. In nature, microorganisms do not exist alone; they are part of tiny ecosystems. There is expected to be diverse signaling and cross-feeding going on between organisms that will elicit production of novel compounds. Although the longstanding argument against this type of research has been that getting a stable mixed culture is almost impossible, it may provide a means for exploiting the true potential of the consortia as a whole. An example of using co-cultivation of two microorganisms producing related products has been suggested as a suitable way towards diversification of microbial structures *(81)*.

4. Genetically Modifying Strains to Produce "Unnatural Microbial Natural Products"

4.1. Expressing the Heterologous Metagenome in a Surrogate Host

Culturable organisms provide only a finite pool of secondary metabolites *(82)*. One approach to maximize the diversity of natural-product extract libraries has been to access the DNA directly from uncultured microorganisms. DNA can be isolated directly from an environmental sample, digested into large fragments with restriction enzymes, and cloned into an artificial vector *(14)*. The vector is then transformed into a surrogate host *(83)*. Environmental DNA libraries can be prepared with large fragments of DNA from a wide range of uncultivated bacteria within an environmental sample *(36)*. This is described as screening the metagenome, the genomes of the total microbiota in an environmental sample *(84)*. The recombinant approach thus obviates the need for culturing diverse microorganisms and provides a relatively unbiased sampling of the vast untapped genetic diversity present in various microenvironments. As an additional advantage, the genes encoding a product of interest are already isolated and can be analyzed using the tools of bioinformatics, thus providing a potential boost to the efforts of analytical chemists to identify the product. Furthermore, the possibility of regulating the expression of isolated environmental gene clusters or combining them with genes for other pathways to obtain new compounds could furnish a further advantage over traditional natural-product discovery methodologies *(113)*. However, it must be noted that these biosynthetic and regulatory genes could be dormant in the host, and optimal induction conditions may be required for the production of novel natural products.

Advances in DNA-sequencing and bioinformatics technologies now make it possible to rapidly identify the clusters of genes that encode bioactive compounds and to make computer predictions of chemical structure based on gene sequence information

(85). A high-throughput genome-scanning method has recently been developed that allows discovery of metabolic loci, independent of expression. Genome sequence tags (GSTs) are genes involved in natural-product biosynthesis. These GSTs are used as probes to screen for the presence of these genes within a clonal library. Any clone that contains a GST can then be screened for novel natural-product gene clusters. More than 450 natural-product clusters have been identified in this manner *(85)*.

4.2. Gene Mixing (Combinatorial Biosynthesis)

The traditional way to diversify unnatural microbial natural products is by random mutagenesis or by culturing microbes with nonnatural precursors. However, the discovery in prokaryotes that the genes for natural products are usually clustered, made it possible to clone an entire pathway into a vector *(14)*. Many natural-product genes are modular and produce multifunctional enzymes. They have a high degree of plasticity. By interchanging and moving around genes within these clusters, hybrid enzymes can be produced that are capable of synthesizing an unlimited set of new molecules *(82)*.

The modular nature of many secondary metabolite genes provides an ideal system to genetically engineer formation of unnatural microbial natural products by incorporating genes from different pathways. An example of such an approach involves the polyketide synthases (PKSs), which are large, multi-domain enzymes that produce polyketides including antibacterials, immunosuppressants, and cholesterol-lowering agents *(86)*. PKSs are encoded by a cluster of continuous genes and have a linear modular organization of similar catalytic domains that build and modify a polyketide backbone *(87)*. Microbial genes can be engineered to produce enzymes with novel catabolic activities *(82)*. The cloning of biosynthetic pathway genes from *Streptomyces* allowed the production of novel compounds by mixing the antibiotic systems of different antibiotic-producing strains *(72)*. Novel compounds were produced by gene transfer between strains producing the isochromanquinone antibiotics actinorhodin, granaticin, and medermycin *(88)*. This pioneering work has been developed by many others for the production of novel enzymes and unnatural natural products *(83,89,90)*. These modular PKS clusters have been manipulated through introduction of different loading domains that specify a branched chain or cyclic substrate and direct inactivation or insertion of individual catalytic domains to produce new enzymes *(87)*. Combinatorial biosynthesis has also been used effectively to generate novel compounds and enzymes in type I and type II PKS systems. This type of approach can be applied to many other modular enzyme systems *(91,92)*.

Genetic engineering and pathway modification will undoubtedly be important in strain optimization in the future. These methods provide a targeted approach for construction of novel pathways and in turn the potential of novel natural products *(93)*.

5. Monitoring Diversity by Dereplicating Microbial Strains and Chemical Extracts

The cost of screening microbial natural products is high, and false positives waste the limited resources available for isolation and structural characterization. In order to increase the output, it is important to spend more time upfront on the prescreening of strains and extracts. In order to focus on the extracts of most interest as quickly as possible, and to avoid repeatedly isolating the same common natural products, efficient dereplication process are of the utmost importance. One may argue that high-throughput screening, in which thousands of extracts can be screened in little time, makes the use of duplicate cultures redundant. However, we must consider that screening a more diverse set of cultures will increase the diversity of the extract library and in turn increase the chance of finding a "hit." Dereplication of the extracts can be done either at the level of the microorganism or the chemical extracts. Database-linking of microbial genetic taxonomy with extract diversity is extremely important (**Fig. 1**). Cluster analysis of an unknown sample and comparison of its taxonomy and extract chromatography to internal databases as well as published literature will provide valuable information to determine whether an extract and/or activity is novel or not.

5.1. Characterization and Selection of Microbial Strains

Ecopia BioSciences Inc. developed an automated genomics platform that predicts the chemical structures of natural products by reading the sequences of the gene clusters that direct their synthesis. By surveying the genome, all of the natural products that a microorganism can make are identified before fermentation studies begin, and the downstream production and purification strategies are specifically tailored to isolate likely new chemical entities (NCEs) and avoid the re-isolation of known compounds. The integration of new genomics technologies greatly increases the efficiency of discovery and makes it possible to build a robust pipeline of NCEs from a small collection of microorganisms, providing a new paradigm for natural-product discovery *(85)*. They scan the genomes of selected microorganisms that were reported to produce known, structurally diverse natural products, to build a database of gene clusters covering the full range of natural-product chemical diversity without sequencing entire genomes. This enables a strategy to prescreen in mini-scale for one microbe of interest in many more media and growth conditions, and look only for a specific compound property, such as molecular weight, ultraviolet (UV) absorbance, and lipophilicity predicted by the specific gene clusters. Then scale-up technologies could be used to identify and purify the specific compounds. The knowledge of the biosynthetic pathway of the compounds will guide rational design or mutagenesis to improve their yields.

Any library of microorganisms is likely to have a high number of duplicate strains. Although identical strains from the same site may be excluded on isolation, many strains of the same species may be collected over time from a wide range of collection sites. Although these strains may be useful later in strain optimization, they are redundant and costly in the initial screening phase and decrease the probability of finding novel compounds. Thus, it is important to dereplicate the culture collection. However, one must consider that strains of the same subspecies may produce different compounds.

Many methods can be used to dereplicate cultures, but molecular techniques are especially well suited to this type of analysis. However, bacterial taxonomists have not yet reached a consensus for defining the fundamental criteria of biological diversity to the species *(94)*. Prokaryotes exchange chunks of their genomes too frequently to make any meaningful species definition *(21)*. An accurate definition of a fungal species is also problematic. Certainly, for all types of microbes, the basic unit is the "ecotype" *(94)*, also called the "geovar" *(20)*. This is the reason why dereplication of strains has

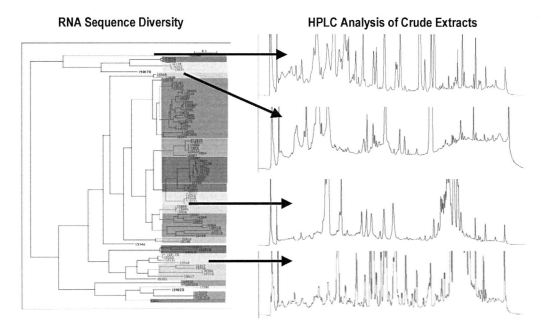

Fig. 1. Database linking microbial genetic taxonomy with extract diversity. The microbial genetic tree is generated based on ribosomal RNA sequence analysis. Chemical extract chromatography is linked to the taxonomy data by a bio-informatics approach.

to be considered only within populations isolated from a particular geographical/ecological region. A strain of *Streptomyces hygroscopicus* isolated from California is not necessarily identical to one isolated from China. Indeed, more than 200 secondary metabolites have been isolated from various *Streptomyces hygroscopicus* strains.

When building a library of microorganisms for use in HTS, thousands of strains are isolated and have to be chosen and characterized. The first step is careful morphological observation. This allows the cultures to be separated into taxonomic groups. Further speciation can be done using molecular or biochemical methods. Biochemical culture-dependent techniques such as fatty acid analysis, pyrolysis mass spectrometry *(95)*, and FT-IR analysis *(96)* were developed initially for clinical isolates and cannot be easily applied to environmental samples that require prolonged growth periods. Changes in the media, incubation temperature, and growth period can alter the profile of the organism, and hence results can be compared efficiently to one another only within one experiment. However, if the culture conditions can be standardized, the use of pyrolysis mass spectrometry analysis can reflect similar clustering of taxonomic groups as molecular methods *(97)*.

The morphogenic groups can be separated further using molecular methods such as restriction fragment length polymorphism (RFLP), which could differentiate strains to the subspecies level *(98–100)*. Ribosomal genes, including the intragenic spacer regions, have been used routinely to differentiate both fungal and actinomycete strains. Thousands of sequences are available in GenBank and the Ribosomal Database Project that can be used to phylogenetically identify interesting organisms. The pitfalls of relying on polymerase chain reaction (PCR)-based rRNA analysis as a measure of microbial diversity in environmental samples have been emphasized *(101)*. Sequencing of other molecular markers, although costly, does however allow identification to the species level. When used in combination with RFLP, strains can be separated to the subspecies level *(100)*.

5.2. Characterization and Choice of Microbial Extracts

Strains of the same species may generate different chemistry in the same media. The first physical characteristics of unknown natural products are determined during the chemical extraction and concentration steps. Solvent-based and acid-base partitioning experiments can help define hydrophobicity and types of functional groups of the natural-product structures.

The most effective selection method from the metabolic aspect is the chemical profile analysis by high-performance liquid chromatography (HPLC) and thin-layer chromatography (TLC) data *(102)*. Identical HPLC retention time or TLC R_f values may not tell you whether two compounds are exactly the same, but different values definitely indicate they are different. Micro-scale extraction procedures have been developed *(103)* and can be automated. The first selection criterion is the metabolic creativity of strains, as the number of peaks revealed by HPLC with detectors of evaporative light scattering detection (ELSD), vs photodiode array (PDA), chemiluminescent nitrogen (CLND), and time-of-flight mass spectrometry (TOFMS) *(104)*. Nielsen et al. proposed a method for identification and confirmation of chemical compound classification based on single- or multiple-wavelength chromatographic profiles *(102)*. Chromatographic matrices from analysis of previously identified samples are used for generating reference chromatograms, and new samples are compared with all chromatograms by calculating resemblance indices *(111)*. In addition, the method allows identification of characteristic sample components by local similarity calculations:

http://www.esainc.com/products/HPLC/Optical_Detectors/esa_ChromaChem.html

Detection from ELSD is based on the universal ability of particles to cause photon scattering when they traverse the path of a polychromatic beam of light. The liquid effluent from HPLC is first nebulized, and the resultant aerosol mist containing the analyte particles is directed through a light beam. A signal is generated that is proportional to the mass present, and is independent of the presence or absence of chromophores, fluorophores, or electroactive groups. Since essentially every compound can be separated by HPLC or micro-HPLC and detected by ELSD, the ESA ChromaChem is equipment that every chromatography laboratory should have. It is a mass-sensitive device, which provides a response directly proportional to an analyte's mass in the sample. The presence of functional groups or chromophores is not necessary for detection. Relative amounts of compounds can be easily assessed by evaporative light-scattering technology. Any nonvolatile analyte can be detected, and gradient elution can be employed to optimize the separation. Aqueous as well as solvent-based mobile phases can be used to detect compounds that are not generally "seen" by other detection techniques. This detector can be used in conjunction with mass spectrometers to provide a complete analysis of the sample. The ChromaChem's unique nebulization system and temperature-controlled drift tube provides sensitivity, reliability, and reproducibility.

The unit's small footprint requires a minimum of bench space, allowing use under space-limited conditions.

The basic methods to compare microbial extracts are HPLC-DAD (diode-array detection) and HPLC-MS. Researchers at Eli Lilly developed a rapid (about one sample per min) surrogate measure of microbial secondary metabolite production computed from the electrospray mass spectra of samples injected directly into a spectrometer *(105,106)*.

The development of a multi-channel mass-spectrometry interface has allowed analysis at high-throughput level. In most cases, LC-MS (liquid chromatography-mass spectrometry) is the most sensitive method for obtaining dereplication information about a compound. A recent development is an eight-way fully automated parallel LC-MS-ELSD system for the analysis of natural products *(107)*. LC-NMR (liquid chromatography nuclear magnetic resonance) should become operational in the near future, allowing the on-line identification of organic molecules *(108,109)*. LC-NMR, although it has lower sensitivity than LC-MS, provides a powerful tool for rapid identification of known compounds and identification of structural classes of novel compounds. LC-NMR is especially useful in instances where the data from LC-MS are incomplete or do not allow confident identification of the active component of a sample.

For strains and chemical tracking, an in-house database has to be built to integrate with commercially available ones such as Antibase database (Wiley Publishers, 2003) or the Dictionary of Natural Compounds (Chapman & Hall, London). There is no single technique that gives 100% confidence to differentiate any two natural-product chemical profiles, but computer-enhanced structural determination methods could integrate various spectral data and raise confidence.

5.3. Chemical Dereplication to Prevent Repeated Discovery

Key elements in the success of a natural-product discovery program are quick identification of bioactive compounds, early elimination of known or unwanted metabolites, and rapid determination of the structure of novel compounds. Dereplication strategies use analytical techniques and database searching to determine the identity of an active compound at an early stage. Dramatic improvements have been achieved during the past years mainly due to the impact of combinatorial chemistry. Natural-product chemistry has to take advantages of these recent developments. The final separation-purification procedures are not discussed here since the procedures are complex and depend on the characteristics of the targeted compounds; this is beyond the scope of this chapter.

In the search for new microbial natural products, we have to consider the frequency of re-isolation of already-described metabolites from microbial cultures. Rough estimates suggest that we have isolated only a minute fraction of the compounds that exist in nature. Full identification of a natural product should be done only after partial purification to determine whether this type of compound is already known or has potential as a useful drug.

Natural-product samples need to be normalized by concentration or weight. Common "interfering" groups of compounds such as detergent-like or toxic compounds should be removed. Samples should be grouped into related chemical classes and then prioritized for further fractionation. The hit profile coupled with genetic and morphological characterization of the strains will build an increasing level of confidence in a putative structure.

The first physical characteristics of unknown natural products are determined during the chemical extraction and concentration steps. Partitioning experiments with solvents exhibiting a range of polarities and pH values will shed light on hydrophobicity and charged functional groups of the natural products.

If the natural-product extract contains a reported commercially available compound, the sample and the reference standard should be co-injected for TLC and HPLC. Identical retention times (HPLC) or Rf values (TLC) may not tell you that they are exactly the same, but different values definitely indicate they are different. If a standard is not available, the hypothesis could be tested by employing a physical test such as MS, looking for ions of the same approximate molecular weight. The chromatographic behavior could also reveal the nature of the compounds of interest, such as logP, surface area, and dipole moment.

Intelligent screening approaches towards microbial natural products are also required. One of the major limiting factors in the drug-discovery industry is that pharmaceuticals have been traditionally designed to target individual factors in a disease system, but diseases are complex in nature and vulnerable at multiple attack points. Therefore, a systematic novel synergistic drug-screening approach based on a multifactorial principle is urgently needed. Many drugs could be more effective at a reduced dosage if low dosages of other synergistic compounds are introduced simultaneously. Many marketed traditional medicines have demonstrated great efficacy and safety profiles in their long history. However, when efforts were made to purify a single molecule, the activity often was lost. SynerZ Pharmaceuticals Inc. has developed a drug-discovery approach consonant with the systems biology framework, and complementary to the target-based approach. Synergistic co-drugs from natural products will enable existing drugs to be more effective and contribute to our better understanding of multiple pathways to cure disease.

Ketoconazole is commonly used to treat *Candida* infections. However, at clinical doses, ketoconazole is associated with important toxic side effects, including hepatitis. In addition, resistant strains often emerge during long-term or prophylactic treatment as a result of the necessarily high concentrations of drug required. The concentration of ketoconazole alone at 0.01 µg/mL gave only about 20% inhibition of growth (**Fig. 2**). When ketoconazole was tested at 1 µg/mL, it gave 90% inhibition of growth. However, the combination of ketoconazole at 0.01X with F0101604 achieved about 95% inhibition (better than 100-fold the ketoconazole amount used here) and the mode of action was cidal, showing a synergistic effect of the two components rather than an additive effect. The natural product SNZ101 purified from F0101604 not only improved the efficacy of a much reduced dosage of ketoconazole, but also broadened its spectrum on drug-resistant strains and reduced its side effects (data not shown). It is clear that natural product F0101604 would be disregarded in conventional screening technology for antifungal lead discovery, because by itself it failed to show any growth inhibition on fungal pathogen *Candida parapsilosis* ATCC 22019.

6. Closing Remarks

One prerequisite to natural-product discovery that remains paramount is the range and novelty of molecular diversity. Currently, natural products are going through a

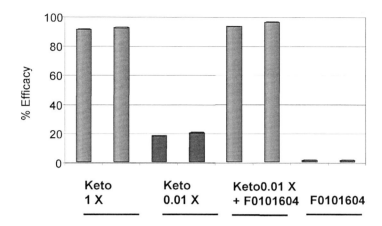

Fig. 2. Synergistic effect of F0101604 with a low dosage of ketoconazole (X = 1 µg/mL). Equal amounts (10^4/mL) of Candida parapsilosis ATCC 22019 are cultured in Mueller-Hinton (MH) broth with Alamar Blue dye in the presence and absence of a subclinical concentration of ketoconazole. Samples are treated as labeled in duplicate. Fluorescence reading after overnight incubation at 35°C in a moist chamber is measured at Ex 544 nm and Em 590 nm, and converted to percentage of growth inhibition.

phase of reduced interest in the drug-discovery field *(112)*. However, new developments may reverse this negative perception.

The systematic exploitation of selected ecosystems, combined with the development of new techniques and media for isolation of novel microorganisms, will allow the collection of representative strains from a large part of the micropopulation. This maximized biodiversity will deliver chemical diversity for an ecosystem.

The direct expression of environmental DNA in heterologous surrogate hosts is progressing. There is a need for rapid and sensitive detection and characterization of new metabolites as well as their gene clusters.

Physiological manipulation should be based on experimental design and measurement of secondary metabolism. Co-cultures will give novel insight into secondary metabolism and require the development of new vessels for stable mixed-culture fermentation.

Gene mixing coupled with the genetic engineering power of PKS, for example, will allow the generation of hybrid or "unnatural microbial natural products."

Total synthesis of natural products with interesting biological activities is paving the way for preparation of new and improved analogs. Combinatorial chemistry permits the selection of the best drug from a large number of candidates. Beyond synthesis and evaluation of organic molecules, a number of new bioorganic methods are emerging on the horizon.

In natural-product chemistry, the rapid and accurate differentiation of chemical compound profiles is based on on-line measurement by LC-ELSD, DAD, MS, and NMR. Automated comparisons of the metabolite profiles of microorganisms can be used as a valuable method for building libraries of natural products for drug discovery *(114)*.

Today, more than 30,000 diseases are clinically described, but less than one-third of these can be treated symptomatically, and only a few can be cured *(110)*. New chemical entities as therapeutic agents are urgently desired. Natural products can play a main role in drug discovery. New strategies for natural products-based drug discovery will increase chemical diversity and reduce redundancy *(115)*. Maximizing the discovery of new compounds and minimizing the re-evaluation of already known natural products will be crucial.

Acknowledgments

Thousands of people have contributed to our understanding of microbial natural products for drug discovery in the past century, and I acknowledge their efforts even though they may not all be cited in this chapter. I am indebted to members of my laboratory, past and present, who have made several of the major contributions cited in this review. I gratefully thank Springer-Verlag GmbH for granting permission that for this chapter, I could expand the contents of my previously invited review "Diversifying microbial natural products for drug discovery," published in Appl Microbiol Biotechnol (2003) 62:446–458 (DOI 10.1007/s00253-003-1381-9).

References

1. Demain AL. Pharmaceutically active secondary metabolites of microorganisms. Appl Microbiol Biotechnol 1999;52:455–463.
2. Chen J, Zheng XF, Brown EJ, Schreiber SL. Identification of an 11-kDa FKBP12-rapamycin-binding domain within the 289-kDa FKBP12-rapamycin-associated protein and characterization of a critical serine residue. PNAS 1995;92:4947–4951.
3. VanMiddlesworth F, Cannell RJP. Dereplication and partial identification of natural products. In: Cannell RJ, ed. Methods In Biotechnology, vol. 4: Natural Product Isolation. Humana Press, Inc., Totowa, NJ: 1998:279–327.
4. Urizar NL, Liverman AB, Dodds DT, et al. A natural product that lowers cholesterol as an antagonist ligand for FXR. Science 2002;296:1703–1706.
5. Grabley S, Thiericke R. The impact of natural products on drug discovery. In: Grabley S, Thiericke R, eds. Drug Discovery From Nature. Springer, New York: 1999:3–37.
6. Che Y, Gloer J, Koster B, Malloch D. Decipinin A and decipienolides A and B: new bioactive metabolites from the coprophilous fungus *Podospora decipiens*. J Nat Prod 2002;65:916–919.
7. Auerbach D, Thaminy S, Hottiger MO, Stagljar I. The post-genomic era of interactive proteomics: facts and perspectives. Proteomics. 2002;2:611–623.
8. Fernandes P. Molecular recognition: identifying compounds and their targets. J Cell Biochem 2001;137:1–6.
9. Verdine G. The combinatorial chemistry of nature. Nature 1996;384 (Supp):11–13.
10. Breinbauer R, Vetter IR, Waldmann H. From protein domains to drug candidates-natural products as guiding principles in the design and synthesis of compound libraries. Angew Chem Int Ed Engl 2002;41:2879–2890.
11. Czaran TL, Hoekstra RF, Pagie L. Chemical warfare between microbes promotes biodiversity. PNAS 2002;99:786–790.
12. Harvey A. Strategies for discovery drugs from previously unexplored natural products. Drug Discov Today 2000;5:294–300.
13. Firn RD, Jones CG. The evolution of secondary metabolism—a unifying model. Mol Microbiol 2000;37:989–994.
14. Handelsman J, Rondon M, Brady S, Clardy J, Goodman R. Molecular biological access to the chemistry of unknown soil microbes: a new frontier for natural products. Chem Biol 1998;5:R245–R249.

15. Monaghan RL, Polishook JD, Pecore VJ, Bills GF, Nallin M, Omstead S. Discovery of novel secondary metabolites from fungi—is it really a random walk through a random forest? Can J Bot 1995;73:S925–S931.
16. Strobel G. Rainforest endophytes and bioactive products. Crit Rev Biotechnol 2002;22:315–333.
17. Shrestha K, Strobel G, Shrivastava SP, Gewali M. Evidence for paclitaxel from three new endophytic fungi of Himalayan yew of Nepal. Planta Med 2001;67:374–376.
18. Foissner W. Notes on the soil ciliate biota (Protozoa, Ciliophora) from the Shimba Hills in Kenya (Africa): diversity and description of three new genera and ten new species. Biodivers Conserv 1999;8:319–389.
19. Bull AT. Clean technology: industry and environment, a viable partnership? Biologist (London) 2000;47:61–64.
20. Staley J, Gosink J. Poles apart: biodiversity and biogeography of sea ice bacteria. Annu Rev Microbiol 1999;53:189–215.
21. Doolittle W. Phylogenic classification and the universal tree. Science 1999;284:2124–2128.
22. Tulp M, Bohlin L. Functional versus chemical diversity: is biodiversity important for drug discovery? Trends Pharmacol Sci 2002;23:225–231.
23. Jensen PR, Fenical W. Strategies for the discovery of secondary metabolites from marine bacteria: ecological perspectives. Annu Rev Microbiol 1994;48:559–584.
24. Proksch P, Edrada RA, Ebel R. Drugs from the sea-current status and microbiological implications. Appl Microbiol Biotechnol 2002;59:125–134.
25. Mincer TJ, Jensen PR, Kauffman CA, Fenical W. Widespread and persistent populations of a major new marine actinomycete taxon in ocean sediments. Appl Environ Microbiol 2002;68: 5005–5011.
26. Feling RH, Buchanan GO, Mincer TJ, Kauffman CA, Jensen PR, Fenical W. Salinosporamide A: a highly cytotoxic proteasome inhibitor from a novel microbial source, a marine bacterium of the new genus *Salinospora*. Angew Chem Int Ed Engl 2003;42:355–357.
27. Whitman WB, Coleman DC, Wiebe WJ. Prokaryotes: the unseen majority. PNAS 1998;95: 6578–6583.
28. Davidson SK, Allen SW, Lim GE, Anderson CM, Haygood MG. Evidence for the biosynthesis of bryostatins by the bacterial symbiont "Candidatus endobugula sertula" of the bryozoan *Bugula neritina*. Appl Environ Microbiol 2001;67:4531–4537.
29. Zinniel D, Lambrecht P, Harris NB, et al. Isolation and characterization of endophytic colonizing bacteria from agronomic crops and prairie plants. Appl Environ Microbiol 2002;68:2198–2208.
30. Castillo UF, Strobel GA, Ford EJ, et al. Munumbicins, wide-spectrum antibiotics produced by *Streptomyces* NRRL 30562, endophytic on *Kennedia nigriscans*. Microbiology 2002;148: 2675–2685.
31. Ananda K, Sridhar K. Diversity of endophytic fungi in the roots of mangrove species on the west coast of India. Can J Microbiol 2002;48:871–878.
32. Tan RX, Zou WX. Endophytes: a rich source of functional metabolites. Nat Prod Rep. 2001;18(4):448–459.
33. Wei ZM, Laby RJ, Zumoff CH, et al. Harpin, elicitor of the hypersensitive response produced by the plant pathogen *Erwinia amylovora*. Science 1992;257:85–88.
34. Rikkinen J, Oksanen I, Lohtander K. Lichen guilds share related cyanobacterial symbionts. Science 2002;297:357.
35. Ahmadjian V. Lichens. Annu Rev Microbiol 1965;19:1–20.
36. Courtois S, Cappellano CM, Ball M, et al. Recombinant environmental libraries provide access to microbial diversity for drug discovery from natural products. Appl Environ Microbiol 2003;69:49–55.
37. Kellenberger E. Exploring the unknown: the silent revolution of microbiology. EMBO Rep 2001;2:5–7.

38. Hugenholtz P, Goebel BM, Pace NR. Impact of culture-independent studies on the emerging phylogenetic view of bacterial diversity. J Bacteriol 1998;180:4765–4774.
39. Davies J. Millennium bugs. Trends Biochem Sci 1999;24:M2–M5.
40. Hunter-Cevera J, Belt A. Isolation of cultures. In: Demain AL, Davies J, eds. Manual of Industrial Microbiology and Biotechnology. American Society for Microbiology, Washington DC: 1999:3–20.
41. Bruns A, Cypionka H, Overmann J. Cyclic AMP and acyl homoserine lactones increase the cultivation efficiency of heterotrophic bacteria from the central Baltic Sea. Appl Environ Microbiol 2002;68:3978–3987.
42. Kaeberlein T, Lewis K. Isolating "uncultivable" microorganisms in pure culture in a simulated natural environment. Science 2002;296:1127–1129.
43. Brewer DG, Martin SE, Ordal ZJ. Beneficial effects of catalase or pyruvate in a most-probable-number technique for the detection of *Staphylococcus aureus*. Appl Environ Microbiol 1977;34:797–800.
44. Kalish H, Camp JE, Stepien M, Latos-Grazynski L, Balch AL. Reactivity of mono-meso-substituted iron(II) octaethylporphyrin complexes with hydrogen peroxide in the absence of dioxygen. Evidence for nucleophilic attack on the heme. J Am Chem Soc 2001;123:11,719–11,727.
45. Ferenci T. Adaptation to life at micromolar nutrient levels: the regulation of *Escherichia coli* glucose transport by endoinduction and cAMP. FEMS Microbiol Rev 1996;18:301–317.
46. Zengler K, Toledo G, Rappe M, et al. Cultivating the uncultured. PNAS 2002;99:15,681–15,686.
47. Connon SA, Giovannoni SJ. High-throughput methods for culturing microorganisms in very-low-nutrient media yield diverse new marine isolates. Appl Environ Microbiol 2002;68:3878–3885.
48. Demain AL. Induction of secondary metabolism. Int Microbiol 1998;1:259–64.
49. Demain AL. Microbial natural products: alive and well in 1998. Nat Biotechnol 1998;16:3–4.
50. Horinouchi S. A microbial hormone, A-factor, as a master switch for morphological differentiation and secondary metabolism in *Streptomyces griseus*. Front Biosci 2002;7:2045–2057.
51. Betina V. Bioactive secondary metabolites of microorganisms. Progr Ind Microbiol 1994;30:5–14.
52. Luchese R, Harrigan W. Biosynthesis of aflatoxin—the role of nutritional factors. J Appl Bacteriol 1993;74:5–14.
53. Boeck L, Christy KL. Production of anticapsine by *Streptomyces griseoplanus*. Appl Microbiol Biotechnol 1971;21:1075–1079.
54. Aharonowitz Y. Nitrogen metabolite regulation of antibiotic biosynthesis. Annu Rev Microbiol 1980;34:209–233.
55. Abbanat D, Maiese W. Biosynthesis of the pyrroindomycins by *Streptomyces rugosporus* LL-42D005; characterization of nutrient requirements. J Antibiot 1999;52:117–126.
56. Gotoh T, Nakahara K, Hashimoto M, et al. Studies on a new immunoacrive peptide, FK-156. II Fermentation, extraction and chemical and biological characterization. J Antibiot 1982;35:1286–1292.
57. Shimada N, Hasegawa S, Harada T, Tomisawa T, Fuji A, Takita T. Oxetanocin, a novel nucleoside from bacteria. J Antibiot 1986;39:1623–1625.
58. Weinberg E. Secondary metabolism: regulation by phosphate and trace elements. Folia Microbiol 1978;23:496–504.
59. Barberel S, Walker J. The effect of aeration upon secondary metabolism of microorganisms. Biotechnol Genet Eng Rev 2000;17:281–323.
60. Pfefferle C, Theobald U, Gurtler H, Fiedler H. Improved secondary metabolite production in the genus *Streptosporangium* by optimization of the fermentation conditions. J Biotechnol 2001;23:135–142.

61. Schimana J, Gebhardt K, Holtzel A, et al. Arymomycins A and B, bew biaryl-bridged lipopeptide antibiotics produced by *Streptomyces* sp. Tu6075. I Taxonomy, fermentation, isolation and biological activities. J Antibiot 2002;55:565–570.
62. Saitoh K, Tenmyo O, Yamamoto S, Furumai T. Pramicidin S, a new pramicidin analog. I Taxonomy, fermentation and biological activities. J Antibiot 1993;46:580–588.
63. Gerth K, Bedorf N HG, Irschik H, Reichenbach H. Epothilons A and B: antifungal and cytotoxic compounds from *Sorangium cellulosum* (Myxobacteria). Production, physico-chemical and biological properties. J Antibiot 1996;49:560–563.
64. Calvo AM, Wilson RA, Bok JW, Keller NP. Relationship between secondary metabolism and fungal development. Microbiol. Mol Biol Rev 2002;66:447–459.
65. Robinson T, Singh D, Nigam P. Solid-state fermentation: a promising microbial technology for secondary metabolite production. Appl Microbiol Biotechnol 2001;55:284–289.
66. Sandor E, Szentirmai A, Paul GC, Thomas CR, Pocsi L, Karaffa L. Analysis of the relationship between growth, cephalosporin C production, and fragmentation in *Acremonium chrysogenum*. Can J Microbiol 2001;47:801–806.
67. Chen G, Wang YS, Li X, Waters B, Davies J. Enhanced production of microbial metabolites in the presence of dimethyl sulfoxyde. J Antibiot 2000;53:1145–1153.
68. Minas W, Bailey JE, Duetz W. Streptomycetes in micro-cultures: growth, production of secondary metabolites, and storage and retrieval in the 96-well format. Antonie Van Leeuwenhoek 2000;78:297–305.
69. Trujillo M, H.U. Gremlich, J.J. Sanglier. Selection strategy of traditional microorganisms for pharmacological screenings. Dev Ind Microb 1997;33: 35–42.
70. Schiewe HJ, Zeeck A. Cineromycins, gamma-butyrolactones and ansamycins by analysis of the secondary metabolite pattern created by a single strain of *Streptomyces*. J Antibiot 1999;52:635–642.
71. Reichenbach H. Myxobacteria, producers of novel bioactive substances. J Ind Microbiol Biotechnol 2001;27:149–156.
72. Hopwood D. Forty years of genetics with *Streptomyces*: from in vivo through in vitro to silico. Microbiology 1999;145:2183–2202.
73. Kupfer D, Reece CA, Clifton SW, Roe BA, Prade RA. Multicellular ascomycetous fungal genomes contain more than 8000 genes. Fungal Genet Biol 1997;21:364–372.
74. Pradella S, Hans A, Spoer C, Reichenbach H, Gerth K, Beyer S. Characterisation, genome size and genetic manipulation of the myxobacterium *Sorangium cellulosim* So ce56. Arch Microbiol 2002;178:484–492.
75. Bentley S, Chater KF. Complete genome sequence of the model actinomycetes *Streptomyces coelicolor* A(3)2. Nature 2002;417:141–147.
76. Omura S, Ikeda H, Ishikawa J, et al. Genome sequence of an industrial microorganism *Streptomyces avermitilis* deducing the ability of producing secondary metabolites. Proc Natl Acad Sci USA 2001;98:12,215–12,220.
77. Silakowski B, Kunze B, Muller R. Multiple hybrid polyketide synthase/non-ribosomal peptide synthetase gene clusters in the myxobacterium *Stigmatella aurantiaca*. Gene 2001;275: 233–240.
78. Pierceal, W., L. Zhang, and D. Hughes. Affinity capillary electrophoresis analyses of protein-protein interactions in target-directed drug discovery. In Haian Fu (ed), "Methods in Molecular Biology, vol 261: Protein-Protein Interactions", Humana, Totowa, NJ: 2003;187–197.
79. Bode HB, Bethe B, Hofs R, Zeeck A. Big effects from small changes: possible ways to explore nature's chemical diversity. Chembiochem 2002;3:619–627.
80. Wieling J, Dijkstra H, Mensink CK, et al. Chemometrics in bioanalytical sample preparation. A fractionated combined mixture and factorial design for the modelling of the recovery of five

tricyclic amines from plasma after liquid-liquid extraction prior to high-performance liquid chromatography. J Chromatogr 1993;629:181–199.
81. Degenkolb T, Heinze S, Schlegel B, Strobel G, Grafe U. Formation of new lipoaminopeptides, acremostatins A,B, and C, by co-cultivation of *Acremonium* sp. Tbp-5 and mycogene rosea DSM 12973. Biosci Biotechnol Biochem 2002;66:883–890.
82. Kennedy J, Hutchinson CR. Nurturing nature: engineering new antibiotics. Nat Biotechnol 1999;17:538–539.
83. Stokes HW, Holmes AJ, Nield BS, et al. Gene cassette PCR: sequence-independent recovery of entire genes from environmental DNA. Appl. Environ. Microbiol. 2001;67:5240–5246.
84. Rondon MR, August PR, Bettermann AD, et al. Cloning the soil metagenome: a strategy for accessing the genetic and functional diversity of uncultured microorganisms. Appl Environ Microbiol 2000;66:2541–2547.
85. Zazopoulos E, Huang K, Staffa A, et al. A genomics-guided approach for discovering and expressing cryptic metabolic pathways. Nature Biotechnol 2003;21:187–190.
86. Xue Q, Ashley G, Hutchinson CR, Santi DV. A multiplasmid approach to preparing large libraries of polyketides. PNAS 1999;96:11,740–11,745.
87. McDaniel R, Thamchaipenet A, Gustafsson C, et al. Multiple genetic modifications of the erythromycin polyketide synthase to produce a library of novel "unnatural" natural products. PNAS 1999;96:1846–1851.
88. Hopwood DA, Malpartida F, Kieser HM, et al. Production of "hybrid" antibiotics by genetic engineering. Nature 1985;314:642–644.
89. Seow K, Meurer G, Gerlitz M, Wendt-Pienkowski E, Hutchinson C, Davies J. A study of iterative type II polyketide synthases, using bacterial genes cloned from soil DNA: a means to access and use genes from uncultured microorganisms. J Bacteriol 1997;179:7360–7368.
90. Christiansen G, Fastner J, Erhard M, Borner T, Dittmann E. Microcystin biosynthesis in Planktothrix: genes, evolution, and manipulation. J Bacteriol 2003;185:564–572.
91. Walsh CT. Combinatorial biosynthesis of antibiotics: challenges and opportunities. Chembiochem 2002;3:125–134.
92. Rix U, Fischer C, Remsing LL, Rohr J. Modification of post-PKS tailoring steps through combinatorial biosynthesis. Nat Prod Rep 2002;19:542–580.
93. Zhang YX, Perry K, Vinci VA, Powell K, Stemmer WP, del Cardayre SB. Genome shuffling leads to rapid phenotypic improvement in bacteria. Nature 2002;415:644–646.
94. Cohan FM. Bacterial species and speciation. Syst Biol 2001;50:513–524.
95. Goodfellow M, Freeman R. Curie-point pyrolysis mass spectrometry as a tool in clinical microbiology. Zentralbl Bakteriol 1997;285:133–156.
96. Bastert J, Korting HC, Traenkle P, Schmalreck AF. Identification of dermatophytes by Fourier transform infrared spectroscopy (FT-IR). Mycoses 1999;42:525–528.
97. Brandao PF, Torimura M, Kurane R, Bull AT. Dereplication for biotechnology screening: PyMS analysis and PCR-RFLP-SSCP (PRS) profiling of 16S rRNA genes of marine and terrestrial actinomycetes. Appl Microbiol Biotechnol 2002;58:77–83.
98. Brandao PF, Clapp JP, Bull AT. Discrimination and taxonomy of geographically diverse strains of nitrile-metabolizing actinomycetes using chemometric and molecular sequencing techniques. Environ Microbiol 2002;4:262–276.
99. Vermis K, Vandekerckhove C, Nelis HJ, Vandamme PA. Evaluation of restriction fragment length polymorphism analysis of 16S rDNA as a tool for genomovar characterisation within the *Burkholderia cepacia* complex. FEMS Microbiol Lett 2002;214:1–5.
100. Schloter M, Lebuhn M, Heulin T, Hartmann A. Ecology and evolution of bacterial microdiversity. FEMS Microbiol Rev 2000;24:647–660.

101. von Wintzingerode F, Bocker S, Schlotelburg C, et al. Base-specific fragmentation of amplified 16S rRNA genes analyzed by mass spectrometry: a tool for rapid bacterial identification. PNAS 2002;99:7039–7044.
102. Nielsen NP, Smedsgaard J, Frisvad JC. Full second-order chromatographic/spectrometric data matrices for automated sample identification and component analysis by non-data-reducing image analysis. Anal Chem 1999;71:727–735.
103. Smedsgaard J. Micro-scale extraction procedure for standardized screening of fungal metabolite production in cultures. J Chromatogr A 1997;760:264–270.
104. Yurek DA, Branch DL, Kuo MS. Development of a system to evaluate compound identity, purity, and concentration in a single experiment and its application in quality assessment of combinatorial libraries and screening hits. J Comb Chem 2002;4:138–148.
105. Higgs RE, Zahn JA, Gygi JD, Hilton MD. Rapid method to estimate the presence of secondary metabolites in microbial extracts. Appl Environ Microbiol 2001;67:371–376.
106. Zahn JA, Higgs RE, Hilton MD. Use of direct-infusion electrospray mass spectrometry to guide empirical development of improved conditions for expression of secondary metabolites from actinomycetes. Appl Environ Microbiol 2001;67:377–386.
107. Cremin PA, Zeng L. High-throughput analysis of natural product compound libraries by parallel LC-MS evaporative light scattering detection. Anal Chem 2002;74:5492–5500.
108. Bobzin SC, Yang S, Kasten TP. LC-NMR: a new tool to expedite the dereplication and identification of natural products. J Ind Microbiol Biotechnol 2000;25:342–345.
109. Bobzin SC, Yang S, Kasten TP. Application of liquid chromatography-nuclear magnetic resonance spectroscopy to the identification of natural products. J Chromatogr B Biomed Sci Appl 2000;748:259–267.
110. Schultz M, Tsaklakidis C. Nach Chem Tech Lab 1997;45:159–165.
111. Garcia JB, Tormo JR. HPLC Studio: a novel software utility to perform HPLC chromatogram comparison for screening purposes J Biomol Screen 2003;8(3):305–315
112. Newman DJ, Cragg GM, Snader KM. Natural products as sources of new drugs over the period 1981–2002. J Nat Prod 2003;66(7):1022–1037
113. Martinez A, Kolvek SJ, Tiong Yip CL, et al. Genetically modified bacterial strains and novel shuttle BAC vectors for constructing environmental libraries and detecting heterologous natural products in multiple expression hosts. Appl Environ Microbiol 2004;70:2452–2463.
114. Tormo JR, García JB, DeAntonio M, et al. A method for the selection of production media for actinomycete strains based on their metabolite HPLC profiles. J Ind Mic Biotech 2003;30: 582–588.
115. Knight V, Sanglier JJ, DiTullio D, et al. Diversifying microbial natural products for drug discovery. Appl Microbiol Biotech 2003;62:446–458.
116. Woese CR. Bacterial evolution. Microbiol Rev 1987;51:221–271.

3
Automated Analyses of HPLC Profiles of Microbial Extracts

A New Tool for Drug Discovery Screening

José R. Tormo and Juan B. García

Summary

Despite the fact that natural products have historically been a prolific source of new compounds, pharmaceutical drug discovery programs have moved away from natural products in favor of synthetic approaches. However, the abundance of synthetic compounds with similar functional groups and, therefore, limited chemical diversity has renewed interest in nature as a good resource for finding new ideas to be applied to the design of the next generation of drugs.

One of the main issues for drug discovery programs based on microbial natural products is how to obtain the maximum potential from microbial strains in terms of chemical diversity of the metabolites produced. Several approaches are now available for enhancing the production and diversity of secondary metabolites from wild-type microorganisms. Automated comparisons of the metabolite profiles of microorganisms can be used as a valuable method for building libraries of natural products for drug discovery. Specific computer analyses of high-performance liquid chromatography chromatograms from organic extracts of fermented microorganisms can be used as a tool for increasing chemical diversity of collections, media improvement, evaluation of natural products libraries, and even determination of taxonomic correlations. Examples of what can be done using some of the new generation of software tools to compare profiles of secondary metabolites include the evaluation of extraction solvents and fermentation formats for the design of natural-product collections, and even the determination of relationships among strains from different origins.

Key Words: HPLC; automation; metabolite profiles; chemical diversity; extraction solvent; fermentation format; fermentation media; actinomycetes; fungi; taxonomy.

1. Learning from Experiences

In recent years, pharmaceutical companies have focused their efforts on finding new leads from combinatorial synthetic libraries. The large number of compounds that can be generated and tested is enticing, and there is a perceived incompatibility of natural products with some of the refined detection techniques commonly used in modern high-throughput screening *(1)*. However, current evidence shows that combining small elements does not necessarily lead to new chemical entities or real diversity *(2)*. Since Nature has been creating new chemical structures in a very efficient manner over millions of years, the large variety of bioactive compounds obtained from microorganisms actually makes them the most specialized synthetic laboratories in practice *(3)*.

From: Natural Products: Drug Discovery and Therapeutic Medicine
Edited by: L. Zhang and A. L. Demain © Humana Press Inc., Totowa, NJ

We should therefore learn from our experiences and give natural products another chance. Why not approach natural-products discovery from a renewed and modern perspective? For example, a variety of new and modern techniques have been developed, including recent initiatives that use natural products as the building blocks for creating combinatorial libraries, and even using purified natural products for creating such libraries *(4,5)*. With the goal of benefiting from such new approaches, we analyzed and improved the capability of microorganisms to produce secondary metabolites, incorporating lessons learned from past experiences in drug discovery. The efforts of our laboratory resulted in the design of a high-quality natural-products library (HQ-NPL), which we believe offers some clear advantages over traditional natural-products libraries.

2. Development of the HPLC Studio Tool

The preparation of natural-products samples for drug discovery usually begins with the fermentation of microorganisms. A historically accepted approach to increasing the chances of finding new drugs has been to maximize chemical diversity by broadening the biodiversity of the microbes, both geographically and taxonomically *(6)*. In addition, it is well established that the production of secondary metabolites by microorganisms is also highly dependent on the components of the fermentation media *(7,8)*. Theoretically, obtaining the maximum yield of secondary metabolites from a strain would entail growing each microorganism in the maximum possible number of media. The combination of source diversity multiplied by diversity of the fermentation conditions should produce the largest number of secondary metabolites per strain. However, increasing the number of sources, microorganisms, and the number of fermentations per microorganism and/or the number of extracts per fermentation are all quite labor intensive, and may be redundant. So the advantages gained thereby need to be balanced against practical constraints.

Chemometrics, i.e., automated measurement of the chemical diversity of extracts from fermentation broths, is a key tool with which to balance research and investment. Automated measurement has been accomplished in our laboratory by creating a software tool that, among other features, selects from a large array of small-scale fermentation conditions those few conditions that provide maximum chemical diversity and quantity of material *(9,10)*. The media selected from these analyses can be used to grow each strain in large scale to generate material for creating a HQ-NPL.

In practice, the automated selection process includes four parts: fermentation of each microorganism in a battery of small-scale production media, extraction of those fermentations, chemical analyses, and final automated treatment of the data to select the most appropriate components (*see* **Fig. 1**). The detailed process will be described in the following subheadings. After this chemometric analysis and selection, each microorganism is grown on a large scale to meet screening requirements. We can thus take advantage of the ability of microorganisms to produce different secondary metabolites depending on the nature of the fermentation medium. The selection of those media in which broader chemical diversity is obtained is achieved through computer analyses of the high-performance liquid chromatography (HPLC) chromatograms of extracts, ultimately leading to an increase in the quality of a natural-products library for drug discovery screening.

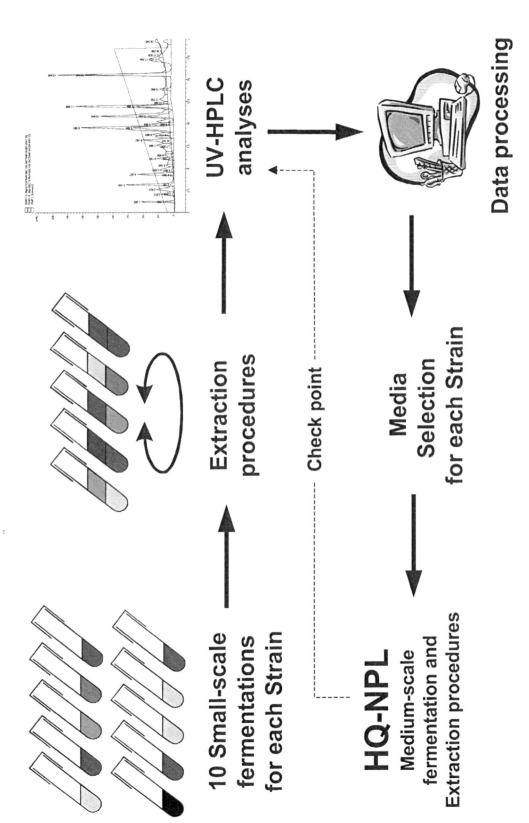

Fig. 1. General overview of the process.

2.1. Prescreening a Small-Scale Microbial Fermentation

Fungi and actinomycetes were isolated from soil and plant samples collected in Costa Rica, Mexico, Panama, South Africa, and Spain. Seed cultures were incubated at 220 rpm, 28°C and 70% relative humidity for 72 h *(11)*. A portion (0.5 mL) of each inoculum was inoculated into ten glass tubes (25 × 150 mm) containing 10 mL of different production media. The over 3000 fermentation tubes were incubated in racks at an inclination of 35° in the same conditions for 21 d for fungal strains and 7 d for actinomycetes, before harvesting. Initially, we decided to obtain a reasonable number of small-volume fermentation media per microorganism type, either fungal or actinomycete. The composition of the production media derived from prior experience with different nutritional sources *(10,12)*. The range of media compositions included as many different carbon and nitrogen sources as possible to amplify the scope of secondary metabolites produced. Glucose, lactose, fructose, maltose, dextrin, glycerol, soluble starch, and the complex sources corn meal, wheat meal, millet meal, and cane molasses were the carbon sources. Yeast extract, primary yeast, peptone, ardamine pH, NZ Soy BL, pharmamedia, distiller-soluble peptonized milk, and meat extract were the nitrogen sources selected (*see* ref. *10* for media details). Extraction of the fermentations was performed with 7 mL organic solvents (*see* **Subheading 3.1.** for details), stirring for 1 h at room temperature. Only the upper half of the organic layers was processed to avoid interface interference during automation procedures. After drying under nitrogen, the residues were dissolved in 1 mL HPLC-grade MeOH and filtered through 0.2-µm membranes to obtain a final 7X concentration (*see* **Fig. 2**).

2.2. Select the Best Sets of Fermentation Conditions

An automated procedure was developed to analyze a large number of samples and select the best sets of fermentation conditions. The HPLC vials were bar coded (**Fig. 2**) by querying an internal database to reduce manipulation errors. Each label referred to a database number, which identified the microorganism, the fermentation medium, and the extraction procedure. This allowed the possibility of comparing different preparation methods such as extraction with methanol, acetone, or methylethylketone (MEK). Each sequence of vials included extracts of unfermented media with which to establish a control baseline. These samples, as well as blank methanol injections, were also used for quality control and to normalize between different batches of microorganisms.

Sets included ten unfermented media, eight strains fermented in each of those fermentation media (80 vials), and methanol blank samples (chemical blanks) for each microorganism as a control for the process. Each one was identified by an HPLC batch file, so that the software could identify how many samples were injected, the database ID of each sample, blank control samples, corresponding blank media samples, and where in the hard drive of the computer the analysis data files generated by the HPLC ChemStation software were stored *(13)*. Moreover, the software can link the fermentation extracts with their corresponding blank media by querying databases, and include the possibility of doing it manually for special nonlabeled samples if necessary. The software also allows backups of the chromatograms in a portable computer, storing the format for later processing if needed *(9)*.

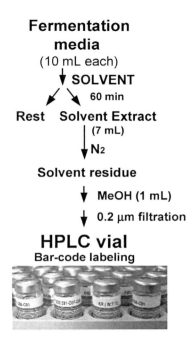

Fig. 2. Small-volume sample management procedures.

2.3. Chemical Analyses

Analyses relied on HPLC reverse-phase gradient chromatography, as it provides complex information about chemical diversity for comparing organic extracts. The detector system was a diode-array HPLC detector, recording simultaneously at 210 nm and 280 nm (14). Diode-array HPLC gradient characterizations of the reconstituted extracts were performed on a ZORBAX Rx-C8 4.6 mm × 250 mm column. A 10 to 100% gradient of acetonitrile in water with a flow-rate gradient of 0.9 to 1.2 mL/min was programmed on an 1100 HP Agilent ChemStation, using a constant temperature of 20°C during each 20-min analysis. Trifluoracetic acid (TFA, 0.01%) was added for pH control (see ref. 10 for more details).

The chemical data obtained for each organic extract, filed in an individual folder by the ChemStation software, consisted of a graph at both 210 and 280 nm vs time (in windows metafile format), and two tables of detected peaks at each of these wavelengths in MS-Excel file format. The Agilent ChemStation software determined the presence of peaks according to standard parameters such as the line slope, described in detail in the reference manual of the Agilent 1100 series of HPLC spectrometers (13).

In the absence of a commercially available program with which to compare more than two complex HPLC chromatograms, we initially relied on simple visual examination to devise the setup of the new software tool.

2.4. HPLC Studio Software Generalities

An application created especially for the project was used for data processing. The software allowed the combination of individual chromatograms and their comparison with great versatility *(9)*. For each strain, the application selected the media showing both the highest chemical diversity (as measured by the number of peaks) and the highest uniqueness (as measured by the presence of peaks not found in other chromatograms) (*see* the automated prioritization of the 10 media where the encoded strain F098,432 was fermented, in **Fig. 3**). This approach provided an automated way to combine data from all the fermentation conditions obtained for a single strain and thereby estimate its maximum chemical diversity (*see* **Fig. 3**, column 12 for cumulative data on chemical diversity).

The chromatograms obtained for a particular microorganism could be processed by the user by selecting the ID numbers from some or all of the fermentation conditions in the injection sequence. In that way, different conditions of the same fermentation medium, and any other data generated by the ChemStation software, could be analyzed.

Assuming that each peak detected at 210 nm represented at least one metabolite, other data processes were needed before the comparisons of different fermentation media could be carried out. Initially, chromatograms from the fermented microorganism were compared with the ones from the nonfermented media in order to remove background peaks caused by media components. Some degree of variability occurred in the HPLC analysis of complex extracts due to several factors, such as the composition of the mixture, the quantity injected, the presence of compounds with very close retention times, and the detector resolution. A resolution time value, determined by visual comparison of different co-injections of mixtures of highly complex microbial extracts, was evaluated for those as ± 0.03 min (1.5 times the resolution of an Agilent HPLC [0.02 min] and 0.065 min in absolute value) *(10)*.

A peak detection threshold was chosen to avoid small peaks and baseline fluctuations. This simplification reduced the data of the chromatogram template based on an absolute area value or a percentage of the total area of the chromatogram. A relative 2% cut-off value of the total area was used for the strains studied. Early-eluting solvent fronts and late-eluting re-equilibration peaks were likewise eliminated from consideration.

The data processing creates a chromatogram for each strain, a so-called chromatogram template, obtained by sequentially combining the chromatograms for each strain. For example, two chromatograms are combined, removing any two peaks with a difference in retention times smaller than the given resolution time (assumed identical). Following that, the software sequentially adds in the other chromatograms. We found that the order in which the chromatograms were added into the template did not affect the final result. All individual data were kept, and average retention times were calculated each time with previously accumulated data. It should be noted that all the data needed for characterizing the diversity in each fermentation medium were present in the chromatogram template: peaks present in only one medium appeared in the template as they appeared in the original chromatogram (with retention times, areas, and extract of origin), whereas peaks present in several media appeared in the template with their mean retention times, all individual areas, and extracts of origin. Thus all peaks present in the chromatogram of a given medium could be obtained from the chromatogram template.

HPLC STUDIO FINAL RESULTS

F-098,432

Resolution Time:	0.065 min.		Criteria Diversity/Area:	75% / 25 %
Retention Time by	Average of times			2 %
From	4 to 19	minutes	Area Selection Parameters	0 Units

Order	Chr # (Letter)	F #	Medium (days)	Extract	Final %	# Peaks	Coinc Peaks	Diversity %	Area %	Area Sum	Diversity Sum	Total Sum
1	2(B)	F-098,432-C06-C01	FR23(7 d.)	MEK	31.12	15	15	36.59	14.73	14.73	36.59	31.12
2	5(E)	F-098,432-C08-C01	GPA(7 d.)	MEK	17.89	12	8	19.51	13.04	27.77	56.10	49.02
3	8(H)	F-098,432-C11-C01	MPG(7 d.)	MEK	12.46	13	5	12.20	13.24	41.01	68.29	61.47
4	6(F)	F-098,432-C09-C01	KHC(7 d.)	MEK	8.31	12	3	7.32	11.27	52.28	75.61	69.78
5	4(D)	F-098,432-C07-C01	GOT(7 d.)	MEK	8.12	12	3	7.32	10.52	62.80	82.93	77.90
6	10(J)	F-098,432-C13-C01	RAM2(7 d.)	MEK	7.60	11	3	7.32	8.45	71.25	90.24	85.50
7	9(I)	F-098,432-C12-C01	PV8(7 d.)	MEK	5.13	9	2	4.88	5.88	77.13	95.12	90.62
8	1(A)	F-098,432-C04-C01	CLA(7 d.)	MEK	4.73	9	1	2.44	11.62	88.75	97.56	95.36
9	7(G)	F-098,432-C10-C01	KR(7 d.)	MEK	3.71	7	1	2.44	7.52	96.27	100.00	99.07
10	3(C)	F-098,432-C05-C01	DNPM(7 d.)	MEK	0.93	4	0	0.00	3.73	100.00	100.00	100.00

Fig. 3. High-performance liquid chromatography (HPLC) Studio report obtained for the encoded strain F-098432.

3. HPLC Studio Applications

The processed data yielded information concerning the relative diversity and number of metabolites produced in each of the fermentations. These chemometric measurements led us to choose which samples should be included in an HQ-NPL.

3.1. Selecting the Best Extraction Solvent

One of the first decisions in the design of a method for creating a library of natural products is the selection of the extraction solvent. The most commonly used solvents in our laboratory are methanol, acetone, and methyl-ethyl-ketone (MEK) *(15)*. We decided to compare the differences among these three solvents in a study using 69 fungal and 77 actinomycete strains. Fermentations (50 mL) in AD_2M_2 medium for fungi and DNPM medium for actinomycetes were aliquoted (10 mL) and extracted in small volumes with methanol (10 mL), acetone (10 mL), or MEK (14 mL) (*see* ref. *10* for media components).

Just by counting the number of peaks between 4 and 19 min, the results indicated that there were no significant differences between acetone and methanol, although acetone yielded a slightly higher number of metabolites per isolate. Interestingly, MEK was ultimately selected as the solvent producing the most complex extracts; in fact it extracted compounds that were extracted by acetone or methanol as well as other compounds with less polarity (**Fig. 4**). The HPLC Studio application was then used to prioritize the extraction solvents for each strain, based simply on number of nonredundant compounds extracted (*see* **Fig. 5**). MEK was the solvent of choice (greater number of nonredundant peaks extracted) in about two-thirds of the cases, followed by acetone, which was the best in just 20% of the cases.

With unknown mixtures, it is impossible to quantitate extracted compounds in an absolute sense; nevertheless, we used total ultraviolet (UV) absorbance (intensity) to estimate the extraction efficacy for each of the solvents. By this measure, MEK was the least effective, while acetone performed well, particularly with actinomycetes (*see* **Fig. 6**). If all three solvents had extracted the same amount of material, they would each produce a balanced value of 33.3% of total area counts. However, MEK values were about 22% for fungi and 15.6% for actinomycetes. This is not surprising, because acetone and methanol are miscible with water, whereas an initial solid/liquid extraction and subsequent liquid/liquid partition occurs for MEK (inmiscible with water). However, in a qualitative sense, MEK reduced the abundance of polar chemicals and enriched the extracts with those of intermediate polarity. Given that the primary factor to consider when creating an HQ-NPL is chemical diversity in samples, MEK is the preferred solvent, even though it extracts a smaller amount of material. A quantitative loss can be adjusted later by increasing the fermentation volume by 15%, or by concentrating samples.

3.2. Selecting the Best Fermentation Format

We also compared fermentations performed in different-sized vessels. Four systems were examined: the classic 50-mL flasks, 10-mL colony tubes, 10-mL EPA vials, and 2.7-mL wells in 24-well plates. Twenty-two organisms (10 fungi in DNPM and 12 actinomycetes in AD_2M_2 medium) were fermented in these four formats and extracted with MEK. The results (**Fig. 7**) indicated that all fermentation formats performed

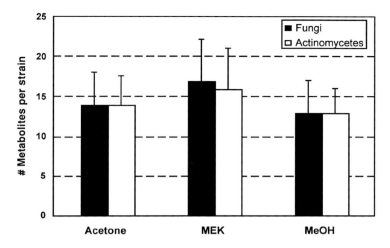

Fig. 4. Average number of metabolites per strain determined for 10 fungal and 12 actinomycete strains as a function of extraction solvent. Data were based on comparing chemical diversity observed by high-performance liquid chromatography analysis at 210 nm.

Solvent Prioritization (%)		Acetone	MEK	MeOH
Fungi	1st selection	23	64	13
	2nd selection	0	36	64
	3rd selection	77	0	23
Actinomycetes	1st selection	21	67	12
	2nd selection	67	33	0
	3rd selection	12	0	88

Fig. 5. Prioritization of the organic extraction solvents determined with the high-performance liquid chromatography (HPLC) Studio tool for 69 fungal strains and 77 actinomycetes. Data were based on prioritizing the chemical diversity observed by HPLC analysis at 210 nm.

equally well as measured by the average number of compounds produced. However, it is possible to obtain extracts with the same number of secondary metabolites but with different compositions.

In order to determine similarity coefficients of extracts based solely on retention times, the commercial software BioNumerics™ was used for cluster analysis and representation of the dendrograms. HPLC profiles were compared by neighbor-joining analysis using the Dice similarity coefficient. The results indicated that, in a significant number of cases, all or almost all the extracts from the same strain clustered within the same group (results not shown).

As a measure of quality control, once the computer selected the three best media, some isolates were re-fermented on a large scale, and the new extracts were compared with the original ones (*see* **Fig. 8**). Identical chromatograms were observed in 95% of the cases. The remainder showed only minor differences between small- and medium-scale formats.

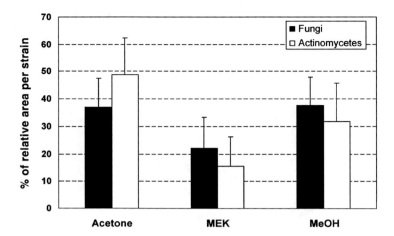

Fig. 6. Relative Areas determined with the high-performance liquid chromatography (HPLC) Studio tool for 69 fungal strains and 77 actinomycetes extracts vs the organic extraction solvent. Data were based on prioritizing the area observed by 210 nm HPLC traces.

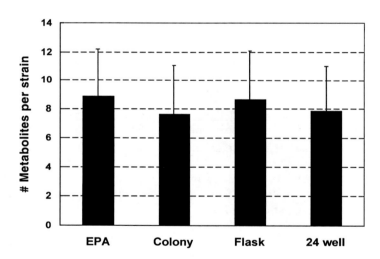

Fig. 7. Average number of metabolites per strain determined for four different fermentation formats obtained for 10 fungal strains and 12 actinomycete extracts. Data were based on comparing chemical diversity pairs observed by high-performance liquid chromatography analysis at 210 nm.

Having confirmed reproducibility, we selected the EPA fermentation vials for the fermentation format. Based on the easy extraction automation of the EPA vials, they appeared to be the best choice overall for small-scale growth of microbial strains.

3.3. Determination of the Relative Chemical Diversity for a Fermentation Condition

As mentioned earlier, HPLC analysis can be used to characterize the relative chemical diversity present in an organic extract obtained from microbial fermentations *(14)*.

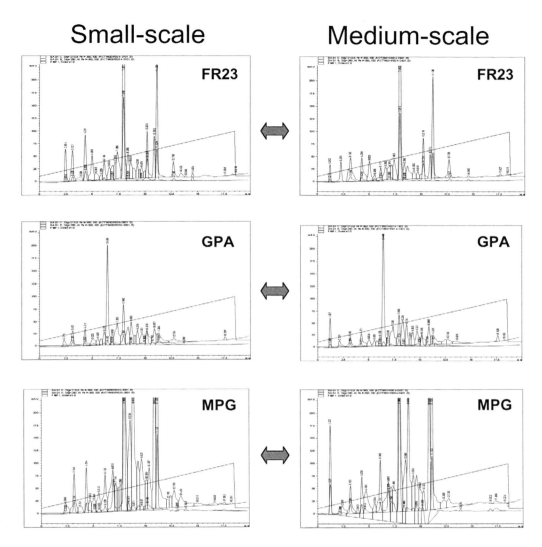

Fig. 8. Comparison between small- and medium-volume high-performance liquid chromatography (HPLC) chromatograms for the three extracts selected by the HPLC Studio software for the F-098432 strain.

The diversity characterization begins by having the HPLC Studio software select the medium with the highest number of peaks. The number of peaks relative to the number of peaks in the chromatogram template indicates its diversity as a percentage. Peaks that are present in the best fermentation medium are then virtually removed from the template prior to the second step. Peaks in each of the other media are counted only if present in the partially depleted template, allowing the second-best medium (that which contributes to most of the remaining peaks) to be determined. This process is repeated until all the media are ranked, establishing a diversity characterization value for each extract (*see* a practical example in ref. *10*).

3.4. Ensuring a Relative Extracted Quantity for a Fermentation Condition

The overall objective was to select not only the fermentation medium resulting in the highest chemical diversity, but also to ensure that sufficient quantity of material is generated for an HQ-NPL. Thus the HPLC UV area was used in addition to the number of peaks in creating the final prioritization list. The user can determine the relative importance of this contribution for each comparison by deliberately selecting a 0 to 100 percentage. Each extract received a final ranking by combining both quantity and chemical diversity.

Introducing a quantity parameter affected the order of prioritization that was obtained solely based on characterization of diversity. As the accumulated diversity percentage is also dependent on the order of accumulation, a recalculation of the diversity characterization percentages for the final order was needed when areas were incorporated in the prioritization. That allowed us to determine the real percentage of the total of diversity that was attained by performing the final prioritization, both for diversity and quantity of accumulated values.

3.5. Selecting the Best Overall Fermentation Conditions for a Single Strain

The computer report generated (*see* **Fig. 3**) can then be used as a guide for determining how many extracts would provide enough value to be included in an HQ-NPL. In that way, we could measure whether the inclusion of any given medium would dramatically increase the total diversity. It is also possible to determine the number of extracts needed to reach a predetermined level of relative diversity and/or quantity.

When ranking by the quantity parameter alone, six medium-scale fermentations were needed per strain to reach a 75% value for diversity. When only the diversity parameter was used, the number of medium-scale fermentations per strain decreased to four. On the other hand, the accumulated areas (quantity values) showed that the first three to four media prioritized covered half of the cumulative of the 10 fermentation conditions *(10)*.

With a prioritization balance set at diversity/quantity = 75/25, the difference achieved between three and four media was not significant enough to justify four medium-scale fermentations. Including more kinds of microorganisms in the medium-scale production would be a preferred option to amplify chemical diversity *(10)*. Using only the first three media prioritized by the HPLC Studio software in medium-scale for additional microorganisms would be the most balanced choice. Such a selection would allow the creation of an HQ-NPL with more strains instead of preparing more medium-scale fermentations per microorganism.

3.6. Selecting the Best Overall Fermentation Conditions for a Set of Strains

Chemometric data for each fermentation medium could also be used to determine its usefulness for a specific set of microorganisms. With small populations of strains, the fermentation conditions that resulted in the greatest diversity and the greatest quantity for each strain did not overlap in most of the cases, and none of the media stood out in terms of their selection frequency *(10)*.

Figure 9A,B shows, respectively, the selection frequencies for each of the 11 fermentation media when used with a representative subset of 50 actinomycetes, and prioritized either by the presence or absence of secondary metabolites (diversity) or by their accumulated areas (quantity). Divergence from the 45° line indicates its difference from the average. The most productive media are above the line, the poorest ones, below. From the graph, we can infer that for this set of strains, the best media in terms of diversity were GOT, FR23, and MPG, whereas in terms of quantity the best media were MH, MPG, and FR23. On the other hand, the poorest media in terms of diversity were PV8 and RAM2, and clearly PV8 in terms of quantity *(10)*. Such comparisons of similar microorganisms can be used to select optimal fermentation conditions and establish a good panel of fermentation media for the preparation of an HQ-NPL.

In terms of taxonomy, a large part of our fermentation program was devoted to actinomycetes. Therefore, we asked whether genera-specific responses to fermentations could be recognized *(16)*. Several fermentation media were more frequently selected for each family of strains, throughout the range of all fermentation conditions. Only the families of *Micromonosporaceae*, *Nocardiaceae*, *Pseudonocardiaceae*, and *Streptomycetaceae* were represented with statistically relevant numbers of strains. None of the media selection frequencies observed for these four families reached 75%. The low values indicated again that there was not a preferred medium for most of the strains of these families, and confirmed the necessity of performing the prioritization study for each strain or family of strains for maximum exploitation of metabolite production *(10)*.

3.7. Comparing Secondary Metabolites Profiles vs Taxonomy

Historically, it has been widely accepted that increasing the chances of finding new drugs relies on the phylogenetic diversity and exploitation of different environments and geographic areas. Relying solely on taxonomy and strain origin is no guarantee of uniqueness, however. It is well known that unrelated microorganisms can produce the same compounds, and that closely related strains do not always produce a similar pattern of secondary metabolites *(17)*.

Microbial isolation and selection has traditionally relied on the expertise of the microbiologist to recognize, on the basis of morphological and genetic characteristics, the uniqueness of new strains obtained from the environment. In the last four decades, thousands of strains have been tested in industrial screening programs, and scientists inevitably test many repeats. According to the new philosophy of natural-products screening, the prioritization of strains to be included in an HQ-NPL is a key factor for the success of a screening effort. A number of approaches have been adopted to avoid screening duplicates: standardized chemotaxonomical analyses, molecular fingerprinting, DNA sequencing of marker regions, and metabolic profiling *(18,19)*. According to our methodologies, by combining the metabolites of a given strain in a large variety of fermentation media, virtual chromatograms obtained with the HPLC Studio tool give scientists a good representation of its metabolic potential.

Figure 10 shows the average number of secondary metabolites (in terms of different retention times) obtained for 20 representative actinomycete strains, classified according to their country of origin. Strains from different geographical origins behaved similarly. The degree of overlap among secondary metabolites produced by several strains was unpredictable. The number of different compounds obtained by combining all the

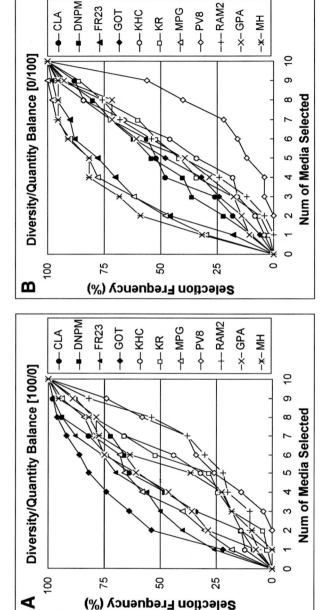

Fig. 9. High-performance liquid chromatography Studio software prioritization for the fermentation media in a representative subset of 50 actinomycete strains vs software prioritization conditions of chemical diversity (**A**) or quantity (**B**). The selection frequency indicates, as a percentage, the number of strains for which a fermentation condition was prioritized with respect to the total number of strains tested.

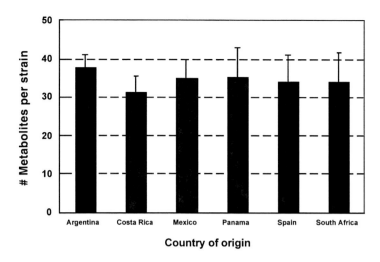

Fig. 10. Average number of different metabolites obtained from 20 random actinomycete strains from each country of origin.

strains of a region was determined in order to evaluate the overall metabolic potential diversity for each geographic location. When a subset of 20 randomly selected strains from each country were added, no significant differences in total number of metabolites per strain were observed among countries (approx 500 different secondary metabolites were detected in each case by the HPLC Studio tool). BioNumerics™ software was also used for cluster analysis of the HPLC chromatograms and graphical representation of dendrograms. For these phenetic analyses, HPLC data matrices generated with the HPLC Studio software were compared with BioNumerics™ in a pairwise manner using the Dice similarity coefficient and UPGMA *(20)*.

Three groups of 20 strains from three different countries (Mexico, Spain, and South Africa) were selected to see whether we could recognize correlations between taxonomic affiliation and metabolic profiles (**Fig. 11A–C**). Dendrograms generated after cluster analyses showed that, in general, strains belonging to the same genus clustered together with a high percentage of similarity, indicating that taxonomically related strains may have some degree of similarity in their metabolic pathways. However, in some cases, certain strains were unique in their metabolic profile, as they failed to cluster with the other representatives of the same taxon. That was the case, for instance, for the *Saccharothrix* strain F095473, isolated from a Mexican soil (**Fig. 11A**), or for four *Actinoplanes* isolates from South Africa that clustered separately with the other *Actinoplanes* (**Fig. 11C**).

3.8. Comparing Secondary Metabolite Profiles vs Geographical Origin

We also wanted to examine the metabolic diversity of microorganisms isolated from different countries. To address the question of whether increasing the geographical diversity of the strains results in a parallel increase of the overall chemical diversity, we performed a cluster analysis of the HPLC chromatograms produced by strains of *Streptomyces* and *Nocardioform* isolated from different countries (**Fig. 12A, B**, respec-

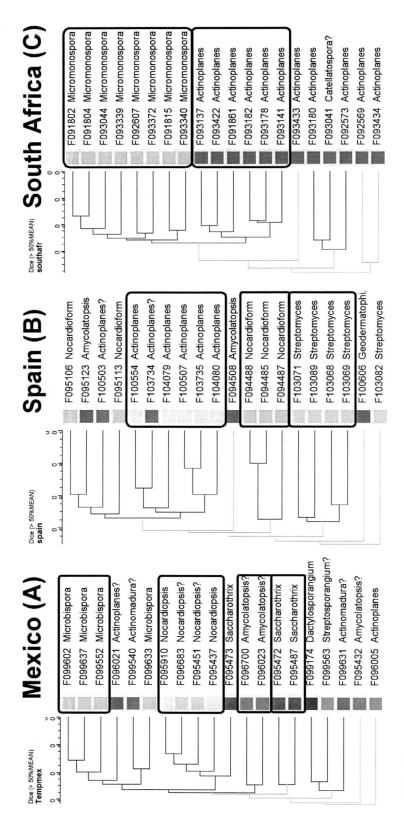

Fig. 11. Dendrograms showing the relationships among 20 actinomycete strains from Mexico, Spain, and South Africa, based on high-performance liquid chromatography profiles (UPGMA, Dice >50%Mean). The taxonomy strains at the genus level are indicated.

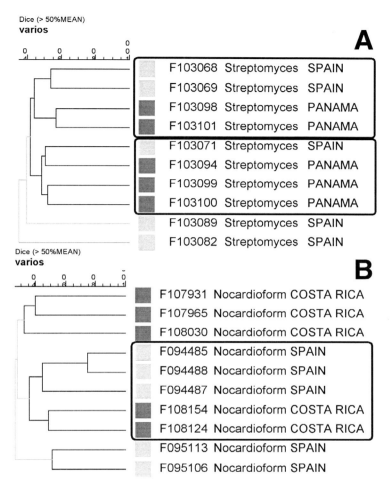

Fig. 12. Dendrograms showing the relationships between some *Streptomyces* and *Nocardioform* isolates from Panama, Costa Rica, and Spain, on the basis of their high-performance liquid chromatography profiles (UPGMA, Dice >50%Mean).

tively). We observed that several cultures of the same taxon clustered together with a high percentage of similarity, suggesting that identical or similar profiles were produced by these strains, independently of their origin. This result confirmed the well-known fact that many types of strains are ubiquitous, especially within the genus *Streptomyces*, and can be found in very different environments. There are also many similar secondary metabolite pathways present in phylogenetically unrelated strains *(17)*. However, it was observed that not all the strains were found within a cluster at the established cut-off point, and some strains of a given taxon also showed unique metabolic profiles.

4. Conclusions

This study shows that the number and quantity of secondary metabolites obtained from microorganisms is measurable, and that the number of medium-scale fermentation conditions needed to encompass suitable chemical diversity can be estimated. The development of a new computational tool, the HPLC Studio software, gave us a valuable instrument with which to attempt the rational design of a collection of natural products, allowing us to focus on and direct the efforts needed to reach a certain threshold of quantity and chemical diversity in a natural-products library.

Likewise, we were able to set a numerical value on the ability of newly isolated strains to produce a range of different secondary metabolites. Some limited correlations were observed between taxonomy and metabolite production. Phylogenetically related strains were grouped into clusters according to their metabolite production, regardless of their country of origin.

All these data were treated by computer-assisted statistical software. Integrating such a tool in the design of a library of natural products derived from microorganisms provided us with additional criteria for the analysis of the metabolites obtained, and hence gave us rational criteria for the design of high-quality natural-products libraries with increased possibilities for finding new biologically active and pharmacologically relevant compounds.

Acknowledgments

The authors are grateful to I. González, G. Bills, and P. Hernández, from the Microbiology and the Sample Management Departments of CIBE, to M. Goetz from the Chemistry Department of Merck Research Laboratories of Rahway, and to F. Peláez, Director of CIBE-MRL-Merck, Sharp & Dohme de España S.A., for their efforts and collaboration during the preparation of this manuscript.

References

1. Oprea TI, Davis AM, Teague SJ, Leeson P. Is there a difference between leads and drugs? A historical perspective. J Chem Inf Comput Sci 2001;41:1308–1315.
2. Lipinski CA, Lombardo F, Dominy BW, Freeney PJ. Experimental and computational approaches to estimate solubility and permeability in drug discovery and development settings. Adv Drug Deliv Rev 1997;23:3–25.
3. Peláez F, Genilloud O. Microorganisms for Health Care, Food and Enzyme Production. 1. Discovering new drugs from microbial natural products. Research Signpost Ed, Kerala, India. 2003;1–22.
4. Müller-Kuhrt L. Putting nature back into drug discovery. Nature Biotech 2003;21:602.
5. Bindseil KU, Jakupovic J, Wolf D, Lavayre J, Leboul J, van der Pyl D. Pure compound libraries; a new perspective for natural product based drug discovery. Drug Disc Today 2001;6:840–847
6. Yarbrough GG, Taylor DP, Rowlands RT, Crawford MS, Lasure L. Screening microbial metabolites for new drugs: theoretical and practical issues. J Antibiot 1993;46:535–544.
7. Cordell GA, Shin YG. Finding the needle in the haystack. The dereplication of natural product extracts. Pure Appli Chem 1999;71:1089–1094.
8. Monaghan RL, Polishook JD, Pecore VJ, Bills GF, Nallin-Omstead M, Streicher SL. Discovery of novel secondary metabolites from fungi—is it really a random walk through a random forest? Can J Bot 1995;73:S925–S931.
9. García JB, Tormo JR. HPLC Studio: a new software utility to perform HPLC chromatogram comparison for screening purposes. J Biomol Screen 2003;8:305–315.
10. Tormo JR, García JB, DeAntonio M, et al. A method for the selection of production media for actinomycete strains based on their metabolite HPLC profiles. J Ind Mic Biotech 2003;30:582–588.

11. Platas G, Collado J, Martínez H, Arrese M, Peláez F, Díez MT. Implementation of conditions of the inoculum stage for *Streptrosporangium* cultures. Microbiologia SEM 1997;13:193–200.
12. Platas G, Peláez F, Collado J, Martinez H, Diez MT. Nutritional preferences of a group of *Streptosporangium* soil isolates. J Biosci Bioen 1999;88:269–275.
13. Hewlett Packard, Agilent. 1995. Understanding Your ChemStation, 5th ed. San Diego, CA, USA.
14. Julian RK, Jr, Higgs RE, Gygi JD, Hilton MD. A method for quantitively differentiating crude natural extracts using high-performing liquid chromatography-electrospray mass spectrometry. Anal Chem 1998;70:3249–3254.
15. Schmid I, Sattler I, Grabley S, Thiericke R. Natural products in high throughput screening: automated high-quality sample preparation. J Biomol Screen 1999;4:15–25
16. Stackebrandt E, Rainey FA, Ward-Rainey NL. Proposal for new hierarchic classification system, *Actinobacteria* classis nov. Int J Syst Bacteriol 1997;47: 479–491.
17. Vilella D, Sánchez M, Platas G, et al. Inhibitors of farnesylation of Ras from a microbial natural products screening program. J Ind Mic Biotech 2000;25:315–327
18. Vandamme P, Pot B, Gillis M, de Vos P, Kersters K, Swings J. Polyphasic taxonomy, a consensus approach to Bacterial Systematics. Microbiol Rev 1996;60:407–438.
19. Fiedler HP. Biosynthetic capacities of actinomycetes. 1. Screening for secondary metabolites by HPLC and UV-visible absorbance spectral libraries. Nat Prod Lett 1993;2:119–128.
20. Sneath PH, Sokal RR. Taxonomic structure. In: Numerical Taxonomy (Kennedy D and Park RB, ed), WH Freeman & Co., San Francisco, USA, 1973, pp. 230–234.

4
Manipulating Microbial Metabolites for Drug Discovery and Production

C. Richard Hutchinson

Summary

Kosan Biosciences was founded on the principle that the chemical diversity of microbial metabolites, which for decades have been a rich source of natural-product drug leads and therapeutically important drugs, can be increased by altering the function of the genes and enzymes that govern the production of these metabolites. In particular, the predictable relationship between the structure and function of the modular type of microbial polyketide synthases has enabled genetic manipulation ("engineering") of the producing organism for production of novel forms of various classes of naturally occurring compounds, such as macrolide antibiotics (erythromycins and FK520) and certain antitumor agents (epothilone D and the geldanamycins). This resembles the approach used by medicinal chemists who synthesize analogs and derivatives of lead compounds in an attempt to improve upon existing drugs or find new ones. Expression of the native or engineered polyketide synthase genes, as well as others that govern metabolite formation, in heterologous hosts is an important aspect of developing commercial systems for drug production. This chapter highlights advances made by Kosan Biosciences and some other groups in this area of natural-products drug research.

Key Words: Antibiotics; anticancer drugs; biosynthesis genes; drugs; erythromycin; epothilone; FK520; geldanamycin; polyketide; polyketide synthase; rapamycin; secondary natural products.

1. Introduction

1.1. Approach

For more than six decades, the chemical diversity of microbial metabolites has made them a rich source of natural-product drug leads and therapeutically important drugs. Notwithstanding the fact that compounds with new structures continue to be uncovered regularly *(1,2)*, there is a pervasive belief that natural products are a waning source of new drugs. Certainly, there has been a precipitous decline in new chemical entities that have been registered with the US Food and Drug Administration (FDA) over the past 5 yr, including ones from natural sources. On the other hand, Craig et al. *(2)* have noted that for the areas of cancer and infectious diseases, 60% to 75% of the drugs used to treat these diseases are of natural origin, based on the numbers of new drugs approved during the period 1983–2002. Part of the explanation for the decline lies in the de-emphasis of natural-product discovery programs by large pharmaceutical companies over the past 10 yr. However, it is likely that inherent limitations in the ability of traditional

From: *Natural Products: Drug Discovery and Therapeutic Medicine*
Edited by: L. Zhang and A. L. Demain © Humana Press Inc., Totowa, NJ

screening methods to access the full range of chemical diversity in nature has also been a contributing factor.

One approach to overcome this dearth of naturally derived drug leads involves the identification and use of the genes governing the biosynthesis of natural products to create new compounds as well as to improve the economics of their production. Research into the genetics of microbial secondary metabolite production since 1980 has provided a wealth of knowledge about its genetic and biochemical basis *(3,4)*. Such information, together with the widely held belief that the enzymes of secondary metabolic pathways are less fastidious overall than their primary metabolism counterparts, has stimulated the growth of a research area called "combinatorial biosynthesis." It encompasses studies of the properties of nonnatural combinations of biosynthesis genes with the aim of exploiting the enzymes' sloppiness for the production of new natural products *(5,6)*. Because this research is largely being done with bacteria, it falls within the larger field of the metabolic engineering of culturable microorganisms.

1.2. Polyketides and Polyketide Synthases

Microbes have the capacity to produce a multitude of structurally complex chemicals by adapting processes of primary metabolism for self-defense and cell signaling purposes. For instance, long-chain fatty acids are made to provide components of cell membranes and to store energy generating potential as triacylglycerides, whereas enzymes called polyketide synthases (PKSs) evolved the capability of making a vast number of compounds known as polyketides from the same substrates used by fatty-acid synthases, and largely by the same type of biochemistry *(7)*. Examples of two types of PKSs and their associated polyketides are illustrated in **Figs. 1** and **2**. Type II PKSs (**Fig. 1**) consist of a collection of largely monofunctional proteins that catalyze the formation of typically polycyclic aromatic compounds from acetate and malonate only. Type I PKSs (**Fig. 2**), in contrast, use large multifunctional proteins to make polyoxygenated, aliphatic compounds from several different kinds of acyl-coenzyme A substrates.

The so-called modular PKSs *(8–10)* are featured in this chapter because these enzymes have enabled the most fruitful genetic engineering route to structural variants (analogs) *(11,12)* of polyketides that are important therapeutic drugs, like the antibacterial erythromycin A (**Fig. 2**) or experimental agents like 17-allylamino-17-demethoxy geldanamycin (17-AAG), which currently is undergoing clinical trials as an antitumor drug (discussed later). Three other types of PKSs are not discussed here; for reviews, *see* refs. *7,13*.

A modular PKS consists of large, multifunction proteins, each with different combinations of domains, called modules, that function like the constituent biochemical activities of fatty-acid synthases. For 6-deoxyerythronolide B synthase (DEBS), the PKS that forms the backbone of the erythromycins and is encoded by the three *eryAI-III* genes *(8,14)*, 6-deoxyerythronolide B (6dEB) is produced by the successive condensation of one propionyl-coenzyme A (CoA) and six 2-methylmalonyl-CoA molecules (**Fig. 2**). The three subunits of DEBS have two modules, each of which contains the activities needed for one cycle of polyketide chain elongation, as illustrated by the structures of the six enzyme-bound intermediates in **Fig. 2**. Every module contains a ketosynthase (KS), an acyltransferase (AT), and an acyl carrier protein (ACP) domain,

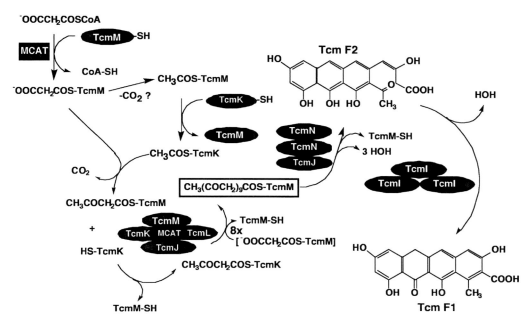

Fig. 1. An illustration of the mechanism of the type II polyketide synthase (PKS) involved in the biosynthesis of tetracenomycin F1 and C *(68)*. The PKS consists of individual protein subunits that act in concert to assemble the acetate starter unit (produced by decarboxylation of enzyme-bound malonate) and 9 chain extender units into an TcmM-bound decaketide by an iterative process involving a malonyl-CoA:ACP acyltransferase and the proteins TcmJ, TcmK, TcmL and TcmM. The latter is cyclized to Tcm F2 by the TcmN enzyme, with assistance by TcmJ, then Tcm F2 is cyclized once more by TcmI to form Tcm F1. The latter intermediate is converted to tetracenomycin C by tailoring enzymes.

which together catalyze a two-carbon extension of the chain between a starter unit (e.g., propionyl-CoA or some enzyme-bound acylthioester intermediate) and a chain extender unit (e.g., 2-methylmalonyl CoA). The AT domains are specific for 2-methylmalonyl CoA except for the AT in the didomain loading module, which prefers propionyl CoA. After each two-carbon unit condensation, the oxidation state of the β-carbon is either retained as a ketone (e.g., module 3), or modified to a hydroxyl, methenyl, or methylene group by the presence of a ketoreductase (KR) (e.g., module 2), a KR + a dehydrase (DH), or a KR + DH + an enoyl reductase (ER) (e.g., module 4), respectively. Polyketide assembly is often terminated by a thioesterase (TE) domain at the C-terminus of the final module (module 6) to release the product from the protein. For DEBS, release is coincident with the intramolecular cyclization that results in the formation of 6dEB, a typical polyketide macrolactone product. Synthesis of 6dEB is followed by C6-hydroxylation (*eryF*) to yield erythronolide B. Addition of the sugar L-mycarose (via thymidine diphospho(TDP)-mycarose) yields 3-*O*-α-mycarosyl-erythronolide B, and the addition of desosamine (via TDP-desosamine) yields erythromycin D. The two sugars are produced by independent pathways (not shown) but controlled by the genes designated *eryB* (mycarose) and *eryC* (desosamine). The final

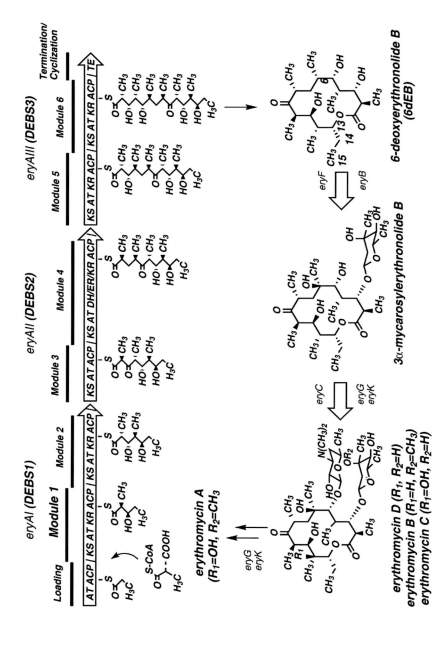

Fig. 2. An illustration of the mechanism of the type I modular polyketide synthase involved in the biosynthesis of 6dEB. Each of the 6-deoxyerythronolide B synthase (DEBS) subunits is represented by a broad, open arrow containing the relevant domains in each module. Key enzyme-bound intermediates of carbon chain assembly are shown bound to the acyl carrier protein domains. Assembly begins at the loading didomain of the first DEBS subunit, upon attachment of propionate which then reacts with one equivalent of bound 2-methylmalonate, obtained from its CoA ester. Further equivalents of 2-methylmalonyl-CoA are used by DEBS to produce 6dEB, as explained in the text. 6dEB is then converted to the erythromycin A-D glycosides by tailoring enzymes.

steps are hydroxylation of erythromycin D to yield erythromycin C by a second P450 enzyme (*eryK*) and *O*-methylation of the mycarosyl residue (*eryG*) to yield the cladinosyl moiety in erythromycin A and B.

In effect, the AT specificity and the types of catalytic domains within a module serve as codes for the structure of each two-carbon unit; the order of the modules in a PKS specifies the sequence of the distinct two-carbon units, and the number of modules determines the length of the polyketide chain. Variations in the acyl-CoA substrates used by a modular PKS, the number of domains within a module and the number of modules in the PKS are responsible for establishing the first set of structural characteristics of the polyketide; differences in the kinds of biochemical transformations the compound produced by the PKS undergoes are dictated by the products of the "tailoring enzymes" (*eryB*, *eryC*, *eryF*, *eryG*, and *eryK* for the erythromycins) and establish the final structure. Consequently, engineering a microorganism to produce novel polyketides can involve altering only the PKS genes or both them and the tailoring genes. These ideas are illustrated below.

1.3. Research and Development Strategy at Kosan Biosciences

To expedite drug discovery and the early stage development process, Kosan has chosen to identify microbial polyketides known to have some type of biological activity and therapeutic potential, then use genetic engineering to alter the compound's structure in ways that could improve its utility (e.g., increasing potency or correcting some remediable limitation in its pharmacology, pharmacokinetics, or pharmacodynamics) or simply the economics of its large-scale production. As far as possible, the changes sought by gene manipulation are based on knowledge of the structure-activity relationships and molecular modeling of drug:target interactions. We do not make libraries of polyketides to screen for activity against some biological target as the route to identification of lead molecules. This could become feasible, however, if suitable advances in PKS engineering are made through the approaches currently being explored, as explained in **Subheading 5.2.**

1.4. Other Approaches to Discovering New Drugs From Natural Products

The time-honored method of screening extracts of culturable microorganisms for biological activities of interest, followed by purification of the responsible metabolites to identify lead compounds, when necessary supported by an ensuing analog synthesis program, has fallen out of favor in large pharmaceutical companies. Nonetheless, several chapters are devoted to this topic in this volume. Genetics-based approaches, such as (1) scanning the genome of micoorganisms by high-throughput DNA sequencing for novel sets of secondary metabolism genes that can be caused to be expressed through manipulation of growth conditions or (2) constructing gene libraries from the DNA of uncultured microorganisms and expressing the clusters of secondary metabolism genes identified by DNA sequencing in heterologous hosts, are gaining favor and thus are covered elsewhere in this book. It remains to be seen whether either approach can uncover drug leads at a pace that will sustain their pursuit commercially.

2. The Macrolide Antibiotic Paradigm
2.1. Erythromycins and Their Analogs

Structure–function studies of DEBS carried out over the past 13 years has created a paradigm for modular PKSs that has guided drug discovery research based on the engineering of any modular PKS. The pioneering work of Katz and co-workers on the *eryA* genes *(8)* and the initial discovery that DEBS consists of multifunctional proteins *(14)* pointed the way to making numerous 6dEB analogs through domain mutation or replacement experiments *(15)*. In some cases, the 6dEB analogs were converted into erythromycin analogs by conducting the work in the erythromycin producing organism *Saccharopolyspora erythraea*, or by using DEBS-null mutants of this organism to bioconvert the 6dEB analogs initially made in a heterologous host organism such as *Streptomyces coelicolor (16)*. Analogs of 6dEB and the erythromycins have been produced that originate from nonnatural starter units *(17,18)*, which lack one or more branched methyl groups *(15,19–21)*, have one methyl group replaced with an ethyl group *(22)*, have different oxidation states at certain hydroxy-bearing positions *(23,24)*, are hydroxylated at different positions *(20,21)*, or have double bonds in place of normally saturated positions *(25)*. Compound libraries containing approximately 50 6dEB analogs have been produced also *(26)*. A subset of these compounds was then converted to their 5-*O*-desosaminyl glycosides *(27)*. Novel erythromycin-like compounds have also been made by adding a module to DEBS *(15,28)*. Representative examples of such analogs are shown in **Fig. 3**.

There are underlying assumptions in the above work: (1) any structural change effected by engineering the PKS domain governing some step in the assembly process will not affect the subsequent steps in the process severely enough to obviate formation of the analog of the normal PKS product; and (2) any structural change effected by engineering or inactivating a specific tailoring enzyme gene will also not obviate formation of the analog of the complete polyketide. The second assumption has been tested to some extent for erythromycin *(29)* and other polyketides *(30)*, and the results of the work summarized above have established certain boundaries on what can be expected for the genetic engineering of modular PKSs in general.

The success of such experiments has often been viewed by the relative amount of product produced by the engineered vs wild-type microorganisms under comparable fermentation conditions. This can vary from 0.1% to nearly 90% of the normal amount, with lower yields seen more often than ones close to that of the wild-type system. Changes in the functionality present in different positions of 6dEB have been achieved with a wide range of success (i.e., yield), and it has not been possible to explain the failures observed or predict with confidence which additional changes could be achieved. Nevertheless, the successful production of the erythromycin analogs achieved to date clearly validates the overall approach that was built on the first successful experiments reported by Donadio et al. in 1991 *(8)*. Furthermore, attempts are being made through carefully designed and executed in vitro experiments with the individual modules of DEBS and its intact subunits to define the parameters that govern the relative activity of engineered modular PKSs; e.g., *see* ref. *31)*.

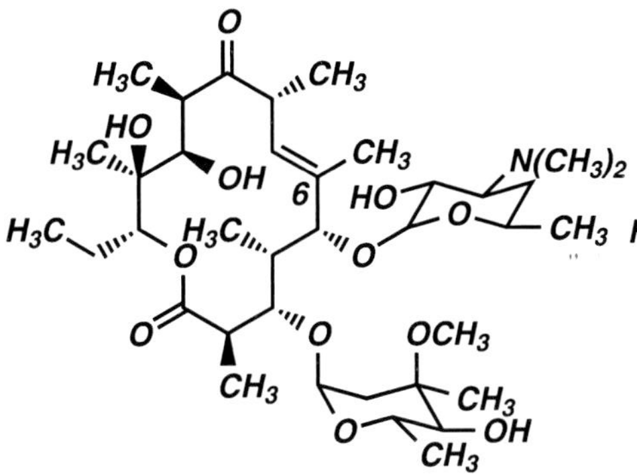

6,7-didehydro-6-deoxyerythromycin 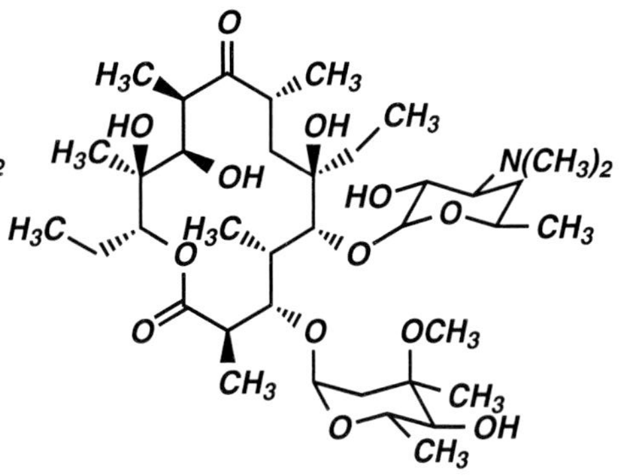   *6-desmethyl-6-ethyl-erythromycin A* 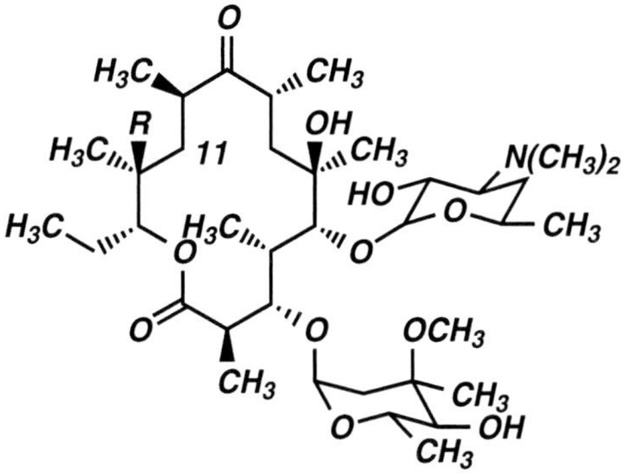   *11-deoxyerythromycins*

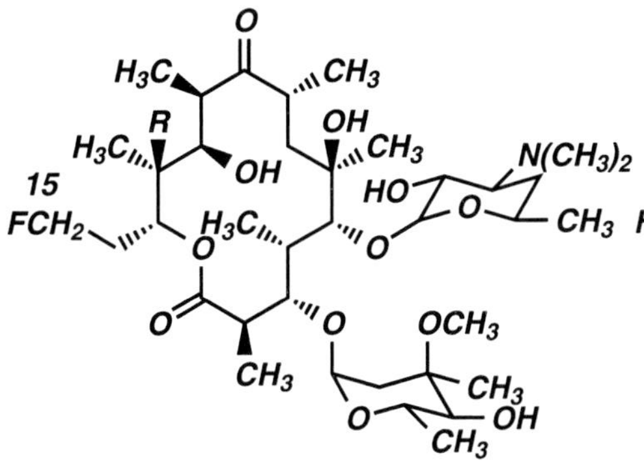

15-fluoroerythromycins 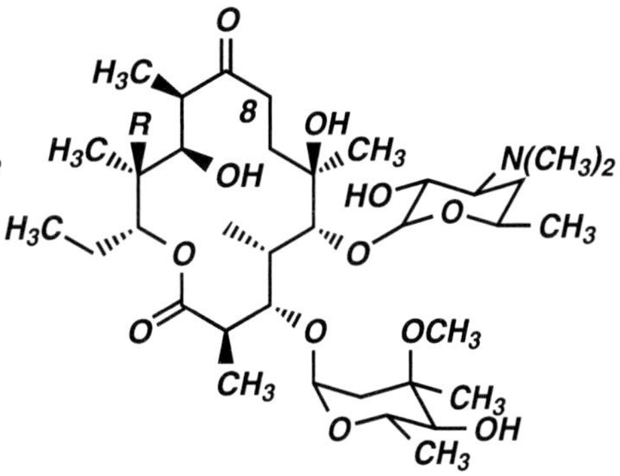   *8-desmethylerythromycins* 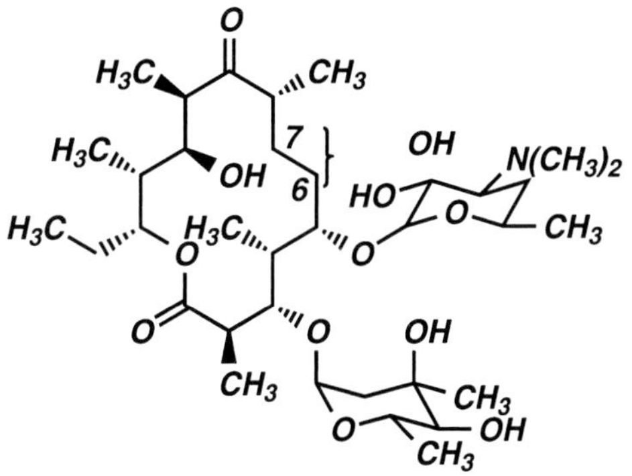   *6-desmethylerythromycin D*

Fig. 3. Examples of six different erythromycins produced by engineered 6-deoxyerythronolide B synthase genes. R = H or OH. The "OH" for 6-desmethylerythromycin D indicates that hydroxylation can take place either at the 6 or 7 position of 6-desmethyl-6-dEB.

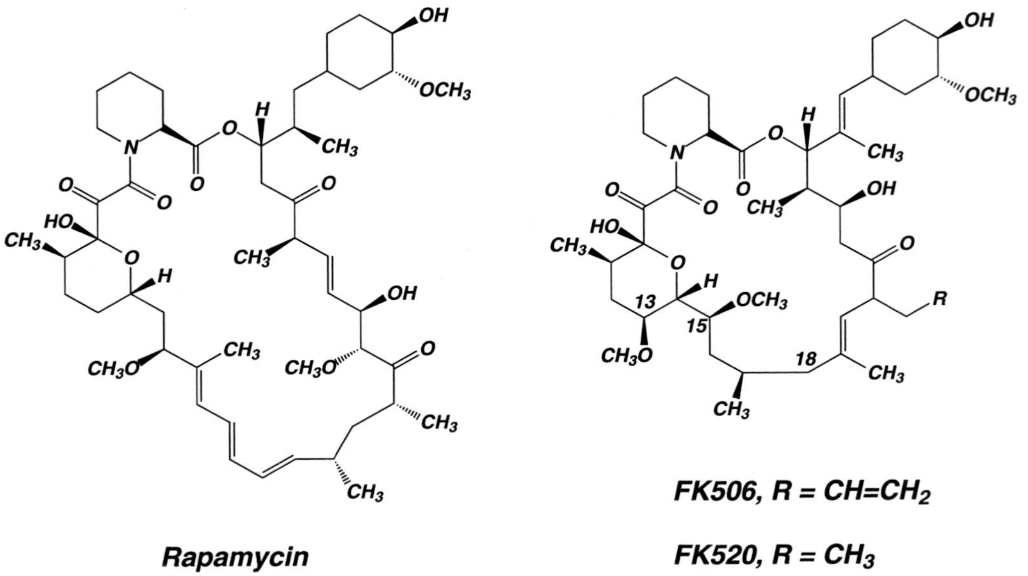

Fig. 4. The structures of the immunosuppressants, rapamycin, FK506 and FK520. The methoxy groups of FK520 at positions 13 and/or 15 have been replaced with hydrogen, methyl or ethyl groups as discussed in the text.

2.2. Engineering Other Macrolactone PKSs
2.2.1. FK520

Three macrolactone antibiotics FK506, FK520, and rapamycin (**Fig. 4**), different from erythromycin and other antibacterial macrolides, are used therapeutically in organ transplantation for their immunosuppressive property. Five analogs of FK520, representing substitutions of the methoxy groups positions 13 and 15 with ethyl, hydrogen, or methyl groups, have been produced by appropriate AT domain exchange experiments directed at the relevant PKS genes *(32)*. The 13-hydrogen analog of FK520 retained the in vitro properties typical of the immunosuppressive FK506 and FK520, whereas the 13-methyl analog displayed greatly reduced suppression of T-cell activation and, along with its 18-hydroxy derivative (prepared chemically), has apparent nerve growth stimulatory activity *(32)*. Five different rapamycin analogs have been produced by the same approach and are under evaluation at Kosan Biosciences (S. Bondi, J. Kennedy, L. Tang, S. Ward, and C. R. Hutchinson, unpublished work).

2.2.2. Epothilones

Epothilones B and D are polyketide macrolactones from subspecies of the myxobacterium *Sorangium cellulosum (33)*, which are undergoing clinical trials as anticancer drugs *(34)*. They inhibit cytoskeletal formation by increasing the stability of microtubules during their formation, which interferes with formation of the mitotic spindle and results in potent arrest of cancerous as well as normal cell growth at the G2/M transition of cell development *(35)*. Because epothilone D seems to have a better therapeutic index than epothilone B, at least in animals *(36)*, this compound was

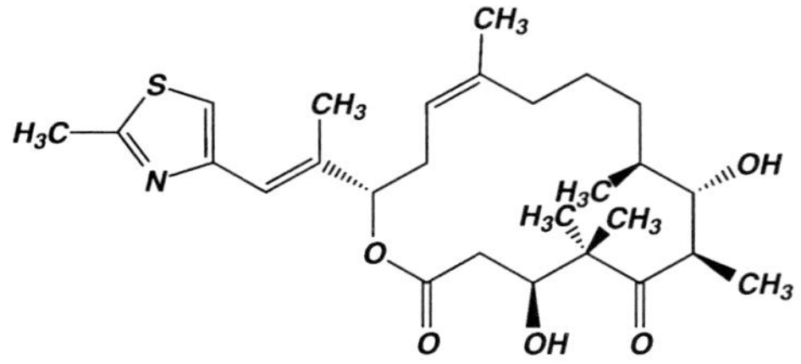

Manipulating Microbial Metabolites 85

[Epothilone D structure]
Epothilone D

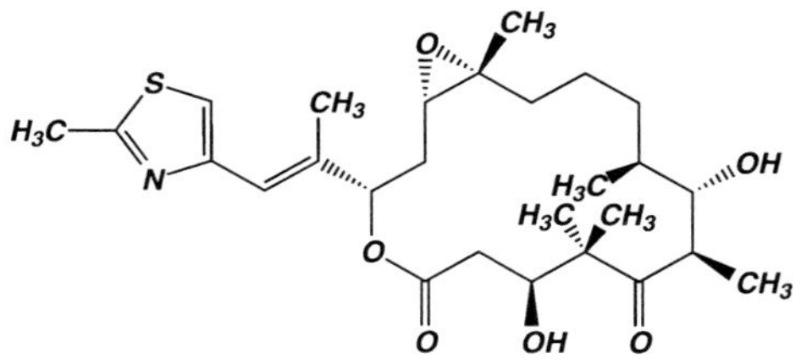

[Epothilone B structure]
Epothilone B

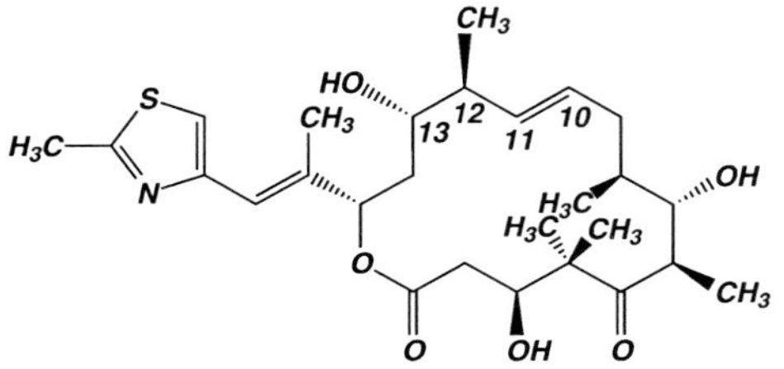

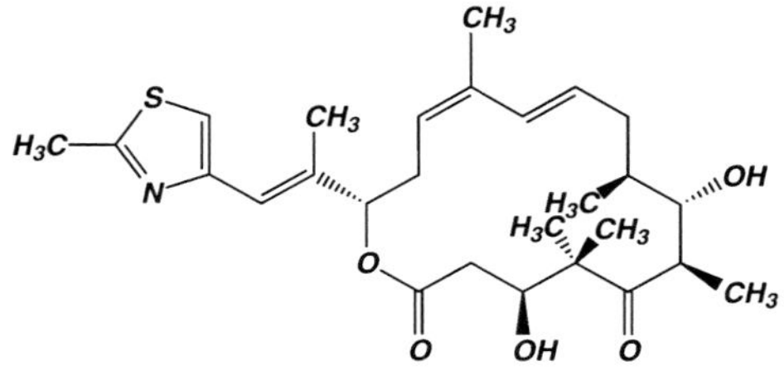

[(E)-10,11-didehydro-12,13-dihydro-13-hydroxyepothilone D structure, with positions 10, 11, 12, 13 labeled]
(E)-10,11-didehydro-12,13-dihydro-13-hydroxyepothilone D

[(E)-10,11-didehydroepothilone D structure]
(E)-10,11-didehydroepothilone D

Fig. 5. The structures of epothilone B and D, and novel analogs of epothilone D that have been produced by engineered epothilone polyketide synthase genes.

selected for development at Kosan Biosciences and is in early Phase II trials as KOS-862. A commercial process for its large-scale production has been established that relies on expression of the cloned epothilone PKS genes *(37,38)* in *Myxococcus xanthus* *(39)*. The convenience of performing gene engineering experiments in this organism has facilitated making several analogs of epothilone D *(40)*; the formation of two of these, 10,11-didehydro-12,13-dihydro-13-hydroxyepothilone D *(40)* and 10,11-didehydro-epothilone D *(41)* (**Fig. 5**), by inactivation of the DH and ER domains, respectively, in module 5 of the epothilone PKS, has shed light on the mechanism of formation of the unusual *cis*-12,13 C=C *(40)*.

3. Geldanamycin Analogs by Gene Engineering

Geldanamycin and the closely related herbimycin A (**Fig. 6**) were initially discovered by virtue of their weak antibacterial and antifungal *(42)* or plant-growth inhibitory *(43)* properties. Interest in these benzoquinone ansamycins increased greatly upon the discovery of the antitumor properties of geldanamycin *(44)* and herbimycin A *(45)*. Neckers and co-workers *(46)* showed in 1994 that their principal cellular target is Hsp90, a ubiquitous and abundant protein chaperone of mammalian cells *(47)*. Geldanamycin competes with ATP for the ATP-binding site of Hsp90 and, when bound, inhibits the ATP-dependent functions of this protein. A particular function is the ability of Hsp90 to chaperone nascent protein kinases that are critical components of signal transduction pathways, especially those in certain cancer cells *(47)*. In the presence of geldanamycin or herbimycin A, the immature kinases undergo rapid degradation, as a consequence of ubiquitination and subsequent catabolism by the proteosome, and the

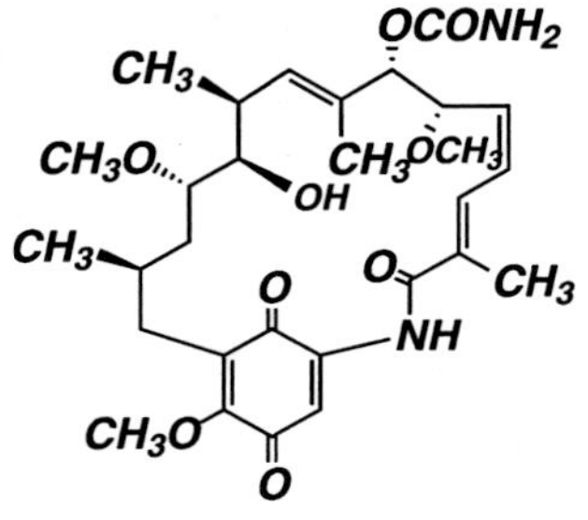

Geldanamycin

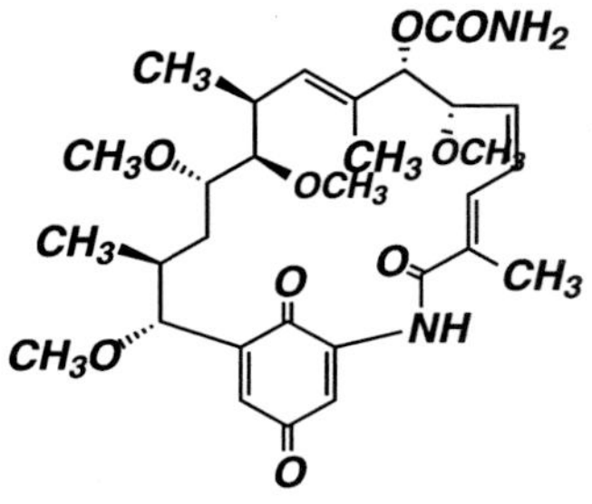

Herbimycin A

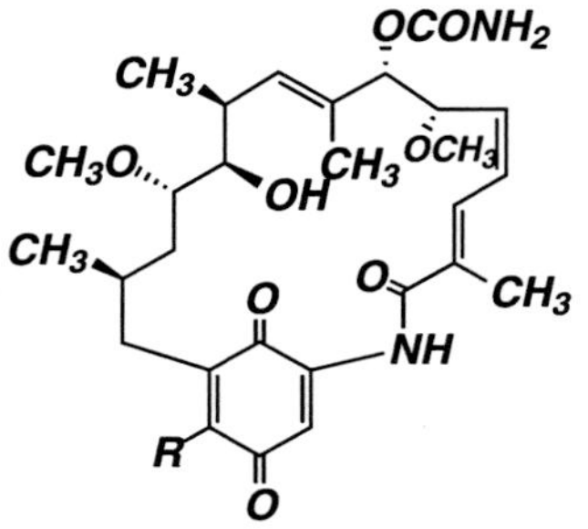

17-AAG, R = CH$_2$=CHCH$_2$-NH

17-DMAG, R = (CH$_3$)$_2$NCH$_2$CH$_2$-NH

Fig. 6. The structures of geldanamycin, herbimycin A and the 17-substituted geldanamycins 17-AAG and 17-DMAG discussed in the text.

levels of the mature kinases become depleted. This can result in a cytostatic effect on a cancer cell or in some cases apoptosis and cell death.

The discovery that Hsp90 and one or more of its protein kinase cohorts are overproduced in several types of human cancers has led to considerable interest in geldanamycin and its analogs as potential anticancer drugs *(48,49)*. Many geldanamycin analogs have been produced by replacement of the C17 *O*-methoxy group with substituted amines *(50)*. 17-AAG (**Fig. 6**), is currently undergoing phase I and early-stage phase II clinical trials *(49,51)*; and a more water-soluble analog, 17-dimethylaminoethylamino-17-demethoxygeldanamycin (17-DMAG) (**Fig. 6**), has completed preclinical development *(52)*.

Additional geldanamycin analogs are being sought in an attempt to ameliorate the hepatotoxicity that is characteristic of geldanamycin, although this is lessened by the 17-amino substitutions noted above *(53,54)*. One approach is to make demethyl and demethoxy analogs by PKS gene engineering, which might show increased potency, or nonquinone analogs by inactivation of the tailoring genes, which might have an improved therapeutic index, because the quinone moiety is believed to facilitate formation of free radicals in vivo *(53,54)*. The geldanamycin biosynthesis genes were cloned and characterized from *Streptomyces hygroscopicus* NRRL 3602 *(55)* to enable genetic engineering experiments; then recombinant strains carrying the *gdmA* PKS genes with AT domain replacements in modules 4, 5, or 7 of the *gdmA2* and *gdmA3* PKS genes *(55)* were created that produced novel geldanamycin analogs like the ones shown in **Fig. 7** (Bianka Hadatsch, Z. Hu, R. McDaniel, K. Patel, M. Piagentini, A. Rascher, L. Vetcher, and C. R. Hutchinson, unpublished work). Inactivation of the *gdmM* monooxygenase gene led to formation of the phenol analog shown; the similar compound formed by the module 7 AT mutant must reflect its inability to serve as a substrate for the GdmM enzyme (A. Rascher and C. R. Hutchinson, unpublished work). Such compounds can be investigated as leads for drug development.

4. Polyketide Genes From the Metagenome of Marine Organisms
4.1. Marine Natural Products As Drug Leads

Cytotoxic, antibacterial, antifungal, anti-inflammatory compounds and many other types of natural products have been isolated from different kinds of marine organisms (*see* ref. 56 and references therein). However, the supply of such potentially valuable drug leads has customarily been hampered by an inadequate source, as a result of the limited availability or difficult accessibility of the producing organism, which usually is very difficult to grow even by aquaculture. It is widely believed that microbial symbionts make marine-derived compounds like the polyketides that are commonly found in terrestrial microorganisms *(57,58)*. Consequently, it should be possible either to isolate pure cultures of these microbes to enable production of the compounds in large amount or to clone the production genes from the metagenome of the marine organism, if a producing microbe cannot be isolated. Once they are cloned, it is feasible to express the genes in different microbial hosts as the means to determine what is produced and whether it is the desired drug *(58)*. Limitations to achieving this goal could be inappropriate expression devices, lack of a unique substrate for a pathway enzyme, or autotoxicity of the produced compound to the host.

Kosan is interested in pursuing the development of antitumor drugs from marine sponges and, as a consequence, has investigated the metagenome of *Discodermia dissoluta* (R. Gadkari, C. Reeves, A. Schirmer, and C. R. Hutchinson, unpublished work). Libraries containing many thousands of clones were made in fosmid and bacterial artificial chromosome (BAC) vectors, using total DNA from the sponge and from fractions enriched for unicellular or filamentous microorganisms, then screened for the presence of the KS domains of modular PKS genes by polymerase chain reaction (PCR) analysis and by direct hybridization methods using suitable gene probes. Typical results are summarized in **Table 1**.

These data firmly establish that (1) a diverse range of KS homologs can be cloned and characterized in total DNA from the sponge samples or from bacterially enriched cell fractions, (2) bacterial genes can make up the majority of clonable genes from the metagenome of the sponge, and (3) in the case studied, multimodular PKS genes can be comparatively rare whereas PKS-NRPS genes can be quite common in the filamentous bacterial cells. It is unclear why the sponge genes, i.e., eukaryotic DNA that is presumed to come from either sponge cells or eukaryotic symbionts like diatoms and dinoflagellates, made up such a small percentage of the types of genes clonable from the total sponge DNA preparations. A possible explanation is related to the fact that the

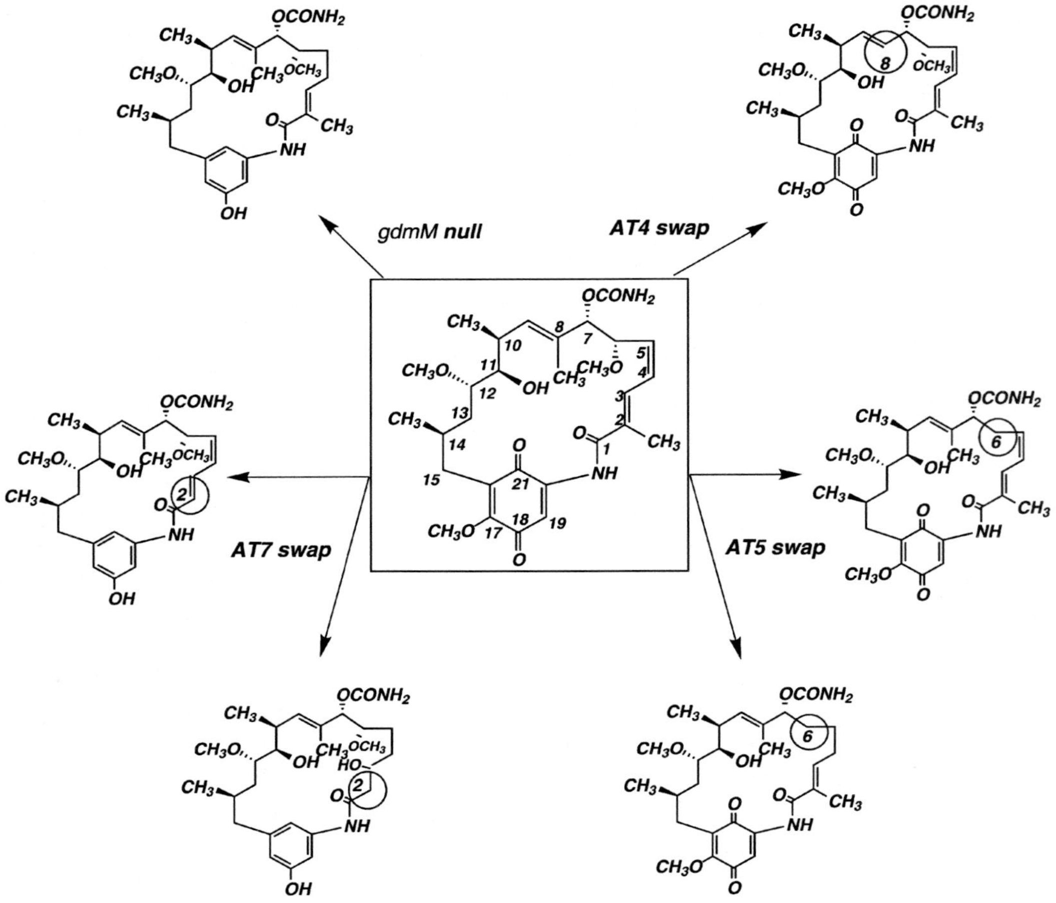

Fig. 7. Structure of the geldanamycin analogs made by engineered *gdmA2A3* polyketide synthase genes or by the strain with a disrupted *gdmM* gene. The positions converted to demethyl or demethoxy positions by acyltransferase swaps in the three different domains of *gdmA2* or *gdmA3* are encircled.

results of cytological staining studies show that the *D. dissoluta* tissue contains at least 100 times more microbial cells than sponge cells; if for some reason the bacterial DNA is more readily clonable than sponge-cell DNA, then the preponderance of the former type of genes is understandable.

5. Future Directions for Engineering Drug Production by Bacteria

5.1. Escherchia coli *As the Host for Large-Scale Production*

New types of microbial hosts are being studied as alternatives to the widely used filamentous actinomycetes bacteria for the commercial production of polyketides and other types of natural products. An example of the use of a common laboratory strain of a myxobacterium to produce epothilone D is presented above. *Escherichia coli* is another possibility, but, unlike the myxobacteria and actinomycetes, it does not contain all the substrates commonly used by a modular PKS, or some of the necessary cofac-

Table 1
Distribution of Modular Polyketide Synthase (PKS) Gene Cluster in Metagenomic Libraries of *Discodermia dissoluta*

Library made from:	Unicellular bacterial enrichment	Filamentous bacterial enrichment	Total sponge
Fosmids screened[a]	64,000	36,000	55,000
PKS positive fosmids[a]	465 (0.7%)	185 (0.5%)	375 (0.7%)
Fosmids with bacterial inserts	>90%	>60%	>80%
PKS fosmids endsequenced	145	148	168
Fosmids with 2 PKS ends[b]	2%	1%	2%
Fosmids with 1 PKS end[c]	29%	47%	41%
Fosmids with no PKS end[d]	68%	52%	57%
PKS fosmids with an NRPS[e] end	0	12%	1%

[a] By hybridization with sponge-derived KS probe pools at low stringency
[b] Contains part of large PKS gene cluster (>5 modules)
[c] Contains start or end of PKS gene cluster of unpredictable size
[d] Contains small PKS gene cluster (1–5 modules)
[e] nonribosomal peptide synthetase

tors. Nevertheless, attempts are underway to introduce these features into typical K12-derived *E. coli* strains and to adapt common expression tools to handle the 5- to 40-kb modular PKS genes. To date, *E. coli* strains have been engineered to contain the genes that convert fed propionate to its CoA form and turn propionyl CoA into 2-methylmalonyl CoA without catabolism of propionate to methylcitrate *(59,60)*, which enable formation of 2-methylmalonyl CoA from fed glucose via the tricarboxylic acid cycle (but only if a precursor of vitamin B12 is added to the fermentation) *(61)*, and which convert the apo form of the modular PKSs into their pantotheinylated, enzymatically active forms *(59)*. An investigation of the variables in expression of the substrate supply, modification, and PKS genes that affect the amount of 6dEB produced by such strains when grown in shake flasks *(62)* has set the stage for optimization of a high cell-density, fed-batch fermentation-based process that has already achieved 1.1 g/L titers of 6dEB *(63)*. Further research is likely to make *E. coli* as attractive as *M. xanthus* or even typical actinomycetes for production of relatively simple polyketides—i.e., ones like epothilone D for which the PKS product does not undergo extensive tailoring reactions that would require introduction of multigene clusters for complex metabolic pathways.

5.2. Creation of Novel PKS Genes

An exciting possibility is on the horizon—that novel polyketides could be manufactured by unusual combinations of the basic building block of modular PKSs: a module that contains all the information for one round of carbon chain extension for a given substrate. Research into the factors governing the choice of chain extension substrate and the ability of the incoming, PKS-tethered substrate to react with this 2-carboxyacylthioester, once it is loaded onto the ACP domain of a module, is beginning to establish some boundary conditions; *see* ref. *64* and references therein. Earlier studies had shown that the productive interaction of two separate modules, when presented with typical acylthioester substrates, was greatly influenced by unique, short amino acid sequences, called "linkers," found at both ends of the protein and even between modules within multimodular proteins *(65–67)*. Consequently, it is logical to assume that collections of individual modules containing appropriate linkers could be screened in bimodular combinations for their ability to convert a given acylthioester starter unit and 2-carboxyacylthioester chain extension substrate to some product. Productive bimodular combinations (defined by kinetic parameters and/or relative yields as a function of particular starter and extender substrates) could then be challenged with a collection of additional modules, and so forth to define the ones that made certain desired products or just any one of a set of theoretically possible products. The groundwork to expedite such research has been laid at Kosan Biosciences; thus, the time is ripe to exploit how multimodular PKS genes themselves must have evolved, in the attempt to expand the repertoire of therapeutically useful natural products nature has provided.

References

1. Cragg GM, Newman DJ, Snader KM. Natural products in drug discovery and development. J Nat Prod 1997;60:52–60.
2. Newman DJ, Cragg GM, Snader KM. Natural products as sources of new drugs over the period 1981–2002. J Nat Prod 2003;66:1022–1037.
3. Hopwood DA. Forty years of genetics with *Streptomyces*: from *in vivo* through *in vitro* to *in silico*. Microbiology 1999;145:2183–2202.
4. Baltz RH. Genetic manipulation of antibiotic-producing *Streptomyces*. Trends Microbiol 1998;6:76–83.
5. Hutchinson CR, McDaniel R. Combinatorial biosynthesis in microorganisms as a route to new antimicrobial, antitumor and neuroregenerative drugs. Curr Opin Investig Drugs 2001;2: 1681–1690.
6. Reeves CD. The enzymology of combinatorial biosynthesis. Crit Rev Biotechnol 2003;23:95–147.
7. Staunton J, Weissman KJ. Polyketide biosynthesis: a millennium review. Nat Prod Rep 2001;18:380–416.
8. Donadio S, Staver MJ, McAlpine JB, Swanson SJ, Katz L. Modular organization of genes required for complex polyketide biosynthesis. Science 1991;252:675–679.
9. Rawlings BJ. Type I polyketide biosynthesis in bacteria (Part B) (1995 to mid-2000). Nat Prod Rep 2001;18:231–281.
10. Rawlings BJ. Type I polyketide biosynthesis in bacteria (Part A—erythromycin biosynthesis) (1994 to 2000). Nat Prod Rep 2001;18:190–227.
11. Khosla C, Gokhale RS, Jacobsen JR, Cane DE. Tolerance and specificity of polyketide synthases. Annu Rev Biochem 1999;68:219–253.
12. Liou GF, Khosla C. Building-block selectivity of polyketide synthases. Curr Opin Chem Biol 2003;7:279–284.
13. Austin MB, Noel JP. The chalcone synthase superfamily of type III polyketide synthases. Nat Prod Rep 2003;20:79–110.
14. Cortes J, Haydock SF, Roberts GA, Bevitt DJ, Leadlay PF. An unusually large multifunctional polypeptide in the erythromycin-producing polyketide synthase of *Saccharopolyspora erythraea*. Nature 1990;348:176–178.
15. Del Vecchio F, Petkovic H, Kendrew SG, et al. Active-site residue, domain and module swaps in modular polyketide synthases. J Ind Microbiol Biotechnol 2003;30:489–494.

16. Carreras C, Frykman S, Ou S, et al. *Saccharopolyspora erythraea*-catalyzed bioconversion of 6-deoxyerythronolide B analogs for production of novel erythromycins. J Biotechnol 2002;92:217–228.
17. Jacobsen JR, Hutchinson CR, Cane DE, Khosla C. Precursor-directed biosynthesis of erythromycin analogs by an engineered polyketide synthase. Science 1997;277:367–369.
18. Pacey MS, Dirlam JP, Geldart RW, et al. Novel erythromycins from a recombinant *Saccharopolyspora erythraea* strain NRRL 2338 pIG1. I. Fermentation, isolation and biological activity. J Antibiot 1998;51:1029–1034.
19. Ruan X, Pereda A, Stassi DL, et al. Acyltransferase domain substitutions in erythromycin polyketide synthase yield novel erythromycin derivatives. J Bacteriol 1997;179:6416–6425.
20. Starks CM, Rodriguez E, Carney JR, et al. Isolation and characterization of 7-hydroxy-6-demethyl-6-deoxyerythromycin D, a new erythromycin analogue, from engineered *Saccharopolyspora erythraea*. J Antibiot 2004;57:64–67.
21. Petkovic H, Lill RE, Sheridan RM, et al. A novel erythromycin, 6-desmethyl erythromycin D, made by substituting an acyltransferase domain of the erythromycin polyketide synthase. J Antibiot 2003;56:543–551.
22. Stassi DL, Kakavas SJ, Reynolds KA, et al. Ethyl-substituted erythromycin derivatives produced by directed metabolic engineering. Proc Natl Acad Sci USA 1998;95:7305–7309.
23. Reid R, Piagentini M, Rodriguez E, et al. A model of structure and catalysis for ketoreductase domains in modular polyketide synthases. Biochemistry 2003;42:72–79.
24. Rodriguez E, Hu Z, Ou S, Volchegursky Y, Hutchinson CR, McDaniel R. Rapid engineering of polyketide overproduction by gene transfer to industrially optimized strains. J Ind Microbiol Biotechnol 2003;30:480–488.
25. Donadio S, McAlpine JB, Sheldon PJ, Jackson M, Katz L. An erythromycin analog produced by reprogramming of polyketide synthesis. Proc Natl Acad Sci USA 1993;90:7119–7123.
26. McDaniel R, Thamchaipenet A, Gustafsson C, Fu H, Betlach M, Ashley G. Multiple genetic modifications of the erythromycin polyketide synthase to produce a library of novel "unnatural" natural products. Proc Natl Acad Sci USA 1999;96:1846–1851.
27. Tang L, McDaniel R. Construction of desosamine containing polyketide libraries using a glycosyltransferase with broad substrate specificity. Chem Biol 2001;8:547–555.
28. Rowe CJ, Bohm IU, Thomas IP, et al. Engineering a polyketide with a longer chain by insertion of an extra module into the erythromycin-producing polyketide synthase. Chem Biol 2001;8:475–485.
29. Gaisser S, Reather J, Wirtz G, Kellenberger L, Staunton J, Leadlay PF. A defined system for hybrid macrolide biosynthesis in *Saccharopolyspora erythraea*. Mol Microbiol 2000;36:391–401.
30. Rix U, Fischer C, Remsing LL, Rohr J. Modification of post-PKS tailoring steps through combinatorial biosynthesis. Nat Prod Rep 2002;19:542–580.
31. Watanabe K, Wang CC, Boddy CN, Cane DE, Khosla C. Understanding substrate specificity of polyketide synthase modules by generating hybrid multimodular synthases. J Biol Chem 2003;278:42,020–42,026.
32. Revill PW, Voda J, Reeves CR, et al. Genetically engineered analogs of ascomycin for nerve regeneration. J Pharm Exper Ther 2002;302:1278–1285.
33. Gerth K, Bedorf N, Hoefle G, Irschik H, Reichenbach H. Antibiotics from gliding bacteria. 74. Epothilons A and B: antifungal and cytotoxic compounds from Sorangium cellulosum (Myxobacteria): production, physico-chemical and biological properties. J Antibiot 1996;49:560–563.
34. Altmann KH. Epothilone B and its analogs—a new family of anticancer agents. Mini Rev Med Chem 2003;3:149–158.
35. Myles DC. Emerging microtubule stabilizing agents for cancer chemotherapy. Annu Rep Med Chem 2002;37:125–132.

36. Chou T-C, O'Connor OA, Tong WP, et al. The synthesis, discovery, and development of a highly promising class of microtubule stabilization agents: curative effects of desoxyepothilones B and F against human tumor xenografts in nude mice. Proc Natl Acad Sci USA 2001;98:8113–8118.
37. Tang L, Shah S, Chung L, et al. Cloning and heterologous expression of the epothilone gene cluster. Science 2000;287:640–642.
38. Molnar I, Schupp T, Ono M, et al. The biosynthetic gene cluster for the microtubule-stabilizing agents epothilones A and B from *Sorangium cellulosum* So ce90. Chem Biol 2000;7:97–109.
39. Lau J, Frykman S, Regentin R, Ou S, Tsuruta H, Licari P. Optimizing the heterologous production of epothilone D in *Myxococcus xanthus*. Biotechnol Bioeng 2002;78:280–288.
40. Tang L, Ward S, Chung L, et al. Elucidating the mechanism of cis double bond formation in epothilone biosynthesis. J Amer Chem Soc 2004;126:46–47.
41. Arslanian RL, Tang L, Blough S, et al. A new cytotoxic epothilone from modified polyketide synthases heterologously expressed in *Myxococcus xanthus*. J Nat Prod 2002;65:1061–1064.
42. DeBoer C, Meulman PA, Wnuk RJ, Peterson DH. Geldanamycin, a new antibiotic. J Antibiot 1970;23:442–447.
43. Omura S, Iwai Y, Takahashi Y, et al. Herbimycin, a new antibiotic produced by a strain of *Streptomyces*. J Antibiot 1979;32:255–261.
44. Sasaki K, Yasuda H, Onodera K. Growth inhibition of virus transformed cells in vitro and antitumor activity in vivo of geldanamycin and its derivatives. J Antibiot 1979;32:849–851.
45. Uehara Y, Murakami Y, Suzukake-Tsuchiya K, et al. Effects of herbimycin derivatives on src oncogene function in relation to antitumor activity. J Antibiot 1988;41:831–834.
46. Whitesell L, Mimnaugh EG, De Costa B, Myers CE, Neckers LM. Inhibition of heat shock protein HSP90-pp60v-src heteroprotein complex formation by benzoquinone ansamycins: essential role for stress proteins in oncogenic transformation. Proc Natl Acad Sci USA 1994;91:8324–8328.
47. Richter K, Buchner J. Hsp90: chaperoning signal transduction. J Cell Physiol 2001;188:281–290.
48. Neckers L. Hsp90 inhibitors as novel cancer chemotherapeutic agents. Trends Mol Med 2002;8:S55–S61.
49. Goetz MP, Toft DO, Ames MM, Erlichman C. The Hsp90 chaperone complex as a novel target for cancer therapy. Ann Oncol 2003;14:1169–1176.
50. Schnur RC, Corman ML, Gallaschun RJ, et al. Inhibition of the oncogene product p185erbB-2 in vitro and in vivo by geldanamycin and dihydrogeldanamycin derivatives. J Med Chem 1995;38:3806–3812.
51. Egorin MJ, Zuhowski EG, Rosen DM, Sentz DL, Covey JM, Eiseman JL. Plasma pharmacokinetics and tissue distribution of 17-(allylamino)-17-demethoxygeldanamycin (NSC 330507) in CD2F1 mice. Cancer Chemother Pharmacol 2001;47:291–302.
52. Egorin MJ, Lagattuta TF, Hamburger DR, et al. Pharmacokinetics, tissue distribution, and metabolism of 17-(dimethylaminoethylamino)-17-demethoxygeldanamycin (NSC 707545) in CD2F1 mice and Fischer 344 rats. Cancer Chemother Pharmacol 2002;49:7–19.
53. Kelland LR, Sharp SY, Rogers PM, Myers TG, Workman P. DT-Diaphorase expression and tumor cell sensitivity to 17-allylamino, 17-demethoxygeldanamycin, an inhibitor of heat shock protein 90. J Natl Cancer Inst 1999;91:1940–1949.
54. Tudor G, Gutierrez P, Aguilera-Gutierrez A, Sausville EA. Cytotoxicity and apoptosis of benzoquinones: redox cycling, cytochrome c release, and BAD protein expression. Biochem Pharmacol 2003;65:1061–1075.
55. Rascher A, Hu Z, Viswanathan N, et al. Cloning and characterization of a gene cluster for geldanamycin production in *Streptomyces hygroscopicus* NRRL 3602. FEMS Microbiol Lett 2003;218:223–230.

56. Faulkner DJ. Marine natural products. Nat Prod Rep 2002;19:1–48.
57. Haygood MG, Schmidt EW, Davidson SK, Faulkner DJ. Microbial symbionts of marine invertebrates: opportunities for microbial biotechnology. J Mol Microbiol Biotechnol 1999;1:33–43.
58. Hildebrand M, Waggoner LE, Lim GE, Sharp KH, Ridley DP, Haygood MG. Approaches to identify, clone and express symbiont bioactive metabolite genes. Nat Prod Rep 2004;21:122–142.
59. Pfeifer BA, Admiraal SJ, Gramajo H, Cane DE, Khosla C. Biosynthesis of complex polyketides in a metabolically engineered strain of *E. coli*. Science 2001;291:1790–1792.
60. Pfeifer B, Hu Z, Licari P, Khosla C. Process and metabolic strategies for improved production of *Escherichia coli*-derived 6-deoxyerythronolide B. Appl Environ Microbiol 2002;68:3287–3292.
61. Dayem LC, Carney JR, Santi DV, Pfeifer BA, Khosla C, Kealey JT. Metabolic engineering of a methylmalonyl-CoA mutase-epimerase pathway for complex polyketide biosynthesis in *Escherichia coli*. Biochemistry 2002;41:5193–5201.
62. Murli S, Kennedy J, Dayem LC, Carney JR, Kealey JT. Metabolic engineering of *Escherichia coli* for improved 6-deoxyerythronolide B production. J Ind Microbiol Biotechnol 2003;30:500–509.
63. Lau J, Tran C, Licari P, Galazzo JL. Development of a high cell-density fed-batch bioprocess for the heterologous production of 6-deoxyerythronolide B in *Escherichia coli*. J. Biotechnol 2004;110:95–103.
64. Kumar P, Li Q, Cane DE, Khosla C. Intermodular communication in modular polyketide synthases: structural and mutational analysis of linker mediated protein-protein recognition. J Amer Chem Soc 2003;125:4097–4102.
65. Gokhale RS, Tsuji SY, Cane DE, Khosla C. Dissecting and exploiting intermodular communication in polyketide synthases. Science 1999;284:482–485.
66. Wu N, Tsuji SY, Cane DE, Khosla C. Assessing the balance between protein-protein interactions and enzyme-substrate interactions in the channeling of intermediates between polyketide synthase modules. J Amer Chem Soc 2001;123:6465–6474.
67. Broadhurst RW, Nietlispach D, Wheatcroft MP, Leadlay PF, Weissman KJ. The structure of docking domains in modular polyketide synthases. Chem Biol 2003;10:723–731.
68. Hutchinson CR. Biosynthetic Studies of daunorubicin and tetracenomycin C. Chem Rev 1997;97:2525–2536.

5
Improving Drug Discovery From Microorganisms

Chris M. Farnet and Emmanuel Zazopoulos

Summary

Microorganisms remain unrivalled in their ability to produce bioactive small molecules for drug development. However, the core technologies used to discover microbial natural products have not evolved significantly over the past several decades, resulting in a shortage of new drug leads. Advances in DNA-sequencing and bioinformatics technologies now make it possible to rapidly identify the clusters of genes that encode bioactive compounds and to make computer predictions of chemical structure based on gene sequence information. These structure predictions can be used to identify new chemical entities and provide important physicochemical "handles" that guide compound purification and structure confirmation. Industrialization of this process provides a model for improving the efficiency of natural-product discovery. The application of advanced genomics and bioinformatics technologies is now poised to revolutionize natural-product discovery and lead a renaissance of interest in microorganisms as a source of bioactive compounds for drug development.

Key Words: Natural products; genomics; drug discovery; bioinformatics; actinomycetes; dereplication; fermentation; structure elucidation.

1. Introduction

Microorganisms produce some of the most important medicines ever developed. They are the source of lifesaving treatments for bacterial and fungal infections (e.g., penicillin, erythromycin, streptomycin, tetracycline, vancomycin, amphotericin), cancer (e.g., daunorubicin, doxorubicin, mitomycin, bleomycin), transplant rejection (e.g., cyclosporin, FK-506, rapamycin), and high cholesterol (e.g., statins such as lovastatin and mevastatin) (**Fig. 1**). Microbial natural products are notable not only for their potent therapeutic activities, but also for the fact that they frequently possess the desirable pharmacokinetic properties required for clinical development. The drugs shown in **Fig. 1** are just a few of the many microbial natural products that reached the market without any chemical modifications required, a testimony to the remarkable ability of microorganisms to produce drug-like small molecules. Indeed, the potential to hit a "home run" with a single discovery distinguishes natural products from all other sources of chemical diversity and fuels the ongoing efforts to discover new compounds.

Traditionally, the search for new natural products has started by growing microorganisms in the laboratory and testing the fermentation broths for bioactivity. However, we now know that microorganisms have many natural-product gene clusters that they do not readily express when grown in the laboratory *(1–3)*. So despite decades of fermentation-broth screening, it is likely that a vast supply of bioactive microbial com-

From: *Natural Products: Drug Discovery and Therapeutic Medicine*
Edited by: L. Zhang and A. L. Demain © Humana Press Inc., Totowa, NJ

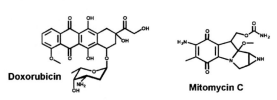

Fig. 1. A few of the landmark medicines produced by microorganisms.

pounds remains to be discovered. This untapped potential provided the impetus for us to develop an automated genomics platform that predicts the chemical structures of natural products by reading the sequences of the gene clusters that direct their synthesis. By surveying the genome, we can identify all of the natural products that a microorganism can make before fermentation studies begin, and specifically tailor the downstream production and purification strategies to isolate likely new chemical entities (NCEs) and avoid the re-isolation of known compounds. The integration of new genomics technologies greatly increases the efficiency of discovery and makes it possible to build a robust pipeline of NCEs from a small collection of microorganisms, providing a new paradigm for natural-product discovery. Here, we describe the core technologies behind the discovery platform developed at Ecopia BioSciences Inc. and present the discovery of a new antifungal agent as a case study to illustrate the power of the genomics-guided approach.

2. Genomics-Guided Natural-Product Discovery
2.1. Genome Scanning Technology

The development of a discovery platform that can predict chemical structures from gene sequences required the development of several new technologies and resources, including a means to efficiently isolate and sequence natural-product gene clusters from microbial genomes; a large reference database of gene clusters linked to the structures of the compounds they encode; and specialized computer applications that can detect correlations between gene sequence and chemical substructure elements. We developed a high-throughput genome-scanning method to sequence natural-product gene clusters without sequencing entire genomes (3). Our strategy was to scan the genomes of selected microorganisms that were reported to produce known, structurally diverse natural products and to build a database of gene clusters covering the full range of natural-product chemical diversity (the Ecopia Decipher® database). The approach proved to be successful, as in all cases the gene clusters corresponding to known natural products were identified by deductive analysis, providing an important training set for chemical structure predictions. The most striking finding, however, was the large number of unexpected gene clusters found in these previously well-studied microorganisms. Genome scanning of approximately 60 actinomycete strains revealed some 700 natural product gene clusters, or an average of a dozen gene clusters per organism. This number exceeds by at least a factor of ten the number of natural products that would have been detected from these organisms by traditional screening approaches (**Fig. 2**). It is now clear that many gene clusters are expressed only under certain growth conditions. Furthermore, even when they are expressed, some gene clusters produce compounds only at very low levels, below the limit of detection of conventional screening methods. This may explain why so many natural products have eluded detection in the past, as it was common practice to screen only one to three growth conditions for each strain.

2.2. Genomics-Guided Discovery Platform

To capitalize on the wealth of gene clusters revealed by genome scanning, we developed a genomics-guided discovery platform designed to rapidly identify clusters encoding likely NCEs and target the compounds for purification (**Fig. 3**). A suite of specialized software and computer applications predicts the structures of compounds encoded by new gene clusters via automated comparisons with known clusters in the database. These structure predictions identify possible NCEs and provide important physicochemical "handles" (including molecular weight, ultraviolet [UV] absorbance, lipophilicity, and other properties) that are then used to detect the desired compound in fermentation broths. To fully exploit their potential, each microorganism is grown in as many as 50 different fermentation media in order to maximize the probability that each of its gene clusters will be expressed. A number of custom-made analytical tools are used to simultaneously display and analyze mass spectroscopy, UV, and bioactivity data generated from extracts prepared from all the growth conditions used. These tools make it possible to detect compounds that may be produced only rarely in fermentation broths, or at very low levels, and to identify those compounds whose properties match the gene-cluster predictions. The structure information provides practical handles to

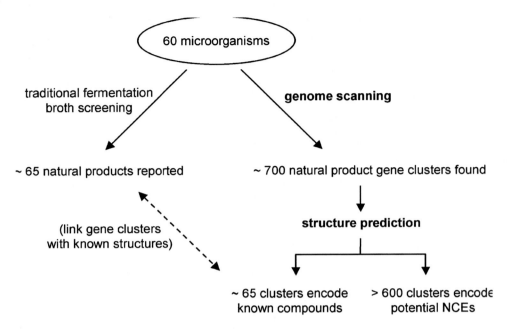

Fig. 2. Genome scanning reveals a vast supply of undiscovered natural products.

guide the purification of targeted compounds and greatly facilitates the final structure elucidation by spectroscopic methods.

The tremendous power of genomics-guided discovery is driven by a database and computational platform that "learns" from previous results and improves with each new compound discovered. In the final step of the discovery cycle, the confirmed structure of a new molecule is linked in the Decipher® database to the gene cluster that encodes it, thus enhancing the ability of the system to make future correlations between genes and chemical structures. Even the "rediscovery" of known compounds adds valuable new information to the database, as the genetic blueprint for each structure identifies new genes-to-molecules correlations. In addition, all of the chemical and biological data generated during the fermentation, extraction, purification, and bioactivity screening stages are fed back into the database and integrated with the genomics information. Sophisticated bioinformatics applications are then able to identify relationships between the diverse data sets that can be used to guide the production and purification of targeted metabolites—for example, by defining the fermentation media that are likely to support the expression of a gene cluster and by identifying purification schemes that have proved successful with isolations from the specific medium and for similar structure types.

3. Genomics-Guided Discovery: A Case Study
3.1. Genome Scanning of Streptomyces aizunensis

The untapped potential of microorganisms to produce bioactive NCEs is illustrated by our experience with the actinomycete *Streptomyces aizunensis*, a producer of the antibiotic bicyclomycin. The bicyclomycin gene cluster was targeted for isolation because the compound contains some unusual functional groups and the genes required

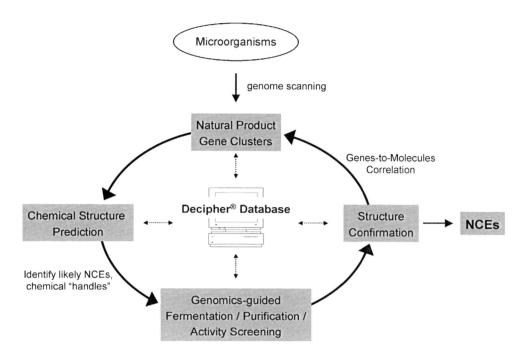

Fig. 3. The genomics-guided discovery platform developed at Ecopia, a new paradigm for natural product discovery. The platform identifies all of the natural-product gene clusters in a genome, predicts the chemical structures of the compounds they encode, and targets compounds that are likely to be NCEs for production and purification.

to make these kinds of structures were not known. Surprisingly, genome scanning of the *S. aizunensis* genome identified 11 natural-product gene clusters in addition to the bicyclomycin cluster, even though this organism produced only bicyclomycin in published fermentation screening studies *(4,5)*. One of the additional gene clusters was predicted to encode a compound similar to the known antibiotic streptothricin, based on computer-based comparisons to other clusters in the database. This information proved to be valuable, as a streptothricin-like compound was indeed detected during subsequent fermentation experiments. Knowing in advance that this compound would be produced made it very easy to identify and circumvent, while purifying other compounds from the fermentation broths. More importantly, the structure predictions generated for the remaining 10 gene clusters did not match any compounds present in databases of known natural products, indicating that they are likely to encode NCEs. Each of these clusters presented an exciting opportunity for new compound discovery. In the following sections, we demonstrate how automated gene sequence analysis was used to predict the structure of an NCE encoded by one of these gene clusters and how genomics information was used to guide the purification of the compound.

3.2. Automated Analysis of Gene Clusters and Chemical Structure Prediction

All gene clusters identified by genome scanning enter a fully automated analysis cascade of specialized software applications that identify open reading frames (ORFs), assign functions to each gene in the cluster, and predict chemical substructure elements. The output of the automated analysis is displayed in an interactive graphical user interface designed to allow scientists to quickly assess the accuracy of the computer predictions and to determine whether a cluster is likely to encode a known compound or an NCE. The automated analysis of one of the *S. aizunensis* gene clusters, designated 023D, is shown in **Fig. 4**. In this cluster, the computer assigned 35 ORFs to protein families based on homology comparisons to proteins in the Decipher database. The disposition of these ORFs in the cluster is shown in window A of **Fig. 4**, where each ORF carries a four-letter code indicating the protein family to which it was assigned. Nine of the ORFs in the 023D cluster were designated as polyketide synthases (PKSs). PKSs and other multimodular protein families (such as nonribosomal peptide synthetases) are further processed by an automated software application that parses the proteins into individual enzymatic domains. Each domain sequence is then compared to a series of protein models of active domains to identify domains that are likely to be nonfunctional. Additional computer scripts are also invoked when particular domains are encountered. For example, the substrate specificity of each acyltransferase (AT) domain is readily assigned by a phylogenetic comparison to AT domains of known specificity, while a similar analysis of thioesterase (TE) domains very effectively distinguishes domains that generate linear polyketide products from those that catalyze the formation of cyclic products. The result is an automated "domain string" (displayed in **Fig. 4**, window C) that captures the structure of the polyketide backbone in a line notation that can then be translated into a chemical structure prediction (**Fig. 4**, window B). The automated analysis of the 023D PKS system predicted a long, linear polyketide chain bearing polyene chromophores. While many cyclic (macrolide) polyene natural products are known, linear polyenes remain relatively rare. Chemical substructure searches using the predicted polyketide backbone identified no similar structures in natural-products databases, providing the first indication that the 023D gene cluster encoded a NCE.

3.3. Correlating Genes With Chemical Substructures

The "family string" generated by the analysis cascade (shown in window D of **Fig. 4**) provides a representation of the cluster that is used in an automated search of the database to identify gene clusters with similar families. The structures of the compounds linked to these clusters are then compared to identify common structural elements. For example, three gene clusters in the Decipher database contain the ADSN, AYTP, and CALB families found in the 023D cluster. Structure analysis of the corresponding compounds identified a single common structural element, a 2-amino-3-hydroxycyclopentenone (C_5N) group in amide linkage to a polyketide carboxylate (**Fig. 5**, upper). Inspection of the computer-predicted function of each family suggested a plausible pathway for the biosynthesis of a C_5N group from glycine and 5-aminolevulinic acid. Thus, the presence of these three genes in a cluster provides a marker for the presence of this functional group in a natural product. Similarly, computer analysis correlated four families in the 023D cluster with the presence of a four-carbon, amine-containing (C_4N) polyketide starter unit (**Fig. 5**, middle) and five families with a 6-deoxyhexose sugar moiety (**Fig. 5**, lower). In both cases the predicted functions of the families suggested likely biosynthetic pathways and strongly supported

Drug Discovery From Microorganisms

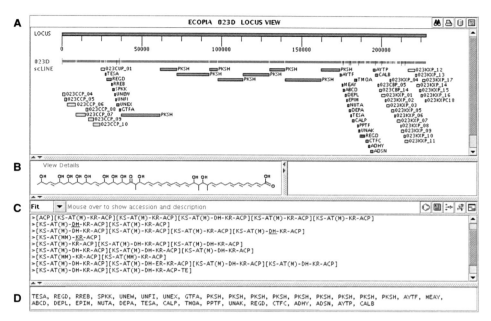

Fig. 4. Automated analysis of the 023D gene cluster: (**A**) overview of the automated gene finding and family calling; (**B**) predicted structure of the polyketide backbone; (**C**) automated domain string; (**D**) automated family string.

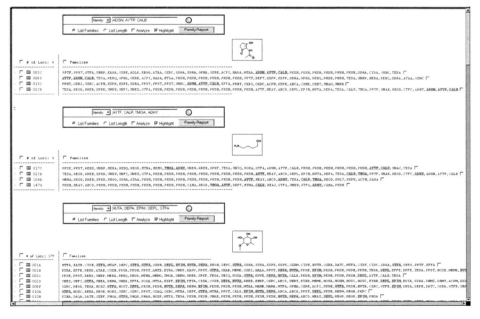

Fig. 5. Correlating genes and structures. An automated application identifies gene clusters having similar gene families (highlighted). The compounds produced by these gene clusters are then compared to identify common structural elements (boxed). This analysis identified three genes that correlate with the C_5N structure element (upper), four genes that correlate with the C_4N polyketide starter group (middle), and five genes that correlate with a 6-deoxyhexose group (lower).

the gene-structure correlations. The family- and domain-string notations demonstrate the utility of reducing gene-sequence information to a series of computable properties, or "genetic descriptors," that a computer can learn to associate with chemical-structure descriptors. As the number of sequenced gene clusters in the Decipher database climbs, the ability to predict chemical structure directly from gene sequence is increasingly refined.

The results of the automated analysis cascade provided a very precise prediction of the structure of the compound encoded by the 023D gene cluster, as shown in **Fig. 6**. Tools for substructure searching are also integrated into the discovery platform and allow a scientist to quickly assess whether a predicted structure has already been reported, providing an early opportunity for *in silico* dereplication. For example, substructure searches of the AntiBase database (Wiley Publishers, 2003) of over 30,000 microbial natural products revealed only 64 products that contain the C_5N group. This example illustrates how even a small amount of structure information can greatly limit the number of structures that need to be considered as candidate products. More importantly, the addition of a second structure element to the query returned no hits from the database, providing further evidence for the novelty of the compound encoded by the 023D gene cluster (**Fig. 6**).

3.4. Finding the Needle in a Haystack: Genomics-Guided Purification

Having strong evidence for an NCE, the compound encoded by the 023D gene cluster was targeted for purification. The structure prediction immediately identified physicochemical properties or "handles" that could be used to guide the purification of the compound. For example, the compound was predicted to have a molecular mass in excess of 1290 Daltons (Da) and a distinctive UV spectrum imparted by the pentaene chromophore. To ensure that the gene cluster was expressed, *S. aizunensis* was grown in more than 50 different fermentation media in 25-mL shake-flask cultures. Methanol extracts of each culture were subjected to high-performance liquid chromatography (HPLC)-UV-mass spectrometry (MS) analyses, and metabolites were monitored using a specially designed system that makes it possible to analyze HPLC fractions simultaneously across all the different media conditions (**Fig. 7**, upper). An overview of the MS traces showed that the profile of metabolites varied considerably from medium to medium. The interface shown in **Fig. 7** is fully interactive, so that the underlying chemical data can be rapidly searched using queries that incorporate multiple physicochemical parameters. For example, a mass filtering function allows the user to search all fractions for masses within a particular range. A search for metabolites having a mass greater that 1290 Da identified a single peak that appeared only in some fermentation media but not in others (**Fig. 7**, middle). Clicking on any fraction pops up a new window that displays the full spectral data set for that fraction. When this was done for one of the peak fractions from the mass filtered search, the data revealed molecular ions consistent with a major isotope of mass 1296.7 Da and a UV absorption spectrum characteristic of a pentaene (**Fig. 7**, lower inset), fully consistent with the structure predictions generated by computer analysis. Thus, the chemical data provided strong evidence that the 1296.7-Da metabolite corresponded to the compound encoded by the 023D gene cluster. In addition to the chemical data, aliquots of each HPLC fraction are routinely tested for antimicrobial activity against a panel of bacterial and fungal patho-

Predicted Structure:

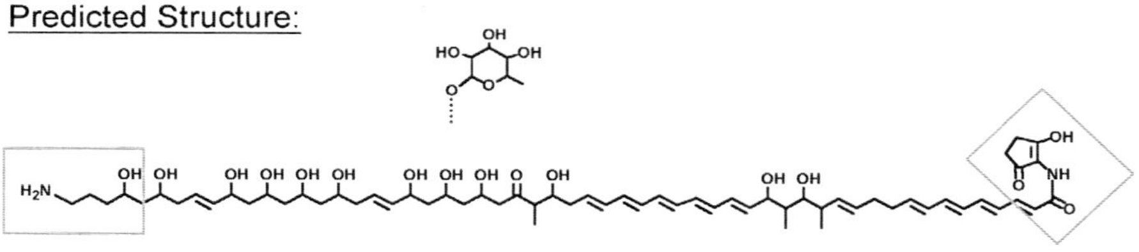

- substructure searching: novel compound
- "handles" for identification and purification:
  - mass > 1,290 Da
  - distinctive UV spectrum

Fig. 6. Computer-generated prediction of the structure of the compound encoded by the 023D gene cluster. The boxed portions indicate examples of substructure elements that can be used to search against a database of known natural products.

gens, and the bioassay results can also be displayed along with the chemical data in the graphical interface. In **Fig. 7** (middle), the screening data for *Candida albicans* is shown, as the fractions containing the 1296.7-Da compound exhibited a potent antifungal activity in the bioassay screens.

Normally, the purification of a natural product from a complex fermentation mixture presents a formidable task that may take months to achieve. In this case, however, having the mass, UV, and bioactivity data in hand made the subsequent purification of the 1296.7-Da compound straightforward, and this was accomplished in a matter of days. Similarly, the elucidation of a complex natural-product structure using the standard analytical techniques can be exceedingly difficult, but in this case it was greatly facilitated by the *in silico* structure prediction. The final structure was confirmed by multidimensional NMR spectrometry, and proved to be entirely consistent with the structure prediction generated by gene-sequence analysis (**Fig. 8**). The new compound, named ECO-02301, displayed potent in vitro activity against numerous fungal pathogens, including *Aspergillus fumigatus* and azole-resistant strains of *Candida albicans*, and was shown to be efficacious in a mouse model of disseminated candidiasis, where treatment of infected mice resulted in a statistically significant increase of the median survival as compared to nontreated animals. ECO-02301 thus represents an exciting new chemical class of natural product and a promising agent for development as a treatment for serious fungal infections.

4. Summary

The discovery of ECO-02301 is one of the early successes of the genomics-guided platform developed at Ecopia and the new paradigm for natural-product discovery. The discovery of this compound from *S. aizunensis* provides direct evidence that the capacity for microorganisms to produce natural products has been greatly underestimated, and that exciting new natural products can be discovered from microorganisms that were already screened using traditional approaches.

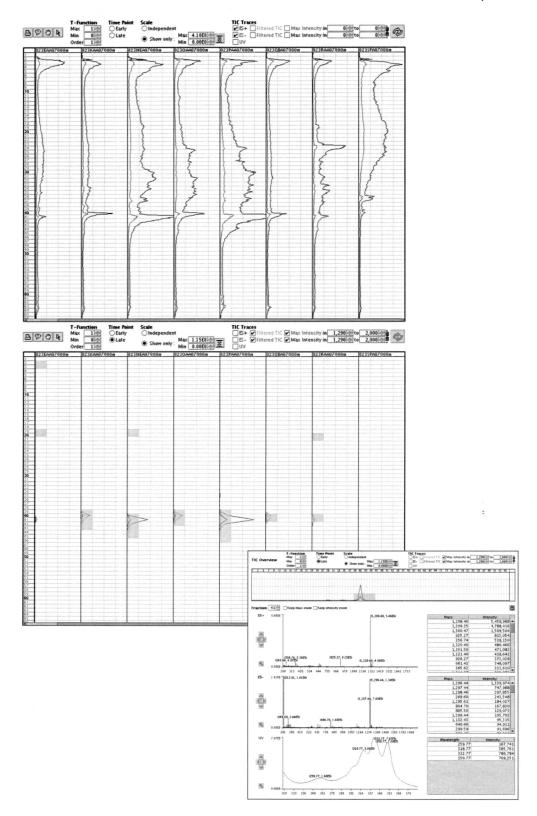

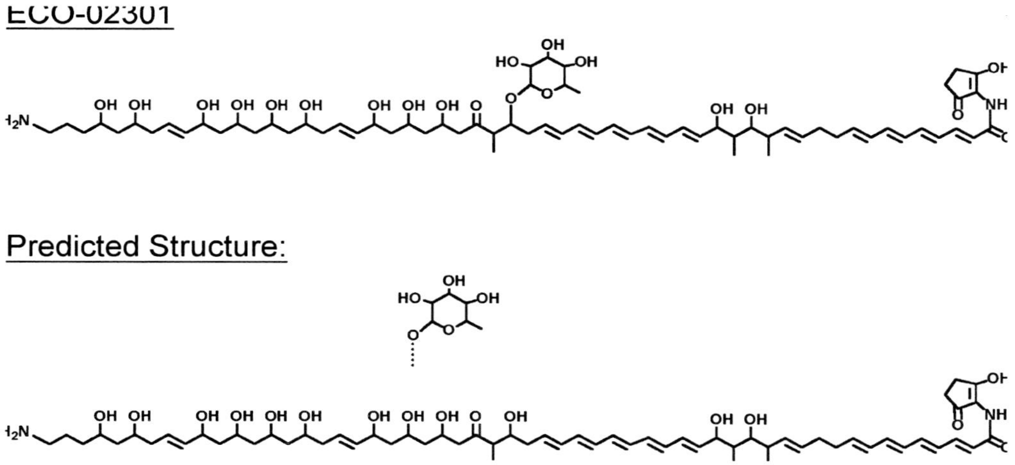

ECO-02301

Predicted Structure:

Fig. 8. The structure of ECO-02301, a new class of antifungal agent. The confirmed structure matches the computer-generated prediction.

Critical to the success of genomics-guided discovery are the database and data-mining tools that enable comparative analysis of gene clusters and the prediction of chemical structures from gene-sequence information. With each new gene cluster added to the Decipher database, the structure-prediction capabilities of the platform become increasingly accurate. The integration of specialized applications to search for metabolites simultaneously across multiple growth conditions makes it possible to correlate the metabolome of an organism with the gene clusters discovered by genome scanning and to identify metabolites that are likely to be NCEs.

Given that only a tiny fraction of the microbial world has been explored, the potential for discovering new natural products is virtually unlimited. Increasingly, new compounds will be discovered at the computer, and bioinformatics technologies will be used to tailor strategies for their production and purification, thus overcoming many of the technical hurdles previously associated with natural-product discovery. With the introduction of powerful new genomics technologies, the remarkable fifty-year track record of microorganisms in producing landmark medicines is now poised to extend well into the new millennium.

Fig. 7. (*opposite page*) Finding the predicted compound in fermentation broths. Upper, parallel high-performance liquid chromatography (HPLC)-mass spectrometry analyses of extracts of *S. aizunensis* grown in multiple fermentation media. Each column represents a different fermentation medium, and each row in a column represents a single HPLC fraction. All fifty media can be analyzed simultaneously by scrolling across the window. Middle, the same interface, filtered for metabolites having a mass greater than 1290 Da, showing a single peak found in only some fermentation media; the *Candida albicans* screening data are also shown in this view (yellow highlights) as an antifungal activity correlated with the peak fractions. Clicking on a peak fraction pulls up a new window (lower inset) showing detailed spectroscopic data consistent with the predicted compound.

Acknowledgments

The authors would like to acknowledge Drs. Jim McAlpine and Brian Bachmann for their contributions to the discovery of ECO-02301.

References

1. Bentley SD, Chater KF, Cerdeno-Tarraga AM, et al. Complete genome sequence of the model actinomycete *Streptomyces coelicolor* A3(2). Nature 2002;417:141–147.
2. Omura S, Ikeda H, Ishikawa J, et al. Genome sequence of an industrial microorganism *Streptomyces avermitilis*: deducing the ability of producing secondary metabolites. Proc Natl Acad Sci USA 2001;98:12,215–12,220.
3. Zazopoulos E, Huang K, Staffa A, et al. A genomics-guided approach for discovering and expressing cryptic metabolic pathways. Nat Biotechnol 2003;21:187–190.
4. Miyamura S, Ogasawara N, Otsuka H, et al. Antibiotic No. 5879, a new water-soluble antibiotic against Gram-negative bacteria. J Antibiotics 1972;25:610–612.
5. Miyamura S, Ogasawara N, Otsuka H, et al. Antibiotic 5879 produced by *Streptomyces aizunensis*, identical with bicyclomycin. J Antibiotics 1973;26:479–484.

6
Developments in Strain Improvement Technology
Evolutionary Engineering of Industrial Microorganisms Through Gene, Pathway, and Genome Shuffling

Stephen B. del Cardayré

Summary

The development of an economically viable production processes is a significant hurdle in the commercialization of natural products. A primary method of achieving this goal is through strain engineering. Evolutionary engineering has been practiced for decades in the form of classic strain improvement. The process of genetic diversification and functional screening has now become a powerful means of improving the function of diverse biological systems from genes and enzymes to whole genomes. Gene shuffling is a method for effecting genetic diversification in evolutionary engineering programs that incorporates recombination into the evolutionary algorithm. This approach dramatically accelerates the process of directed evolution and is arguably the most robust method for the purposeful manipulation of biological structure and function. This chapter reviews the theory and practice of gene shuffling-mediated evolutionary engineering in the context of commercial strain improvement. Described are examples of the improvement of commercial natural-product fermentation processes through the shuffling of individual genes, metabolic pathways, and whole genomes.

Key Words: Fermentation; directed evolution; shuffling; natural products; Streptomyces; metabolic engineering.

1. Introduction

Despite the medical and commercial success of the class of compounds known as natural products, the investment in the discovery and the commercialization of new natural products has significantly decreased since its peak in the mid to late twentieth century. Explanations for this decline include the decrease in the discovery of new classes of natural compounds having interesting new activities, the advent of combinatorial chemistry, the desire of medicinal chemists to have rapid access to significant quantities of relatively pure compounds for chemical studies, and the time and expense of developing an economically viable process for the production of compounds for clinical and commercial use. Although many pharmaceutical companies have divested of their internal natural-products programs, smaller biotechnology companies are picking up this effort and breathing new life into the field of natural-product discovery. New technologies include methods to identify and express cryptic secondary metabolic pathways *(1)*, isolate the biosynthetic potential of "unculturable" organisms *(2–4)*, the prospecting of marine microorganisms *(5–7)*, and the various combinatorial approaches to exploit the metabolic potential of natural-product biosynthesis—i.e., combinatorial

From: *Natural Products: Drug Discovery and Therapeutic Medicine*
Edited by: L. Zhang and A. L. Demain © Humana Press Inc., Totowa, NJ

biology *(8)*, combinatorial biosynthesis *(9,10)*, and combinatorial biocatalysis *(11,12)*. Although these technologies are successfully expanding accessible chemical diversity, the financial hurdles of developing and commercializing newly discovered compounds remain. Sufficient quantities of pure compound must be available for development and clinical studies, and a final approved compound must be economically accessible to the consumer and profitable for the producer. The goal of this chapter is to discuss technologies aimed at the improvement of process economics in the production of natural products; we will focus specifically on the application of evolutionary engineering technologies for the improvement of fermentation-based bioprocesses. The theory and practice of gene-shuffling technology shall be introduced, followed by commercially relevant examples of its application to the improvement of a specific gene in a native biosynthetic pathway, of engineered biosynthetic pathways, and of a whole fermentation strain genome, each for the production of commercial natural products.

1.1. Economics of Production

A comprehensive discussion of the economics of fermentation processes has been described previously (ref. *13*, and references therein). Simply, the cost of goods (COG) for producing a fermented compound is primarily associated with its production in the fermenter and isolation from the resulting culture. These costs include raw materials, energy, capital investments, time, and labor, and are directly offset by the overall isolated yield of compound by the process. COG can thus be estimated as COG ($/kg) = $(C_f + C_i)/yP_f$, where C_f is the cost associated with the fermentation, C_i is the cost associated with product isolation, P_f is the mass of product produced by the fermentation, and y is the percent yield of the isolation process. Accordingly, the economics of production can be improved by either decreasing the costs associated with production and isolation (C_f and Ci) or increasing the total process isolated yield (P_f and/or y). Achieving these economic goals is generally realized through iterations of process and biocatalyst (strain) engineering. Process engineering is the manipulation of the physics and chemistry of the fermentation and isolation process (the environment to which the biological catalyst and its products are exposed), whereas strain engineering is the genetic manipulation of the biological catalyst to improve its performance under fermentation process conditions. The technologies described in this chapter are new tools to dramatically facilitate the strain-engineering aspect of fermentation process improvement.

1.2. Strain Improvement

In fermentation processes, a whole cell biological catalyst, such as a bacterium, fungus, or mammalian cell culture, is propagated in a medium composed of raw materials, which over time are converted to biomass, the product of interest, and byproducts. The total mass of product produced in the process is proportional to the mass of biocatalyst produced, the specific activity of the biocatalyst at different times throughout the fermentation, the time the biocatalyst is catalytically active, and the volume of the reactor. Since both the volume of the reactor and the time it is running are key contributors to the costs of the process, an important metric of process efficiency is the amount of product produced in a given volume over time, which is expressed as *volumetric productivity* ($g^{-1}L^{-1}d^{-1}$). A process that runs for 10 d and results in 10 g/L of product has a volumetric productivity of 1 gL^{-1}d^{-1}. An economically ideal process would be short, based on inexpensive raw materials, produce a large amount of product and little biomass or byproducts, and be run under conditions that support an equally inexpensive isolation process. Strain engineering is aimed at achieving this ideal through the genetic manipulation of the biocatalyst.

The primary goal of strain engineering for the production of natural products is the improvement of volumetric productivity. Most natural strain isolates produce secondary metabolites at very low levels (10–100 mg/L), while mature commercial strains can make significantly more than 10 g/L *(14)*. There are in addition other performance criteria that may be targeted—for example, the purity of the products produced, ability to use inexpensive feedstocks, stability, low viscosity, and tolerance to optimal process conditions. In regards to purity, secondary metabolites often result from complex biosynthetic pathways that produce not only the compound of interest but also structurally related "shunt products" that are difficult to remove in the isolation process. Decreasing the levels of these compounds in the fermentation can dramatically simplify the isolation process, result in an increase in isolation yield, and a corresponding decrease in the cost of the final product.

Rationally engineering a commercially useful cellular phenotype is usually not simple. A great deal of information, genetic tools, and a strong model of the structure and function of the target cell are required. Most industrial organisms are significantly less well characterized then model academic systems (e.g., there are few biochemical pathway, genetic, or genomic data) and lack the technical prerequisites described above. This is especially true for emerging industrial organisms (such as those producing a new, promising natural product). This is not to say that excellent examples of successful metabolic engineering of industrial microorganisms do not exist. They do *(15–19)*. However, they predominantly are for well-characterized systems.

The primary approach to improving industrial organisms is classic strain improvement. This proven method of random mutagenesis and screening for improved strains remains the industry standard for commercial strain improvement, and has been thoroughly reviewed *(20–23)*. The beauty of this approach is its simplicity. It requires no molecular tools, metabolic models, or biochemical or genomic information, only an effective mutagen and an accurate screen for the desired phenotype. However robust, classic strain improvement is time and labor intensive, and can be the economic hurdle that limits commercial realization of a promising product or biological process. Even for well-characterized organisms, this approach remains the industrial method of choice for strain improvement. Indeed, this "black box" approach to genetic improvement is the foundation of evolutionary engineering.

2. Evolutionary Engineering and DNA Shuffling

Natural evolution is a recursive process of genetic diversification and functional selection (**Fig. 1**). Evolutionary engineering is a controlled process of evolution where both the genetic diversification and the functional selection are controlled in the lab. Evolutionary engineering has emerged as a complementary tool to rational design, and has proven to be one of the most reliable tools by which the function of biological systems can be purposefully manipulated. The approach does not rely on the application of structure-function relationships, but rather applies nature's own design algo-

Fig. 1. Evolution is a recursive process of genetic diversification and functional selection. "Evolutionary engineering" or "directed evolution" is an application of the evolutionary algorithm for the purposeful manipulation of biological systems in which the recursive cycles of genetic diversification and functional screening are controlled in the laboratory.

rithm. Large populations of genetically diverse individuals are generated and then screened for a desired function, with those having acquired the desired function surviving to parent the next population of diverse offspring (just as in classical strain improvement). This recursive approach to biological systems engineering asks the question "which structure supports a desired function?" as opposed to "does this preconceived structure support a desired function?" In this way, the functions of complex biological systems are routinely tailored without relying on rational structural designs or simplified models.

2.1. Genetic Diversification

Next to the classical breeding of plants and animals, classical strain improvement is perhaps the oldest application of evolutionary engineering. This process relies on recursive random mutagenesis of entire living genomes. In the late 1980s and early 1990s, the directed evolution of single genes emerged *(24–26)*, in which genes were randomly mutated by error-prone polymerase chain reaction (PCR) or direct synthesis, and then screened for improved function. This very successful approach turned out to be a major step in the ability to routinely improve the function of biological macromolecules. However, this process of sequential random mutagenesis and screening, like classic strain improvement, lacks recombination, which is a key component and catalyst of the evolution algorithm. In nature, recombination, through sexual mating and horizontal gene transfer, facilitates both the amplification and the consolidation of the useful genetic information within a population. Recombination between members of a population allows the genetic potential of the entire population to be exploited in the evolution process, with useful mutations from individuals congregating and deleterious mutations being segregated over time. In essence, recombination facilitates information exchange in a learning system. In the absence of recombination, these benefits are lost. Individuals carrying useful mutations within a population cannot benefit from each other (through breeding) and must evolve independently (randomly acquiring new mutations). Mutations arise linearly and may need to be rediscovered. Useful genetic diversity can be discarded, as only one or a few improved genetic variants from a population can advance to further rounds of diversification. Finally, deleterious mutations can accumulate over time, eroding the benefit of accumulated beneficial mutations. As a result, evolution in the absence of recombination can be significantly slower and less efficient (**Fig. 2a**).

In 1994, Stemmer introduced DNA-shuffling technology *(27,28)*. DNA shuffling provides a means to breed a population of DNA fragments in vitro or in vivo (**Fig. 2b**). The shuffling reaction produces from a starting population of genes a combinatorial library of new genes containing permutations and combinations of the genetic diversity originating in the starting population (e.g., if a population of genes contain 10 random mutations that affect functional improvement, there are $2^{10} = 1024$ possible combinations of these mutations that could be generated through their shuffling). Unlike the results of classical breeding, the progeny of shuffled libraries can have genetic material from all of the parents that were shuffled (as opposed to only two in the case of classical breeding). These shuffled libraries in general contain new genes that have significant functional improvements over the parental sequences. Because the library is not a library of random mutations, which primarily are deleterious or neutral in function, but rather a library of highly functional combinatorial mutants, improved genes can be identified from screening a relatively small sample of the library. It has since been demonstrated that DNA shuffling in combination with high-throughput screening (HTS) dramatically accelerates the process of evolutionary engineering and is applicable to a diversity of biological phenomena of both commercial and academic interest.

DNA shuffling as a technology has significantly evolved since it was first reported in 1994. Typical approaches to gene shuffling now include multi-gene shuffling (i.e., the shuffling of homologous genes from different species) *(29,30)*, synthetic shuffling *(31)*, and *in silico* shuffling *(32,33)*, all of which can be further enhanced using statistical and bioinformatic tools. Successful applications are numerous in both industry and academia, and include improved enzyme activity, selectivity, and stability, improved flux through metabolic pathways, and improved operators, vectors, antibodies, protein therapeutics, and vaccines. A recent advance in shuffling technology is its application to whole genomes *(34,35)*. Genome shuffling, which shall be discussed in detail in **Subheading 5.1.**, is an improved means of effecting classic strain improvement that incorporates recombination into the mutagenesis and screening algorithm, allowing the strain improvement process to proceed at an accelerated pace.

2.2. Screening

The two components of evolution are genetic diversification and functional selection. The process of functional selection can be the greatest technical challenge of any evolutionary engineering effort, since in the lab "selection" generally is effected through a screen. If a screen does not measure the commercial performance criteria, it may identify only laboratory novelties, but never a truly improved variant that can be commercialized. As most HTS relies on the common 96-well plate format, it is almost impossible to develop a screen that replicates commercial process conditions. Accordingly, a great deal of industrial evolutionary engineering relies on quantitatively understanding commercial process conditions and the desired performance criteria of the catalyst being evolved, and then devising methods to mimic these at small scale such that real performance can be measured.

A screen has two components: the reaction that needs improvement, such as a biocatalytic conversion or fermentation, and the analysis of that reaction, such as the quantification of the product and its purity in the reaction mixture. The reaction itself is

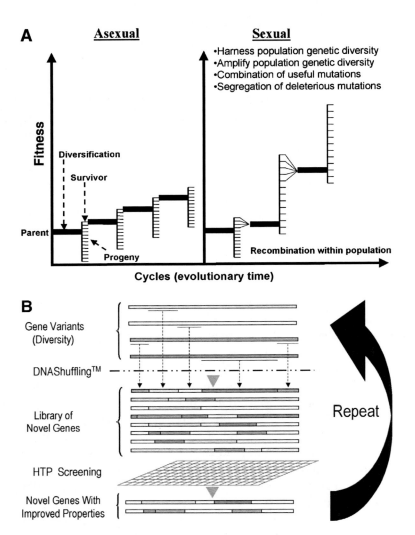

Fig. 2. (A) Recombination is a catalyst of the evolutionary process. Asexual evolution is a recursive process of random mutagenesis and functional selection in which improved mutants must acquire new mutations randomly in a *linear* fashion, i.e., 10 rounds of mutagenesis and selection will result in about 10 new mutations. Sexual evolution incorporates recombination into this process and results in the exchange of information within a population, such that improved mutants within a population can acquire new useful mutations from each other. Recombination amplifies the useful genetic diversity of a population in combinatorial manner, i.e., if there are 10 useful mutations in a population there are $2^{10} = 1024$ combinations of those mutations that can result from their recombination. Recombination also harnesses the genetic diversity of a population by allowing mutations to become linked to each other and thereby distributed more broadly throughout a selected population. Useful mutations are combined in an *exponential* manner, i.e., 3 rounds of recombination can result in the accumulation of $2^3 = 8$ mutations. Finally, recombination provides a means for useful mutations to become segregated from deleterious mutations, which would be difficult to lose by random mutagenesis. Accordingly, sexual evolution proceeds faster than asexual evolution. (B) DNA shuffling combines the practices of classical breeding and molecular biology to allow one to breed a population of DNA molecules in vitro. In one shuffling format, genes are randomly fragmented and then reassembled in a primerless polymerase chain reaction. During each thermal cycle, single-

the most difficult to design, as it should mimic commercial conditions as closely as possible. An exact replication of the physical and chemical parameters under which a process is run should be scaled down as much as possible—in general, to <1 mL of reaction mixture. These reaction conditions must then be accurately reproduced for each of the genetic variants being compared, such that any variation in the measured productivity of a reaction reflects the performance of the catalyst, not a variation in the reaction conditions. Indeed, the variability of the system limits the ability to identify statistically relevant improvements. Analysis of the completed reactions is generally easier to solve, assuming that the right equipment is available. Most analytical methods are now amenable to high-throughput applications. UV/Vis and fluorescence spectroscopy are routinely used for the analysis of 96-well-plate-mediated reactions. Mass spectroscopy has become a very sensitive high-throughput workhorse, with MS/MS providing very precise chemical data *(36)*. Unfortunately, most chromatographic methods, such as HPLC and gas chromatography (GC), remain only medium throughput (approx 200 samples per day per column) for most applications. Ninety-six parallel capillary electrophoresis, however, is now proving to be a powerful means of quickly separating and quantifying the components in complex mixtures, such as natural-product fermentation broths and mixtures of enantiomers *(37)*. Once quantitative data are produced, however, rigorous statistical analyses should be applied to identify whether any genetic variants are truly improved *(23)*.

2.3. High-Throughput Fermentation

In the case of fermentation strain improvement, generating a quality screen is a particularly difficult challenge. Commercial fermenters can range from 1 to 200 m^3, supporting cell growth to very high density. The media are complex and often contain solids and oils. Most fermentations are controlled at specific temperature, pH, and dissolved oxygen, and may have controlled feeding of important raw materials and nutrients. The resulting fermentation broths are a complex mixture of compounds that generally require significant processing before pure compound is isolated. Traditionally, strain improvement has relied on small-scale fermentations carried out in shake flasks or test tubes and then analyzed chromatographically. However, these methods are now being adapted to 96-well format *(34)*. Dispensing particulate medium containing oils into 96-well plates is feasible using automated liquid handling, such as a Multimek (Beckman-Coulter) outfitted with wide-bore pipet tips (Robbins), which can uniformly dispense media containing large particles. Often buffers incorporated into the media can provide needed pH control without undesired effects on culture perfor-

---

(Fig. 2. continued) stranded fragments from one gene hybridize to homologous regions of other genes and prime the DNA polymerization reaction, resulting in DNA fragments containing genetic material from at least two of the original genes. After numerous thermal cycles, complete chimeric genes are synthesized, with the final population representing a combinatorial library of the original starting genetic diversity. This library is then screened for new genes having desired properties. The process is then reiterated until the specific performance is achieved.

mance. Inoculation of seed cultures can be achieved through automated colony pickers, such as the Q-bot (Genetix, UK), transfer and assays can be accomplished using automated liquid-handling systems and robotics, and fermentations can be carried out in 96-, 48-, and 24-well plates vigorously agitated in controlled-environment shakers, which maintain uniform temperature and high humidity to prevent evaporation. The challenge is primarily for mycelial organisms, where high cell densities result in viscous cultures that require high oxygen transfer rates. Research to identify conditions for optimal delivery of oxygen has been described *(38–40)*. In addition, new technologies are emerging to support 96-well fermentations with controlled dO, pH, and temperature *(41)*; however, these have not yet reached commercial launch.

3. Gene Shuffling and Enzyme Evolution

There are countless examples of the application of gene shuffling to the improvement of enzymes in the primary literature, and these are routinely reviewed *(42,43)*. Industrial performance criteria for enzymes that have been improved by these methods include expression, stability, catalytic tolerance to diverse and extreme environments, catalytic activity, and substrate selectivity, including alterations in chemical, regiochemical, and enantiomeric selectivity. Most literature examples describe the improvement of in vitro enzyme activity. However, for enzymes involved in the biosynthesis of a fermented natural product, these same characteristics expressed in vivo can significantly influence the volumetric productivity of a fermentation and isolation process *(44)*. Targeting biosynthetic enzymes for evolutionary engineering is a method of rational metabolic engineering that requires knowledge of the biosynthetic pathway, limiting steps, genes encoding the targeted enzymes, and genetic tools for their manipulation. Below are examples of how DNA shuffling was applied in this manner to improve the production of natural products.

3.1. Gene Shuffling of aveC for Improved Commercial Production of Doramectin

Avermectin and its analogs are biosynthesized by the actinomycete *Streptomyces avermitilis* and are today the most widely used drugs in animal health and agriculture, with current worldwide sales exceeding $1 billion. *S. avermitilis* normally produces eight distinct but closely related avermectins. Through strain improvement efforts via random mutagenesis and the use of exogenously supplied fatty acids, the avermectin analog doramectin (CHC-B1) was developed. Doramectin, which is sold commercially as Dectomax™, is co-produced during fermentation with the undesired analog CHC-B2. CHC-B1 and CHC-B2 are structural homologs, which differ only by the hydration of the C22-C23 methylene bond (**Fig. 3**), and are derived from avermectin B1 and B2. CHC-B2 must be removed through the isolation process. An economic improvement in the process would be achieved if the CHC-B1:CHC-B2 ratio were improved without compromise of the final CHC-B1 titer.

*ave*C was identified as a gene that affected the ratio of avermectin B1 and B2 produced in fermentations *(45,46)*. The gene was cloned by complementation from a classically improved strain of *S. avermitilis* that demonstrated an altered B1:B2 ratio in fermentation broths. The *ave*C gene has no close homologs or known function at this

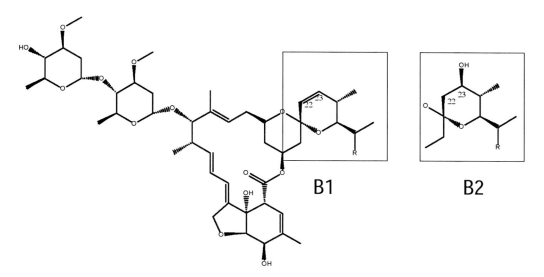

Fig. 3. Structure of avermectins. Avermectin B1 and B2 differ from each other only by the chemistry of the C22–C23 bond. B1 has a *cis* double bond, whereas B2 has this bond hydrated at the C23 position. Doramectin is derived from B1. Avermectin R = C_2H_5, doramectin R = C_6H_{11}.

time. *S. avermitilis* disrupted in *aveC* produces only background levels of avermectin. Stutzman-Engwall et al. investigated whether DNA shuffling of the *aveC* gene might result in new variants that would confer an ability to produce improved ratios of avermectins B1 and B2, and hence a corresponding improvement in the production of CHC-B1:CHC-B2 in commercial doramectin processes *(36)*.

The *aveC* gene was first randomly diversified by mutagenic PCR, and the gene variants were cloned and expressed in *S. avermitilis*. Extracts from the variants, grown in high-throughput solid-phase fermentations containing authentic fermentation media components were then analyzed by an HTP MS/MS assay that quantified the total avermectin B1 and B2 produced. From this library, mutant *aveC* genes were identified that conferred upon *S. avermitilis* an improved B1:B2 ratio. Whereas the wild-type *aveC* gene resulted in a B1:B2 ratio of 1:1.6, an evolved *aveC* containing two mutations (D48E/A89T) resulted in a B1:B2 ratio of 1:0.4, a fourfold improvement that also resulted in a 40% increase in B1 production under commercial conditions (**Fig. 4**).

Additional evolution was pursued to further increase the B1:B2 ratio. Using the *aveC* (D48E/A89T) gene as backbone, three iterations of DNA semi-synthetic shuffling and HTS were carried out. From a population of significantly improved variants, an evolved *aveC* mutant containing 10 mutations was identified that resulted in a B1:B2 ratio of 1:0.07, a 23-fold improvement (**Fig. 4**) *(47)*! The significance of this study is dramatic. A gene whose contribution to the biosynthesis of a natural product remains unknown, was engineered to effect a >20-fold improvement in the commercial performance criteria. This milestone in the engineering of natural-product biosynthetic pathways emphasizes the industrial importance of evolutionary engineering and in particular DNA shuffling as an engineering tool.

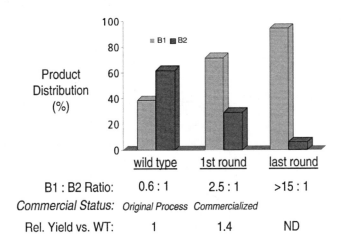

Fig. 4. Improvements in doramectin process. Initial production strains of *S. avermitilis* produced a CHC-B1:CHC-B2 ratio of 0.6:1. A new strain carrying an *aveC* gene improved by evolutionary engineering produced a B1:B2 ratio of 2.5:1, and has been commercialized. After three additional rounds of *aveC* shuffling, a new strain was generated that produced a ratio B1:B2 of 15:1, a 23-fold improvement *(36,47)*.

3.2. Gene Shuffling to Enhance Production of Shikimate From Escherichia coli

Shikimic acid is an intermediate in the aromatic amino acid and vitamin pathway, a chemical intermediate in the synthesis of the anti-influenza drug Tamilfu®, and its biosynthetic pathway contributes intermediates to a variety of natural products, such as rifamycin, rapamycin, and tacrolimus (FK506). The first committed step in shikimate biosynthesis in *Escherichia coli* is the feedback-regulated condensation of phosphoenolpyruvate (PEP) and D-erythrose-4-phosphate (E4P) by 3-deoxy-D-*arabino*-heptulosonate (DAHP) synthase. The shikimate pathway must compete with the PEP-dependent carbohydrate phosphotransferase (PTS) system for microbial transport and phosphorylation of glucose. Frost and colleagues hypothesized that improved flux through the shikimate pathway could be achieved if the condensation of pyruvate with E4P could be engineered to replace DAHP synthase as the first committed step in the shikimate pathway. To this end, Ran et al. demonstrated that 2-keto-3-deoxy-6-phosphogalactonate (KDPGal) aldolase could catalyze the condensation of pyruvate and E4P, and support the in vitro synthesis of 3-dehydroshikimic acid when incubated with pyruvate, E4P, 3-dehydroquinate synthase, and 3-dehydroquinate dehydratase. They then demonstrated that *dgo*A (which encodes KDPGal aldolase) from both *E. coli* and *Klebsiella pneumoniae* complemented in part *E. coli* strains having the genes encoding all isoforms of the DAHP synthase knocked out *(48)*.

To improve flux through this engineered shikimate pathway, the *dgo*A genes were improved by evolutionary engineering *(48)*. Both the *E. coli* and the *K. pneumonia dgo*A genes were improved through two rounds of error-prone PCR mutagenesis followed by one round of DNA shuffling. Improved genes were identified by a nutritional selection having increased stringency in each cycle of directed evolution (i.e., that required *dgo*A variants to support growth on progressively lower concentrations of shikimate pathway downstream products). The best-improved genes contained four to nine amino acid changes, demonstrated four- to eightfold increase in DAHP synthase activity and a 7- to 30-fold decrease in KDPGal aldolase activity relative the parental enzymes. To compare the in vivo flux through the shikimate pathway, the encoding constructs of the improved and parental genes were expressed in *E. coli* NR7 (*E. coli* KL3 with all DAHP synthase isozymes genetically inactivated). 3-Dehydroshikimic acid production from these strains was then monitored in controlled stirred-tank fermentations. While the *K. pneumoniae dgo*A gene supported only background production of 3-dehydroshikimate, its best progeny supported 8.3 g/L. Similarly, the *E. coli dgo*A gene supported 2 g/L, while its best progeny supported 12 g/L. This is an elegant example of the combination of rational metabolic engineering and evolutionary engineering to dramatically improve the flux to a natural product, and represents a new paradigm for overcoming PEP availability.

3.3. Pathway Shuffling and Evolution

An important goal in natural-product production is the engineering of new metabolic pathways into fermentation organisms. Often a desired compound is not produced by a known organism, but a pathway comprised of host and recombinantly expressed genes can be constructed that enable the desired biosynthesis. Although this approach has met certain success at creating new strains making desirable compounds, often the compound is not produced at commercially viable levels. In these cases, it is likely that the engineered pathway contains the metabolic bottleneck, and the flux through that pathway needs to be improved. Again, evolutionary engineering is a pragmatic approach to improving performance. The DNA encoding a part or the entire metabolic pathway can be subjected to DNA shuffling, and host organisms expressing the pathway variants can then be screened for improved compound production. Evolution in this context allows the individual genes to become better expressed, the encoded recombinant enzymes to become optimized to function in the new host environment in a more balanced fashion, and catalytically limiting enzymatic steps to be improved. This approach has been successfully demonstrated in the adaptation and improvement of several metabolic pathways, including the arsenate detoxification pathway from *Staphylococcus aureus* expressed in *E. coli (49)*, an engineered pathway for production of adipyl-7-amino deacetoxycephalosporanic acid (Tobin et al., unpublished results), the mevalonic acid pathway from *Saccharomyces cerevisiae* expressed in *E. coli* for improved fermentative terpene production (Chatterjee et al., unpublished data), and for the production of an intermediate in a new synthesis of the flavor compound strawberry furanone *(50)*.

4. Genome Shuffling and Organism Evolution

Despite successes in the rational and targeted engineering of fermentation strains, the primary means of improving even well-characterized and mature industrial organisms remains classical strain improvement. This is because most whole-cell phenotypes are by nature complex, and it is difficult to model or identify which genes in the dynamic cellular machine may be limiting. By evolving an entire genome, no assumptions as to what is limiting are made, and an organism can be improved through the evolution of its entire genome. As described previously, classical strain improvement suffers from being time and labor intensive. In addition, it is an asexual process, which has the associated limitations of accumulating beneficial and difficult to lose deleterious mutations in a linear fashion. In 2001, Zhang et al. introduced whole-genome shuffling *(34)*. By incorporating recombination into the recursive cycles of mutagenesis and screening of classical strain improvement, Zhang et al. were able to overcome some of the limitations of the classical method and dramatically accelerate the rate of strain improvement.

4.1. Genome Shuffling

Genome shuffling combines the advantage of multiparental crossing that is allowed by DNA shuffling with the recombination of entire genomes normally associated with conventional breeding. While in vitro recombination of genes (<10 kb) is routine, it is not for whole genomes (>>1 Mb). In vivo methods for effecting recombination primarily are based on natural mechanisms and result in low recombination efficiencies ($<10^{-5}$). Hopwood *(51)* and Baltz *(52)* described the application of protoplast fusion technology as an efficient means of obtaining efficient recombination (>10%) between strains of *Streptomyces*. Hopwood and Wright later demonstrated that a single protoplast fusion could also result in recombination between up to four strains of *Streptomyces coelicolor*, although with very low efficiency (approx 10^{-6}) *(53)*. It was proposed that this tool could be used to breed cell lines resulting from divergent lineages in a strain-improvement program *(54,55)*. However, the technology never became popular in this regard, likely because a single fusion between divergent lines resulted in new strains that performed poorly relative to their parents. Indeed this is what would be expected, as the beneficial mutations from divergent cell lines would be diluted as a result of a single cross between distantly related strains of a lineage—i.e., if two strains, each having 10 unique mutations, were crossed, the average recombinant would still have 10 mutations, but 5 mutations would originate from each line. The majority of mutations would have to be complementary for improvement to be expected from a significant portion of the population resulting from a cross. Zhang et al. applied the fact that shuffling is a recursive process of pair-wise recombination; in vitro shuffling is a process of recursive pairwise recombination between DNA molecules that occurs during each thermal cycle of the shuffling reaction *(27)*. Taking advantage of the high pairwise recombination efficiencies of protoplast fusion, they demonstrated that recursive protoplast fusion (**Fig. 5**) of a population of different cells resulted in the shuffling of their genomes. This was verified by recursively fusing the same four multiply marked strains of *S. coelicolor* used by Hopwood and Wright *(50)*. The recursively fused population, relative to the singly fused population, had a 10^5-fold increase in recombinants having genetic markers from the four original parents *(34)*. This significant increase in "complex progeny" suggested that one could shuffle a population of genetically diverse improved strains, such as improved siblings from a classical improvement program, and, through screening a small population, identify improved strains containing mutations from many of the parental breeding stock. An example of the successful application of this approach to strain improvement is described in **Subheading 5.2.**

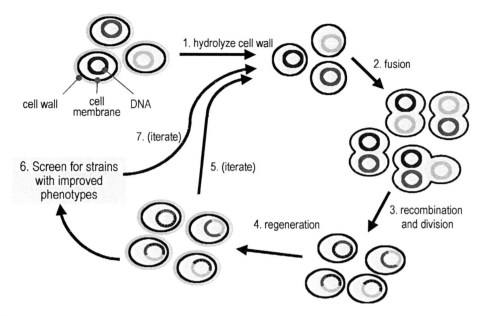

Fig. 5. Genome shuffling by recursive protoplast fusion. A population of cells to be shuffled is identified, and protoplasts from each strain are prepared. These are then combined and fused as a mixed population and allowed to regenerate. In general, 10% of the population of regenerated cells are pair-wise recombinants, i.e., they have genetic information from any two parents; but very few cells have genetic information from three or more parents. Recursive fusion of the regenerated protoplasts allow the 10% recombinants from each round to recombine with different parents or with each other, such that a population regenerated from recursive fusions can contain >2% cells that have genetic information from any four parents, hence representing a genome-shuffled library (34). When applied to strain development, the starting population of cells would be strains that each have some desired but genetically distinct property.

4.2. Improved Production of Tylosin by Genome Shuffling of Streptomyces fradiae

Tylosin (**Fig. 6**) is a complex polyketide antibiotic sold for veterinary health and nutrition that is isolated from the fermentation broths of *S. fradiae*. The biosynthesis of tylosin requires a biosynthetic cluster of approx 100 kb in addition to numerous genes distributed throughout the *S. fradiae* genome. Eli Lilly initiated in the early 1960s an extensive strain improvement program to improve tylosin production from *S. fradiae*. The program began with strain SF1, a stable derivative of the natural isolate obtained through natural selection. This strain was taken through 20 rounds of classical strain improvement (CSI) for improved tylosin production, resulting in the commercial production strain SF21.

Zhang et al. employed this system to demonstrate that genome shuffling could accelerate the strain-improvement process. They propogated 22,000 MNNG-induced mutants of SF1 in 96-well two-stage mini-fermentations. The mutants were screened for improved tylosin production using a two-tier spectrophotometric and HPLC assay. Eleven strains were identified and used as breeding stock for genome shuffling by

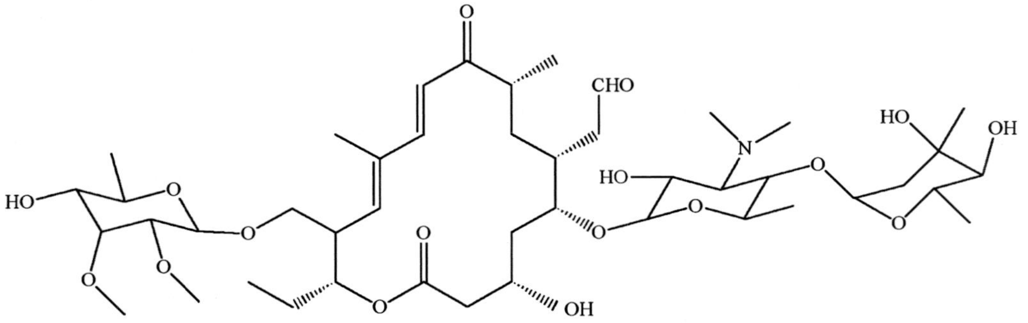

Fig. 6. Tylosin structure.

recursive protoplast fusion (**Fig. 5**). After two rounds of genome shuffling and screening only 2000 mutants, the group identified new shuffled strains that produced tylosin at levels equivalent to SF21 in shake-flask cultures (**Figs. 7, 8**). A molecular and morphological comparison of GS1 and GS2 to SF1 and SF21 demonstrated that GS1 and GS2 were morphologically similar to SF1, but shared a functionally similar but structurally distinct mutation in a regulatory gene (*tyl*Q) shared by SF21. This study demonstrated that the shuffling of the improved population resulting from one round of CSI could result in strains that performed as well as those resulting from 20 rounds of CSI. As a practical comparison, this represents the difference between 20 years and over a million screens for CSI as compared to 1 yr and less than 25,000 screens for genome shuffling (**Fig. 8**). The study also demonstrated that the diverse mutations discovered early in a strain-improvement program are important, and that only a few are actually required to result in significant improvement. In this example, the genome-shuffled strains would have had no more than 11 mutations (and likely less). Indeed, this may explain the similar morphology of SF1 and the shuffled strains GS1 and GS2 as compared to the much smaller SF21 (**Fig. 8**).

This study demonstrates that commercially significant strain improvement can be achieved in a single year, a timeline that significantly decreases the resource and economic burden normally associated with strain improvement. From the perspective of developing new natural products, this technology may decrease the titer hurdle that developing new strains must face to be commercialized. Regarding the further improvement of mature strains, it suggests that the shuffling of archived strains, such as those divergent strains from a lineage that were not ultimately pursued, may be a powerful means of taking advantage of previous screening efforts and accessible diversity to generate new and significantly improved strains. Indeed genome shuffling may enable the scenario originally proposed by Hopwood and Chater *(54)*, in which the mutations from divergent lineages could be combinatorially explored through protoplast fusion.

5. Summary

Natural products remain an important and lucrative source of chemical and functional diversity. New technologies aimed at further harnessing the potential of these compounds are still at their infancy, and their promise for generating useful new mol-

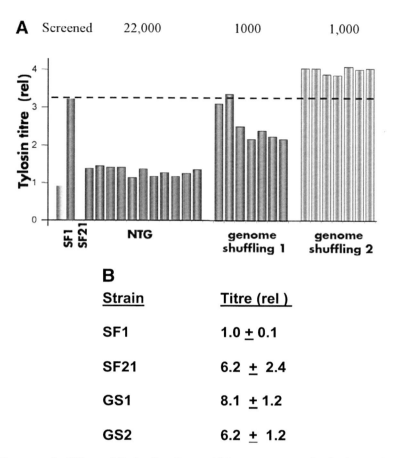

Fig. 7. Genome shuffling of *S. fradiae* for rapid improvement of tylosin production, from Zhang et al. *(34)*. (**A**) A naïve strain of *S. fradiae*, SF1, was mutagenized with NTG, and 22,000 derived mutants were screened by high-throughput fermentation (300 µL) for improved tylosin production. Eleven of the improved strains identified were recombined by one round of genome shuffling (**Fig. 5**), and 1000 individuals from the shuffled library were screened for further improvements. Seven new strains that demonstrated significant improvements over the NTG mutants were then recombined by another round of genome shuffling. Screening of this population identified another seven strains that had even further improvements. These supported tylosin production in the small-scale fermentation that was superior to a commercial production strain, SF21, that was derived from more than 20 yr of classical strain improvement. (**B**) Relative tylosin production in shake-flask fermentations by strains from the second round of genome shuffling, GS1 and GS2, as compared to SF1 and SF21.

ecules remains. However, new natural products will continue to face the economic hurdles and impatient timelines of the pharmaceutical industry. The challenge to produce significant quantities of compound quickly for development and clinical studies, and to economically produce the final product, remains the responsibility of strain and process engineers. Evolutionary engineering is now a proven and robust tool for the purposeful manipulation of biological systems, and provides an important new tool for the strain engineer. Gene shuffling in combination with high-throughput screening is

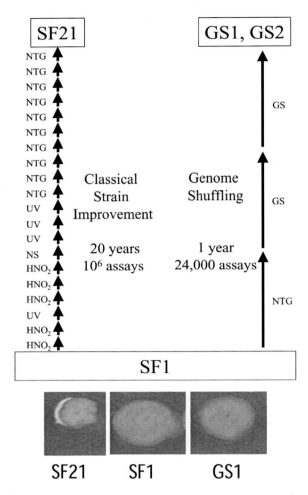

Fig. 8. Comparison of *S. fradiae* improvement by genome shuffling and classical strain improvement. SF21 was produced from strain SF1 by 20 rounds of classic strain improvement involving over a million screens and 20 yr of effort. GS1 and GS2 were also generated from SF1 by one round of classic strain improvement and two rounds of genome shuffling involving 24,000 screens and 1 year. Colony morphologies of SF1, SF21, and GS1 highlight the difference in mutational loads of SF21 and GS1 relative to SF1.

arguably the most effective means of achieving successful and efficient evolutionary engineering of genes, enzymes, metabolic pathways, and whole genomes. The technology has already realized notable commercial success in the improvement of established natural-product-producing microorganisms. Might this technology facilitate a resurgence of natural products in the drug discovery and development process? Hopefully so.

References

1. Zazopoulos E, Huang K, Staffa A, et al. A genomics-guided approach for discovering and expressing cryptic metabolic pathways. Nat Biotechnol 2003;21:187–190.

2. Miao V, Coeffet-LeGal MF, Brown D, Sinnemann S, Donaldson G, Davies J. Genetic approaches to harvesting lichen products. Trends Biotechnol 2001;19:349–355.
3. Seow KT, Meurer G, Gerlitz M, Wendt-Pienkowski E, Hutchinson CR, Davies J. A study of iterative type II polyketide synthases, using bacterial genes cloned from soil DNA: a means to access and use genes from uncultured microorganisms. J Bacteriol 1997;179:7360–7368.
4. Wang GY, Graziani E, Waters B, et al. Novel natural products from soil DNA libraries in a streptomycete host. Org Lett 2000;2:2401–2404.
5. Mikhailov VV, Kuznetsova TA, Eliakov GB. [Bioactive compounds from marine actinomycetes]. Bioorg Khim 1995;21:3–8.
6. Haygood MG, Schmidt EW, Davidson SK, Faulkner DJ. Microbial symbionts of marine invertebrates: opportunities for microbial biotechnology. J Mol Microbiol Biotechnol 1999;1:33–43.
7. Wagner-Dobler I, Beil W, Lang S, Meiners M, Laatsch H. Integrated approach to explore the potential of marine microorganisms for the production of bioactive metabolites. Adv Biochem Eng Biotechnol 2002;74:207–238.
8. Schloss PD, Handelsman J. Biotechnological prospects from metagenomics. Curr Opin Biotechnol 2003;14:303–310.
9. Cropp TA, Kim BS, Beck BJ, Yoon YJ, Sherman DH, Reynolds KA. Recent developments in the production of novel polyketides by combinatorial biosynthesis. Biotechnol Genet Eng Rev 2002;19:159–172.
10. Reeves CD. The enzymology of combinatorial biosynthesis. Crit Rev Biotechnol 2003;23:95–147.
11. Michels PC, Khmelnitsky YL, Dordick JS, Clark DS. Combinatorial biocatalysis: a natural approach to drug discovery. Trends Biotechnol 1998;16:210–215.
12. Rich JO, Michels PC, Khmelnitsky YL. Combinatorial biocatalysis. Curr Opin Chem Biol 2002;6:161–167.
13. Reisman RB. Economics. In: Demain AL, Davies JE (eds), *Manual of Industrial Microbiology and Bitechnology*. Second edition. American Society of Microbiology, Washington, DC: 1999:273–288.
14. Aharonowitz Y, Cohen G. The microbiological production of pharmaceuticals. Sci Am 1981;245:141–152.
15. Askenazi M, Driggers EM, Holtzman DA, et al. Integrating transcriptional and metabolite profiles to direct the engineering of lovastatin-producing fungal strains. Nat Biotechnol 2003;21:150–156.
16. Baltz RH, McHenney MA, Cantwell CA, Queener SW, Solenberg PJ. Applications of transposition mutagenesis in antibiotic producing streptomycetes. Antonie Van Leeuwenhoek 1997;71:179–187.
17. de Graaf AA, Eggeling L, Sahm H. Metabolic engineering for L-lysine production by *Corynebacterium glutamicum*. Adv Biochem Eng Biotechnol 2001;73:9–29.
18. Paradkar AS, Mosher RH, Anders C, et al. Applications of gene replacement technology to *Streptomyces clavuligerus* strain development for clavulanic acid production. Appl Environ Microbiol 2001;67:2292–2297.
19. Ohnishi J, Mitsuhashi S, Hayashi M, et al. A novel methodology employing *Corynebacterium glutamicum* genome information to generate a new L-lysine-producing mutant. Appl Microbiol Biotechnol 2002;58:217–223.
20. Queener SW, Lively DH. Screening and selection for strain improvement. In: Demain AL, Solomon NA (eds), *Manual of Industrial Microbiology and Biotechnology*. American Society of Microbiology, Washington, DC:1986;155–169.
21. Baltz RH. Mutagenesis in *Streptomyces* spp. In: Demain AL, Solomon NA (eds), *Manual of Industrial Microbiology and Biotechnology*. American Society for Microbiology, Washington, DC: 1986;184–190.

22. Baltz RH. Molecular genetic approaches to yield improvement in actinomycetes. In: Strohl WR (ed), *Biotechnology of Antibiotics*. Vol 82. second ed. Marcel Dekker, New York: 1997:49–62.
23. Vinci VA, Byng G. Strain improvement by non-recombinant methods. In: Demain AL, Davies JE (eds), *Manual of Industrial Microbiology and Biotechnology*. 2 ed. ASM, Washington, DC: 1999;103–113.
24. Ellington AD, Szostak JW. Selection in vitro of single-stranded DNA molecules that fold into specific ligand-binding structures. Nature 1992:355:850–852.
25. Bartel DP, Szostak JW. Isolation of new ribozymes from a large pool of random sequences [see comment]. Science 1993;261:1411–1418.
26. Chapman KB, Szostak JW. In vitro selection of catalytic RNAs. Curr Opin Struct Biol 1994;4:618–622.
27. Stemmer WP. DNA shuffling by random fragmentation and reassembly: in vitro recombination for molecular evolution. Proc Natl Acad Sci USA 1994;91:10,747–10,751.
28. Stemmer WP. Rapid evolution of a protein in vitro by DNA shuffling. Nature 1994;370:389–391.
29. Crameri A, Raillard SA, Bermudez E, Stemmer WP. DNA shuffling of a family of genes from diverse species accelerates directed evolution. Nature 1998;391:288–291.
30. Ness JE, Welch M, Giver L, et al. DNA shuffling of subgenomic sequences of subtilisin. Nat Biotechnol 1999;17:893–896.
31. Ness JE, Kim S, Gottman A, et al. Synthetic shuffling expands functional protein diversity by allowing amino acids to recombine independently. Nat Biotechnol 2002;20:1251–1255.
32. Govindarajan S, Ness JE, Kim S, Mundorff EC, Minshull J, Gustafsson C. Systematic variation of amino acid substitutions for stringent assessment of pairwise covariation. J Mol Biol 2003;328:1061–1069.
33. Gustafsson C, Govindarajan S, Emig R. Exploration of sequence space for protein engineering. J Mol Recognit 2001;14:308–314.
34. Zhang YX, Perry K, Vinci VA, Powell K, Stemmer WP, del Cardayre SB. Genome shuffling leads to rapid phenotypic improvement in bacteria. Nature 2002;415:644–646.
35. Patnaik R, Louie S, Gavrilovic V, et al. Genome shuffling of *Lactobacillus* for improved acid tolerance. Nat Biotechnol 2002;20:707–712.
36. Stutzman-Engwall K, Conlon S, Fedechko R, et al. Engineering the *aveC* gene to enhance the ratio of doramectin to its CHC-B2 analogue produced in *Streptomyces avermitilis*. Biotechnol Bioeng 2003;82:359–369.
37. Reetz MT. An overview of high-throughput screening systems for enantioselective enzymatic transformations. Methods Mol Biol 2003;230:259–282.
38. Minas W, Bailey JE, Duetz W. Streptomycetes in micro-cultures: growth, production of secondary metabolites, and storage and retrieval in the 96-well format. Antonie Van Leeuwenhoek 2000;78:297–305.
39. Duetz WA, Ruedi L, Hermann R, O'Connor K, Buchs J, Witholt B. Methods for intense aeration, growth, storage, and replication of bacterial strains in microtiter plates. Appl Environ Microbiol 2000;66:2641–2646.
40. Duetz WA, Witholt B. Effectiveness of orbital shaking for the aeration of suspended bacterial cultures in square-deepwell microtiter plates. Biochem Eng J 2001;7:113–115.
41. Maharbiz MM, Holtz WJ, Howe RT, Keasling JD. Microbioreactor arrays with parametric control for high-throughput experimentation. Biotechnol Bioeng 2004;85:376–381.
42. Powell KA, Ramer SW, Del Cardayre SB, et al. Directed evolution and biocatalysis. Angew Chem Int Ed Engl 2001;40:3948–3959.
43. Giver L, Arnold FH. Combinatorial protein design by in vitro recombination. Curr Opin Chem Biol 1998;2:335–338.

44. del Cardayre S, Powell K. DNA shuffling for whole cell engineering. In: Vinci VA, Parekh SR (eds), *Handbook of Industrial Cell Culture*. Humana, Totowa, New Jersey: 2003;465–482.
45. Ikeda H, Omura S. Avermectin biosynthesis. Chem Rev 1997;97:2591–2610.
46. Stutzman-Engwall KJ, Katoh Y, McArthur HAI, Stutzman-Engwall KJ, Katoh Y, McArthur HAIStutzman-Engwall KJ, Katoh Y, McArthur HAIs; Pfizer, assignee. *Streptomyces avermitilis* gene directing the ratio of B1:B2 avermectins. US patent US6248579. June 19, 2001.
47. Stutzman-Engwall K, Conlon S, Fedechko R, et al. Semi-synthetic DNA shuffling of *aveC* leads to improved industrial scale production of Doramectin by *Streptomyces avermitilis*. Metab Eng 2004, in press.
48. Ran N, Draths KM, Frost JW. Creation of a shikimate pathway variant. J Am Chem Soc 2004, in press.
49. Crameri A, Dawes G, Rodriguez E, Jr., Silver S, Stemmer WP. Molecular evolution of an arsenate detoxification pathway by DNA shuffling. Nat Biotechnol 1997;15:436–438.
50. Newman L, Garcia H, Hudlickey T, Selifonov S. Directed evolution of the dioxygenase complex for the synthesis of furanone flavor compounds. Tetrahedron 2004;60:729–734.
51. Hopwood DA, Wright HM, Bibb MJ, Cohen SN. Genetic recombination through protoplast fusion in *Streptomyces*. Nature 1977;268:171–174.
52. Baltz RH. Genetic recombination in *Streptomyces fradiae* by protoplast fusion and cell regeneration. J Gen Microbiol 1978;107:93–102.
53. Hopwood DA, Wright HM. Bacterial protoplast fusion: recombination in fused protoplasts of *Streptomyces coelicolor*. Mol Gen Genet 1978;162:307–317.
54. Hopwood DA, Chater KF. Fresh approaches to antibiotic production. Philos Trans R Soc Lond B Biol Sci 1980;290:313–328.
55. Hopwood DA. The many faces of recombination. In: Sebek OK, Laskin AI (eds), *Genetics of Industrial Microorganisms*. Washington, DC: American Society for Microbiology. 1979:1–9.

PART III
SPECIFIC GROUPS OF DRUGS

7
The Discovery of Anticancer Drugs From Natural Sources

David J. Newman and Gordon M. Cragg

Summary
Since the early 1940s, the search for agents that may treat or ameliorate the scourge of cancer has involved all aspects of chemistry and pharmacology. Throughout these years, natural products have played an extremely important role as first the major source of drugs used for direct treatment, as scaffolds upon which chemists would practice their skill, and now as modulators of specific cellular pathways in the tumor cell. Even today, over 60% of the 140 plus agents currently available in Western medicine can trace their provenance to a natural-product source.

Key Words: Natural products; cancer drug discovery.

1. Introduction

The original antitumor drugs that went into clinical usage before and after WWII were synthetic agents that were originally based upon the so-called nitrogen mustards used as blister gases during WWI, and effectively acted as acylating agents modifying the bases within the DNA helices of both normal and cancerous cells.

Following WWII and the initial investigations of the microbial world for antibiotics, came the realization that a significant number of microbial products with nominal antibiotic activity were also effective against eukaryotic cells from both fungi and mammals. In addition, there was some anecdotal evidence that plant secondary metabolites also demonstrated activity against mammalian cells, including tumor cells. Thus, from roughly the early 1950s to date, the systematic investigation of plant and microbial products for their potential as antitumor agents has been carried out worldwide, with a large amount of the funding coming from the US National Cancer Institute in one form or another.

To put these results into perspective, of the 140-plus drugs that have been approved worldwide from the early 1940s through the end of 2002 for the treatment of the manifold diseases that fall under the generic heading of cancer, the majority are "other than true synthetics." What this statement means is that of the 126 small molecules identified by excluding biologicals, vaccines, and hematological/immunological stimulants, the number of nonsynthetic agents is either 77 (62%) or 85 (67%), depending upon the use of definitions. If one uses the simple definitions listed in **Table 1** (i.e., N, ND, and S*) then the lesser figure applies, whereas if the "NM" category is included, the greater figure applies.

From: *Natural Products: Drug Discovery and Therapeutic Medicine*
Edited by: L. Zhang and A. L. Demain © Humana Press Inc., Totowa, NJ

**Table 1
Definitions of Drug Sources**

Major categorization
  "N"   Natural product.
  "ND"   Derived from a natural product and is usually a semi-synthetic modification.
  "S"   Totally synthetic drug, often found by random screening/modification of an existing agent.
  "S*"   Made by total synthesis, but the pharmacophore is/was from a natural product.

Minor categorization
  "NM"   Natural product mimic

In **Table 2**, in addition to showing the formal designation as to pharmacophoric source, we also show the actual biological source (plant or microbe) of the "other than synthetic molecules."

2. Prokaryotes (Other Than Marine-Derived)

2.1. Approved Agents

If one inspects **Table 2**, then the initial important involvement of the formal microbial world in cancer chemotherapy can be seen, with 21 of 126 (or 17%) of the formal natural products/natural product-derived categories from a chemical-source perspective falling into this biological category.

The first of the microbial compounds to be used systematically was named generically as "Dactinomycin." Chemically, it is named D-actinomycin C_1, and the common usage is actinomycin D (Structure 1). This was first introduced in the early 1960s and has been the subject of very significant chemical synthetic and semisynthetic programs in many parts of the world since its introduction. However, to date, no derivatives have been approved for clinical use outside of experimental protocols. In recent years, there has been a series of reports that actinomycin D has activities related to perturbation of the signal transduction cascade(s) distinct from its well-known intercalating ability, and it will be interesting to see whether these activities rejuvenate interest in this class of molecules (1–5).

One of the most important classes of compounds derived from the order *Actinomycetales* (and this is the formal source of most microbial-derived materials) is that known generically as the anthracyclines, with one of the most useful being daunorubicin (Structure 2) and its derivative doxorubicin (adriamycin) (Structure 3), which even today is still a major component of the treatment regimen for breast cancer. Although there have been many similar molecules isolated and described in the literature, it is doxorubicin and its more modern derivatives such as epirubicin (Structure 4), pirirubicin (Structure 5), idarubicin (Structure 6), and, more recently, valrubicin (Structure 7) that have been approved for cancer treatment. The mechanism of action (MOA) of these molecules, aside from their formal identification as intercalators into the DNA helix, is now known to be inhibition of topoisomerase II, one of the important enzymes in the replication pathway of DNA during cell cycling.

Table 2
All Small Molecule Anticancer Drugs (1940s–2002)
(Organized Alphabetically by Generic Name Within Chemical Source)

Generic name	Year introduced	Chem-source	Bio-source
aclarubicin	1981	N	M
actinomycin D	Pre-1981	N	M
angiotensin II	1994	N	A
arglabin	1999	N	M
asparaginase	Pre-1981	N	M
BEC	1989	N	P
bleomycin	Pre-1981	N	M
daunomycin	Pre-1981	N	M
doxorubicin	Pre-1981	N	M
masoprocol	1992	N	P
mithramycin	Pre-1981	N	M
mitomycin C	Pre-1981	N	M
paclitaxel	1993	N	P
pentostatin	1992	N	M
peplomycin	1981	N	M
solamargine	1987	N	P
streptozocin	Pre-1981	N	M
testosterone	Pre-1981	N	A
vinblastine	Pre-1981	N	P
vincristine	Pre-1981	N	P
alitretinoin	1999	ND	P
amrubicin hydrochloride	2002	ND	M
cladribine[a]	1993	ND	M*
cytarabine ocfosfate[a]	1993	ND	M*
docetaxel	1995	ND	P
dromostanolone	Pre-1981	ND	A
elliptinium acetate	1983	ND	P
epirubicin hydrochloride	1984	ND	M
estramustine	Pre-1981	ND	A
ethinyl estradiol	Pre-1981	ND	A
etoposide	Pre-1981	ND	P
etoposide phosphate[b]	1996	ND	P
exemestane	1999	ND	A
fluoxymesterone	Pre-1981	ND	A
formestane	1993	ND	A
fulvestrant	2002	ND	A
gemtuzumab ozogamicin	2000	ND	M
hydroxyprogesterone	Pre-1981	ND	A
idarubicin hydrochloride	1990	ND	M
irinotecan hydrochloride	1994	ND	P
medroxyprogesterone acetate	Pre-1981	ND	A
megesterol acetate	Pre-1981	ND	A
methylprednisolone	Pre-1981	ND	A

(*continued*)

Table 2 (Continued)

Generic name	Year introduced	Chem-source	Bio-source
methyltestosterone	Pre-1981	ND	A
miltefosine	1993	ND	A
mitobronitol	Pre-1981	ND	A
pirarubicin	1988	ND	M
prednisolone	Pre-1981	ND	A
prednisone	Pre-1981	ND	A
teniposide	Pre-1981	ND	P
testolactone	Pre-1981	ND	A
topotecan hydrochloride	1996	ND	P
triamcinolone	Pre-1981	ND	A
triptorelin	1986	ND	A
valrubicin	1999	ND	M
vindesine	Pre-1981	ND	P
vinorelbine	1989	ND	P
zinostatin stimalamer	1994	ND	M
aminoglutethimide	1981	S	
amsacrine	1987	S	
arsenic trioxide	2000	S	
bisantrene hydrochloride	1990	S	
busulfan	Pre-1981	S	
camostat mesylate	1985	S	
carboplatin	1986	S	
carmustine	Pre-1981	S	
chlorambucil	Pre-1981	S	
chlortrianisene	Pre-1981	S	
cis-diamminedichloroplatinum	Pre-1981	S	
cyclophosphamide	Pre-1981	S	
dacarbazine	Pre-1981	S	
diethylstilbestrol	Pre-1981	S	
flutamide	1983	S	
fotemustine	1989	S	
heptaplatin /SK-2053R	1999	S	
hexamethylmelamine	Pre-1981	S	
hydroxyurea	Pre-1981	S	
ifosfamide	Pre-1981	S	
levamisole	Pre-1981	S	
lobaplatin	1998	S	
lomustine	Pre-1981	S	
lonidamine	1987	S	
mechlorethanamine	Pre-1981	S	
melphalan	Pre-1981	S	
mitotane	Pre-1981	S	
mustine hydrochloride	Pre-1981	S	
nedaplatin	1995	S	
nilutamide	1987	S	

(continued)

Table 2 (Continued)

Generic name	Year introduced	Chem-source	Bio-source
nimustine hydrochloride	Pre-1981	S	
oxaliplatin	1996	S	
pipobroman	Pre-1981	S	
porfimer sodium	1993	S	
procarbazine	Pre-1981	S	
ranimustine	1987	S	
sobuzoxane	1994	S	
thiotepa	Pre-1981	S	
triethylenemelamine	Pre-1981	S	
uracil mustard	Pre-1981	S	
zoledronic acid	2000	S	
aminogluethimide	Pre-1981	S*	P
capecitabine	1998	S*	M*
carmofur	1981	S*	M*
cytosine arabinoside	Pre-1981	S*	M*
doxifluridine	1987	S*	M*
enocitabine	1983	S*	M*
floxuridine	Pre-1981	S*	M*
fludarabine phosphate	1991	S*	M*
fluorouracil	Pre-1981	S*	M*
gemcitabine hydrochloride	1995	S*	M*
goserelin acetate	Pre-1981	S*	A
leuprolide	Pre-1981	S*	A
mercaptopurine	Pre-1981	S*	M*
methotrexate	Pre-1981	S*	M
mitoxantrone hydrochloride	1984	S*	M
tamoxifen	Pre-1981	S*	A
thioguanine	Pre-1981	S*	M*
bexarotene	2000	S*/NM	P
raltitrexed	1996	S*/NM	M
temozolomide	1999	S*/NM	M
anastrozole	1995	S/NM	
bicalutamide	1995	S/NM	
camostat mesylate	1985	S/NM	
fadrozole hydrochloride	1995	S/NM	
gefitinib	2002	S/NM	
imatinib mesilate	2001	S/NM	
letrazole	1996	S/NM	
toremifene	1989	S/NM	

[a]Marine involvement, see text.
[b]Prodrug (not counted).
M = Microbe; P = Plant; A = Animal; M* = Marine involvement (nucleoside).
N, ND, S, S*, S*/NM, S/NM; *see* **Table 1** in text.

That there is still potential for doxorubicin, in spite of its cardiotoxicity, is demonstrated by the number of delivery techniques that have been either tested in clinical trials, are currently in clinical trials, or have been approved for use as methods of delivery of this agent. These have included monoclonal antibodies bearing a doxorubicin warhead directed towards a particular epitope (predominately in leukemia treatment) and liposomal delivery systems involving both conventional (one is in clinical use in the USA) and "stealth" liposomes containing doxorubicin. In addition, doxorubicin is a favorite warhead for polymeric delivery systems. A group that has shown the utility of this approach is that of Duncan, with the reported clinical trials of PK1 and PK2 under the auspices of the United Kingdom cancer research consortia *(6)*.

Another series of extremely important molecules, also from the *Actinomycetales*, are the family of glycopeptolide antibiotics known as the bleomycins (Structure 8) (particularly A_2; Blenoxane®) and, initially, the closely related structural class, the phleomycins. These molecules were originally reported by Umezawa's group at the Institute of Microbial Chemistry in Tokyo, and developed as antitumor agents by Bristol-Myers. Their "original" MOA was elucidated by Hecht et al. (MIT and the University of Virginia), who demonstrated that a metal ion (Cu^{2+} or Fe^{2+}) was required to activate the oxidative breakage of the DNA helix, once binding to the helices had occurred. For most of the years since the introduction of Blenoxane® into clinical use, the primary target was thought to be DNA. However, in a recent series of investigations using both bleomycin and synthetic modifications of the molecule, Hecht has discovered that RNA interacts with the bleomycins at much lower concentrations than those required for any effect on DNA, thus opening up the possibility that the major site of action is, in fact, inhibition of t-RNA, not DNA (Hecht, personal communication). Irrespective of their exact mechanism, these agents are major products in the clinician's armamentarium.

Other microbial products that are still in use include the mitosanes, such as mitomycin C, the glycosylated anthracenone, mithramycin, streptozotocin, and pentostatin, and, most importantly, calicheamicin. The latter compound (Structure 9) is possibly the most potent antitumor compound to be approved for clinical use. It is a compound with in vitro activity at the sub-picomolar level, and for a significant number of years, it languished as it was just too toxic to pursue, in spite of its exquisite activity *(7)*. It was also the progenitor of a new chemical class, the endiynes, which, on activation, undergo an unprecedented rearrangement and interaction with DNA to produce double-stranded DNA cleavage. The actual in vivo MOA of these molecules is still open to debate, but modern techniques are now being employed in elucidating the possible pathways involved *(8)*. In 2000, Wyeth-Ayerst (originally Lederle, now Wyeth) gained approval for Mylotarg®, where the compound is linked to a specific monoclonal antibody that is directed against chronic myologenous leukemia *(9)*.

2.2. Microbial Agents in Clinical Trials

In addition to the agents listed above, there are some extremely interesting structural classes of agents isolated from microbes, or modified from such agents, in various phases of clinical trials.

There are perhaps two major structural types that evince significant interest on the part of both chemists and cell biologists. These are the structures based on the

Anticancer Drugs From Natural Sources

Structures 1–9.

epothilones (Structure 10), from the extremely prolific *Myxomycetales*, and the indolocarboxazoles such as staurosporine (Structure 11), from the *Actinomycetales*.

2.2.1. Epothilones

The epothilones are of great interest as potential antitumor agents due to their MOA being the same as that of paclitaxel (*vide infra*), though having at first glimpse, quite a different topology. Molecular modeling, however, has shown that there are significant common structural features in the two basic molecules *(10–12)*. Originally, the epothilones were difficult to obtain in any quantity, and a significant amount of time and effort was spent by both academics and industry in efforts to synthesize both epothilones A and B and their precursors, C and D. The major impetus behind all of these studies was the realization that the epothilones were active against paclitaxel-resistant cell lines *(13)*. Bristol-Myers also synthesized the aza-derivative (Structure 12) and currently have it and epothilone B in phase II clinical trials.

During this time period, the group at Kosan licensed the work of Danishevsky on epothilone D, which he had synthesized at Memorial Sloan-Kettering (MSK), and cloned and expressed the polyketide gene cluster that produced epothilones A/B. Removal of the terminal gene for the P_{450} enzyme that epoxidized the macrolidic double bond in the precursor epothilones C/D, and transfer to a different host, produced crystalline epothilone D from a 100-L fermentation *(14)*. This material is now undergoing phase II clinical trials in conjunction with MSK and Roche. There have been significant efforts expended on the total synthesis of this class of compounds *(15,16)*.

2.2.2. Indolocarbazoles

The indolocarboxazoles first came into prominence with the identification of staurosporine and its simple derivative UCN-01 (Structure 13) as potential inhibitors of components of the eukaryotic cell cycle and of phosphokinase C *(17)*. The differences in the potential mechanisms of action of staurosporine and simple derivatives vs synthetic analogs were shown by a comparison with the synthetic compound, SB-218078 (Structure 14). Although not a natural product, the compound SB-218078 recently reported by workers *(18)* at SmithKline Beecham (now Glaxo SmithKline) is definitely derived from the indolocarbazole base structure; a reduced furan ring replaces the substituted pyranose sugar found in staurosporine, and the maleimide ring of the didemnimides replaces the reduced pyranone of staurosporine. This compound demonstrates at least a 15-fold selectivity for Chk1 inhibition vs Cdk1 (*aka* cdc2), and a 65-fold selectivity over PKC (IC_{50} values of 15, 250, and 1000 nM respectively). In contrast, staurosporine is active against all three kinases at the 5–8 nM level, and UCN-01 has a 15-fold selectivity vs Cdk1, but has comparable activity against PKC (IC_{50} values of 7, 100, and 4 nM respectively).

These workers also confirmed the findings of Shao et al. *(19)* as to the synergistic effect of UCN-01 on camptothecin cytotoxicity, and demonstrated that SB-218078 had comparable activity, implying that the mechanism of this abrogation with UCN-01 is due to Chk1 inhibition. Concomitantly with this work, Graves et al. *(20)* reported that UCN-01 inhibits Chk1, thus causing the phosphatase Cdc25c to be dephosphorylated, and hence allowing activation of Cdk1 (Cdc2) and abrogating the G2 checkpoint.

Currently UCN-01 is still in phase I clinical trials. For further discussions as to the possibilities of this basic structural type as inhibitors of the cell cycle, the relatively recent review of Newman et al. *(21)* should be consulted, together with the review of Dancey and Sausville *(22)* on the potential of protein kinase inhibitors in general as candidate antitumor agents.

The compounds related to the cytotoxin rebeccamycin (Structure 15) are extremely interesting from a mechanistic aspect, as through relatively simple modifications of the base indolo-carbazole, one can generate molecules that have both topoisomerase I and topoisomerase II inhibitory activity. The findings have led to the synthesis of water-soluble analogs, such as NB-506 (Structure 16) and second-generation products, all of which show excellent preclinical in vivo activity. The topoisomerase I activity of this class of compounds has been extensively studied *(23)*, and one of the most interesting discoveries in this structural class has been the simple analog R-3 (Structure 17). This compound was first reported as a topoisomerase I inhibitor in 1997 *(24)*, and not only inhibits topoisomerase I but also completely inhibits the phosphorylation of SF2/ASF, a member of the SR protein family, in the absence of DNA. This compound is the only compound so far reported that inhibits the protein kinase activity of topoisomerase I, an activity first reported in 1996 *(25)* and reviewed in 1997 *(26)*. A derivative (Structure 18) whose generic name is edotecarin, though previously known as J-107088, is currently in phase III trials with Pfizer for therapy of brain carcinomas. If it is ultimately commercialized, it will be the first non-camptothecin-based topoisomerase I inhibitor.

Simple alteration of the substitution pattern around the indolocarbazole moiety of rebeccamycin gives a molecule known as BMS-27557 or NSC 655649 (Structure 19), which now has topoisomerase II rather than topoisomerase I selectivity. This compound is now in phase II trials with Exelixis for a variety of carcinomas.

2.2.3. Geldanamycin Derivatives

It is noteworthy that the first signal transduction modulator, other than a formal Cdk or PKC inhibitor, to enter clinical trials is a microbial product. Thus 17-allylamino-geldanamycin (17-AAG) (Structure 20) entered phase I trials with the NCI and Kosan (the licensee) in 2001. This compound binds at the major ATP-binding site of the protein chaperone HSP-90. Although now it is known not to be suitable for use a single agent in humans, it appears to be effective when used in combination with cytotoxins such as paclitaxel. Very recently *(27)*, it was shown that the HSP-90 of BT474 breast tumor cells binds 17-AAG much more strongly (approx100-fold higher affinity) than that in "normal" cells (human renal epithelial cells and dermal fibroblasts). That there are further avenues to explore in the inhibition of the protein chaperones can be seen from the recent publications of Neckers and Ivy *(28)* on HSP90 and its potential, and of Beliakoff et al. *(29)* on the potential for use of HSP90 inhibitors in dealing with hormone-refractory breast cancers.

3. The Plant Kingdom

3.1. Approved Agents

Plants have a long history of use in the treatment of cancer *(30)*, though many of the claims for the efficacy of such treatment should be viewed with some skepticism because cancer, as a specific disease entity, is likely to be poorly defined in terms of folklore and traditional medicine *(31)*. The first agents to advance into clinical use were the so-called *vinca* alkaloids, vinblastine and vincristine (Structures 21, 22), isolated from the Madagascar periwinkle, *Catharanthus roseus*, which was used by various cultures for the treatment of diabetes. These drugs were first discovered during an investigation of the plant as a source of potential oral hypoglycemic agents *(31)*. What is also of interest is that vincristine is in fact a heterologous dimer of two alkaloids that apart are not active as cytotoxins, but chemically combined, the compound acts as a tubulin interactive agent. Recently, two semisynthetic analogs of vincristine, vinorelbine and vindesine, were approved for clinical use *(32)*.

Podophyllotoxin was isolated as the active antitumor agent from the roots of various species of the genus *Podophyllum*. These plants possessed a long history of medicinal use by early American and Asian cultures, including the treatment of skin cancers and warts *(31)*. Although podophyllotoxin is still used as a topical treatment for warts (USP formulary), it was too toxic for use as an injectable antitumor agent. Following exten-

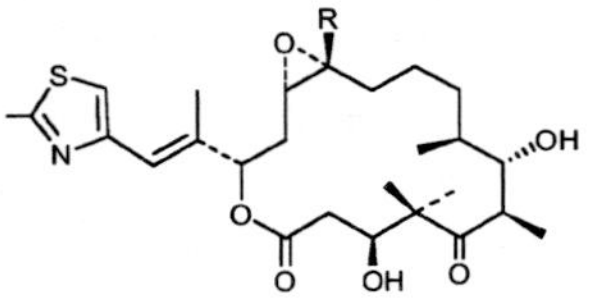

10a Epothilone A; R = H
10b Epothilone B; R = CH3

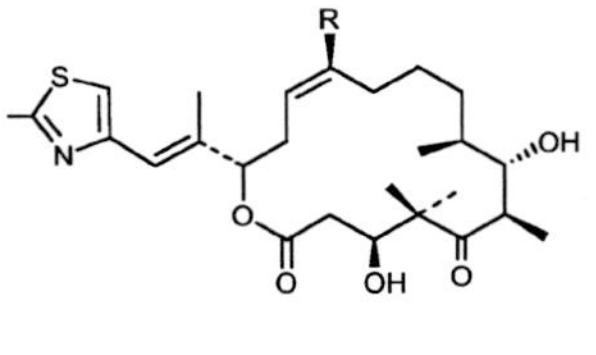

10c Epothilone C; R = H
10d Epothilone D; R = CH3

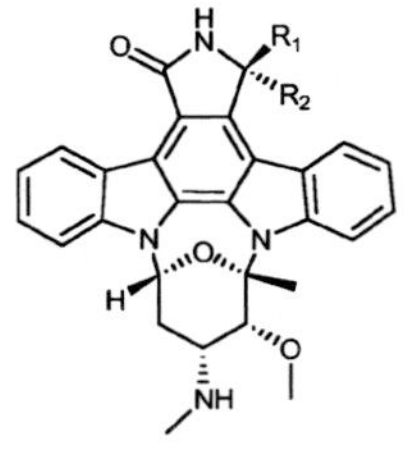

11 Staurosporine R₁ = R₂ = H
13 UCN 01 R₁ = H; R₂ = OH

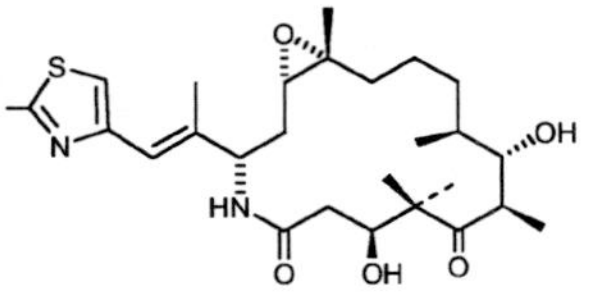

12 16-Azaepothilone B

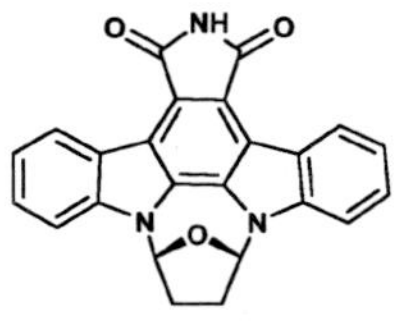

14 SB 218078

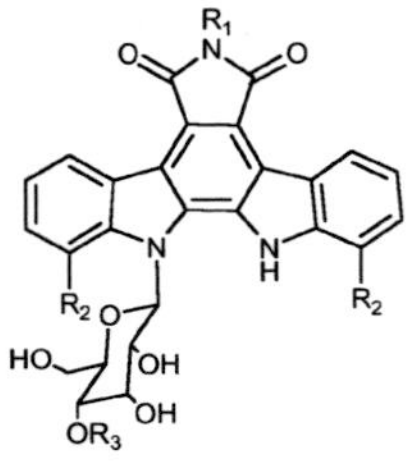

15 Rebeccamycin R₁ = H; R₂ = Cl; R₃ = CH₃
16 NB 506 R₁ = NHCHO; R₂ = OH; R₃ = H
17 R 3 R₁ = OH; R₂ = H; R₃ = CH₃
19 BMS 27557 R₁ = CH₂CH₂N(CH₂CH₃)₂; R₂ = Cl; R₃ = CH₃

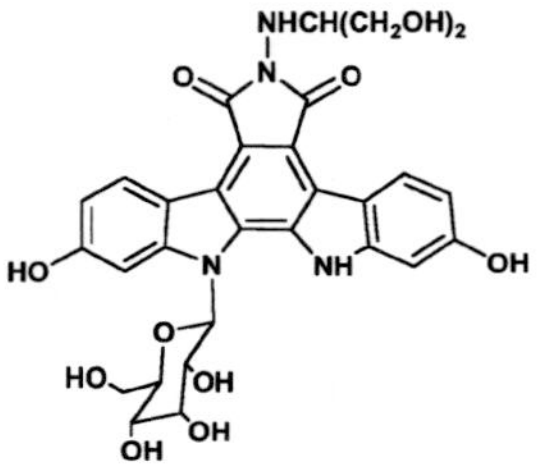

18 Edotecarin

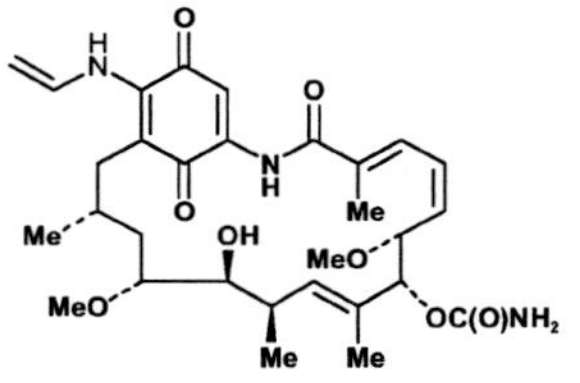

19 17-Allylaminogeldanamycin

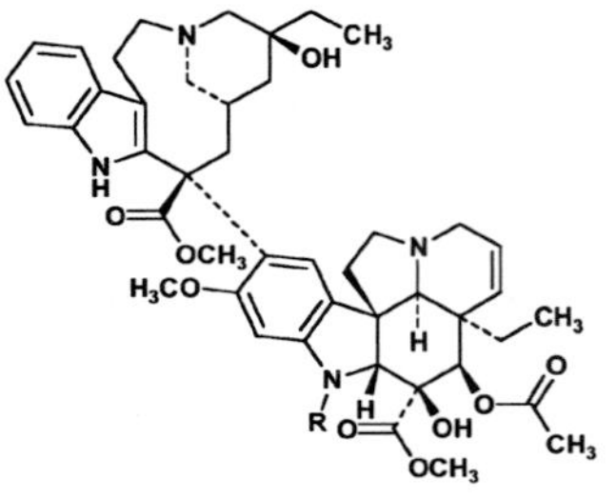

21 Vinblastine R = CH₃
22 Vincristine R = CHO

Structures 10–22

sive semisynthetic modifications of the isomeric natural product, epipodophyllotoxin, by Sandoz (now part of Novartis), two clinically approved agents, etoposide and teniposide (Structures 23, 24), were developed.

More recent additions to the armamentarium of plant-derived chemotherapeutic agents are the taxanes and camptothecins. Taxol® (Structure 25) *(33)* (generic name paclitaxel) was initially isolated from the bark of *Taxus brevifolia*, collected in Washington State as part of a random collection program by the U.S. Department of Agriculture for the National Cancer Institute (NCI) *(34)*. The use of various parts of *T. brevifolia* and other *Taxus* species (e.g., *canadensis, baccata*) by several Native American tribes for the treatment of some non-cancerous conditions has been reported *(30)*, while the leaves of *T. baccata* are used in the traditional Asiatic Indian (Ayurvedic) medicine system *(35)*, with one reported use in the treatment of "cancer" *(30)*. Paclitaxel, along with several key precursors (the baccatins), occurs in the leaves/needles of various *Taxus* species, and the ready semi-synthetic conversion of the relatively abundant baccatins to paclitaxel, as well as to active paclitaxel analogs, such as docetaxel *(36)*, has provided a major, renewable natural source of this important class of drugs.

Paclitaxel was a pivotal discovery from a mechanistic aspect in the treatment of cancer, as it was the first compound identified that caused the stabilization of microtubules in the mitotic cycle. As a result of such stabilization, the cell stops division and usually undergoes programmed cell death (the apoptotic cascade). Although both colchicines and *vinca* alkaloids were known to interact with microtubules, it became obvious *(37,38)* that paclitaxel had an entirely different mechanism of action, effectively "freezing" the tubulin polymerization/depolymerization cycle. Paclitaxel entered clinical use in 1993 and docetaxel was approved in 1995, initially for treatment of refractory ovarian cancer and then breast cancer, though now the taxanes are being used in a large number of cancer treatments.

It is rare for a single chemistry group to first identify a clinical agent with an undescribed MOA, and then discover another totally different chemotype with a previously unknown MOA. However, Wall and Wani, the discoverers of paclitaxel, were also responsible for the identification, isolation, and purification of camptothecin (Structure 26). This compound, which they isolated from the Chinese ornamental tree *Camptotheca acuminata (39)*, was entered into clinical trials as its sodium salt by the NCI in the late 1970s, but it turned out to be too toxic due to reformation of the lactone ring in the bladder. Subsequent work under a National Cooperative Drug Discovery Program (NCDDG) involving the NCI, Johns Hopkins University, and SmithKline Beckman (now Glaxo SmithKline) led to the synthesis of the semisynthetic derivative topotecan (Structure 27), which was approved in the United States in 1996. Another semi-synthetic derivative, irinotecan (Structure 28), was approved roughly 18 mos earlier in Japan. Until the identification of the indolocarbazole derivatives referred to earlier (**Subheading 2.2.2.**), the camptothecins were the only chemotype reported to consistently inhibit topoisomerase I activity.

Elliptinium acetate (Structure 29), a derivative of ellipticine isolated from several genera of the *Apocynaceae* family, was approved in the early 1980s in France as a treatment for breast cancer. Though there is still some usage, it is probable that various taxanes are utilized in preference to this agent.

3.2. Plant Agents in Clinical Trials

A significant number of plant-sourced agents are still in clinical trials at one level or another. Some are being investigated as direct cytotoxins, whereas others are being studied from the aspect of their potential role as inhibitors of particular cell-cycle enzymes, proteins, or pathways.

3.2.1. Homoharringtonine

Homoharringtonine (Structure 30), from the Chinese tree *Cepahlotaxus harringtonia* var *drupacea* (Sieb and Zucc), is still in clinical trials against various leukemias in the West, but is reported as being used in China as an anticancer agent *(40)*.

3.2.2. Flavonoids

An early report of a natural product compound class that ultimately led to Cdk inhibitors was that of Ranelleti et al. *(41)* on the antitumor effect of quercetin (Structure 31).

Structures 23–34

This compound from the class known as flavonoids can be thought of as an ATP mimic, where the flat bicyclic chromone ring is an isostere of adenine. Quercetin showed a dose-dependent and reversible inhibition of tumor cell proliferation over the range from 10 nM to 10 µM. Subsequent cell-cycle analysis demonstrated a block at the G_0/G_1 interphase, consistent with Cdk inhibition, though it should be noted that this compound has been reported to inhibit topoisomerase II; however, a close analog, myricetin (Structure 32), has shown an IC_{50} close to 10 µM vs Cdk2 *(42)*.

Flavopiridol (Structure 33) was made by the Indian subsidiary of Hoechst (now Aventis) following the isolation and synthesis of the plant-derived natural product rohitukine (Structure 34). It was the most active of approx 100 analogs when assayed against Cdks and showed about 100-fold more selectivity compared to its activity vs tyrosine kinases. It showed roughly comparable activity in the 100–400 nM range for IC_{50} values, depending upon the specific Cdk. It was the first compound at NCI identified as a potential antitumor agent that subsequently was proven to be a relatively specific Cdk inhibitor *(43)*. The initial report *(44)* on its Cdk 2 inhibitory activity was made in 1994, followed by data demonstrating antitumor activity by Czech et al. *(45)* in 1995. Flavopiridol is currently in phase III clinical trials as an inhibitor of cyclin-dependent kinase 2 (Cdk2), both as a single agent and as a modulator in combination with other agents, particularly paclitaxel and cis-platinum. It has been reported to lead to partial and/or complete remissions in a number of phase I patients, leading to phase II studies in patients with paclitaxel-resistant tumors *(46)*. There have been a number of relatively recent reports of this agent and the various combinations with other drugs and drug candidates, a significant number of which are either natural products or derived from natural products *(22,47,48)*.

3.2.3. Combretastatins

The combretastatins, derived from *Combretum caffrum*, are a family of stilbenes which act as anti-angiogenic agents, causing vascular shutdown in tumors and resulting in tumor necrosis *(49)*. A water-soluble analog, combretastatin A-4 phosphate (Structure 35), has shown promise in early clinical trials and is currently in phase II. These relatively simple compounds, which interact with tubulin (though they do not stabilize polymerized tubulin), have been the progenitors of a very large number of synthetic molecules that are currently in a variety of testing stages ranging from preclinical to phase I clinical trials. The evolution of a significant number of derivatives is documented in reviews by Li and Sham *(50)* and Cragg and Newman *(51)*.

3.3. Comments on Current Use of Plant Sources

Approximately 40% of the 150 compounds listed in the recent review *(51)* of tubulin-interactive and topoisomerase I inhibitors currently in preclinical and clinical trials, are based upon steroids, taxanes, combretastatins, or camptothecins.

4. Marine-Derived Agents

4.1. Introduction

With the advent of marine natural-product research approximately 50 yr ago, it has become rather obvious that the marine environment produces chemical compounds that have significant activities against mammalian cells. Perhaps the first notable discovery of biologically active compounds from marine sources was reported from 1950 to 1956 by Bergmann et al. *(52–54)* on two compounds, spongouridine and spongothymidine (Structures 36, 37), which they had isolated from the Caribbean sponge, *Cryptotheca crypta*. What was significant about these materials was that they demonstrated for the first time that naturally occurring nucleosides could be found containing sugars other than ribose or deoxyribose.

These two compounds can be thought of as the prototypes of all of the modified nucleoside analogs made by chemists that have crossed the antiviral and antitumor stages since then. Once it was realized that biological systems would recognize the base and not pay too much attention to the sugar moiety, chemists began to substitute the "regular pentoses" with acyclic entities, and with cyclic sugars bearing unusual substituents. These experiments led to a vast number of derivatives that were tested extensively as antiviral and antitumor agents over the next 30-plus years. Suckling *(55)* showed how such structures evolved in the (then) Wellcome laboratories, leading to azidothymidine (AZT) and, incidentally, to Nobel prizes for Hitchens and Elion, though no direct mention was made of the original arabinose-containing leads from natural sources.

Showing that "Nature may follow chemists rather than the reverse, or conversely that it was always there but the natural products chemists were slow off the mark," arabinosyladenine (Ara-A or Vidarabine®) was synthesized in 1960 as a potential antitumor agent *(56)*, was later produced by fermentation *(57)* of *Streptomyces griseus*, and was isolated together with spongouridine *(58)* from a Mediterranean gorgonian (*Eunicella cavolini*) in 1984.

Of the many compounds derived from these early discoveries, some, such as Ara-A, Ara-C, Acyclovir, and later AZT and DDI, have gone into clinical use, but most have simply become entries in chemical catalogs. Though this class of compounds was known to possess antiviral and antitumor activities, synthetic analog studies led to the development of cytosine arabinoside (Ara-C) (Structure 38) as a clinically useful anticancer agent approx 15 yr later.

Currently there are no approved agents that have come directly from the marine environment, but a very significant number of marine-sourced compounds have either been in clinical trials or are currently in clinical or preclinical trials as anticancer agents.

Table 3 shows the current status of compounds from all marine sources that were or are now in clinical trials as antitumor agents, and some that are in preclinical testing. That the marine environment should produce a very large number of exquisitely potent agents is not surprising when it is realized that the majority of marine invertebrates have no methods of physical defense against predators. Aside from mollusks, which tend to have a hard carapace, the other phyla lack any form of outward defense, and thus rely upon "chemical warfare." A probable reason for the presence of extremely active chemotypes is simply that the dilution effect of seawater means that any toxin must be very potent, as it usually has to cross a finite distance to deter any predator.

4.2. Didemnin B, the First Agent to Enter Clinical Trials

Didemnin B (Structure 39) was isolated by Rinehart's group from extracts made of the tunicate *Trididemnum solidum (59)* that demonstrated potent antiviral and cytotoxic activity against P388 and L1210 murine leukemia cell lines. Didemnin B was subsequently advanced into preclinical and clinical trials (phases I and II) under the auspices of the NCI in the very early 1980s as the first defined chemical compound directly from a marine source to go into clinical trials for any major human disease. The compound turned out to be too toxic for use, and trials were officially terminated in the mid-1990s by NCI. However, the experience gained from these efforts was immensely helpful in aiding the trials of other compounds.

Rinehart's group developed methods of large-scale isolation and purification, and total syntheses that permitted significant structure-activity relationships to be derived *(60)*. This enabled materials to be provided to other investigators so that basic biochemical studies could be performed, leading to the identification of a potential mechanism of action for this compound—namely, the binding to elongation factor 1-α reported by Crews et al. *(61)*. Subsequent reports showed that didemnin B binds noncompetitively to palmitoyl protein thioesterase *(62)*, and the following year, it was reported *(63)* that rapamycin inhibited the didemnin-induced apoptosis of human HL-60 cells, perhaps by binding to the FK-506 binding protein(s).

In 2002, Vera and Joullie *(64)* published an excellent review of didemnins as cell probes and targets for syntheses, and also made some reasonable arguments that the dosing schedules used in the early clinical trials may well have been nonoptimal for demonstrating activity as a cytotoxin, rather than as an immunosuppressive/modulator.

Although didemnin B was not successful, a very close chemical relative is currently in clinical trials (see aplidine, **Subheading 4.4.3.**). Recently, Rinehart published an overview of these compounds as part of a discussion of antitumor compounds from tunicates *(65)*.

4.3 Post-Didemnin B Compounds That Entered Clinical Trials

A significant number of compounds from marine sources that have been entered into preclinical and clinical trials since the early 1980s, emanated from programs funded by the US National Cancer Institute. The NCI has supported many of the searches for agents active against cancer, via a variety of mechanisms, ranging from random to targeted shallow-water collections, by both direct contractors and via grants to academics, both individually and as parts of consortia. Thus, NCI has the systems in place for collection, bioactivity determinations, and subsequent testing in animals and humans, with the aim of finding new and potent treatments for cancers. In addition, the National Science Foundation, via grants to academics for basic marine research, has significantly contributed to the discovery process, as the agents that were identified were then tested via the NCI in a significant number of cases.

Because of the extremely long time frame involved in such processes (for example, paclitaxel, took over 20 yr from structural determination and reporting until US Food and Drug Administration [FDA] approval), the compounds that will be discussed fall into two approximate time frames: those from the initial collection programs (which aided the didemnin B discovery), and those that are further back in the system, which have been discovered as a result of the modified NCI screens utilizing the 60-cell line (or functional equivalent) screen from the early 1990s.

4.3.1. Dolastatins

The dolastatins are a series of cytotoxic peptides that were originally isolated in very low yield by Pettit's group *(66–70)* as part of its work on marine invertebrates from the Indian Ocean mollusk, *Dolabella auricularia*. Owing to the potency and mechanism of action of dolastatin 10 (Structure 40), a linear depsipeptide that was shown to be a tubulin-interactive agent binding close to the *vinca* domain in a site where other peptidic agents bound *(71,72)*, the compound entered phase I clinical trials in the 1990s under the auspices of the NCI. Because the natural abundance was so low, Pettit and others developed synthetic methods that provided enough material under cGMP conditions to commence trials *(66)*. Dolastatin 10 progressed through to phase II trials as a single agent, but it did not demonstrate significant antitumor activity in a phase II trial against prostate cancer in humans *(73)*. Similarly, no significant activity was seen in a phase II trial against metastatic melanoma *(74)*.

As a result of the synthetic processes alluded to earlier, many derivatives of the dolastatins have been synthesized, with TZT-1027 (Auristatin PE or Soblidotin) (Structure 41) now in phase II clinical trials in Europe, Japan, and the United States under the auspices of either Teikoku Hormone, the originator, or the licencee, Daiichi Pharma-

Table 3
Status of Selected Marine-Derived Compounds in Clinical and Preclinical Trials (12/03)

Name	Source	Status (Disease)	Comment
Didemnin B	*Trididemnum solidum*	Phase II (Cancer)	Dropped mid-1990s
Bryostatin 1	*Bugula neritina*	Phase II (Cancer)	Now in combination therapy trials; licenced to GPC Biotech in Germany by Arizona State Univ. May be produced by bacterial symbiont
Dolastatin 10	*Dolabella auricularia* (Marine microbe derived; cyanophyte)	Phase II (Cancer)	Many derivatives made synthetically; now known to be a cyanobacterial product. No positive effects in phase II trials so far reported
Cematodin	Synthetic derivative of Dolastatin 15 (Marine microbe derived; cyanophyte)	Phase II (Cancer)	Some positive effects on melanoma pts in phase II. Listed as discontinued.
TZT-1027	Synthetic Dolastatin	Phase II (Cancer)	Also known as Auristatin PE and Soblidotin.
Girolline	*Pseudaxinyssa cantharella*	Phase I (Cancer)	Discontinued due to hypertension
Ecteinascidin 743	*Ecteinascidia turbinata*	Phase III (Cancer) in 2003. EC CPMP did not approve rapid marketing for sarcoma (07-11/2003) but orphan drug designation by EC for ovarian cancer (10/03)	Licenced to Ortho Biotech (J&J). Partial synthesis from a microbial product for production
Aplidine	*Aplidium albicans*	Phase II (Cancer) EC COMP/EMEA approved orphan drug status for ALL (07/2003)	Dehydrodidemnin B
E7389	*Lissodendoryx* sp.	Phase I (Cancer)	Eisai's synthetic halichondrin B derivative.
Discodermolide	*Discodermia dissoluta*	Phase I (Cancer)	Licenced to Novartis by Harbor Branch Oceanographic Institution

(continued)

Table 3 (Continued)

Name	Source	Status (Disease)	Comment
Kahalalide F	*Elysia rufescens*/ *Bryopsis* sp.	Phase II (Cancer) early 2003	Licenced to PharmaMar by Univ. Hawaii. Synthesized by Univ. Barcelona chemists.
ES-285 (Spisulosine)	*Spisula polynyma*	Phase I (Cancer)	*Rho*-GTP inhibitor
HTI-286 (Hemiasterlin derivative)	*Cymbastella* sp.	Phase I (Cancer)	Derivative made by Univ. British Columbia; licensed to Wyeth.
Bengamide derivative	*Jaspis* sp.	Phase I (Cancer)	Bengamide A licenced to Novartis; derivative made that was a Met-AP1 inhibitor, withdrawn 2002
Cryptophycins (also Arenastatin A)	*Nostoc* & *Dysidea arenaria*	Phase I (Cancer)	From a terrestrial cyanophyte, but also from a sponge as Arenastatin A. Synthetic derivative licenced to Lilly by Univ. Hawaii, but withdrawn 2002
KRN-7000	*Agelas mauritianus*	Phase I (Cancer)	An Agelasphin derivative
Squalamine	*Squalus acanthias*	Phase II (Cancer)	Squalamine
AE-941 (Neovastat)	Shark	Phase III (Cancer)	"Defined mixture of <500 kDa from cartilage"
Laulimalide	*Cacospongia mycofijiensis*	Preclinical (Cancer)	Synthesized by many groups in both partial and complete synthetic schemes
Curacin A	*Lyngbya majuscula*	Preclinical (Cancer)	Synthesized, more soluble combi-chem derivatives being evaluated
Vitilevuamide	*Didemnum cucliferum* & *Polysyncraton lithostrotum*	Preclinical (Cancer)	
Diazonamide	*Diazona angulata*	Preclinical (Cancer)	Synthesized and new structure elucidated
Eleutherobin	*Eleutherobia* sp.	Preclinical (Cancer)	Synthesized and derivatives made by combi-chem. Can be made by aquaculture
Dictyostatin-1	*Spongia*	Preclinical (Cancer)	
Sarcodictyin	*Sarcodictyon roseum*	Preclinical (Cancer) (derivatives)	Combi-chem synthesis
Peloruside A	*Mycale hentscheli*	Preclinical (Cancer)	
Salicylihalimides A & B	*Haliclona* sp.	Preclinical (Cancer)	First marine Vo-ATPase inhibitors

ceuticals. Another derivative of dolastatin 15, known as Cematodin (Structure 42) or LU-103793, was entered into phase I clinical trials by the Knoll division of Abbott GMBH for treatment of breast cancer. It progressed into phase II, but trials appear to have been discontinued. By using tritium-labeled dolastatin 15 as a bioprobe, Hamel's group at NCI *(75)* recently reported that the *vinca* domain in tubulin may well be composed of a series of overlapping domains rather than being a single entity, as different levels/types of competition were found when selected tubulin-interactive agents were used to investigate the binding characteristics of the labeled probe.

Similarly to bryostatin (*see* **Subheading 4.4.1.**), there was always a potential question with the dolastatins as to whether or not they were microbial in origin, as peptides with unusual amino acids have been well documented in the literature as coming from the *Cyanophyta*. In the last few years, this supposition has been shown to be fact. Thus, in 1998, workers at the Universities of Guam and Hawaii reported the isolation and purification of simplostatin 1 (Structure 43) from the marine cyanobacterium *Simploca hynoides (76)*. This molecule differed from dolastatin 10 by the addition of a methyl group on the first *N*-dimethylated amino acid. Subsequently, in 2001, the same groups reported the direct isolation of dolastatin 10 from another marine cyanobacterium that was known to be grazed on by *D. auricularia (77)*. Dolastatin 10 was in fact isolated from the nudibranch following feeding of the cyanophyte, thus confirming the original hypothesis (personal communication, Dr. V. J. Paul). Recently, the mechanism of action of symplostatin 1 was shown to be similar to dolastatin 10 but to be somewhat more toxic to mice at comparable doses *(78)*.

4.3.2. Girolline (Girodazole)

This very simple compound, a substituted imidazole (Structure 44), was reported from the sponge *Pseudaxinyssa cantharella (79)*, and was shown by workers at Rhone-Poulenc Rorer (now Aventis) to be an inhibitor of protein synthesis, acting preferentially on the termination step in eukaryotic protein synthesis. This is in contrast to other known protein synthesis inhibitors, such as emetine, homoharringtonine, anguidine, and bruceantin, which generally act at either the initiation or elongation steps *(79)*. Girolline proceeded through to phase I clinical trials in humans, but the trials were stopped due to significant hypertensive effects seen in treated patients.

4.3.3. Bengamide Derivatives

Bengamides A and B (Structures 45, 46) were first reported in 1986 by Crews's group at the University of California, Santa Cruz (UCSC), as antihelminthic compounds, with some antibiotic and cytotoxic activity *(80)*. The number of bengamide analogs isolable from the same sponge was extended to bengamide G *(81,82)*. In a subsequent paper with workers from Novartis, the number of compounds in the group was extended, and their antitumor activities reported *(83)*.

The bengamides were evaluated by Novartis (then Ciba-Geigy). As a result of their intrinsic activities, a synthetic program was put in place that developed a derivative of bengamide A (Structure 47) as a clinical candidate. This compound was shown to be an inhibitor of methionine aminopeptidases and subsequently entered phase I clinical trials in 2000, but was withdrawn in the middle of 2002.

Structures 35–43, 53

4.3.4. Cryptophycins

These compounds were reported from two blue-green algae, initially by a group from Merck in 1990 using *Nostoc* sp. strain MB5357 (ATCC 53789), which was originally isolated from a lichen on a Scottish island; they reported only the antifungal activity, finally deciding not to proceed with development as it was too toxic. Moore's group at the University of Hawaii subsequently identified the same compound *(84)* from a nonmarine cyanophyte, *Nostoc* sp., strain GSV-224, and almost contemporaneously, a similar molecule was reported by Kobayashi et al. from an Okinawan sponge (*see* next paragraph). The Universities of Hawaii and Wayne State then licensed the natural cryptophycins and synthetic derivatives to the Eli Lilly Company for advanced preclinical and clinical development. This led to the selection of a semi-synthetic derivative, cryptophycin 52 (LY355703) (Structure 48), as a phase I clinical candidate in the middle 1990s *(85)*. The routes, both chemical and pharmacological, leading to the choice of this particular derivative were described by Shih and Teicher *(86)* of the Lilly Research Laboratories. The compound progressed towards phase II trials, but in 2002, cryptophycin 52 was withdrawn from trials.

Although the original cryptophycins came from terrestrial cyanophytes, and the clinical candidate came from semisynthetic modifications of the natural product, in 1994 Kobayashi et al. reported that an acetone extract of the Okinawan sponge, *Dysidea arenaria*, had potent cyctoxicity *(87)* and yielded a compound (arenastatin A) that subsequently turned out to be identical to cryptophycin 24 (Structure 49), reported by Moore's group in 1995 *(88)*. A later report from the Japanese group demonstrated that arenastatin A and synthetic analogs also are tubulin-interactive agents similar in activity to the other cryptophycins reported by Moore et al. *(89,90)*.

4.4. Compounds Currently (December 2003) in Clinical Trials

4.4.1. Bryostatin

In 1968, NCI commissioned a large-scale (for those days) collection of the bryozoan, *Bugula neritina*, by Jack Rudloe of the Gulf Specimen Company off the west coast of Florida, and this collection was sent to Pettit's group for workup. The extract was found to be inactive against the L1210 murine leukemia cell line but to give a 68% increase in life span using the same concentration against the P388 murine leukemia cell line *(91)*. Following very significant amounts of work by Pettit and his group, including more collections on a larger scale, significant problems with isolation as a result of dealing with vanishingly small quantities of a very potent agent, and problems related to assay reproducibility, the compound was purified and identified as bryostatin 3 (Structure 50), one of a series of closely related compounds that now number 20.

Subsequent work by Pettit's group identified two other geographic areas where significant (in relative terms) quantities of bryostatin 1 (Structure 51) could be isolated from *B. neritina* colonies. As a result of prodigious efforts on the part of the Pettit group and workers at NCI-Frederick, by 1990 there was enough cGMP-grade material to commence systematic clinical trials. From these studies *(92)*, it was determined that bryostatins bind to the same receptors as the phorbol esters, which are tumor promoters, but have no tumor promoter activity. As a result of this binding, however, the protein kinase C isoforms in various tumor cells are significantly down-regulated, leading to inhibition of growth, alteration of differentiation, and/or death.

To date, bryostatin 1 has been in more than 80 human clinical trials, with more than 20 being completed at both the phase I and phase II levels. There have been some responses to the compound as a single agent, with effects ranging from complete remission (CR) and partial remission (PR) to stable disease (SD). However, the use as a single agent is probably not the optimal usage for this agent *(92,93)*. When combined with cytotoxic drugs such as the *vinca* alkaloids, paclitaxel, fludarabine, cisplatin, and so on, responses in phase I trials have been reported *(92)*, and currently three phase I and five phase II trials are ongoing in a number of centers, using a variety of protocols. Details can be obtained from http://clinicaltrials.gov for these and for other agents in clinical trials. The major side effect appears to be myalgia, but this is treatable by standard supportive therapies.

One very interesting question arising from the search for bryostatin sources was why is the nominal producing organism so ubiquitous, but the number of *B. neritina* colonies that actually produce detectable bryostatin 1–3 levels so low and geographically spread? One possible answer comes from the work of Haygood and her collaborators at the Scripps Institution of Oceanography. Haygood showed that the bryozoan is actually the host to a symbiotic organism that may well be the actual producer of the compound; in an elegant series of experiments, she and her colleagues demonstrated the presence of a putative PKS-I gene fragment in colonies that produced bryostatin, that was absent in non-producers *(94)*. In addition, Davidson and Haygood demonstrated that there are subdivisions within *B. neritina* samples taken from the same sites but at different depths. Thus, at depths greater than 9 m, (the D type), bryostatins 1–3 and minor components are found, whereas at less than 9 m (S type), only the minor derivatives are seen. The symbiotic organisms isolated from each type differ in certain aspects of their respective 16S ribosomal sequences, giving rise to the possibility that the bryozoans are also different taxonomically *(95)*. Reports at the Society for Industrial Microbiology meetings in 2002 and 2003 demonstrated that Haygood and collaborators were pursuing the possibility of transferring this particular PKS fragment to other, more amenable microbes in order to further investigate the possibility of producing bryostatin by fermentative means. At the September 2003 International Marine Biotechnology Conference in Chiba, Japan, Haygood *(96)* reported on the current status of the PKS search, suggesting that this system resembled that reported by Piel *(97)* for the *Paederus* beetle's pseudomonal symbiont PKS, which produces pederine, in that there are no AT domains in the clusters. Further work is ongoing, utilizing "remote" AT domains from another organism.

4.4.2. Ecteinascidin 743

Antitumor activity of extracts from the ascidian, *Ecteinascidia turbinata*, had been reported as early as 1969 by Sigel et al. *(98)*, but it was not until 1990 that the structures of the active components were published simultaneously by Rinehart et al. *(99)* and Wright et al. *(100)*.

The structure of the most stable member of the series, known as Et743 from the absorption maximum, is shown below (Structure 52). The basic structure, without the exocyclic isoquinoline group, is a well-known base structure *(101)* of compounds from both microbes (saframycins, naphthyridinomycins, safracins, quinocarcins) and sponge-derived metabolites (renieramycins, with the latest variation, renieramycin J

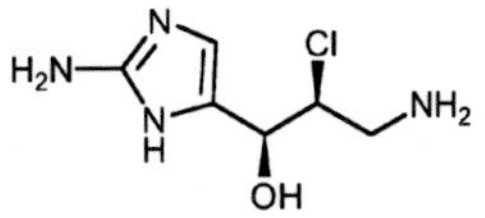

Structures 44–52

44 Girolline / Girodazole
45 Bengamide A; R = H 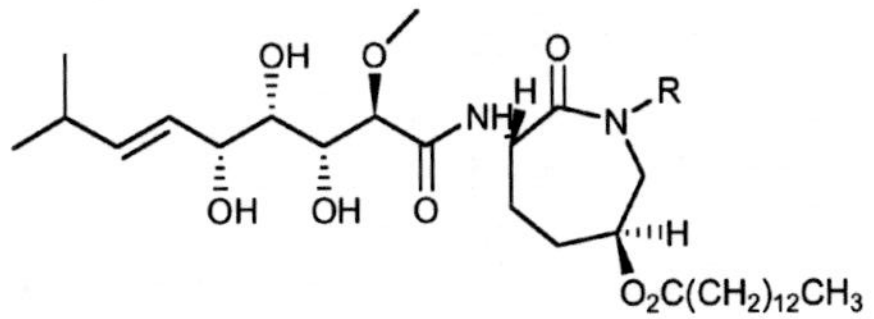
46 Bengamide B; R = CH3
47 LAF-389 / NVP-LAF-389 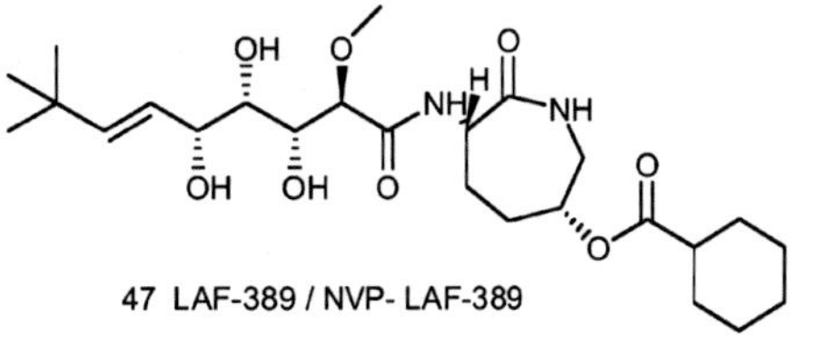
48 Cryptophycin 52; R₁ = R₂ = CH₃; R₃ = Cl 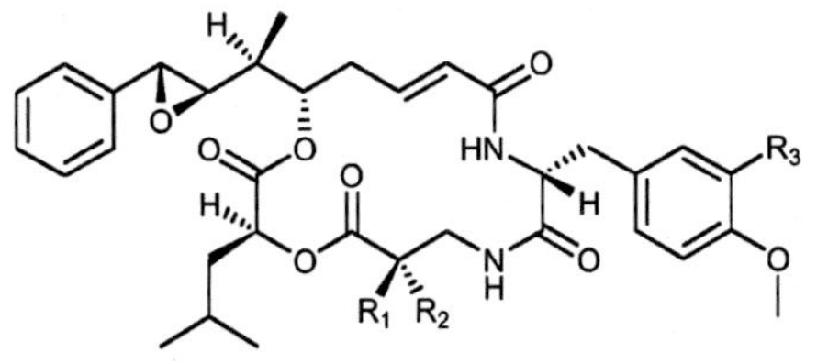
49 Arenastatin A; R₁ = R₂ = R₃ = H
50 Bryostatin 3 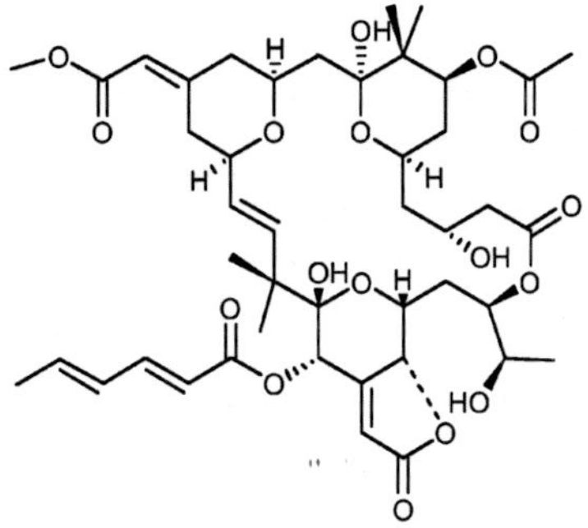
51 Bryostatin 1 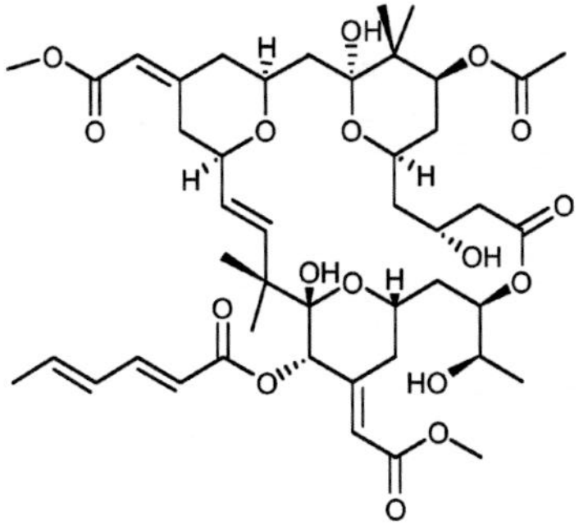
52 Ecteinascidin 743

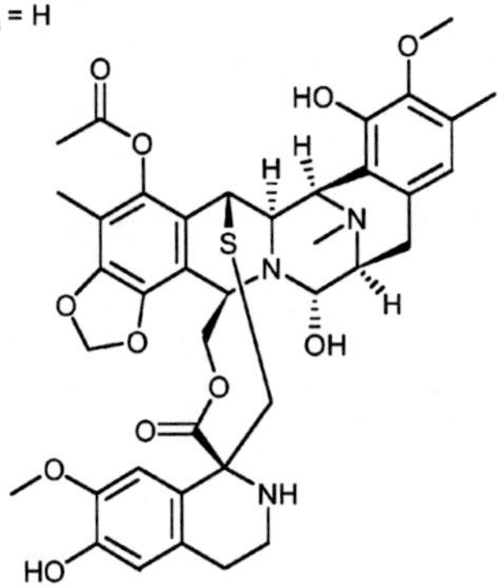

being reported by Oku et al. *(102)*), but the exocyclic substituent was novel, as was the bridging sulfur. The yield from natural sources was very low, and in order to provide enough material to perform basic studies as to the mechanism of action, and in vitro and in vivo animal studies, very significant amounts of the ascidian had to be collected from areas around the Caribbean. The compound was synthesized by Corey *(103)* and, as a result, he made a phthalimide-substituted version known as phthalascidin, which demonstrated reasonable activity in the regular test systems *(104)*.

Et743 was licensed by the University of Illinois to the Spanish company PharmaMar for subsequent development. PharmaMar chemists then performed an elegant semi-synthesis from the marine *Pseudomonas fluorescens* metabolite safracin B, which provided cGMP-grade Et743 from a 21-step synthetic process on a scale large enough to provide enough material for clinical trials *(105)*. This was made possible in spite of a low overall yield of 1.4% because the starting material could be obtained on a large scale by fermentation. The mechanisms of action of Et743 have been shown to include the following at physiologically relevant concentrations: effects on the transcription-coupled nucleotide excision repair process (TC-NER) *(106,107)*; and interaction between the Et743 DNA adduct and DNA transcription factors, in particular the NF-Y factor *(108)*.

The compound was placed into human clinical trials while these mechanisms were being worked out, and so far it has been administered to over a 1000 patients in phase I and phase II trials covering a variety of cancers. Results from the European phase I and pharmacokinetic trials were recently reported by Twelves et al. *(109)*. As a result of these trials, Et743 was preregistered in the EU and granted orphan drug status for sarcoma by the European Commission's Committee for Orphan Medicinal Products (COMP) of the European Agency for the Evaluation of Medicinal Products (EMEA). It was licenced to Johnson and Johnson (Ortho Biotech) in 2001 under the brand name Yondelis®, with the generic name of trabectedin. It is currently in a variety of phase II trials in the United States for the treatment of sarcoma, and is in phase II trials in Europe for this and ovarian cancer. Two reports on the phase II trials were recently published *(110,111)*, giving details of toxicities and response levels predominantly in sarcoma patients. Recently, the EU's Committee for Proprietary Medicinal Products (CPMP) recommended that marketing authorization for advanced soft-tissue sarcoma not be granted for the EU under a rapid orphan drug classification, and this decision was upheld on appeal in November 2003. However, the same agency granted orphan drug status for ovarian cancer in October 2003, so trials are still ongoing.

4.4.3. Aplidine

This compound, formally dehydrodidemnin B (Structure 53), was first reported in a patent application *(112)* in 1991 and then referred to in a paper from Rinehart's group in 1996 on the structure-activity relationships amongst the didemnins *(60)*. The antitumor potential was first reported by PharmaMar scientists *(113,114)* in 1996, and the total synthesis was reported in a patent application *(115)* in 2000.

The compound, generic name "aplidine or dehydrodidemnin B" and a trade name of Aplidin®, was entered into phase I clinical trials in 1999 under the auspices of PharmaMar in Canada, Spain, France, and the UK for treatment of both solid tumors and non-Hodgkin's lymphoma. These were successfully completed with over 200 patients, and demonstrated that a dosage of up to 5 mg.M² was well tolerated in either a 3-h or 24-h infusion every other week *(116)*. The dose-limiting toxicity was muscle pain that was responsive to either dose limitation or addition of carnitine. Interestingly, in the presence of carnitine, the maximum tolerated dose could be increased by 40% to 7 mg.M². Phase II clinical trials are now underway in Europe comparing the two dosage regimens in renal and colon carcinomas, together with an outpatient regimen, and, in July 2003, the European Commission's COMP/EMEA awarded orphan drug status for acute lymphoblastic leukemia (ALL) to Aplidin®.

The precise mechanism of action of this agent is not yet known, but it appears to block vascular endothelial growth factor (VEGF) secretion and block the corresponding VEGF-VEGF-Receptor-1 (also known as *flt-1*) autocrine loop in leukemic cells *(117)*. In addition, effects on *jnk*, *src*, and p38-MAPK, possibly mediated via glutathione depletion, were recently reported *(118)*, with the end result being induction of the apoptotic cascade in MDA-MB-231 cells at aplidine levels of 5 nM, which are below the blood levels achievable in humans. In these experiments, general caspase inhibitors decreased apoptotic efficacy by approx 50%, thus implicating at least two different mechanisms of apoptosis, one via caspases, the other not involving caspase activation.

What is very interesting is that the removal of two hydrogen atoms (i.e., conversion of the lactyl side chain to a pyruvyl side chain) appears to significantly alter the toxicity profile, as this is the only formal change in the molecule when compared to didemnin B, though the comments on dosage regimens under didemnin B (**Subheading 4.2.**) from Vera and Joullie should be taken into account when such comparisons are made in the future.

4.4.4. Halichondrin B (and Derivatives)

Halichondrin B (Structure 54) is one of a series of compounds originally isolated by Uemura et al. in 1985 from the Japanese sponge *Halichondria okadai (119,120)*. Subsequently, a number of sponges from other areas of the Pacific and Indian oceans were reported to contain one or more of these macrolides, including *Axinella* sp. from the Western Pacific *(121)*, *Phakellia carteri* from the Eastern Indian Ocean *(122)*, and *Lissodendoryx* sp. off the East Coast of South Island, New Zealand *(123)*.

In 1992, Kishi's group at Harvard, funded by the NCI, reported that they had successfully synthesized both halichondrin B and norhalichondrin B *(124)*. Working with the US division of the Japanese pharmaceutical company Eisai, Kishi's synthetic schemes were utilized to synthesize a large number of variants of halichondrin B, particularly smaller molecules that maintained biological activity but were intrinsically more chemically stable, due to the substitution of a ketone for the ester linkage in the macrolide ring. Two of these agents were subsequently evaluated by NCI in conjunction with the Eisai Research Institute in the United States, and one, originally ER-086526 (NSC 707389) and now renumbered E7389 (Structure 55), is now in phase I clinical trials in conjunction with the NCI.

Details of the biology and chemistry of this compound and other compounds in the series have recently been published by Eisai scientists, thus demonstrating the power of current synthetic chemistry when applied to a very potent marine-derived natural product *(125,126)*. Using the synthetic techniques described, enough cGMP material, produced by total synthesis, was provided to the NCI for the projected clinical trials.

4.4.5. Discodermolide

This polyhydroxylated lactone (Structure 56) was reported by the Harbor Branch Oceanographic Institution (HBOI) group in 1990 following isolation from the Caribbean sponge *Discodermia dissoluta*, originally collected at a depth of 133 m off the Bahamas *(127)*, with a revision to the stereochemistry being published the following year *(128)*. Originally, the compound was judged to be a new immunosuppressive agent and an incidental cytotoxin, but in 1996, it was reported that discodermolide bound to microtubules more potently than Taxol®, a discovery that confirmed *in silico* studies at the University of Pittsburgh *(129)*. The reports from HBOI led to synthesis of the (-) isomer by Schreiber's group *(130)*. Subsequently, other groups made the same isomer, and then in the late 1990s to 2003, Marshall and Johns *(131)*, Halstead *(132)*, Smith et al. *(133)*, and Paterson et al. *(134)* all reported syntheses that produced varied isomers in good yield. Kilogram amounts are now achievable by total synthesis *(135)*. In the interim, HBOI licensed the compound to Novartis as a preclinical candidate, and it is now in phase I clinical trials as a potential treatment against solid tumors.

The HBOI group are still discovering more derivatives of the natural product, and the same group isolated a marine macrolide, dictyostatin 1 (Structure 57), having some of the structural features of discodermolides, from a deep-water lithistid sponge, by following a tubulin interaction assay rather than relying on cytoxicity *(136)*. This material had previously been reported, but as a cytotoxin, from a Maldavian *Spongia* species by Pettit et al., and in a relatively recent paper, Horwitz's group demonstrated how discodermolide and Taxol may well fit into the same site on tubulin *(137)*.

4.4.6. Kahalalide F

This cyclic depsipeptide (Structure 58) was isolated from the Sacoglossan mollusk *Elysia rufescens*, following grazing by the mollusk on a green macroalga, *Bryopsis* sp. After isolation and identification, it was discovered that the depsipeptide also occurs in the alga, but on a wet-weight basis, the mollusk concentrates the depsipeptides significantly *(138,139)*.

The compound was licensed to PharmaMar by the University of Hawaii in the 1990s, and it entered preclinical testing. Its full MOA has not yet been determined, but it was known to target lysosomes *(140)*, thus suggesting selectivity for tumor cells with high lysosomal activity, such as prostate tumors. The compound was synthesized in a very efficient manner using solid-phase peptide techniques *(141)*, and entered phase I clinical trials in Europe in December 2000 for the treatment of androgen-independent prostate cancer. The compound has now entered phase II trials, predominantly as a treatment for prostate cancer. Recently, Suarez et al. *(142)* demonstrated that kahalalide F induces cell death via "oncosis" (progression of cellular processes leading to necrotic cell death), possibly initiated by lysosomal membrane depolarization.

4.4.7. Spisulosine

In 1999, workers from PharmaMar reported on the initial studies with a molecule known as ES-285 or spisulosine (Structure 59), which was isolated from the marine clam *Spisula polynyma*. They demonstrated that this compound causes a loss of actin stress fibers, which may well be a result of its resemblance to lysophosphatidic acid (LPA) and hence an interaction with the LPA receptor, which is known to modulate the levels of the *Rho* proteins *(143)*. This molecule is currently in phase I trials in Europe under the aegis of PharmaMar.

4.4.8. HTI-286 (Hemiasterlin Derivative)

Hemiasterlin (Structure 60) was originally reported by Kashman's group from the South African sponge *Hemiasterella minor*, an organism that also contained jaspamide and geodiamolide TA *(144)*. This was quickly followed by the report of a group of peptides that exhibited significant cytotoxic activities from a Papua New Guinea sponge that was originally classified as *Pseudoaxinyssa* sp. but, due to a taxonomic revision, the genus was changed to *Cymbastela*. This particular sponge produced a number of peptides, including geodiamolides A to F, hemiasterlin as described by Kashman, and two novel hemiasterlins, A and B, plus other geodiamolides and criamides *(145)*. In 1997, following testing of the hemiasterlin and the A and B derivatives in experiments to determine their mechanism of action, it was discovered that these agents interact with tubulin to produce microtubule depolymerization in a manner similar to that reported for nocodazole and vinblastine *(146)*. Further investigations by Hamel's group using hemiasterlin isolated at NCI *(147)* indicated that this peptide, together with cryptophycin 1 and dolastatin 10, inhibits tubulin assembly and probably is bound at what is being called the "peptide binding site" *(148)*.

In the intervening time, Andersen at the University of British Columbia (UBC) commenced a synthetic program in order to produce the original hemiasterlin using a synthetic scheme that would permit variations on the overall structure in order to determine SAR requirements *(149)*. The hemiasterlins, including the analogs made by Andersen's group (which included HTI-286 (Structure 61) *(150)*, were licensed by the UBC to Wyeth for development as part of an NCDDG program, of which Andersen was a member. Significant amounts of synthetic work were performed around these structures, but the original agent was still superior and is now in phase I clinical trials and is scheduled for phase II *(151)*.

4.4.9. KRN-7000

In 1993, workers at the Kirin Brewery, in conjunction with Higa at the University of the Ryukyus, reported the first isolation of α-galactosylceramides from natural sources, the marine sponge *Agelas mauritianus*. Very interestingly, these compounds, the agelasphins (Structure 62), demonstrated antitumor and potential immunostimulatory activities *(152,153)*. Following these results, and demonstration that these molecules were potent in vivo active agents against the murine B16 melanoma, various derivatives were made, culminating in the production of KRN-7000 (Structure 63) *(154,155)*. This compound entered phase I clinical trials in both Asia and Europe in 2001 for cancer immunotherapy.

4.4.10. Squalamine

This compound, isolated from the common dogfish shark *Squalus acanthias*, collected off the New England coast, was originally reported by Zasloff's group and collaborators from the University of Pennsylvania in 1993, and was shown to be a fairly simple aminosterol (Structure 64) with broad-spectrum antibiotic activity *(156)*. The compound was licensed to Maganin Pharmaceuticals (now Genaera) for development, entered phase I, and now has progressed into phase II clinical trials for non-responding solid tumors, as part of a combination with standard agents, and as primary treatment for advanced ovarian cancer. It went into phase I studies for prostate cancer in 2003.

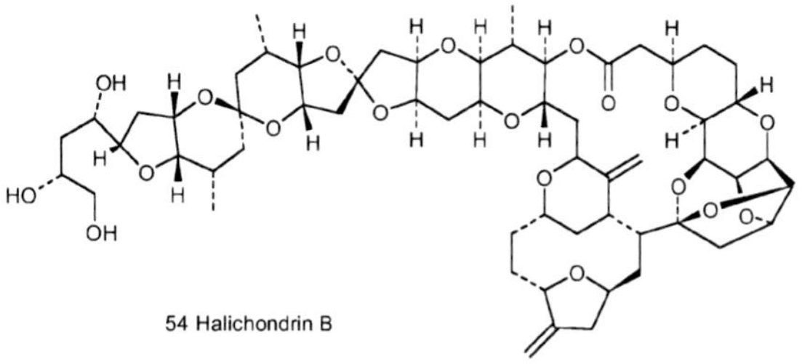

Structures 54-61

54 Halichondrin B
55 E7389 / NSC 707389 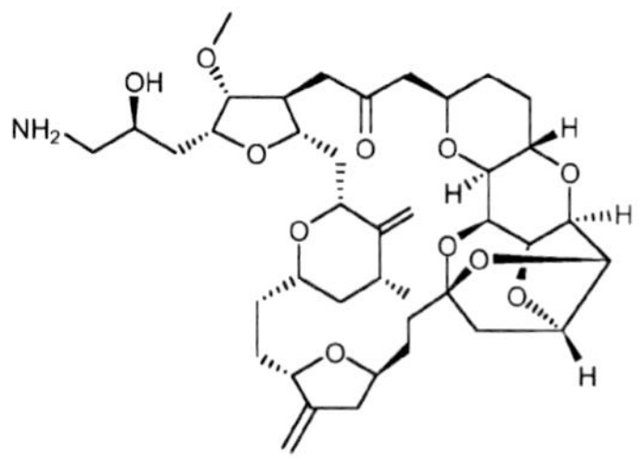
56 Discodermolide 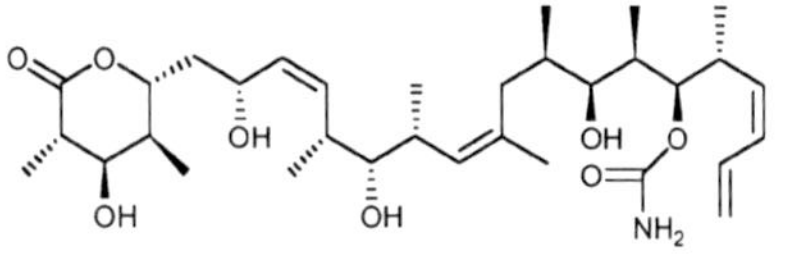
57 Dictyostatin 1 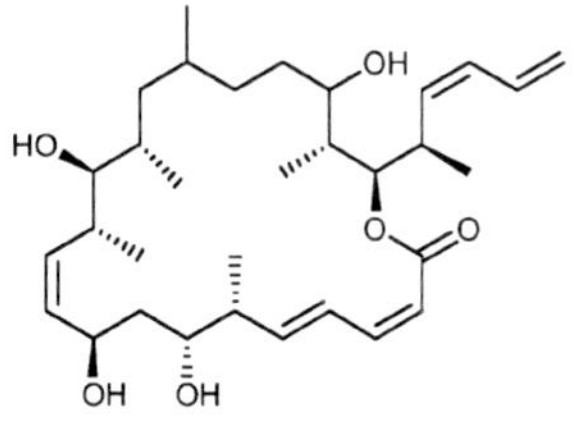
58 Kahalalide F 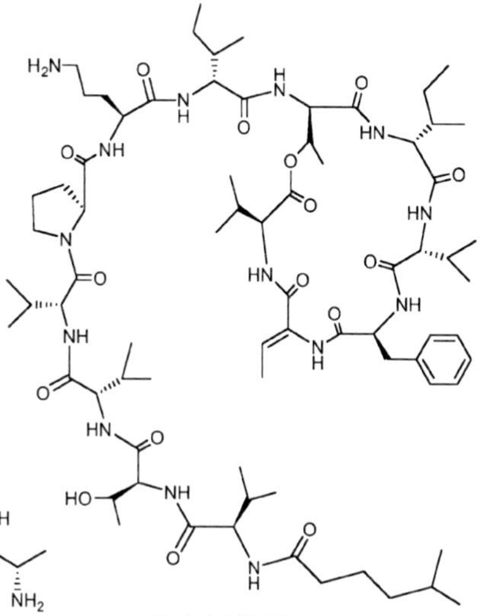
59 Spisulosine / ES-285 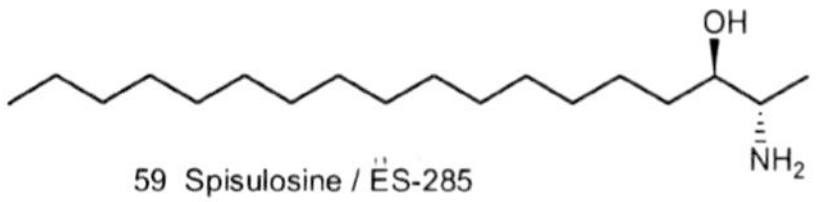
60 Hemiasterlin 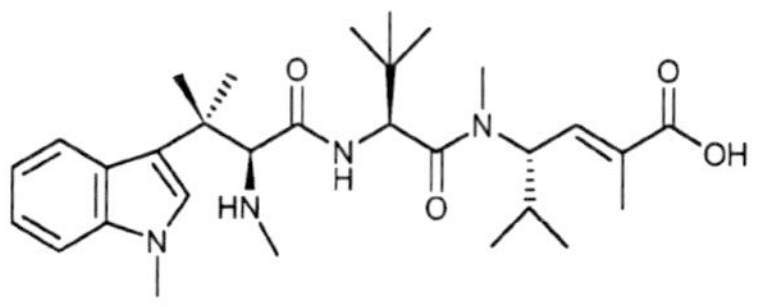
61 HTI-286 / SPA 110

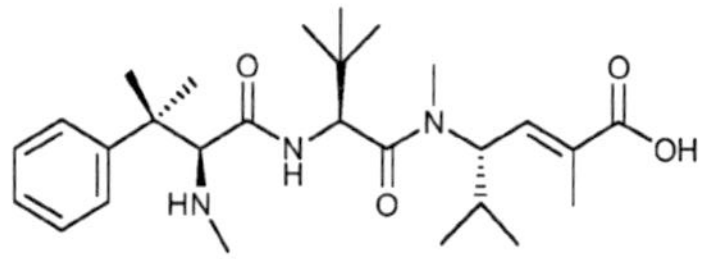

There are reports in the literature that it may well have antiangiogenic activity under some conditions *(157)*.

4.4.11. AE-941 or Neovastat

This is not a true compound, but is probably best thought of as a "defined mixture" in that it is a standardized liquid extract comprising the <500-kDa fraction from shark cartilage. This material is made under cGMP conditions from taxonomically identified shark species harvested under sustainable conditions, and has quality controls that permitted both the FDA and the Canadian equivalent to give approval for clinical trials. The methods of preparation have been published in reasonable detail by Sorbera et al. *(158)*, and the first formal report of angiostatic and antitumor activity was given in 1997 at the American Association for Cancer Research meeting *(159)*.

The preparation is in many phase III clinical trials in Canada, Europe, and the United States, with details of the pivotal studies in phase III renal carcinomas being given in two recent reviews *(160,161)*. The initial clinical results from these studies will be of great interest.

4.5. Selected Antitumor Compounds From Marine Sources in Preclinical Status

4.5.1. Tubulin-Interactive Agents

These compounds include materials that interact with tubulin, both in a manner similar to that of paclitaxel and to the *vinca* alkaloids, with the compound laulimalide (Structure 65) being a very good example as a microtubule-stabilizing agent. Hamel *(71)* indicated that this agent might well bind at a site different from the taxanes, though it is possible that it might also be binding to unpolymerized tubulin or to aberrant polymeric tubulin. Over 10 synthetic routes to laulimalide have been published, plus many more that give methods of synthesis of "subassemblies" of the overall molecule *(162,163)*.

Other agents that intereact with tubulin include curacin A (Structure 66) from the cyanobacterium *Lyngbya majuscula (164)*, vitilevuamide (Structure 67) from the ascidians *Didemnum cuculiferum* and *Polysyncraton lithostrotum (165)* (this compound should be compared to those given in the recent review of bioactive peptides by Janin *(166)*), and diazonamide A (Structure 68) from the ascidian *Diazona angulata (167)* (and its subsequent reassessment and synthesis of a revised structure that has biological activity by Cruz-Monserrate et al. *[168]*).

The sarcodictyins (Structure 69), from the Mediterranean corals *Sarcodictyon roseum* and *Eleutherobia aurea*, were reported by Pietra's group *(169,170)* and bear a close structural resemblance to eleutherobin (Structure 70), which was originally isolated by Fenical's group *(171)* from the Australian octacoral *Eleutherobia* sp. Eleutherobin was reisolated from the Caribbean octacoral *Erythropodium caribaeorum* by Andersen et al. *(172)* and then reported from whole-organism aquaculture by Andersen et al. *(173)*.

Peloruside A (Structure 71), from the New Zealand marine sponge *Mycale hentscheli (174)*, is an interesting case, as the biological activity of this compound induces apoptosis following a G_2-M arrest; its relatively simple structure may well lend itself to synthetic modifications *(175)*. Also, as noted by Ghosh and Kim *(176)*, it has structural

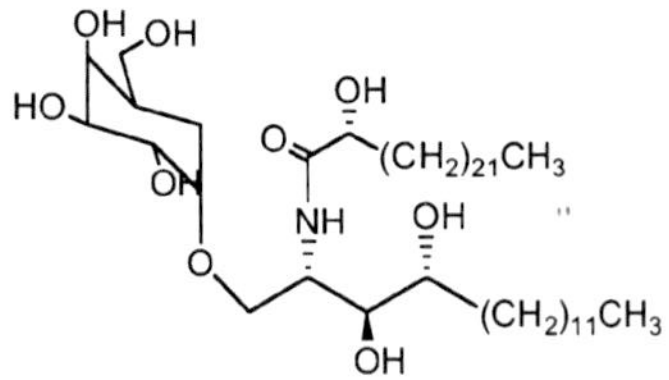

Structures 62–68

62 Agelasphin 7a 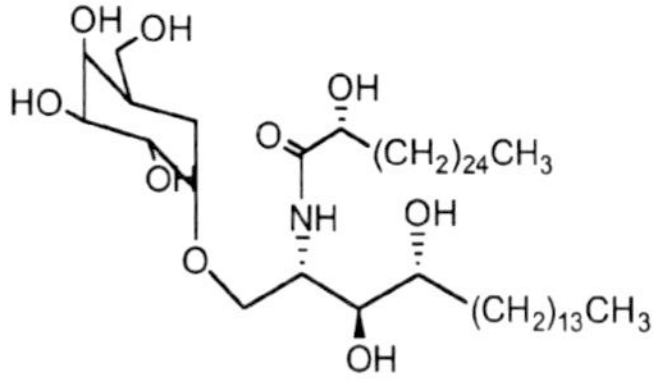
63 KRN 7000 / AGL582 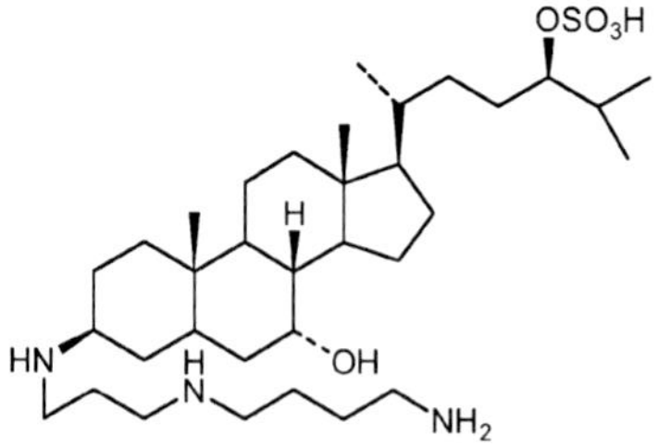
64 Squalamine 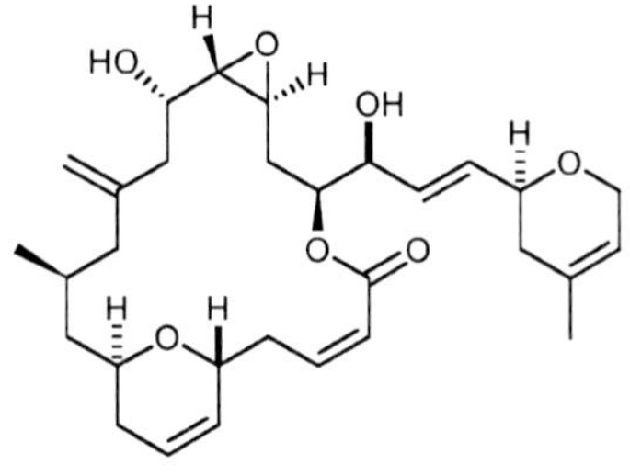
65 Laulimalide / Fijianolide B 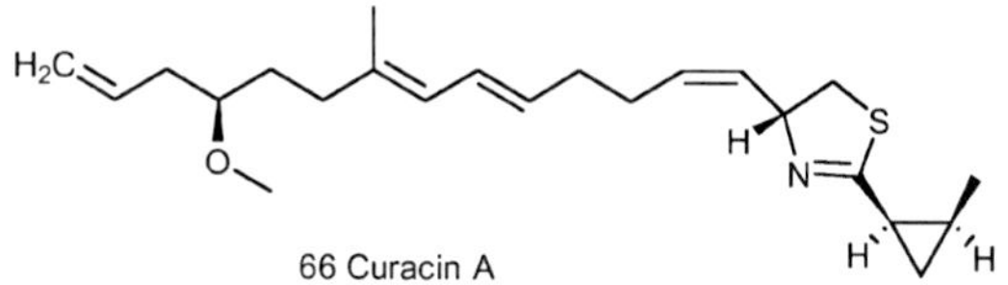
66 Curacin A 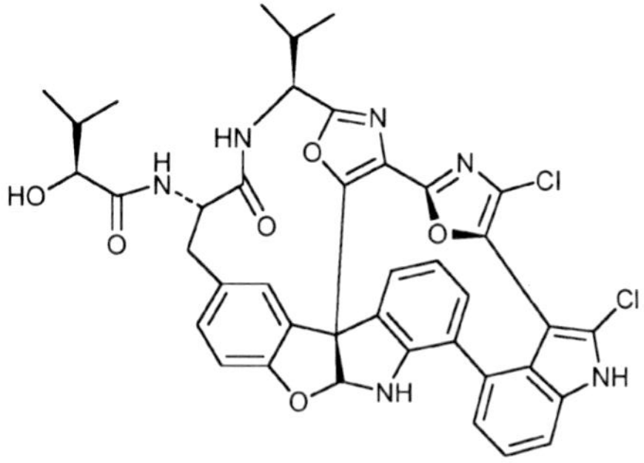
67 Vitilevuamide
68 Diazonamide (Revized)

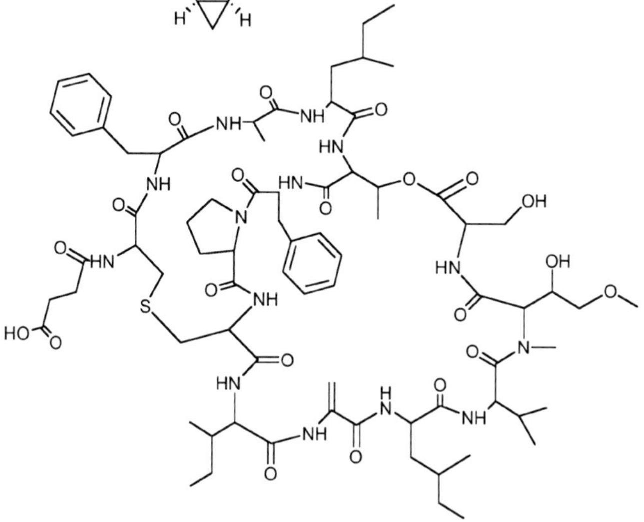

157

similarities to the epothilones. Their partial synthesis may significantly aid in the production of enough material to further evaluate the full potential of this compound.

4.5.2. Vacuolar-ATPase Inhibitors

Vo-ATPases occur throughout eukaryotes, and their prime function is to pump hydrogen ions from one side of a membrane to the other. These particular enzymes perform this function within vacuoles in the cell and are dependent upon ATP for the necessary energy to perform the function.

In 1997, Boyd's group at the NCI reported the isolation of two closely related and very cytotoxic novel macrolide structures, the salicylihalimides A and B (Structures 72, 73), from the Western Australian marine sponge *Haliclona* sp., with GI_{50} values below 1 n*M* for sensitive melanoma lines in the NCI 60 cell line screen *(177)*. These agents did not resemble any profiles from the NCI's 60-cell-line dataset of standard agent responses, but did show similar patterns to the bafilomycin and concanamycin derivatives. The latter compounds are known to exhibit Vo-ATPase inhibitory activity but are too toxic for human use, though bafilomycins have been used for plant protection.

Subsequent work from the same laboratory expanded the range of the structures to include another marine-derived product series, the lobatamides, and both series were shown to be specific for the higher eukaryotic Vo-ATPases but not the fungal equivalents. Subsequently, similar molecules have been isolated from bacteria, fungi, and myxobacteria. Although syntheses have been published, to date, no formal in vivo assays of the original compounds have been performed due to a lack of access to the natural source from Western Australia.

This class of compounds is of interest not only as antitumor agents but also in bone resorption; thus, they may well have utility in osteoporosis. Two recent reviews have been published on these agents *(178,179)*.

5. The Actual Source(s) of Marine Products?

The continuing question of whether symbiotic (and/or commensal) microbes are the actual producers of cytotoxic and other metabolites is being investigated by a variety of groups, and as information evolves, the probability increases that a very significant number of polyketide-derived compounds and non-ribosomally produced peptides are of microbial origin. A recent review by Janin *(166)* shows the structures of a large number of such molecules. A very recent example is the one given by Piel on the production of pederine by the pseudomonal symbiont from *Paederus* beetles referred to earlier (**Subheading 4.4.1.**). In a presentation at the Society for Microbiology meeting in August 2003, Piel gave further information that demonstrated the formation of a very close structural relative of the marine sponge cytotoxins, the onnamides, by this terrestrially derived PKS system.

As can be seen by inspection of the structure of Et743, it bears a close similarity to the antibiotics saframycin B, naphthridinomycin, saframycin, and to sponge metabolites of the renieriamycin class. There are many other agents from marine sources, such as the mycalamides and onnamides, that have structures or partial structures similar to molecules isolated from terrestrial or other marine phyla, and it was posited from such circumstantial evidence that a significant number of ostensibly marine-derived com-

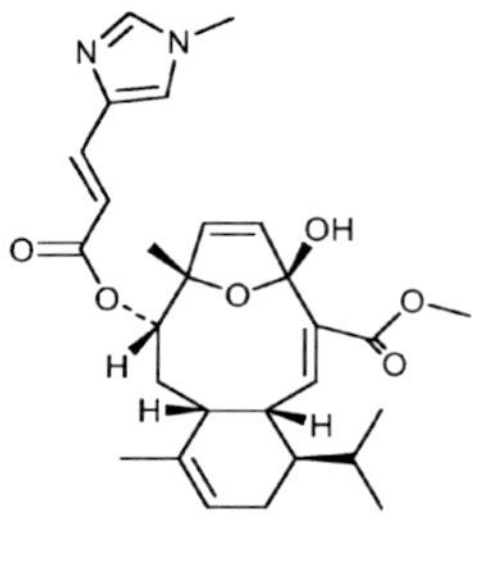

Anticancer Drugs From Natural Sources

69 Sarcodictyin

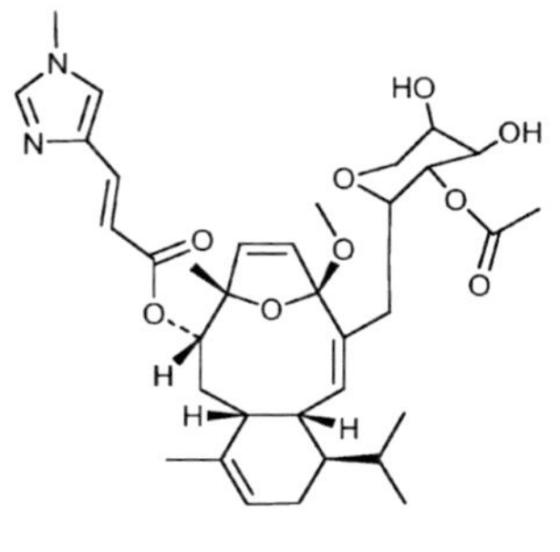

70 Eleutherobin

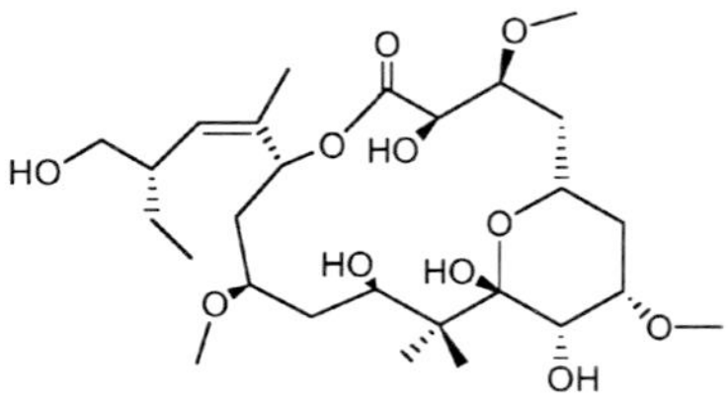

71 Peloruside A

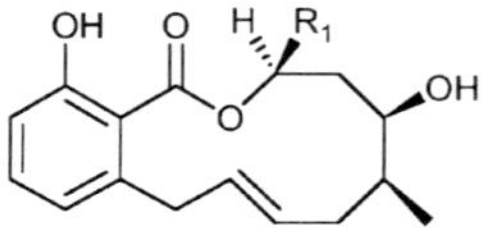

72 Salicylihalimide A; $R_1 =$

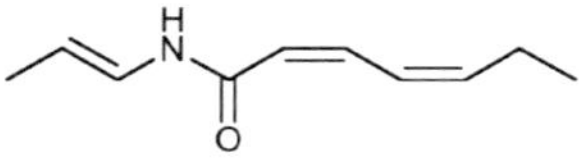

73 Salicylihalimide B; $R_1 =$

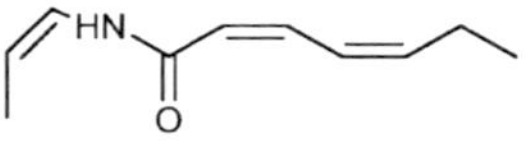

Structures 69–73

pounds, particularly from the *Porifera*, were in fact derived from commensal and/or symbiotic microbes.

However, it was not until the autumn of 2003 that there was other than circumstantial evidence of production, when Hill (University of Maryland) and Hamann (University of Mississippi) reported at the 6th International Marine Biotechnology Conference on their work with a purified commensal streptomycete isolated from a deep-water Indonesian sponge. They demonstrated that the microbe, when grown in a laboratory setting, with some media but not others, produced manzamine A and 8-hydroxymanzamine A, compounds that were isolated from the sponge itself following wild collection *(180,181)*.

6. Conclusions

We have shown that natural products from all sources have had and will continue to have major influences on the identification and production of agents against cancers of all types. Our experience implies that the novel chemotypes directed against cancer have tended to come from microbial and marine sources rather than from vascular plants. This may simply be due to the centuries of exploration of the plant kingdom compared to the other two potential sources of natural products. However, by using the power of directed combinatorial library syntheses based on natural products and/or natural-product-like molecules, the potential to produce agents that have been modified in order to increase their utility as drug candidates has been and continues to be demonstrated. Early examples are the modification of the eleutherobin/sarcodictyin skeletons by Nicolaou's group *(182)*, the potential for production of novel epothilone-based structures, also by Nicolaou's group *(183)*, and the synthesis of natural-product-like libraries by Schreiber's group *(184,185)*.

If one couples to these chemical techniques, the novel structures that are now being reported from total extraction of DNA from soils *(186)*, and the amplification of polyketide/nonribosomal peptide synthetases from a large number of different microbial families, then the horizon is limitless.

Thus, the final point of emphasis is that the drug discovery research community should continue to investigate Nature's resources for leads to novel chemotypes that can then be modified and/or extended using the power of all types of modern chemical and biochemical techniques.

References

1. Kleeff J, Kornmann M, Sawhney H, Korc M. Actinomycin D induces apoptosis and inhibits growth of pancreatic cancer cells. Int J Cancer 2000;86:399–407.
2. Zisman A, Ng C-P, Pantuck AJ, Bonavida B, Belldegrun AS. Actinomycin D and gemcitabine synergistically sensitize androgen-independent prostate cancer cells to apo2l/trail-mediated apoptosis. J Immunotherapy 2001;24:459–471.
3. Bock J, Sazabo I, Jekle A, Gulbins E. Actinomycin D-induced apoptosis involves the potassium channel KV1.3. Biochem Biophys Res Comm 2002;295:526–531.
4. Elliott MJ, Baker JD, Dong YB, Yang HL, Gleason Jr. JF, McMasters KM. Inhibition of cyclin a kinase activity in E2F-1 chemogene therapy of colon cancer. Tumor Biol 2002;23:324–336.
5. Zhao Y, Brown TL, Kohler H, Muller S. Mts-conjugated-antiactive caspase 3 antibodies inhibit actinomycin D-induced apoptosis. Apoptosis 2003;8:631–637.
6. Duncan R. Polymer conjugates for tumour targeting and intracytoplasmic delivery. The EPR effect as a common gateway? Pharm Sci Technol Today 1999;2:441–449.
7. Lee MD, Ellestad GA, Borders DB. Calicheamicins: discovery, structure, chemistry, and interactions with DNA. Acc Chem Res 1991;24:235–243.
8. Watanabe CMH, Supekova L, Schultz PG. Transcriptional effects of the potent enediyne anticancer agent Calicheamicin γI. Chem Biol 2002;9:245–251.
9. Hamann RR, Hinman LM, Hollander I, et al. Gemtuzumab ozogamicin, a potent and selective anti-CD33 antibody-calicheamicin conjugate for treatment of acute myeloid leukemia. Bioconjugate Chem 2002;13:47–58.
10. Day BW. Mutants yield a phamacophore model for the tubulin-paclitaxel binding site. TIPS 2000;21:321–323.
11. Giannakakou P, Gussio R, Nogales E, et al. Common pharmacophore for epothilone and taxanes: molecular basis for drug resistance conferred by tubulin mutations in human cancer cells. Proc Natl Acad Sci USA 2000;97:2904–2909.
12. Gussio R, Fojo TA, Giannakakou P. Reply. TIPS 2000;21:323–324.
13. Kowalski RJ, Giannakakou P, Hamel E. Activities of the microtubule-stabilizing agents epothilones A and B with purified tubulin and in cells resistant to paclitaxel (Taxol®). J Biol Chem 1997;272:2534–2451.
14. Frykman S, Tsuruta H, Lau J, et al. Modulation of epothilone analog production through media design. J Ind Microbiol Biotech 2002;28:17–20.
15. Nicolaou KC, Ritzen A, Namoto K. Recent developments in the chemistry, biology and medicine of the epothilones. Chemical Comm 2001:1523–1535.
16. Wartmann M, Altmann KH. The biology and medicinal chemistry of epothilones. Curr Med Chem Anti-Cancer Agents 2002;2:123–148.

17. Omura S, Sasaki Y, Iwai Y, Takeshima H. Staurosporine, a potentially important gift from a microorganism. J Antibiot 1995;48:535–548.
18. Jackson JR, Gilmartin A, Imburgia C, Winkler JD, Marshall LA, Roshak A. An indolocarbazole inhibitor of human checkpoint kinase (chk1) abrogates cell cycle arrest caused by DNA damage. Cancer Res 2000;60:566–572.
19. Shao RG, Cao CX, Shimizu T, O'Connor PM, Kohn KW, Pommier Y. Abrogation of an S-phase checkpoint and potentiation of camptothecin cytotoxicity by 7-hydroxy-staurosporin (UCN-01) in human cancer cell lines, possibly influenced by p53 function. Cancer Res 1997;57:4029–4035.
20. Graves PR, Yu L, Schwarz JK, et al. The chk1 protein kinase and the cdc25c regulatory pathways are targets of the anticancer agent UCN-01. J Biol Chem 2000;275:5600–5605.
21. Newman DJ, Cragg GM, Holbeck S, Sausville EA. Natural products and derivatives as leads to cell cycle pathway targets in cancer chemotherapy. Current Can Drug Targ 2002;2:279–308.
22. Dancey J, Sausville EA. Issues and progress with protein kinase inhibitors for cancer treatment. Nature Revs Drug Discov 2003;2:296–313.
23. Prudhomme M. Recent developments of rebeccamycin analogues as topoisomerase I inhibitors and antitumor agents. Curr Med Chem 2000;7:1189–1212.
24. Bailly C, Riou J-F, Colson P, Houssier C, Rodrigues-Pereira E, Prudhomme M. DNA cleavage by topoisomerase I in the presence of indolocarbazole derivatives of rebeccamycin. Biochemistry 1997;36:3917–3929.
25. Rossi F, Labourier E, Forne T, et al. Specific phosphorylation of SR proteins by mammalian DNA topoisomerase I. Nature 1996;381:80–82.
26. Tazi J, Rossi F, Labourier E, Gallouzi I, Brunel C, Antoine E. DNA topoisomerase I: a custom-officer at DNA-RNA worlds border? J Mol Med 1997;75:786–800.
27. Kamal A, Thao L, Sensintaffar J, et al. A high-affinity conformation of hsp90 confers tumour selectivity on hsp90 inhibitors. Nature 2003;425:407–410.
28. Neckers L, Ivy SP. Heat shock protein 90. Current Opin Oncology 2003;15:419–424.
29. Beliakoff J, Bagatell R, Paine-Murrieta G, Taylor CW, Lykkesfeldt AE, Whitesell L. Hormone-refractory breast cancer remains sensitive to the antitumor activity of heat shock protein 90 inhibitors. Clin Cancer Res 2003;9:4961–4971.
30. Hartwell JL, Plants used against cancer. Lawrence, MA: Quarterman, 1982.
31. Cragg GM, Boyd MR, Cardellina II JH, Newman DJ, Snader KM, McCloud TG. Ethnobotany and drug discovery: the experience of the US National Cancer Institute. In: Chadwick DJ, Marsh J (eds), Ethnobotany and the Search for New Drugs: CIBA Foundation Symposium, Chichester, UK: Wiley & Sons: 1994:178–196.
32. Newman DJ, Cragg GM, Snader KM. The influence of natural products upon drug discovery. Natural Prod Rep 2000;17:215–234.
33. Kingston DGI. Taxol, a molecule for all seasons. Chemical Comm 2001:867–880.
34. Cragg GM, Schepartz SA, Suffness M, Grever MR. The taxol supply crisis. New NCI policies for handling the large-scale production of novel natural product anticancer and anti-HIV agents. J Nat Prod 1993;56:1657–1668.
35. Kapoor LD, CRC Handbook of Ayurvedic Medicinal Plants. Boca Raton, Florida: CRC Press; 1990.
36. Cortes JE, Pazdur R. Docetaxel. J Clin Oncol 1995;13:2643–2655.
37. He L, Orr GA, Horwitz SB. Novel molecules that interact with microtubules and have functional activity similar to taxol. Drug Discov Today 2001;6:1153–1164.
38. Jordan MA. Mechanism of action of antitumor drugs that interact with microtubules and tubulin. Curr Med Chem-Anti-Cancer Agents 2002;2:1–17.
39. Potmeisel M, Pinedo H, Camptothecins: New Anticancer Agents. Boca Raton, Florida: CRC Press; 1995.

40. Lee K-H. Current developments in the discovery and design of new drug candidates from plant natural product leads. J Nat Prod 2004;67:273–283.
41. Ranelleti FO, Ricci R, Larocca LM, et al. Growth inhibitory effect of quercetin and presence of type-II estrogen-binding sites in human colon-cancer cell lines and primary colorectal tumors. Int J Cancer 1992;50:486–492.
42. Walker DH. Small-molecule inhibitors of cyclin-dependent kinases: molecular tools and potential therapeutics. Curr Top Microbiol Immunol 1998;227:149–165.
43. Sielecki T, Boylan JF, Benfield PA, Trainor GL. Cyclin-dependent kinase inhibitors: useful targets in cell cycle regulation. J Med Chem 2000;43:1–18.
44. Losiewicz MD, Carlson BA, Kaur G, Sausville EA, Worland PJ. Potent inhibition of cdc2 kinase activity by the flavanoid, l86-8275. Biochem Biophys Res Comm 1994;201:589–595.
45. Czech J, Hoffman D, Naik R, Sedlacek H. Anti-tumoral activity of flavone l86-8275. Int J Oncol 1995;6:31–36.
46. Kaubisch A, Schwartz GK. Cyclin-dependent kinase and protein kinase C inhibitors: a novel class of antineoplastic agents in clinical development. Cancer J 2000;6:192–212.
47. Senderowicz AM. Small-molecule cyclin-dependent kinase modulators. Oncogene 2003;22:6609–6620.
48. Dai Y, Grant S. Cyclin-dependent kinase inhibitors. Curr Opin Pharmacol 2003;3:362–370.
49. Holwell SE, Cooper PA, Grosios K, et al. Combretastatin A-1 phosphate, a novel tubulin-binding agent with in vivo anti-vascular effects in experimental tumors. Anticancer Research 2002;22:707–712.
50. Li Q, Sham HL. Discovery and development of antimitotic agents that inhibit tubulin polymerization for the treatment of cancer. Expert Opin Ther Patents 2002;12:1663–1702.
51. Cragg GM, Newman DJ. A tale of two tumor targets: topoisomerase I and tubulin. The Wall and Wani contribution to cancer chemotherapy. J Nat Prod 2004;67(2):232–244.
52. Bergmann W, Feeney RJ. The isolation of a new thymine pentoside from sponges. J Am Chem Soc 1950;72:2809–2810.
53. Bergmann W, Feeney RJ. Contributions to the study of marine products. XXXII. The nucleosides of sponges. J Org Chem 1951;16:981–987.
54. Bergmann W, Burke DC. Contributions to the study of marine products. XI. The nucleosides of sponges. IV. Spongosine. J Org Chem 1956;21:226–228.
55. Suckling ƐJ. Chemical approaches to the discovery of new drugs. Sci Prog Edin 1991;75:323–360.
56. Lee WW, Benitez A, Goodman L, Baker BR. Potential anticancer agents. Xl. Synthesis of the β-anomer of 9-(d-arabinofuranosyl)-adenine. J Am Chem Soc 1960;82:2648–2649.
57. Davis P/ Fermentation of 9-(β-d-arabinofuranosyl)adenine. US Patent 1969, 1159290, 26JUL69.
58. Cimino G, De Rosa S, De Stefano S. Antiviral agents from a gorgonian, *Eunicella cavolini*. Experientia 1984;40:339–400.
59. Rinehart K, Gloer JB, Cook JC. Structures of the didemnins, antiviral and cytotoxic depsipeptides from a Caribbean tunicate. J Am Chem Soc 1981;103:1857–1859.
60. Sakai R, Rinehart KL, Kishore V, et al. Structure-activity relationships of the didemnins. J Med Chem 1996;39:2819–2834.
61. Crews CM, Collins JL, Lane WS. GTP-dependent binding of the antiproliferative agent didemnin to the elongation factor 1a. J Biol Chem 1994;269:15,411–15,414.
62. Meng L, Sin N, Crews CM. The antiproliferative agent didemnin B uncompetitively inhibits palmitoyl protein thioesterase. Biochemistry 1998;37:10,488–10,492.
63. Johnson KL, Lawen A. Rapamycin inhibits didemnin B-induced apoptosis in human HL-60 cells: evidence for the possible involvement of FK506-binding protein 25. Immunol Cell Biol 1999;77:242–248.

64. Vera M, Joullie MM. Natural products as probes of cell biology: 20 years of didemnin research. Med Res Revs 2002;22:102–145.
65. Rinehart K. Antitumor compounds from tunicates. Med Res Rev 2000;20:1–27.
66. Pettit GR. The dolastatins. Fortschr Chem Org Naturst 1997;70:1–79.
67. Bai R, Pettit GR, Hamel E. Dolastatin 10, a powerful cytostatic peptide derived from a marine animal: inhibition of tubulin polymerization mediated through the *vinca* alkaloid binding domain. Biochem Pharmacol 1990;39:1941–1949.
68. Bai R, Friedman SJ, Pettit GR, Hamel E. Dolastatin 15, a potent antimitotic depsipeptide derived from *Dolabella auricularia*: interactions with tubulin and effects on cellular microtubules. Biochem Pharmacol 1992;43:2637–2645.
69. Pettit GR, Kamano Y, Herald CL, et al. The isolation and structure of a remarkable marine animal antineoplastic constituent: dolastatin 10. J Am Chem Soc 1987;109:6883–6885.
70. Pettit GR, Kamano Y, Dufresne C, Cerny RL, Herald CL, Schmidt JM. Isolation and structure of the cytostatic linear depsipeptide dolastatin 15. J Org Chem 1989;54:6005–6006.
71. Hamel E. Interactions of antimitotic peptides and depsipeptides with tubulin. Biopolymers 2002;66:142–160.
72. Bai R, Pettit GR, Hamel E. Binding of dolastatin 10 to tubulin at a distinct site for peptide antimitotic agents near the exchangeable nucleotide and *vinca* alkaloid sites. J Biol Chem 1990;265:17,141–17,149.
73. Vaishampayan H, Glode M, Du W, et al. Phase II study of dolastatin-10 in patients with hormone-refractory metastatic prostate adenocarcinoma. Clin Canc Res 2000;6:4205–4208.
74. Margolin K, Longmate J, Synold TW, et al. Dolastatin-10 in metastatic melanoma: a phase II and pharmacokinetic trial of the California Cancer Consortium. Invest New Drugs 2001;19: 335–340.
75. Cruz-Monserrate Z, Mullaney JT, Harran PG, Pettit GR. Dolastatin 15 binds in the *vinca* domain of tubulin as demonstrated by Hummel-Dreyer chromatography. Eur J Biochem 2003;270:3822–3828.
76. Harrigan GG, Luesch H, Yoshida WY, et al. Symplostatin 1: a dolastatin 10 analogue from the marine cyanobacterium *Symploca hynoides*. J Nat Prod 1998;61:1075–1077.
77. Luesch H, Moore RE, Paul VJ, Mooberry SL, Corbett TH. Isolation of dolastatin 10 from the marine cyanobacterium *Symploca* species VP642 and total stereochemistry and biological evaluation of its analogue symplostatin 1. J Nat Prod 2001;64:907–910.
78. Mooberry SL, Leal RM, Tinsley TL, Luesch H, Moore RE, Corbett TH. The molecular pharmacology of symplostatin 1: a new antimitotic dolastatin 10 analog. Int J Cancer 2003;104:512–521.
79. Ahond A, Bedoya Zurita M, Colin M, et al. La girolline, novelle substance antitumorale extraite de l'eponge *Pseudaxinyssa cantharella*. C R Acad Sci Paris 1988;307:145–148.
80. Quinoa E, Adamczeski M, Crews P, Bakus GJ. Bengamides, heterocyclic anthelminthics from a *Jaspidae* marine sponge. J Org Chem 1986;51:4494–4497.
81. Adamczeski M, Quinoa E, Crews P. Novel sponge-derived amino acids. 5. Structures, stereochemistry, and synthesis of several new heterocycles. J Am Chem Soc 1989;111:647–654.
82. Adamczeski M, Quinoa E, Crews P. Novel sponge-derived amino acids. 11. The entire absolute stereochemistry of the bengamides. J Org Chem 1990;55:240–242.
83. Thale Z, Kinder FR, Bair KW, et al. Bengamides revisited: new structures and antitumor studies. J Org Chem 2001;66:1733–1741.
84. Trimurtulu G, Ohtani I, Patterson GML, et al. Total structures of cryptophycins, potent antitumor depsipeptides from the blue-green alga *Nostoc* sp. strain GSV 224. J Am Chem Soc 1994;116:4729–4737.
85. Sessa C, Weigang-Kohler K, Pagani O, et al. Phase I and pharmacological studies of the cryptophycin analogue LY355703 administered on a single or weekly schedule. Eur J Cancer 2002;38:2388–2396.

86. Shih C, Teicher BA. Cryptophycins: a novel class of potent antimitotic antitumor depsipeptides. Curr Pharm Design 2001;7:1259–1276.
87. Kobayashi M, Aoki S, Ohyabu N, Kurosu M, Wang W, Kitagawa I. Arenastatin A, a potent cytotoxic depsipeptide from the Okinawan marine sponge *Dysidea arenaria*. Tetrahedron Lett 1994;35:7969–7972.
88. Golakoti T, Ogino J, Heltzel CE, et al. Structure determination, conformational analysis, chemical stability studies, and antitumor evaluation of the cryptophycins. Isolation of 18 new analogs from *Nostoc* sp. strain GSV 224. J Am Chem Soc 1995;117:12,030–12,049.
89. Koiso Y, Morita K, Kobayashi M, Wang W, Ohyabu N, Iwasaki S. Effects of arenastatin A and its synthetic analogs on microtubule assembly. Chem-Biol Interact 1996;102:183–191.
90. Morita K, Koiso Y, Hashimoto Y, et al. Interaction of arenastatin A with porcine brain tubulin. Biol Pharm Bull 1997;1997:171–174.
91. Newman DJ. Bryostatin—from bryozoan to cancer drug. In: Gordon DP, Smith AM, Grant-Mackie JA (eds), Bryozoans in Space and Time, Wellington: NIWA, 1996:9–17.
92. Pettit GR, Herald CL, Hogan F. Biosynthetic products for anticancer drug design and treatment: bryostatins. In: Baguley BC, Kerr DJ (eds), Anticancer Drug Development, Academic, San Diego: 2002;203–235.
93. Clamp A, Jayson GC. The clinical development of the bryostatins. Anti-Cancer Drugs 2002;13:673–683.
94. Davidson SK, Allen SW, Lim GE, Anderson CM, Haygood MG. Evidence for the biosynthesis of bryostatins by the bacterial symbiont "*candidatus* endobugula sertula" of the bryozoan *Bugula neritina*. Appl Environ Microbiol 2001;67:4531–4537.
95. Davidson SK, Haygood MG. Identification of sibling species of the bryozoan *Bugula neritina* that produce different anticancer bryostatins and harbor distinct strains of the bacterial symbiont "*candidatus* endobugula sertula". Biol Bull 1999;196:273–280.
96. Haygood MG. The role of a bacterial symbiont in the biosynthesis of bryostatins in the marine bryozoan *Bugula neritina*. Abs Pap 6th Int Mar Biotech Conf 2003:Abs. S5-2A-K.
97. Piel J. A polyketide synthase-peptide synthase gene cluster from an uncultured bacterial symbiont of *Paederus* beetles. Proc Natl Acad Sci USA 2002;99:14,002–14,007.
98. Sigel MM, Wellham LL, Lichter W, Dudeck LE, Gargus J, Lucas AH. Anticellular and antitumor activity of extracts from tropical marine invertebrates. In: Younghen Jr HW (ed), Food-Drugs from the Sea Proceedings, Marine Technology Society, Washington, DC: 1969; 281–294.
99. Rinehart K, Holt TG, Fregeau NL, et al. Ecteinascidins 729, 743, 745, 759A, 759B and 770: potent antitumor agents from the Caribbean tunicate *Ecteinascidia turbinata*. J Org Chem 1990;55:4512–4515.
100. Wright AE, Forleo DA, Gunawardana GP, Gunasekera SP, Koehn FE, McConnell OJ. Antitumor tetrahydroisoquinoline alkaloids from the colonial ascidian *Ecteinascidia turbinata*. J Org Chem 1990;55:4508–4512.
101. Scott JD, Williams RM. Chemistry and biology of the tetrahydroisoquinoline antitumor antibiotics. Chem Rev 2002;102:1669–1730.
102. Oku N, Matsunaga S, van Soest RWM, Fusetani N. Renieramycin J, a highly cytotoxic tetrahydroisoquinoline alkaloid, from a marine sponge *Neopetrosia* sp. J Nat Prod 2003;66: 1136–1139.
103. Corey EJ, Gin DY, Kania RS. Enantioselective total synthesis of ecteinascidin 743. J Am Chem Soc 1996;118:9202–9203.
104. Martinez EJ, Owa T, Schreiber SL, Corey EJ. Phthalascidin, a synthetic antitumor agent with potency and mode of action comparable to ecteinascidin 743. Proc Natl Acad Sci USA 1999;96:3496–3501.

105. Manzanares I, Cuevas C, Garcia-Nieto R, Marco E, Gago F. Advances in the chemistry and pharmacology of ecteinascidins, a promising new class of anticancer agents. Curr Med Chem-Anti-Cancer Agents 2001;1:257–276.
106. Takebayashi Y, Pourquier P, Zimonjic DB, et al. Antiproliferative activity of ecteinascidin 743 is dependent upon transcription-coupled nucleotide-excision repair. Nat Med 2001;7:961–966.
107. Zewail-Foote M, Li V-S, Kohn H, Bearss D, Guzman M, Hurley KH. The inefficiency of incisions of ecteinascidin 743-DNA adducts by the UVRabc nuclease and the unique structural feature of the DNA adducts can be used to explain the repair-dependent toxicities of this antitumor agent. Chem Biol 2001;8:1033–1049.
108. Bonfanti M, La Valle E, Fernandez Sousa-Faro J-M, et al. Effect of ecteinascidin-743 on the interaction between DNA binding proteins and DNA. Anti-Cancer Drug Des 1999;14:179–186.
109. Twelves C, Hoekman K, Bowman A, et al. Phase I and pharmokinetic study of Yondelis® (ecteinascidin-743; Et-743) administered as an infusion over 1 h or 3 h every 21 days in patients with solid tumours. Eur J Cancer 2003;39:1842–1851.
110. van Kesteren C, de Vooght MMM, Lopez-Lazaro L, et al. Yondelis® (trabectedin, Et-743): the development of an anticancer agent of marine origin. Anti-Cancer Drugs 2003;14:487–502.
111. Laverdiere C, Kolb CA, Supko JG, et al. Phase II study of ecteinascidin 743 in heavily pretreated patients with recurrent osteosarcoma. Cancer 2003;98:832–840.
112. Rinehart KL, Lithgow-Bertelloni AM. Novel antiviral and cytotoxic agent. US Patent 1989, GB Appl. 89/22,026.
113. Faircloth G, Rinehart K, Nunez de Castro I, Jimeno J. Dehydrodidemnin B, a new marine derived antitumour agent with activity against experimental tumour models. Ann Oncol 1996;7:34.
114. Urdiales JL, Morata P, Nunez de Castro I, Sanchez-Jimenez F. Anti-proliferative effect of dehydrodidemnin B (DDB), a depsipeptide isolated from Mediterranean tunicates. Cancer Lett 1996;102:31–37.
115. Cuevas C, Cuevas F, Gallego P, et al. Synthetic methods for aplidine and new antitumoral deriv., methods of making and using them. US Patent 2000, 2000/16148, 30JUN2000.
116. Jimeno JM. A clinical armamentarium of marine-derived anti-cancer compounds. Anti-Cancer Drugs 2002;13 (Suppl.1):S15–S19.
117. Broggini M, Marchini SV, Galliera E, et al. Aplidine, a new anticancer agent of marine origin, inhibits vascular endothelial growth factor (VEGF) secretion and blocks VEGF-VEGFR-1 (*flt-1*) autocrine loop in human leukemia cells MOLT-4. Leukemia 2003;17:52–59.
118. Cuadrado A, Garcia-Fernandez LF, Gonzalez L, et al. Aplidin® induces apoptosis in human cancer cells via glutathione depletion and sustained activation of the epidermal growth factor receptor, *src*, *jnk*, and p38 MAPK. J Biol Chem 2003;278:241–250.
119. Uemura D, Takahashi K, Yamamoto T, et al. An antitumor polyether macrolide from a marine sponge. J Am Chem Soc 1985;107:4796–4798.
120. Hirata Y, Uemura D. Halichondrins: antitumor polyether macrolides from a marine sponge. Pure Appl Chem 1986;58:701–710.
121. Pettit GR, Herald CL, Boyd MR, et al. Isolation and structure of the cell growth inhibitory constituents from the Western Pacific marine sponge *Axinella* sp. J Med Chem 1991;34:3339–3340.
122. Pettit GR, Tan R, Williams MD, et al. Isolation and structure of halistatin 1 from the Eastern Indian Ocean marine sponge *Phakellia carteri*. J Org Chem 1993;58:2538–2543.
123. Gravelos DG, Lake R, Blunt JW, Munro MHG, Litaudon MSP. Halichondrins: cytotoxic polyether macrolides. US Patent 1993, EP 0 572 109 A1.
124. Aicher TD, Buszek KR, Fang FG, et al. Total synthesis of halichondrin B and norhalichondrin B. J Am Chem Soc 1992;114:3162–3164.

125. Towle MJ, Salvato KA, Budrow J, et al. In vitro and in vivo anticancer activities of synthetic macrocyclic ketone analogues of halichondrin B. Cancer Res 2001;61:1013–1021.
126. Yu MJ. Structurally simplified analogs of halichondrin B: discovery of E7389, a highly potent anticancer agent. Abs Pap Am Chem Soc 2002;224:238-Medi Part 232.
127. Gunasekera SP, Gunasekera M, Longley RE, Schulte GK. Discodermolide: a new bioactive polyhydroxylated lactone from the marine sponge *Discodermia dissoluta*. J Org Chem 1990;55:4912–4915.
128. Gunasekera SP, Gunasekera M, Longley RE, Schulte GK. Discodermolide: a new bioactive polyhydroxylated lactone from the marine sponge *Discodermia dissoluta*. J Org Chem 1991;56:1346.
129. ter Haar E, Kowalski RJ, Hamel E, et al. Discodermolide, a cytotoxic marine agent that stabilizes microtubules more potently than taxol. Biochemistry 1996;35:243–250.
130. Nerenberg JB, Hung DT, Somers PK, Schreiber SL. Total synthesis of the immunosuppressive agent (-) discodermolide. J Am Chem Soc 1993;115:12,621–12,622.
131. Marshall JA, Johns BA. Total synthesis of (+)-discodermolide. J Org Chem 1998;63:7885–7892.
132. Halstead DP. Total synthesis of (+)-miyakolide, (-)-discodermolide, and (+)-discodermolide (PhD Thesis). Cambridge: Harvard: 1998.
133. Smith III AB, Kaufman MD, Beauchamp TJ, LaMarche MJ, Arimoto H. Gram-scale synthesis of (+)-discodermolide. Org Lett 1999;1:1823–1826.
134. Paterson I, Delgado O, Florence GJ, Lyothier I, Scott JP, Sereinig N. 1,6-asymmetric induction in boron-mediated aldol reactions: application to a practical total synthesis of (+)-discodermolide. Org Lett 2003;5:35–38.
135. Paterson I, Florence GJ. The development of a practical total synthesis of discodermolide, a promising microtubule-stabilizing anticancer agent. Eur J Org Chem 2003:2193–2208.
136. Isbrucker RA, Cummins J, Pomponi SA, Longley RE, Wright AE. Tubulin polymerizing activity of dictyostatin-1, a polyketide of marine sponge origin. Biochem Pharmacol 2003;66:75–82.
137. Martello LA, LaMarche MJ, He L, Beauchamp TJ, Smith III AB, Horwitz SB. The relationship between taxol and (+)-discodermolide: synthetic analogs and modeling studies. Chem Biol 2001;8:843–855.
138. Hamann MT, Scheuer PJ. Kahalalide F. A bioactive depsipeptide from the sacoglossan mollusk *Elysia rufescens* and the green alga *Bryopsis* sp. J Amer Chem Soc 1993;115:5825–5826.
139. Hamann MT, Otto CS, Scheuer PJ, Dunbar DC. Kahalalides: bioactive peptides from a marine mollusk *Elysia rufescens* and its algal diet *Bryopsis* sp. J Org Chem 1996;61:6594–6600.
140. Garcia-Rocha M, Bonay P, Avila J. The antitumoral compound kahalalide F acts on cell lysosomes. Cancer Lett 1996;99:43–50.
141. Lopez-Macia A, Jimenez JC, Royo M, Giralt E, Alberico F. Synthesis and structure determination of kahalalide F. J Am Chem Soc 2001;123:11,398–11,401.
142. Suarez Y, Gonzalez L, Cuadrado A, Berciano M, Lafarga M, Munoz A. Kahalalide F, a new marine-derived compound, induces oncosis in human prostate and breast cancer cells. Mol Cancer Ther 2003;2:863–872.
143. Cuadros R, Montejo de Garcini E, Wandosell F, Faircloth G, Fernandez-Sousa JM, Avila J. The marine compound spisulosine, an inhibitor of cell proliferation, promotes the disassembly of actin stress fibers. Cancer Lett 2000;152:23–29.
144. Talpir R, Benayahu Y, Kashman Y, Pannell L, Schleyer M. Hemiasterlin and geodiamolide TA; two new cytotoxic peptides from the marine sponge *Hemiasterella minor* (Kirkpatrick). Tetrahedron Lett 1994;35:4453–4456.
145. Coleman JE, de Silva ED, Kong F, Andersen RJ, Allen TM. Cytotoxic peptides from the marine sponge *Cymbastela sp*. Tetrahedron 1995;51:10,653–10,662.

146. Anderson HJ, Coleman JE, Andersen RJ, Roberge M. Cytotoxic peptides hemiasterlin, hemiasterlin A and hemiasterlin B induce mitotic arrest and abnormal spindle formation. Cancer Chemother Pharmacol 1997;39:223–226.
147. Gamble WR, Durso NA, Fuller RW, et al. Cytotoxic and tubulin-interactive hemiasterlins from *Auletta* sp. and *Siphonochalina* spp. sponges. Bioorg Med Chem 1999;7:1611–1615.
148. Bai R, Durso NA, Sackett DL, Hamel E. Interactions of the sponge-derived antimitotic tripeptide hemiasterlin with tubulin: comparison with dolastatin 10 and cryptophycin 1. Biochemistry 1999;38:14,302–14,310.
149. Andersen RJ, Coleman JE, Piers E, Wallace DJ. Total synthesis of (-)-hemiasterlin, a structurally novel tripeptide that exhibits potent cytotoxic activity. Tet Lett 1997;38:317–320.
150. Nieman JA, Coleman JE, Wallace DJ, et al. Synthesis and antimitotic/cytotoxic activity of hemiasterlin analogues. J Nat Prod 2003;66:183–199.
151. Loganzo F, Discafani C, Annable T, et al. HTI-286, a synthetic analogue of the tripeptide hemiasterlin, is a potent antimicrotubule agent that circumvents p-glycoprotein-mediated resistance in vitro and in vivo. Cancer Res 2003;63:1838–1845.
152. Natori T, Koezuka Y, Higa H. Agelasphins, novel α-galactosylceramides from the marine sponge *Agelas mauritianus*. Tet Lett 1993;34:5591–5592.
153. Natori T, Morita M, Akimoto K, Koezuka Y. Agelasphins, novel antitumor and immunostimulatory cerebrosides from the marine sponge *Agelas mauritianus*. Tetrahedron 1994;50: 2771–2784.
154. Hoshi A, Castaner J. KRN-7000. Drugs Fut 1996;21:152–153.
155. Motoki K, Kobayashi E, Uchida T, Fukushima H, Koezuka Y. Antitumor activities of α-, β-monogalactosylceramides and four diastereomers of an α-galactosylceramide. Bioorg Med Chem Lett 1995;5:705–710.
156. Moore KS, Wehrli S, Roder H, et al. Squalamine: an aminosterol antibiotic from the shark. Proc Natl Acad Sci USA 1993;90:1354–1358.
157. Hao D, Hammond LA, Eckhardt SG, et al. A phase I and pharmacokinetic study of squalamine, an aminosterol angiogenesis inhibitor. Clin Canc Res 2003;9:2465–2471.
158. Sorbera LA, Castaner RM, Leeson PA. Ae-941. Drugs Fut 2000;25:551–556.
159. Alpert L, Savard P, Ross N, et al. Angiostatic and antitumoral activity of AE-941 (Neovastat®), a molecular fraction derived from shark cartilage. Proc Am Assoc Cancer Res 1997;38:Abs 1530.
160. Gingras D, Boivin D, Deckers C, Gendron S, Bathomeuf C, Beliveau R. Neovastat—a novel antiangiogenic drug for cancer therapy. Anti-Cancer Drugs 2003;14:91–96.
161. Bukowski RM. AE-941, a multifunctional antiangiogenic compound: trials in renal cell carcinoma. Expert Opin Investig Drugs 2003;12:1403–1411.
162. Ahmed A, Hoegenauer EK, Enev VS, et al. Total synthesis of the microtubule stabilizing antitumor agent laulimalide and some nonnatural analogues: the power of Sharpless' asymmetric epoxidation. J Org Chem 2003;68:3026–3042.
163. Mulzer J, Ohler E. Microtubule-stabilizing marine metabolite laulimalide and its derivatives: synthetic approaches and antitumor activity. Chem Rev 2003;103:3753–3786.
164. Gerwick WH, Proteau PJ, Nagle DE, Hamel E, Blokhin A, Slate DL. Structure of curacin A, a novel antimitotic, antiproliferative and brine shrimp toxic natural product from the marine cyanobacterium *Lyngbya majuscula*. J Org Chem 1994;59:1243–1245.
165. Edler MC, Fernandez AM, Lassota P, Ireland CM, Barrows LR. Inhibition of tubulin polymerization by vitilevuamide, a bicyclic marine peptide, at a site distinct from colchicine, the *vinca* alkaloids, and dolastatin 10. Biochem Pharmacol 2002;63:707–715.
166. Janin YL. Peptides with anticancer use or potential. Amino Acids 2003;25:1–40.
167. Lindquist N, Fenical W, Van Duyne GD, Clardy J. Isolation and structure determination of diazonamides A and B, unusual cytotoxic metabolites from the marine ascidian *Diazona chinensis*. J Am Chem Soc 1991;113:2303–2304.

168. Cruz-Monserrate Z, Vervoort HC, Bai R, et al. Diazonamide A and a synthetic structural analog: disruptive effects on mitosis and cellular microtubules and analysis of their interactions with tubulin. Mol Pharmacol 2003;63:1273–1280.
169. D'Ambrosio M, Guerriero A, Pietra F. Sarcodictyin-A and sarcodictyin-B, novel diterpenoidic alcohols esterified by (E)-n(1)-methylurocanic acid-isolation from the Mediterranean stolonifer *Sarcodictyon roseum*. Helv Chim Acta 1987;70:2019–2027.
170. D'Ambrosio M, Guerriero A, Pietra F. Isolation from the Mediterranean stoloniferan coral *Sarcodictyon roseum* of sarcodictyin C, D, E, and F, novel ditepenoidic alcohols esterified by (E)- or (Z)-n(1)-methylurocanic acid-failure of the carbon-skeleton type as a classification criterion. Helv Chim Acta 1988;71:964–976.
171. Lindel T, Jensen PR, Fenical W, Long BH, Casazza AM, Carboni J, Fairchild CR. Eleutherobin, a new cytotoxin that mimics paclitaxel (taxol®) by stabilizing microtubules. J Am Chem Soc 1997;119:8744–8745.
172. Cinel B, Roberge M, Behrisch H, van Ofwegen L, Castro CB, Andersen RJ. Antimitotic diterpenes from *Erythropodium caribaerum* test pharmacophore models for microtubule stabilization. Org Lett 2000;2:257–260.
173. Taglialatela-Scafati O, Deo-Jangra U, Campbell M, Roberge M, Andersen RJ. Diterpenoids from cultured *Erythropodium caribaeorum*. Org Lett 2002;4:4085–4088.
174. West LM, Northcote PT. Peloruside A: a potent cytotoxic macrolide isolated from the New Zealand marine sponge *Mycale* sp. J Org Chem 2000;65:445–449.
175. Hood KA, West LM, Rouwe B, et al. Peloruside A, a novel antimitotic agent with paclitaxel-like microtubule-stabilizing activity. Cancer Res 2002;62:3356–3360.
176. Ghosh AK, Kim J-H. An enantioselective synthesis of the C_1-C_9 segment of antitumor macrolide peloruside A. Tetrahedron Lett 2003;44:3967–3969.
177. Erickson KL, Beutler JA, Cardellina II JH, Boyd MR. Salicylihalimides A and B, novel cytotoxic macrolides from the marine sponge *Haliclona* sp. J Org Chem 1997;62:8188–8192.
178. Beutler JA, McKee TC. Novel marine and microbial natural product inhibitors of vacuolar ATPase. Curr Med Chem 2002;9:1241–1253.
179. Yet L. Chemistry and biology of salicylihalimide A and related compounds. Chem Rev 2003;103:4283–4306.
180. Yousaf M, Rao KV, Gul W, et al. Solving limited supplies of marine pharmaceuticals through the rational and high-throughput modification of high yielding marine natural producer scaffolds. Abs Pap 6th Int Mar Biotech Conf 2003:Abs. S14-13B-13.
181. Kasanah N, Rao KV, Wedge D, Hill RT, Hamann MT. Biotransformation and biosynthetic studies of the manzamine alkaloids. Abs Pap 6th Int Mar Biotech Conf 2003:Abs S14-13B-14.
182. Nicolaou KC, Winssinger N, Vouloumis D, et al. Solid and solution phase synthesis and biological evaluation of combinatorial sarcodictyin libraries. J Am Chem Soc 1998;120:10,814–10,826.
183. Nicolaou KC, Roschangar F, Vourloumis D. Chemical biology of epothilones. Angew Chem, Int Ed Engl 1998;37:2014–2045.
184. Koeller KM, Haggarty SJ, Perkins BD, et al. Chemical genetic modifier screens: small molecule trichostatin suppressors as probes of intracellular histone and tubulin acetylation. Chem Biol 2003;10:397–410.
185. Burke MD, Schreiber SL. A planning strategy for diversity-oriented synthesis. Angew Chem, Int Ed Engl 2004;43:46–58.
186. Brady SF, Chao CJ, Clardy J. New natural product families from an environmental DNA (eDNA) gene cluster. J Am Chem Soc 2002;124:9968–9969.

8
Case Studies in Natural-Product Optimization
Novel Antitumor Agents Derived From Taxus brevifolia *and* Catharanthus roseus

Jian Hong and Shu-Hui Chen

Summary

Taxus brevifolia is a Pacific yew tree that produces Taxol. *Catharanthus roseus* is one of the most well-studied plants in the world, and has provided a large number of alkaloids, including the clinically important antitumor agents vinblastine and vincristine. Both classes of antitumor agents interfere with tubulin-microtubule dynamics through opposite modes of action. They are the most renowned natural products, have been successfully brought to market, and have also served as leads for further optimization. This review highlights the chemical modification of *Taxus* diterpenoids and bisindole *Catharanthus* alkaloids, with an emphasis on structure–activity relationship studies and analog optimizations, leading to the discovery of a new generation of antitumor drugs.

Key Words: Antitumor agents; *Taxus brevifolia*; *Catharanthus roseus*; tubulin; microtubules; taxol; bisindole alkaloids; vinblastine; vincristine; structure–activity relationship.

1. Introduction

Cancers and related diseases are a major cause of human death worldwide. With the development of molecular biology and pharmacology, great progress has been made toward a better understanding of these diseases. Among current cancer treatment options, only chemotherapy has effectively treated systemic cancers. Such chemotherapy relies heavily on the use of natural products and their modified analogs. Among the most effective natural products for cancer chemotherapy, vinblastine and paclitaxel (Taxol®) are well renowned. They have been successfully brought to market and have also served as lead compounds for further optimization.

Microtubules are dynamic and polymeric structures assembled from two structurally similar protein subunits, α- and β-tubulin. Each tubulin contains approx 440 amino acid residues. Microtubules play an important role in cellular activities by providing pathways for cellular transport processes, constructing the cell cytoskeleton, and they are also involved in mitosis. Because cancer cells proliferate more rapidly than normal cells, disrupting the cell-division process to stop the growth of diseased cells is clearly an important goal and the mode of action for many antitumor agents used in chemotherapy *(1)*.

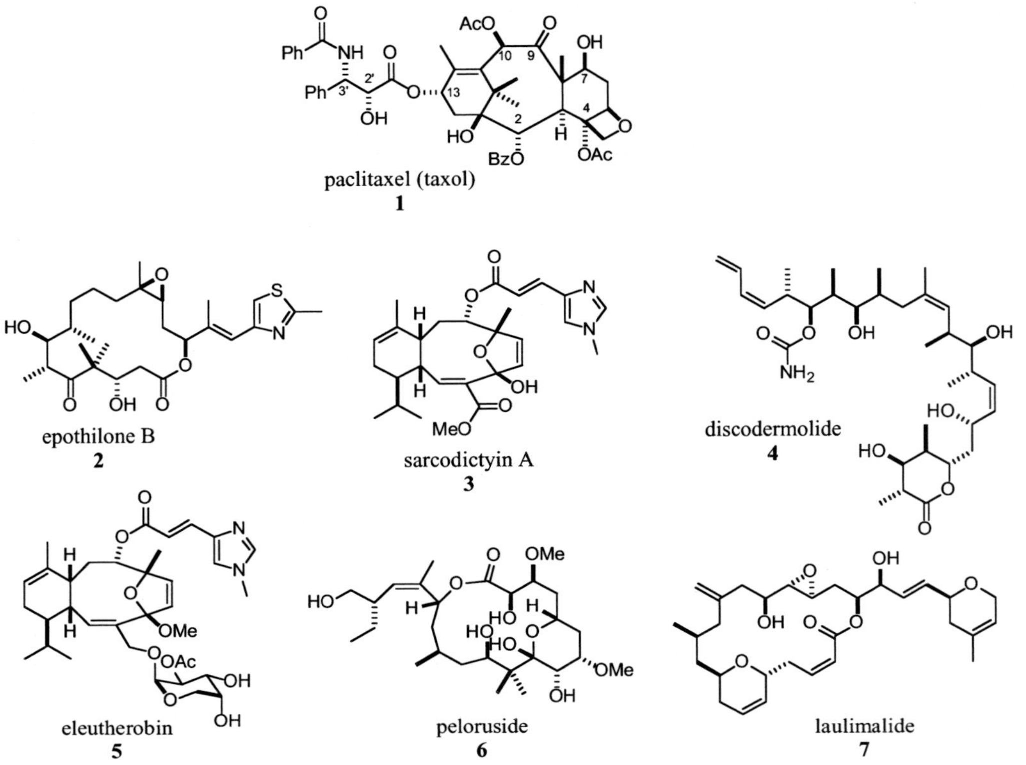

Fig. 1. Selected microtubule stabilizing agents.

Mechanistically, there are two ways of disrupting the tubulin-microtubule equilibrium: (1) treating with agents capable of promoting tubulin polymerization and thus stabilizing microtubules against depolymerization, and (2) treating with agents capable of preventing tubulin polymerization, thus preventing the formation of microtubules.

Several representative tubulin polymerization promoters (type 1 compounds) are shown in **Fig. 1**. These include Taxol, epothilones A and B, eleutherobin, sarcodictyin A, peloruside, laulimalide, and discodermolide *(2)*.

A number of bisindole alkaloids derived from *Catharanthus roseus*, shown in **Fig. 2**, were found to inhibit the tubulin polymerization process (type 2 compounds). *Catharanthus roseus* (L.) G. Don (Apocynaceae), also commonly known as rosy periwinkle, is one of the most thoroughly studied plants in the world. It is indigenous to Madagascar and is cultivated in India, Israel, and the United States *(3)*. To date, more than 100 alkaloids have been isolated from the leaves and roots of this plant, including an extensive series of bisindole alkaloids *(4)*. Of all these, vinblastine and vincristine have become clinically important antitumor agents, marketed as Velban® and Oncovin®, respectively. Vinblastine was first discovered in 1958 by Noble at the University of Western Ontario *(5)*, and independently by Svoboda at the Lilly Research Laboratories *(6)*, as an antiproliferative factor in leaf extracts of the periwinkle plant. Further purification of those extracts led to the discovery of vincristine (leurocristine), leurosidine, and leurosine *(7,8)*.

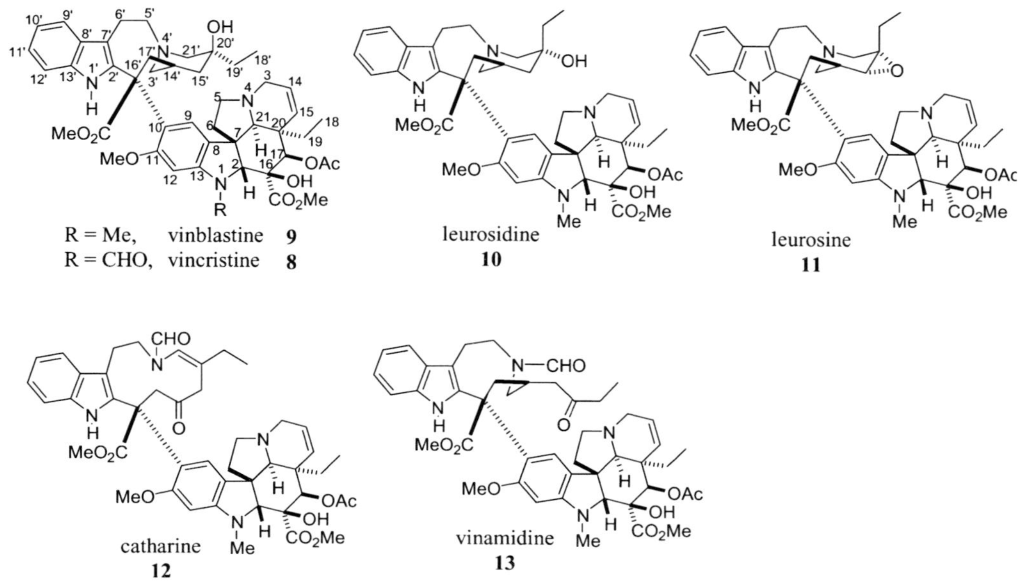

Fig. 2. Selected bisindole alkaloids from *Catharanthus roseus*.

Other well-known tubulin polymerization inhibitors, shown in **Fig. 3**, include maytansine, rhizoxin, cryptophycin 1, ustiloxin A, colchicine, and podophyllotoxine *(9)*. Vinblastine and vincristine block the cell cycle at the G2/M phase. They exert their cytotoxic effect by interfering with the subtle equilibrium between tubulin and microtubules: vinblastine and vincristine form complexes with tubulin at a site different from the site to which colchicine binds, and prevent tubulin polymerization, thus leading to the inhibition of microtubule formation and thereby blocking mitosis *(10)*.

2. Optimization and SAR Studies of Paclitaxel and Analogs From *Taxus brevifolia*

Natural-product drug discovery has been a well-established process, in which microorganisms possess the remarkable ability to produce drug-like small molecules. Many natural products reached the market without any chemical modifications, such as steroids, β-lactams, and erythronolides. Indeed, the potential to hit a "home run" with a single discovery distinguishes natural products from all other sources of chemical diversity and fuels the ongoing efforts to discover new compounds. On the other hand, many natural products present important structural motifs and pharmacophores, which can be optimized to produce valuable drugs with improved properties *(11a–c)*.

Paclitaxel 1, an antimicrotubule agent initially isolated from the bark of *Taxus brevifolia (11d)*, has garnered considerable attention in recent years due to its efficacy in the treatment of various types of cancer, including ovarian, breast, and lung carcinoma *(12)*. In addition, paclitaxel has demonstrated promising activity against Kaposi sarcoma, bladder, prostate, esophageal, head and neck, cervical, and endometrial cancers *(13)*. Paclitaxel has a unique mechanism of action, which distinguishes it from other anticancer agents. The cytotoxic effects of paclitaxel are believed to arise from

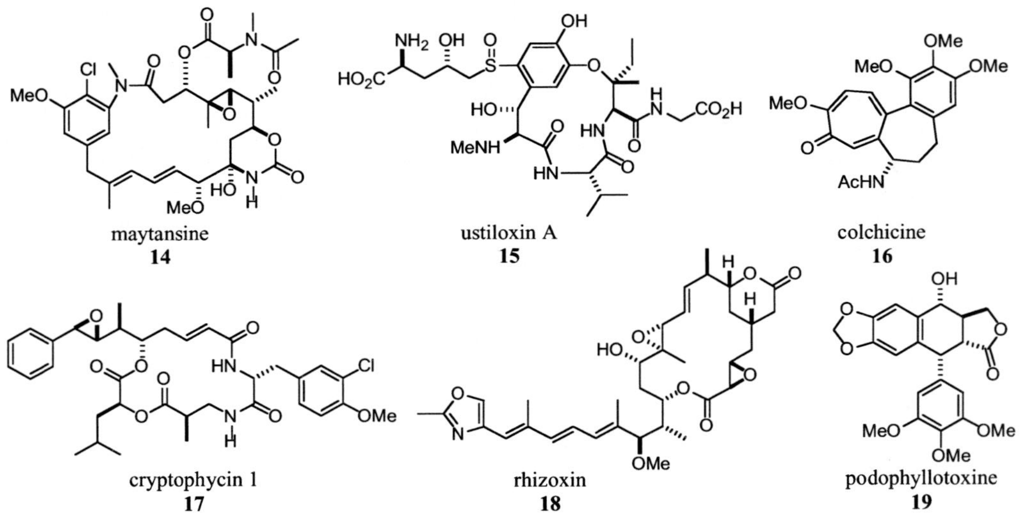

maytansine
14

ustiloxin A
15

colchicine
16

cryptophycin 1
17

rhizoxin
18

podophyllotoxine
19

Fig. 3. Additional tubulin polymerization inhibitors.

its ability to promote tubulin polymerization and stabilize microtubules thus formed even in the absence of cofactors such as GTP. A major consequence of this shift in tubulin-microtubule equilibrium is the inhibition of mitosis *(14)*. The continually expanding therapeutic profile of paclitaxel, coupled with its novel mode of action, has spurred intense research activity on many fronts, including structure–activity relationship (SAR) studies *(15)*. A semisynthetic side-chain analog of paclitaxel, docetaxel (**Fig. 4**, compound 20) was approved by the FDA in 1996 for the treatment of advanced breast cancers *(16)*. Although both paclitaxel and docetaxel have demonstrated impressive antitumor activity, recent reports have shown that the use of these drugs often results in undesirable side effects as well as various drug resistance *(17)*. Therefore, it is necessary to develop better taxanes with improved efficacy and safety profiles against paclitaxel-sensitive and -resistant tumors. With these considerations in mind, we have been involved in an analoging program aimed at discovering better taxanes via both core and side-chain modifications. Results emerging from rather extensive SAR studies at the diterpenoid core clearly show that the functional groups at the northern region of the molecule, namely C-7 *(18)*, C-9 *(19)*, and C-10 *(20)* substituents, contribute relatively little to receptor binding, whereas functionalities at the southern region of paclitaxel, namely C-2 *(21)* and C-4 *(22)* together with oxetane ring *(23)* and C-1 *(23b,24)* are essential elements for tubulin binding and cytotoxicity.

As a result of fruitful research by scientists at Bristol-Myers Squibb, two core-modified paclitaxel analogs—BMS-184476 *(25)* and BMS-188797 *(26)*—were selected, on the basis of their impressive in vitro and in vivo antitumor activities, for human trials against various paclitaxel-resistant tumors. In addition, a C-14 hydroxylated paclitaxel analog, IDN5109 (SB-T-101131) *(27)*, with additional modifications on its side chain, was selected by Bayer Corp. for human clinical trials (**Fig. 4**). More recently, Iimura et al. described an orally active taxotere analog containing morpholinoethyl moiety at the C-10 position *(28)*.

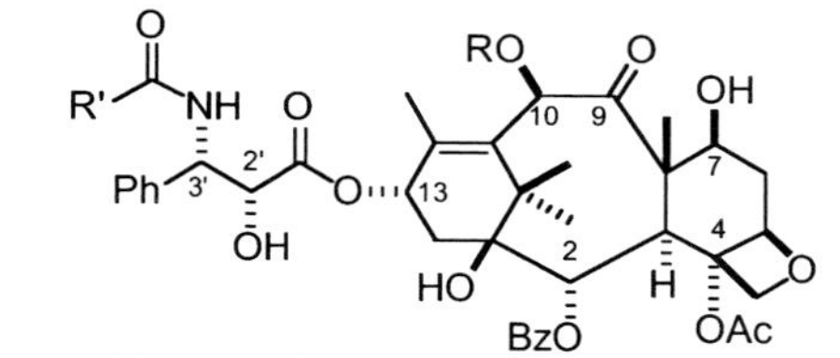

Docetaxel (Taxotere)
R = H, R' = O*t*-Bu
20

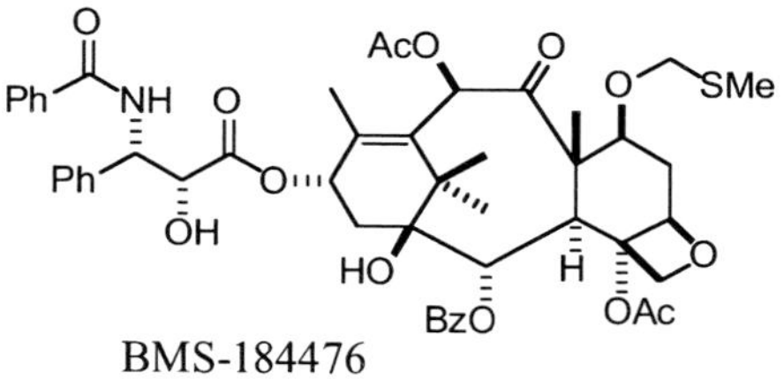

BMS-184476
21

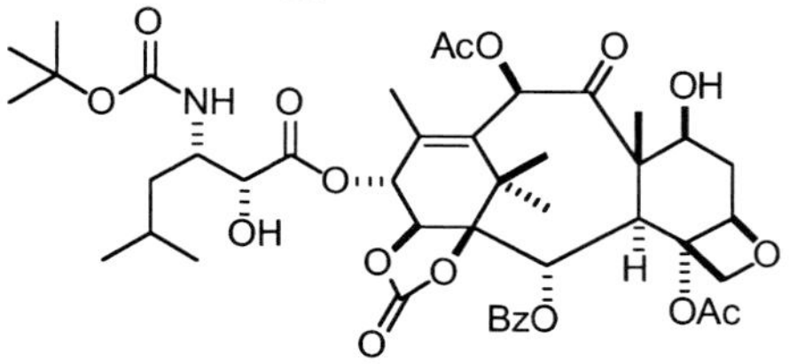

IDN5109 (SB-T-101131)
22

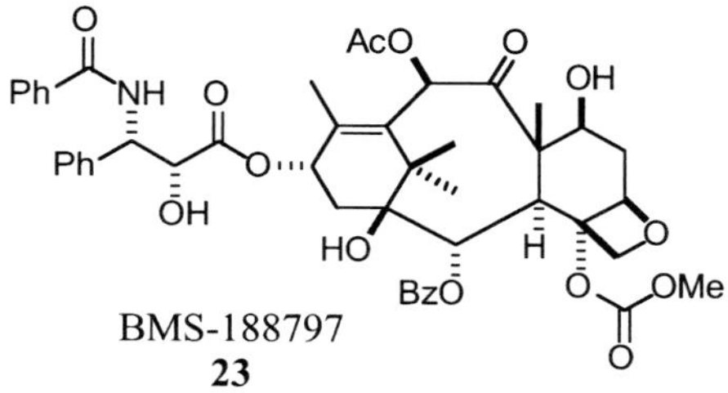

BMS-188797
23

Fig. 4. Other clinically important taxol derivatives

In this chapter, we will highlight the effects of SAR modifications at C-2, C-4, and oxetane (D-ring) on tubulin polymerization as well as on cytotoxicity determined in various taxol-sensitive and -resistant cell lines. In vivo antitumor activities of a few selected compounds will also be discussed.

2.1. C-2 SAR Investigation

2.1.1. General Synthetic Route

A convergent synthetic route for the preparation of C-2 modified taxol analogs was reported by Kingston et al. *(21d)*. As outlined in **Structure 1**, C-2 & C-7 bis-silylated taxol (compound 24) was subjected to a phase-transfer catalyzed debenzoylation condition to yield the corresponding C-2 hydroxyl derivative (compound 25). This intermediate was then condensed with the requisite C-2 acids to provide the respective C-2 ester derivatives (compound 26), which were converted to the final products (compound 27) upon desilylation.

2.1.2. C-2 SAR Trends

In 1993, Chen et al. reported that the C-2 deoxytaxol (compound 27f) was found to be >100-fold less cytotoxic than Taxol against HCT-116 cell lines (a human colon cancer cell line), suggesting that the C-2 benzoate was involved in the intimate binding of taxol with its receptor *(21a)*. Subsequent data from Chen et al. *(21b)* and Kingston et al. *(21c,d)* reported on the cytotoxicity values (IC_{50} n*M* in HCT-116) determined for a set of C-2-substituted benzoates—compounds 27a (>88), 27b (0.70), 27c (3.8), 27d (0.36), and 27e (2.8). Compared with the cytotoxicity obtained with paclitaxel (IC_{50} 2–2.5 n*M*), compounds 27b and 27d were found to be more potent than paclitaxel. In view of the data obtained with compound 27a, it is clear that para-substitution was not tolerated. A separate report from Georg et al. demonstrated that, on the basis of cytotoxicity

Structure 1. Synthesis of C-2 modified taxol analogs.

determined in the B16 melanoma cell line, the C-2 phenyl-bearing analog 27g was about fivefold more potent than its corresponding C-2 saturated counterpart 27h *(21e)*. This result was in good agreement with that disclosed by Ojima and his collaborators on the corresponding taxol analogs *(21f)*. More recently, Kadow and Chen prepared a series of structurally unique C-1/C-2 orthoester derivatives, including compound 27i shown in **Fig. 5**. Interestingly, this novel analog retained good cytotoxicity against HCT-116, with an IC_{50} value of 6 nM *(21g)*. It is also worthwhile to mention that the C-2 *epi*-taxol 27j was devoid of bioactivity in both the tubulin-binding assay and cytotoxicity assay when tested against the HCT-116 cell line *(21h)*.

More recently, Fang et al. showed that the C-2 benzamide bearing derivative 27k was 16-fold less cytotoxic than paclitaxel in KB cells *(21i)*. These recent results provide further evidence for the importance of the characteristics and stereochemistry of the C-2 functionality for the bioactivity of paclitaxel.

Encouraged by the excellent potencies detected in the paclitaxel-sensitive cell line HCT-116, a number of promising C-2 analogs were further evaluated in the paclitaxel-resistant cell line 1A9(PTX22). This cell line expresses an altered β-tubulin that confers 21-fold resistance to paclitaxel. When tested against the 1A9(PTX22) cell line, compounds 27c and 27d showed 15-fold cross-resistance relative to the corresponding sensitive cell line. In contrast to paclitaxel, the C-2 meta-azidobenzoate derivative 27e encountered only fivefold resistance *(21d)*.

2.2. C-4 SAR Studies

2.2.1. General Synthetic Route for C-4 Derivatization

As shown in **Structure 2**, two novel series of C-4 ester and carbonate paclitaxel analogs were synthesized from the key intermediate, 4-deacetyl baccatin derivative 30,

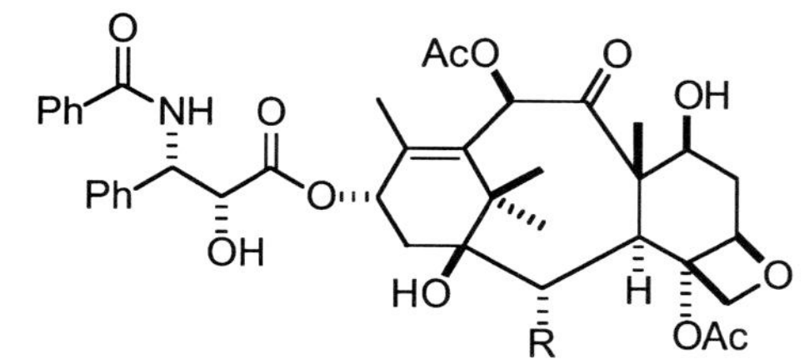

27a R = OC(O)Ph-OMe-p
27b R = OC(O)Ph-OMe-m
27c R = OC(O)Ph-OMe-o
27d R = OC(O)Ph-N_3-o
27e R = OC(O)PhN_3-m
27f R = H

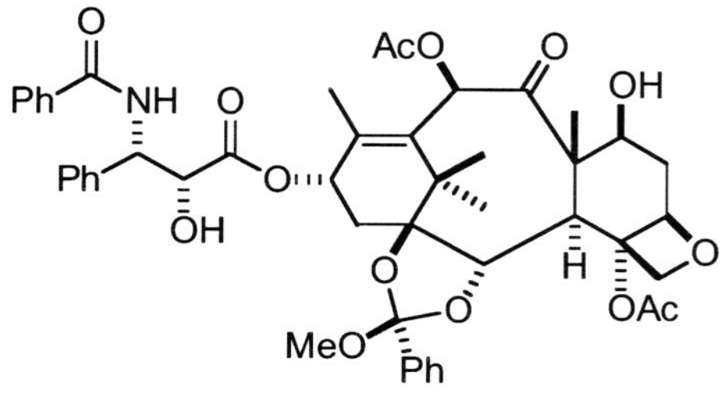

27l C-2 modified analogs

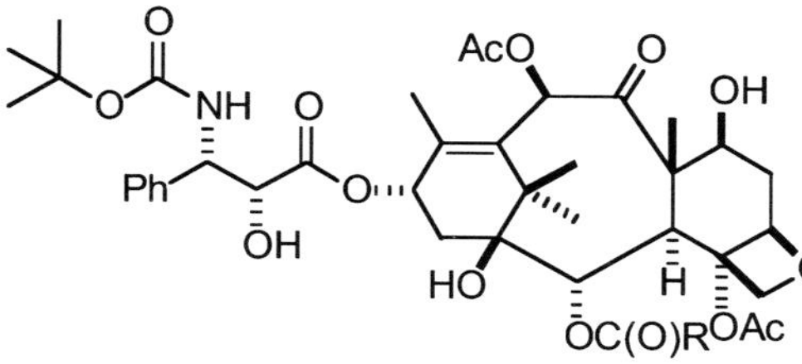

27g R = Ph
27h R = c-hexyl

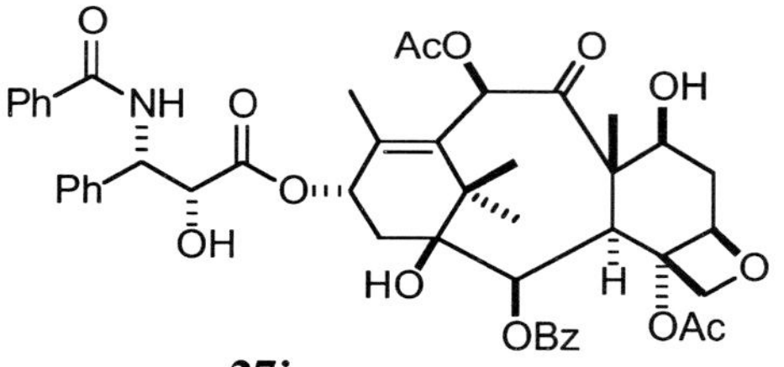

27j

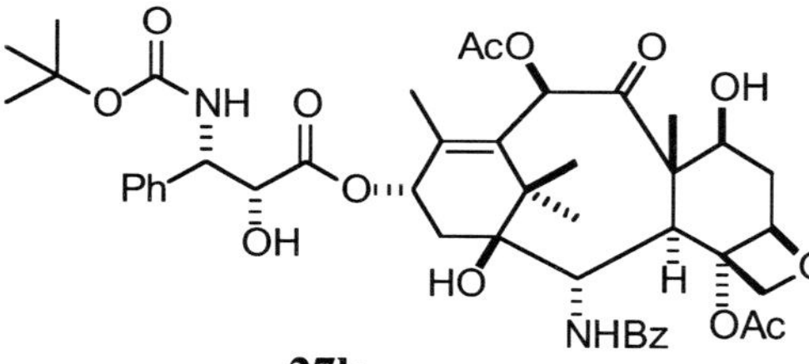

27k

Fig. 5. Representative C-2 modified taxol analogs.

which was in turn prepared according to Chen's protocol (Red-Al/THF/0°C) in two steps from 7,13-bisTES baccatin III (compound 28) *(22a)*. Deprotonation of compound 30 with lithium bis(trimethylsilyl)amide, followed by reaction with an acyl chloride or a chloroformate, afforded the desired C-4 ester baccatin (compound 31) or the C-4 carbonate baccatin (compound 32) in 42–94% yield. These two intermediates were converted, via desilylation (at C-1, C-7 and C-13) and subsequent mono-resilylation, to their corresponding 7-TES baccatin derivatives 33 and 34. Final side-chain attachment (at C-13) onto 33 or 34 led to two novel series of C-4-modified paclitaxel analogs, compounds 36 and 37, respectively *(22a)*.

2.2.2. C-4 SAR Trends

As shown in **Table 1**, the C-4 deacetyl derivative 38 was considerably less potent than paclitaxel in the tubulin polymerization and the cytotoxicity assays *(22b,e)*. In 1996, Chen et al. reported that the C-4 methyl ether analog 39 displayed 18-fold weaker cytotoxicity in comparison to paclitaxel *(22d)*. These results clearly show that the C-4 acyl moiety is an important element in receptor binding.

To further explore the size of the C-4 binding pocket, Chen and his collaborators synthesized and evaluated a large number of C-4 analogs, including various C-4 esters and C-4 carbonates highlighted in **Table 1**. Interestingly, two C-4 benzoyl bearing

Structure 2. Synthesis of C-4 ester and carbonate derivatives.

derivatives (compounds 36a and 36b) were significantly less active than paclitaxel, suggesting that the binding pocket at C-4 is perhaps limited to 6-carbon alkyls *(22b)*.

In sharp contrast to the results mentioned above, all of the aliphatic esters 36c–h exhibited good to excellent activities in the tubulin and the cytotoxicity assays. A rather important trend was observed simply by comparing the cytotoxicity data of the following five C-4 straight chain (two to six carbons) ester analogs: 1 (paclitaxel), 36c, 36d, 36f, and 36h. It is obvious that the 4-butyrate ester 36d was the most potent analog within that subset. This trend seemed to suggest that the four-carbon chain is probably the optimal size for effective receptor binding. In order to further optimize activity, other four-carbon-bearing C-4 esters such as C-4 cyclopropyl ester analog 36e were prepared. The compound 36e was found to be the most potent ester derivative in **Table 1** *(22a,b)*. Interestingly, the C-4 aziridine carbamate derivative 40, a closely related analog of the C-4 ester 36e, was found to be 10-fold less potent than 36e in the tubulin polymerization assay *(22c)*. As also documented in **Table 1**, two C-4 carbonate analogs, 23 and 37b, exhibited two- to fourfold improved potency relative to paclitaxel in both the tubulin binding assay and the cytotoxicity assay against HCT-116 human colon carcinoma cell line *(22b)*.

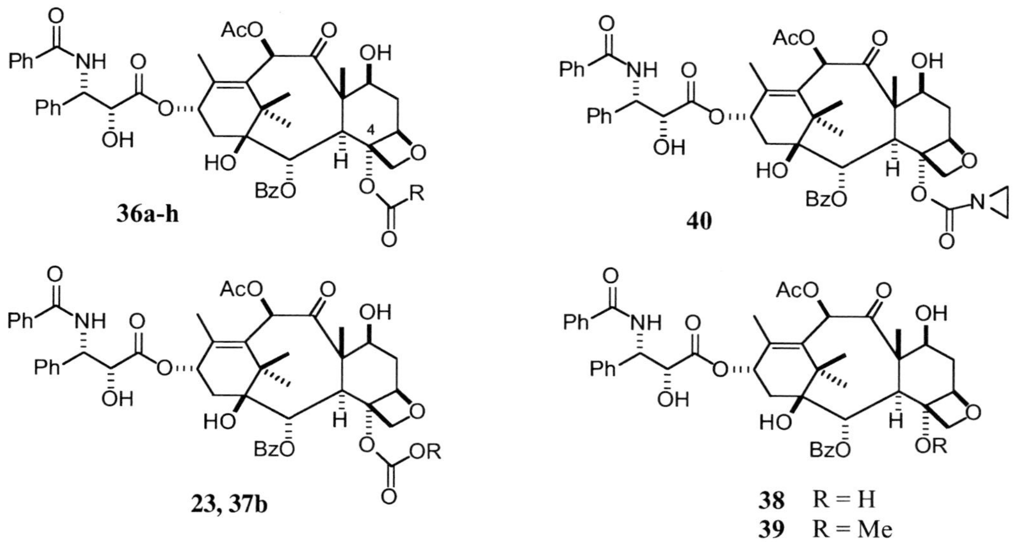

Fig. 6. Representative of C-4 modified paclitaxel analogs.

2.2.3. Side-Chain Modified Novel C-4 Esters and Carbonates

In light of recent reports indicating that replacement of the 3'C-phenyl and 3'N-Bz with their respective 3'C-furyl and 3'N-Boc could lead to paclitaxel side-chain analogs possessing enhanced in vitro activity *(29)*, several potent C-4 esters (36d and 36e) and carbonates (23 and 37b) (**Fig. 6**) were further derivatized into their side-chain analogs for biological evaluation. The in vitro cytotoxicity against HCT-116 and HCT-116(VM46) and in vivo efficacy (*ip*) of these analogs are summarized in **Table 2**.

2.2.3.1. IN VITRO EVALUATION AGAINST SENSITIVE AND RESISTANT HCT-116 CELLS

The general conclusion that emerges from the in vitro data is that side-chain modifications do not seem to have a significant impact on the cytotoxicity of these C-4 esters and C-4 carbonates as listed in **Table 2**. In fact, none of the 3'-C and/or 3'-N replacements gained a clear benefit over the corresponding parent compounds. For example, four pairs of C-4 cyclopropyl esters (compounds 36e, 42–44) and C-4 methyl carbonates (compounds 23, 45), regardless of their side-chain substituents, displayed almost equal potencies in the cytotoxicity assay against the HCT-116 cell line *(22b)*.

Careful analysis of the R/S ratio listed in **Table 2** clearly indicates a moderate increase in size at C-4 resulted in an improved ability to overcome resistance in the HCT-116(VM46) cell line. This cell line is more than 100-fold more resistant to paclitaxel due to overexpression of P-glycoprotein (Pgp), which results in decreased intracellular concentrations of paclitaxel and leads to the multidrug-resistance phenotype (MDR). Although clinical resistance for paclitaxel is not well understood, Pgp overexpression and altered tubulin have been suggested as two possible mechanisms. Remarkably, as shown in **Table 2**, replacement of the C-4 acetoxy moiety in paclitaxel with either cyclopropyl ester (2.4-fold for 36e) or ethyl carbonate (2.9-fold for 37b) effectively overcame Pgp-mediated resistance to paclitaxel. In these two cases, further side-chain modifications did not impact the R/S ratio very much (see data obtained for compounds

Table 1
In Vitro Potencies of Paclitaxel Side-Chain Bearing C-4 Analogs

C-4 Analogs	R	Tubulin Polym. Ratio*	HCT-116 IC$_{50}$ (nM)
Paclitaxel 1	Me	1.0	2.4–4.0
36a	Ph	>60	>1000
36b	p-F-Ph	61	790
36c	Et	1.5	2.0
36d	n-Pr	0.61	1.1
36e	c-Pr	0.24	1.0
36f	n-Bu	1.0	2.0
36g	c-Bu	0.44	1.5
36h	(CH$_2$)$_4$CH$_3$	3.4	6.0
23	Me	0.41	2.0
37b	Et	0.64	1.0

*Ratio obtained in the tubulin polymerization assay, and ratios < 1.0 signify analogs more potent than paclitaxel.

42–44 and 46 for detail). On the other hand, for other C-4 analogs 36d and 23, such side-chain modifications (t-Boc as the 3'-N cap and/or 2-furyl as the 3'-C replacement) had a significant impact on the R/S ratio. The resulting nonpaclitaxel side-chain analogs 41 and 45 exhibited impressive R/S ratios, ranging from 0.5 to 3.6 *(26)*.

2.2.3.2. IN VIVO EVALUATION (M-109 LUNG CARCINOMA)

As shown in **Table 2**, of the 10 C-4 esters and carbonates tested against murine M-109 lung carcinoma *(22b,26c)*, 7 of them were active (T/C > 125%). The best results were obtained with the paclitaxel side chain bearing 4-cyclopropyl ester 36e and one nonpaclitaxel side chain containing 4-methylcarbonate 45, with T/C (%) values of 188 and 183, respectively. It is also worthwhile to note that some of these C-4 analogs were 10–15 times more potent than paclitaxel (as indicated by the dose mg/kg values). These include two C-4 cyclopropyl esters (36e, 43) and two C-4 ethyl carbonate derivatives (37b, 46). However, attempts to improve the in vivo efficacy by means of further side-chain modifications were not successful. For instance, 3'C-furyl/3'-N-Boc bearing C-4 cyclopropyl analogs of 42–44 and 41 possessed reduced activities (as measured by T/C values) as compared with their corresponding parent analogs 36e and 36d. In one case, C-4 methyl carbonate 45 exhibited slightly better in vivo activity than the paclitaxel side-chain-bearing analog 23. No side-chain effect was seen with C-4 ethyl carbonate analogs (37b vs 46). In view of these discrepancies, it is fair to say that side-chain modification provided at best only minimal improvements in in vivo efficacy with the M-109 (*ip*) tumor model *(22b,26c)*.

2.2.3.3. FURTHER IN VIVO EVALUATIONS

Promising C-4 esters and carbonates with demonstrated in vivo efficacy in the M-109 solid tumor model (*ip*) were tested in one or more secondary distal tumor models (iv

Table 2
Non-Paclitaxel Side-Chain-Bearing Analogs

Analog	R_2	R_1	R	HCT-116 IC_{50} (nM)	R/S Ratio	T/C (mg/kg) M-109 (ip)
Taxol	Ph	Ph	Me	2.4-4.0	125	159-228 (60)
36d	Ph	Ph	n-Pr	1.1	31	164 (25)
41	t-BuO	2-Furyl	n-Pr	2.8	0.5	96 (1)
36e	Ph	Ph	c-Pr	1.0	2.4	188 (6)
42	t-BuO	Ph	c-Pr	0.8	0.9	115 (1.6)
43	Ph	2-Furyl	c-Pr	0.9	5.2	132 (3)
44	t-BuO	2-Furyl	c-Pr	1.3	1.3	100 (4)
23	Ph	Ph	OMe	2.0	68	161 (50)
45	t-BuO	2-Furyl	OMe	1.6	3.6	183 (32)
37b	Ph	Ph	OEt	1.0	2.9	142 (6)
46	t-BuO	2-Furyl	OEt	3.0	1.4	144 (4)

administration of drug to a mouse bearing a subcutaneously implanted tumor) in an effort to identify improved paclitaxel(s). To our disappointment, none of these compounds were found to be more efficacious than paclitaxel. It became apparent in the C-4 analog series that an enhanced ability to overcome MDR resistance in vitro did not correlate with expected enhanced in vivo efficacy. This observation parallels the results of Nicolaou's group obtained with C-2 taxoids *(29e)*, as well as our own findings with C-7 paclitaxel derivatives *(18)*. Despite limitations seen with the initial ip/ip M-109 tumor model, the C-4 cyclopropyl ester 36e and the C-4 methyl carbonate derivative 45 demonstrated the best activities. In light of the fact that furyl/Boc side-chain analogs in the C-4 cyclopropyl series (42–44) were inactive in vivo, this suggested that the in vivo activity of the C-4 methyl carbonate series might be more robust in tolerating the incorporation of a more potent side chain. To confirm this hypothesis, we decided to evaluate C-4 methylcarbonate 23 (BMS-188797) in distal tumor models (against L2987 lung carcinoma and HCT/pk colon carcinoma), despite its modest ability to overcome resistance in vitro. To our satisfaction, BMS-188797 demonstrated better efficacy than paclitaxel in these distal tumor models after iv administration *(30)*. On the basis of these data, BMS-188797 was selected for further evaluation in phase I and II clinical trials.

2.3. Oxetane Ring Modification

In order to probe the importance of the intact oxetane ring for bioactivity, several oxetane ring opened analogs (e.g., compounds 47 and 48) as well as oxetane ring oxygen modified analogs (e.g., compounds 49–52) were prepared and evaluated for their tubulin assembly activity and in vitro cytotoxicity against cancer cell lines (*see* **Fig. 7**).

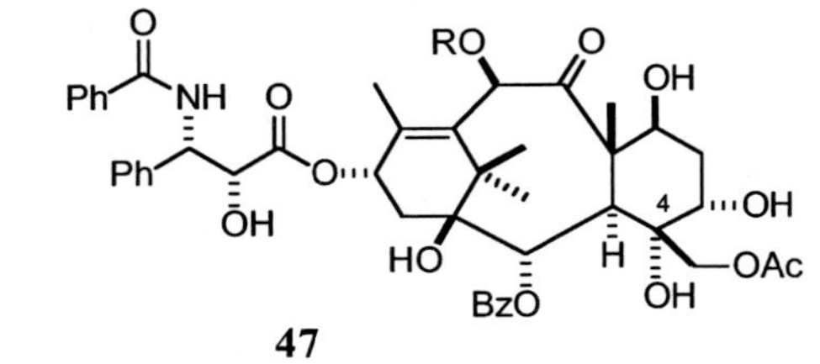

Fig. 7. Oxetane ring-modified taxol or taxotere derivatives.

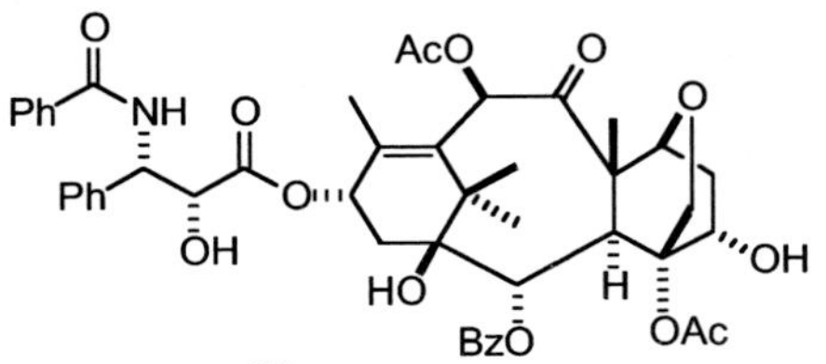

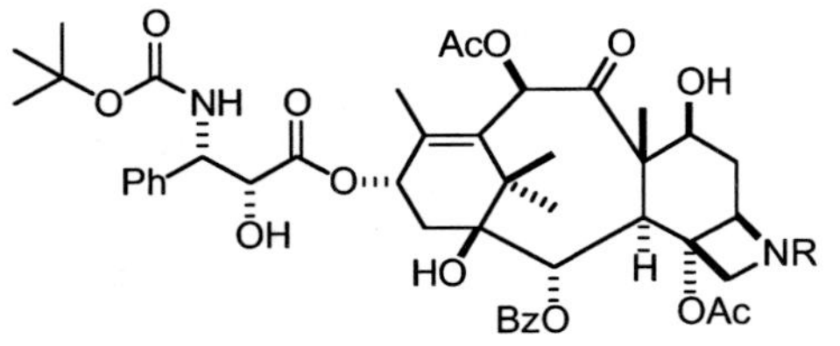

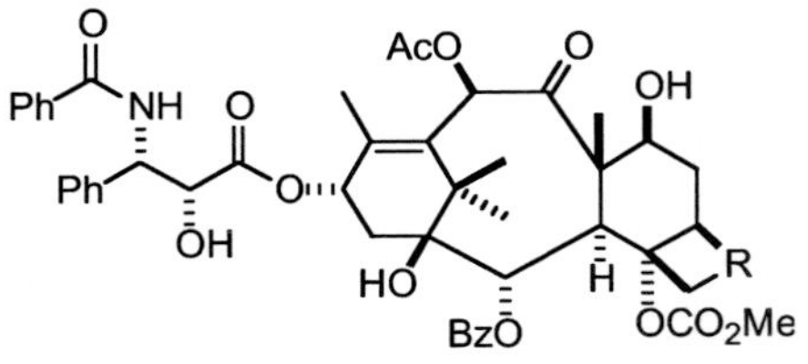

Pioneering work by Kingston showed that D-secotaxol derivatives such as compound 47 were devoid of biological activity *(23a)*. On the basis of this result, it was concluded that the oxetane ring is essential for interaction with microtubules and hence cytotoxicity. Since the key C-4 acetate moiety was absent in compound 47, further research continued to search for compounds bearing both the intact C-4 functionality as well as a modified oxetane ring. In 1993, Chen and his collaborators prepared an interesting bicyclic D-ring containing analog 48 with the intact C-4 acetate, as seen in paclitaxel. Surprisingly, compound 48 also showed significantly reduced bioactivity relative to paclitaxel *(23b)*. The lack of bioactivity observed with compound 48 seemed to suggest that the oxetane ring is required for activity. A recent report from Dubois et al. detailed synthesis and evaluation of two 5(20)-aza-docetaxel (taxotere). When tested against the KB cell line in vitro, both compounds 49 and 50 were found to be inactive. In addition, compound 50 showed 16-fold weaker potency than taxotere, and compound 49 was essentially inactive in the tubulin polymerization assay *(23c)*. The nitrogen atom in the azetidine ring (D-ring) would be protonated at neutral pH and thus would possibly interact with tubulin at a different site from that found with the neutral oxygen atom in taxotere. For these reasons, the role of the oxetane ring on the activity of paclitaxel or docetaxel was still not well understood. In order to pinpoint the role of the oxetane ring oxygen atom on activity, Kingston and his collaborators embarked on the synthesis and comparative evaluation of 5,20-thiapaclitaxel derivative 52, with its corresponding C-4 methyl carbonate paclitaxel analog 51. When tested in the tubulin polymerization and cytotoxicity assays, compound 52 had negligible activity in all assays, with IC_{50} values >1 μM (Burkitt lymphoma) and >2.5 μM (prostate carcinoma). In contrast, compound 51 was clearly superior in activity to paclitaxel and even docetaxel. The IC_{50} values detected with compound 51 were 2 nM (CA46, Burkitt lymphoma), 1 nM (PC3 prostate carcinoma), and 0.2 nM (MCF-7, breast carcinoma).

**Table 3
Biological Activity of Bisindole Alkaloids in the CRFF-CEM
T-Cell Leukemia Cell Line**

Compds	IC$_{50}$ (µg/mL)	Compds	IC$_{50}$ (µg/mL)
vincristine	0.025	vinrosidine	1.0
vinblastine	0.1	vinamidine	7.7
vinleurosine	0.4		

On the basis of these results, Kingston concluded that the D-ring region of the pharmacophore of paclitaxel is very sensitive to steric effects, and the oxygen atom in the oxetane ring is acting as a hydrogen-bond acceptor *(23d)*. It is worthwhile to point out that the similar 5(20)-thiadocetaxel was later prepared by Potier and his co-workers *(23e)*.

3. Optimization and SAR Studies for Designed Bisindole *Catharanthus* Alkaloid Analogs

For several decades, vinblastine and vincristine have become clinically important antitumor agents due to their strong antineoplastic activity against a wide spectrum of human tumors *(31)* (**Table 3**). These two alkaloids, although structurally almost identical, differ markedly in the type of tumors that they affect and in their toxic properties. In particular, vinblastine is used in the treatment of Hodgkin's disease, whereas vincristine is considered the drug of choice in treating acute childhood leukemia, and is more widely used *(32)*.

Vinblastine and vincristine are highly potent drugs, which also have serious side effects, especially on the neurological system *(33)*. Therefore, development of new synthetic analogs with the goal of obtaining more effective and less toxic drugs is desirable *(34)*. This long-standing interest has led to worldwide research programs that continue today, to investigate structural modifications of these alkaloids. Research has mainly focused on the partial or total synthesis of vinblastine analogs from the two parts of the dimeric structure, the *Iboga* and *Aspidosperma* portions, by carbon skeleton modification and functional group transformation. To date, these efforts have led to more than a dozen candidates in clinical evaluation. Optimization and SAR studies of vinblastine structure will continue for the foreseeable future, in order to obtain new generations of effective antitumor drugs.

3.1. SAR Studies on the Velbanamine Portion: Discovery of Vinorelbine and Vinflunine

Although the upper moiety of vinblastine has been less studied due to its lack of functionalities, it is still considered to be a very important region in terms of the potency and novelty of analogs. Modification of the upper velbenamine moiety via the rearrangement of catharanthine is less easily accessible and would require many steps. However, several SAR observations can be deduced from the biological evaluation of various analogs that have been prepared *(35)*.

Structure 3. Potiers synthesis of vinorelbine via second modified Polonovski reaction.

The stereochemical configurations at C16' and C14' in the velbanamine portion are critical. Inversion of the configurations leads to loss of activity. The C16' carbomethoxy group of the velbanamine portion is also important, since the decarboxylated dimer is inactive.

Structural variation at C15'-C20' of the velbanamine portion is well tolerated. Leurosidine (the C20' epimer), leurosine (the epoxide), the C20'-deoxy derivative, the C15'-C20' dehydro-derivative, and C20'-desethyl derivative, all exhibit different inhibition activities for microtubule assembly (36).

Several N4'-oxides have been prepared by oxidation of the parent compounds with peroxide. They have shown significantly reduced antitumor activity as well as reduced tubulin-binding affinity. Other skeletal modifications, such as removal of the D-ring of vinblastine, have been reported. One example, vinamidine (also known as catharinine), with the D-ring cleavage, has resulted in a near complete loss of biological activity.

One of the most successful examples of upper-region modified vinblastine analogs is vinorelbine (compound 57), which retains excellent activity but is less neurotoxic than vincristine. It was prepared by a second Potier-Polonovski reaction with anhydrovinblastine (compound 53). As shown in **Structure 3**, the resulted bisiminium species was hydrolyzed and recyclized to provide skeleton-modified vinorelbine (37–39).

The antitumor activity of navelbine has been evaluated in many murine tumor models. Results have shown that it demonstrated remarkable activity against leukemia models, although it was inactive in carcinoma and fibrosarcoma models. Vinorelbine is currently available worldwide for the treatment of non-small cell lung cancer and of breast cancer. It is orally active, and its use will almost certainly be extended to the treatment of other cancerous diseases.

Super acid chemistry has been applied to prepare vinflunine (compound 58, **Structure 4**), the difluoro-derivative of vinorelbine. This compound has improved in vivo

Case Studies in Natural-Product Optimization

Structure 4. Syntheses of vinflunine and difluorodeoxyvinblastine via super acid chemistry.

n = 1 vinorelbine (**57**)
n = 2 anhydrovinblastine (**53**)

HF-SbF$_5$
CCl$_4$, -40 °C
35 - 40%

n = 1 vinflunine (F 12158) **58**
n = 2 difluorodeoxyvinblastine **59**

antitumor activity over vinorelbine against B16 melanoma and a panel of human tumor xenografts. It is currently undergoing phase I clinical evaluation *(40)*. In a similar fashion, a new family of anhydrovinblastine and vinorelbine analogs was prepared using super acid chemistry. Halogen atoms, ketone, and hydroxy groups are introduced into the upper velbanamine portion *(41)*.

It was discovered that fluorination at C19' position of vinorelbine dramatically increases the in vivo activity, while hydroxylation at the same position results in the total loss of activity. In the anhydrovinblastine series, monohalogenation of the C19' position does not affect the activity, but dihalogenation significantly decreases the activity. In contrast to the above observation, the same modifications greatly decrease the in vivo activity for the vinorelbine series. In addition to the above findings, deacetylation at the C17 position of dihydrovinorelbine and vinflunine resulted in increases of both in vitro cytotoxicity and in vivo activity.

Another strategy to synthesize new upper-region-modified vinblastine analogs has been established as shown in **Structure 5** *(42)*. The preliminary SAR results suggested that the integrity of the velbanamine skeleton in the bisindole alkaloids is important for maintaining their potency. Only one of the tetrahydro-isoquinoline derivatives (compound 64) exerted a marginal activity on the tubulin polymerization inhibition test. Other derivatives lacking the C5'-C6' bond in the C-ring were found to be inactive.

3.2. SAR Studies on the Vindoline Portion

Most of the SAR studies involve the vindoline portion of bisindole alkaloids. This is due to the fact that the presence of various functionalities, such as the C16 and C17 positions, offers good opportunities for derivatizing new analogs. Many functional groups at C16 have been found to be equivalent to the ester "pharmacophore" at this position. These findings have led to the discovery of several new generations of analogs such as vindesine and vinzolidine.

3.2.1. Vindoline's Indole Methyl Derivatives

Vinepidine (LY 119863) (compound 66) is a 20'-deoxy epimer of vincristine *(43)*. As shown in **Structure 6**, it can be prepared from the 20'-*epi*-deoxyvinblastine by oxidation of the vindoline portion's methyl group. In terms of antitumor activity, vinepidine was more efficient than vincristine against lymphosarcoma and leukemia. Although its biochemical pharmacology, antitumor efficacy, and potency led to a phase

Structure 5. A new strategy to synthesize upper-region-modified vinblastine analogs.

Structure 6. Synthesis of vinepidine.

I clinical study, vinepidine's observed neuromuscular toxicity prevented its further clinical development *(44,35)*.

Vinformide (compound 68), also known as *N*-formylleurosine, is an *N*-formyl analog of leurosine. It was prepared from leurosine by oxidation of the vindoline portion's methyl group with various oxidants, including chromic acid *(45)*. Another synthesis started from vincristine reacted with Vilsmeier reagent to produce anhydrovincristine, which was later epoxided to yield vinformide *(46)*. Vinformide was found to be 20 (P388 cell) or 1000 (K562 cell) times less toxic than vincristine, and is useful in the treatment of lymphoma, leukemia, and Hodgkin's disease. On the other hand, in spite of its promising antineoplastic potency, vinformide also exerts an acute cardiotoxic side effect *(47)*.

Other *N*1 derivatives have been reported, and they are shown in **Structure 7**. Many of them have demonstrated certain antitumor activities as well as improved therapeutic index compared with vinblastine. This finding indicates that vindoline's indole methyl group is a useful position to functionalize potentially and provide more potent vinblastine derivatives *(48)*.

3.2.2. Analogs Derived From the C16 Carbomethoxy Group: Discovery of Vindesine and Analogs

Vindesine (LY 99094) (compound 71) is a chemically modified vinblastine analog that differs slightly from vinblastine by having an amide group instead of an ester group at C16. Such minor differences are responsible for the profound alteration in the

Structure 7. Synthesis of vinformide.

Structure 8. Synthesis of vindesine.

antitumor spectrum, potency, and toxicity of these compounds. In particular, vindesine possesses an antitumor spectrum against rodent tumor systems that resembles vincristine, rather than its parent vinblastine, while its neurotoxic potential appears to be less than that of vincristine. Vindesine is prepared from vinblastine or deacetylvinblastine by various methods, as shown in **Structure 8** *(49)*.

The selection of vindesine for clinical evaluation provided an opportunity to explore the consequences of minor structural changes within the "bottom" vindoline portion. Clinical trial data indicated that vindesine is an active oncolytic agent that appears to be less neurotoxic than vincristine. Further studies also indicated that vindesine is active in vincristine-resistant childhood leukemia. When used in combination with cisplatin, vindesine showed efficacy in treatment of non-small cell lung cancer. It is currently available for the treatment of leukemia in several countries.

An extensive series of *N*-substituted vindesine analogs has been prepared from the reaction of deacetyl-vinblastine acid azide (compound 72, **Structure 9**) with the appropriate amines. These *N*-alkylvindesines have reduced activity. In terms of collective antitumor activity, vindesine emerges as the congener with optimum qualities *(50)*.

Structure 9. Preparation of vindesine analogs (*N*-alkylvindesines).

Structure 10. Synthesis of vinblastin-16-oyl amino acid derivatives.

Similarly, a group of 21 vinblastine-C16-carboxyl amino acid derivatives (compound 76, **Structure 10**) were synthesized by linking amino acid carboxylic esters to the vinblastin-C16-carboxyl moiety through an amide linkage. The influence of the nature of the amino acid, ester alkyl chain lengths, the stereoisomerism of the amino acid, and the reacetylation of the hydroxyl group at the C17 position of the vindoline moiety have been studied. Among those 21 congeners, vinblastine-L-Trp-OEt (compound 76b) and vinblastine-L-Ile-OEt (compound 76a) stand out as the best derivatives, based on their antitumor activities. Further studies also showed that the presence of a tryptophan at the C16-carboxyl moiety diminished the toxicity. Therefore derivative vinblastine-L-Trp-Oet (compound 76b), known as vintriptol, was selected for phase I and II clinical trials *(51,52)*.

A series of new amino phosphonic acide derivatives of vinblastine has been synthesized and tested in vitro and in vivo for antitumor activity. All of these compounds were capable of inhibiting tubulin polymerization in vitro. The antitumor activity strongly depended on the stereochemistry of the phosphonates. The most active compounds have the *S*-configuration. One of them, a compound with an isopropyl group (R = *i*-Pr) (compound 78), exhibited a remarkable activity against cancer cell lines both in vitro and in vivo, and was brought into phase I clinical trials *(53)*.

Cephalosporin substituted at the C3' position with the potent oncolytic agent desacetylvinblastine hydrazide was synthesized as a potential prodrug for the treatment of solid tumors. The design of this novel prodrug was based on the knowledge that hydrolysis of a cephalosporin's β-lactam bond can result in the expulsion of the C3' substituent. As shown in **Structure 12**, treatment of these candidate prodrugs (compounds 79–81) with the P99 β-lactamase enzyme efficiently catalyzed their conversion to the free drug form *(54)*.

Structure 11. Synthesis of α-amino phosphonic acid derivatives of vinblastine.

Structure 12. Synthesis of various cephalosporin-vinblastine prodrugs.

Napavin (compound 84, **Structure 13**), a photo-reactive vinblastine derivative, was synthesized from vinblastine. In contrast to vinblastine, which binds noncovalently to the intracellular protein tubulin, napavin was designed to form a covalent bond with target proteins upon irradiation, to prevent its elimination and thus increase the half-life of its action. It was found that as new photo-reactive cytostatic substance, napavin can overcome multidrug resistance of tumor cells *(55)*.

3.2.3. The C16 Spiro-Fused Analogs: Discovery of Vinzolidine and Analogs

Different from the vast majority of the C16 analogs of vinblastine, a group of C16 spiro-fused derivatives represents a novel class of semisynthetic vinblastine analogs. In particular, the C16 spiro-fused oxazolidine-1,3-dione (compound 85a–c) can be prepared in two ways, as shown in **Structure 14**. A number of the bisindole oxazolidinedione analogs have been prepared using these methods. Their ability to

Structure 13. Synthesis of napavin.

Structure 14. Synthesis of vinzolidine analogs.

induce a mitotic block and their antitumor efficacy were examined. Vinzolidine (LY 104208) (compound 85c) stood out as exceptionally active and well absorbed when administered by the oral route. These results have allowed vinzolidine to undergo further clinical studies *(56,57)*.

As shown in **Structure 15**, spirolactone (compound 88) was formed during the reaction of paraformaldehyde with 17-deacetylvinblastine under acidic conditions. Its chemotherapeutic activity was assessed for experimental P338 leukemia. The data showed that this compound had a much better activity than vinblastine *(58)*.

A new series of semisynthetic C16-spiro-oxazolidine-1,3-diones (compound 89a–c) were prepared from 17-deacetoxy-vinblastine using one of the standard methods. As shown in **Table 4**, it was found that compound 89b had lower toxicity than vinblastine and exerted different pharmacological effects. The spiro-oxazolidine ring and the substitution of a formyl group for a methyl group were responsible for the unique pharmacological effects observed. The studies also showed that these compounds displayed their cytotoxic activities at significantly higher concentrations than parent compounds, although their antimicrotubular activities were similar in vitro *(59)*.

3.2.4. The C17 Hydroxyl Group-Derived Analogs and Prodrugs

Vinglycinate (LY 49040) (compound 91) is a glycinate prodrug derived from the C17 hydroxy group of vinblastine. It was prepared from vinblastine as shown in **Structure 16**. Vinglycinate demonstrated measurable activity against several leukemia models and was orally absorbed. Its toxicity and the experimental antitumor spectrum were found to be very similar to that of vinblastine. As a result, vinglycinate was brought

Structure 15. Vinzolidine and analogs.

Table 4
Biological Activity of Vinzolidine Analogs in Various Tests

Compds	R$_1$	R$_2$	In vivo effect P388 Dose (mg/kg)	Cytotoxity MTA IC$_{50}$ (µM)	Anti-tubulin effect IC$_{50}$ (nM)
89a	CH$_3$	CH$_2$CH$_2$Cl	60	0.31	200
89b	CHO	CH$_2$CH$_2$Cl	20	0.31	136
89c	CH$_3$	CH$_2$CH=CH$_2$	40	0.35	185
vinblastine	-	-	-	0.005	248
vincristine	-	-	1	0.013	290

Structure 16. Synthesis of vinglycinate.

Table 5
In Vitro Biological Activity of Several Bisindole Alkaloid Analogs

Compds	R_1	R_2	P3UCLA cells IC_{50} (µg/mL)	CEM cells IC_{50} (µg/mL)
92a	H	$NHNH_2$	<0.0003	0.03
92b	H	OCH_3	0.0003	-
92c	$COCH_3$	OCH_3	-	0.10
92d	$CO(CH_2)_2CO_2H$	OCH_3	0.003	1.4

into clinical studies. Further clinical development was stopped because it showed no improvement over vinblastine in terms of efficacy and toxicity *(60,35)*.

In a fashion similar to the preparation of vinglycinate, a series of 17-(α-aminoacyl) derivatives (compounds 92a–d) have been prepared. As shown in **Table 5**, the antitumor activity of these compounds was examined. It was shown that the larger size of the acyl substitutent at C17 reduced the antitumor activity significantly. Several analogs derived from the C17 hydroxyl group have been prepared, and the activities have been tested. The results in **Table 3** showed the same trend *(61)*.

A series of peptidyl conjugates at the C17 position of vinblastine have been developed for solving the marginally effective and dose-limiting systemic toxicity issues. One of them, compound 95, as shown in **Structure 17**, was evaluated as a prodrug targeted to prostate cancer cells. This compound features an octapeptide segment attached by an ester linkage at the C17 position of vinblastine, and undergoes rapid cleavage by prostate-specific antigen (PSA) ($T_{1/2}$ = 12 min) between the Gln and Ser residues. In nude mouse xenograft studies, it reduced circulating PSA levels by 99% and tumor weight by 85% at a dose just below its MTD *(62)*.

4. Summary and Future Perspectives

Since the first isolation and structural elucidation of *Taxus* diterpenoids and bisindole *Catharanthus* alkaloids, tremendous synthetic and medicinal chemistry efforts have been pursued and disclosed in the literature. Vinblastine and vincristine have served as lead structures for discovery chemistry and lead optimization.

Structure 17. Synthesis of peptidyl prodrugs of vinblastine.

The discovery chemistry of the bisindole alkaloids and their biosynthetic precursors that occur in the same plant has culminated in the development of many novel efficient synthetic routes that are applicable for SAR studies and process development. All these efforts have enriched our knowledge in both synthetic organic chemistry and medicinal chemistry research areas.

The original biosynthetic hypothesis and subsequent synthetic studies led to an understanding of the biosynthetic pathway, allowing us to couple both catharanthine and vindoline to access bisindole alkaloids. As a result of the discovery chemistry, many natural as well as semi-synthetic analogs of vinblastine were prepared via the biosynthetic concepts and newly developed methodologies. In addition to the above development, other areas of pharmacokinetics and pharmacology have been extensively explored, and several allied disciplines have been well integrated. For example, structural–activity relationship studies have led to an understanding of various pharmacophores present in the molecule. Medicinal chemistry also led us to probe the relevant biology, especially mode of action and tubulin-microtubule dynamics. New generations of vinblastine analogs, represented by navelbine, vindesine, and vinzolidine, as well as vinflunine and many others, have been brought into clinical studies, and many of them have been successful in reaching the market.

Future progress will focus on the necessity of developing a three-dimensional structure of the vinblastine binding site, which will benefit our rational drug design. Such a ligand-receptor model will also allow us to probe the tubulin-microtubule dynamics at a molecular level. Meanwhile, a suitable screening model for clinically useful bisindole alkaloids needs to be defined. Other research horizons will arise from the development of bio-conjugates or their hybrids with bisindole alkaloids, as well as from our increasing knowledge of the molecular mechanisms of drug resistance.

With the goal of identifying paclitaxel derivatives retaining in vivo efficacy against paclitaxel-resistant tumors, scientists from many pharmaceutical industries and academic institutions synthesized and evaluated various paclitaxel analogs containing modifications at the core or/and side-chain regions. As a result of this intensive effort, BMS-184476 *(25)*, BMS-188797 *(26)*, and IDN5109 *(27)* were selected, on the basis of their excellent in vitro activity and in vivo activity in a number of animal models, for clinical evaluation against a number of solid tumors. The future success of these novel taxanes in clinical trials should provide clinicians with more treatment options for combating various paclitaxel-resistant cancers. In addition, the success of orally active paclitaxel analogs in clinical trials should reduce the cost of chemotherapy by eliminating the need for premedication, as required for paclitaxel treatment.

In conclusion, the discovery of the *Taxus* and *Catharanthus* family of natural products will continue to provide good opportunities to explore chemical biology, leading to the generation of new clinically useful drugs for the treatment of human cancers.

Acknowledgments

The authors would like to thank Dr. Alan Warshawsky for proofreading of this manuscript. We would also like to thank Yifei Zhou and Xinyi Chen for their support in making this manuscript possible.

References

1. Hyams JS, Lloyd CW (eds). *Microtubules*. Wiley-Liss Inc.: New York, 1994.
2. Nicolaou KC, Hepworth D, King NP, Finlay MRV. Chemistry, biology and medicine of selected tubulin polymerizing agents. Pure Appl Chem 1999;71:989–997.
3. Brossi A, Suffness M (eds), *The Alkaloids*: Antitumor Bisindole Alkaloids From *Catharanthus roseus*. Academic Press, Inc.: 1990; Vol. 37.
4. Sapi J, Massiot G. Bisindole alkaloids. In Saxton JE (ed), *The Monoterpenoid Indole Alkaloids*. John Wiley & Sons, Ltd.: 1994, 523–646.
5a. Noble RL, Beer CT, Cutts JH. Further biological activities of vincaleukoblastine—an alkaloid isolated from vinca rosea (L.). Biochem Pharmacol 1958;1:347–348.
5b. Noble RL, Beer CT, Cutts JH. Role of chance observations in chemotherapy: vinca rosea. Ann NY Acad Sci 1958;76:882–894.
6a. Johnson IS, Wright HF, Svoboda GH. Experimental basis for clinical evaluation of antitumor principles derived from vinca rosea linn. J Lab Clin Med 1959;54:830.
6b. Johnson IS, Wright HF, Svoboda GH, Vlantis J. Antitumor principles derived from vinca rosea Linn. I. Vincaleukoblastine and leurosine. Cancer Res 1960;20:1016–1022.
7. Svoboda GH. Alkaloids of vinca rosea (*Catharanthus roseus*). IX. Extraction and characterization of leurosidine and leurocristine. Lloydia 1961;24:173–178.
8a. Svoboda GH. A note on several new alkaloids from vinca rosea Linn. I. Leurosine, virosine, perivine. J Am Pharm Ass, Sc Ed 1958;47:834.
8b. Svoboda GH, Johnson IS, Gorman M, Neuss N. Current status of research on the alkaloids of vinca rosea Linn. (Catharanthus roseus G. Don). J Pharm Sci 1962;51:707–720.
9. Jordan A, Hadfield JA, Lawrence NJ, McGown AT. Tubulin as a target for anticancer drugs: agents which interact with the mitotic spindle. Med Res Rev 1998;18:259–296.
10. Hamel E. Antimitotic natural products and their interactions with tubulin. Med Res Rev 1996;16:207–231.
11a. Cragg GM, Newman DJ, Snader KM. Natural products in drug discovery and development. J Nat Prod 1997;60:52–60.
11b. Newman DJ, Cragg GM, Snader KM. Natural products as sources of new drugs over the period 1981–2002. J Nat Prod 2003;66:1022–1037.
11c. Tietze LF, Bell HP, Chandrasekhar S. Natural product hybrids as new leads for drug discovery. Angew Chem Int Ed 2003;42:3996–4028.
11d. Wani MC, Taylor HL, Wall ME, Coggon P, McPhall A T. Plant antitumor agents. VI. The isolation and structure of Taxol, a novel antileukemic and antitumor agent from Taxus brevifolia. J Am Chem Soc 1971;93:2325.
12. Rowinsky EK. The development and clinical utility of the Taxane class of antimicrotubule chemotherapy agents. Annu Rev Med 1997;48:353–374.

13. Venook AP, Egorin MJ, Rosner GL, et al. Phase I and pharmacokinetic trial of paclitaxel in patients with hepatic dysfunction: Cancer and Leukemia Group B 9264. J Clin Oncol 1998;16:1811–1819.
14. Schiff PB, Horwitz S B. Taxol stabilizes microtubules in mouse fibroblast cells. PNAS 1980;77:1561–1565.
15a. Kingston DGI, Molinero AA, Rimoldi JM. The taxane diterpenoids. In: *Progress in the Chemistry of Organic Natural Products*. Springer: New York, 1993, 1–206.
15b. Chen SH, Farina V. Paclitaxel chemistry and structure-activity relationships. In: Farina V (ed), *The Chemistry and Pharmacology of Taxol and its Derivatives*. Elsevier Science: Amsterdam, 1995, 165–254.
16a. Gueritte-Voegelein F, Guenard D, Lavelle F, Le Goff M T, Mangatal L, Potier P. Relationships between the structure of taxol analogs and their antitumor activity. J Med Chem 1991;34:992–998.
16b. Bissery M-C, Guenard D, Gueritte-Voegelein F, Lavelle F. Experimental antitumor activity of Taxotere, a taxol analog. Cancer Res 1991;51:4845–4852.
17. Rowinsky EK, Onetto N, Canetta RM, Arbuck SG. Taxol: the first of the taxanes, an important new class of antitumor agents. Sem Oncol 1992;19:646–662.
18. Chen SH, Kant J, Member SW, et al. Taxol structutre-activity relationships: Synthesis and biological evaluation of Taxol analogs modified at C-7. Bioorg Med Chem Lett 1994;4:2223–2228.
19a. Klein LL. Synthesis of 9-dihydrotaxol: a novel bioactive taxane. Tetrahedron Lett 1993;34:2047–2050.
19b. Pulicani JP, Bourzat J-D, Bouchard H, Commercon A. Electrochemical reduction of taxoids: selective preparation of 9-dihydro-, 10-deoxy- and 10-deacetoxy-taxoids. Tetrahedron Lett 1994;35:4999–5002.
19c. Ishiyama T, Iimura S, Yoshino T, et al. New highly active taxoids from 9β-dihydrobaccatin-9,10-acetals. Part 2. Bioorg Med Chem Lett 2002;12:2815–2819.
20a. Kant J, O'Keeffe WS, Chen SH, et al. A Chemoselective approach to functionalize the C-10 position of 10-deacetylbaccatin III. Synthesis and biological properties of novel C-10 taxol analogues. Tetrahedron Lett 1994;35:5543–5546.
20b. Ojima I, Slater JC, Michaud E, et al. Syntheses and SAR of the second-generation antitumor taxoids: exceptional activity against drug-resistant cancer cells. J Med Chem 1996;39:3889–3896.
21a. Chen SH, Wei J-M, Farina V. Taxol SAR: synthesis and biological evaluation of 2-deoxytaxol. Tetrahedron Lett 1993;34:3205–3206.
21b. Chen S H, Farina V, Wei J-M, et al. Structure-activity relationships of taxol: synthesis and biological evalaution of C-2 taxol analogs. Bioorg Med Chem Lett 1994;4:479–482.
21c. Chaudhary AG, Gharpure MM, Rimoldi JM, et al. Unexpectedly facile hydrolysis of the C-2 benzoate group of taxol and syntheses of analogs with increased activities. J Am Chem Soc 1994;116:4097–4098.
21d. Kingston DGI, Chaudhary AG, Chordia MD, et al. Synthesis and biological evaluation of 2-acyl analogues of paclitaxel (taxol). J Med Chem 1998;41:3715–3726.
21e. Boge TC, Himes RH, Vander Velde D G, Georg GI. The effect of the aromatic rings of taxol on biological activity and solution conformation: Synthesis and evalaution of saturated taxol and taxotere analogues. J Med Chem 1994;37:3337–3343.
21f. Duclos O, Zucco M, Ojima I, Bissery M-C, Lavelle F. Structure-activity relationship study on new taxoids. 207th National Meeting of the American Chemical Society; San Diego, CA, American Chemical Society: Washington, DC, 1994; Abstract MEDI 86.
21g. Bristol-Myers Squibb PCT patent application entitled "Ortho-ester analogs of paclitaxol" WO 98/00419 (Jan. 8, 1998).

21h. Chordia MD, Kingston DGI. Synthesis and biological evaluation of 2-epi-paclitaxel. J Org Chem 1996;61:799–801.
21i. Fang W-S, Liu Y, Liu H-Y, Xu S-F, Wang L, Fang Q-C. Synthesis and cytotoxicity of 2α-Amino decetaxel analogs. Bioorg Med Chem Lett 2002;12:1543–1546.
22a. Chen SH, Kadow JF, Farina V, Fairchild CR, Johnston KA. First syntheses of novel paclitaxel (taxol) analogs modified at the C-4 position. J Org Chem 1994;59:6156–6158.
22b. Chen SH, Wei J-M, Long BH, et al. Novel C-4 paclitaxel (taxol) analogs: potent antitumor agents. Bioorg Med Chem Lett 1995;5:2741–2746.
22c. Chen SH, Fairchild C, Long BH. Synthesis and biological evaluation of novel C-4 aziridine-bearing paclitaxel (taxol) analogs. J Med Chem 1995;38:2263–2267.
22d. Chen SH. First syntheses of C-4 methyl ether paclitaxel analogs and the unexpected reactivity of 4-deacetyl-4-methyl ether baccatin III. Tetrahedron Lett 1996;37:3935–3938.
22e. Datta A, Jayasinghe LR, Georg GI. 4-Deacetyltaxol and 10-acetyl-4-deacetyltaxotere: Synthesis and biological evaluation. J Med Chem 1994;37:4258–4260.
22f. Chen SH, Farina V, Vyas DM, Doyle TW. Synthesis of a paclitaxel isomer: C-2-acetoxy-C-4-benzoate paclitaxel. Bioorg Med Chem Lett 1998;8:2227–2230.
23a. Samaranayake G, Magri NF, Jitrangsri C, Kingston DGI. Modified taxol. 5. Reaction of taxol with electrophilic reagents and preparation of a rearranged taxol derivative with tubulin assembly activity. J Org Chem 1991;56:5114–5119.
23b. Chen SH, Huang S, Wei J-M, Farina V. The chemistry of taxanes: reaction of taxol and baccatin derivatives with Lewis acids in aprotic and protic media. Tetrahedron 1993;49:2805–2828.
23c. Marder-Karsenti R, Dubois J, Bricard L, Guenard D, Gueritte-Voegelein F. Synthesis and biological evaluation of D-ring-modified taxanes: 5(20)-azadocetaxel analogs. J Org Chem 1997;62:6631–6637.
23d. Gunatilaka AAL, Ramdayal FD, Sarragiotto MH, Kingston DGI. Synthesis and biological evaluation of novel paclitaxel (taxol) D-ring modified analogues. J Org Chem 1999;64:2694–2703.
23e. Merckle L, Dubois J, Place E, et al. Semisynthesis of D-ring modified taxoids: novel thia derivative of docetaxel. J Org Chem 2001;66:5058–5065.
24. Yuan H, Kingston DGI, Long BH, Fairchild CA, Johnston KA. Synthesis and biological evaluation of C-1 and ring modified A-norpaclitaxel. Tetrahedron 1999;55:9089–9100.
25. Altstadt TJ, Fairchild CA, Golik J, et al. Synthesis and antitumor activity of novel C-7 paclitaxel ethers: discovery of BMS-184476. J Med Chem 2001;44:4577–4583.
26a. Kadow JF, Alstadt T, Chen SH, et al. Discovery of more efficacious analogs of paclitaxel (Taxol) for human clinical evaluation, in Anticancer Agents, Ojima I, Vite GD, Altmann K-H (eds). ACS Symposium Ser. 796; American Chemical Society: Washington, DC, 2001; pp 43–58.
26b. Poss MA, Moniot JL, Trifunovich ID, et al. US Patents 5,808,102 (Sep. 15, 1998) and 6,090,951 (Jul. 18, 2000).
26c. Chen SH. Discovery of a novel C-4 modified 2nd generation paclitaxel analog BMS-188797. In: Guo M, Chen SH, Reiner J, Zhao K (eds), *Frontiers of Biotechnology & Pharmaceuticals*, Vol. 3. Science Press, New York: 2002; 157–171.
27. Distefano M, Scambia G, Ferlini C, et al. Anti-proliferative activity of a new class of taxanes (14β-hydroxy-10-deacetylbaccatin III derivatives) on multidrug-resistance-positive human cancer cells. Int J Cancer 1997;72:844.
28. Iimura S, Uoto K, Ohsuki S, et al. Orally active docetaxel analogue: synthesis of 10-deoxy-10-C-morpholinoethyl docetaxel analogues. Bioorg Med Chem Lett 2001;11:407–410.
29a. Ojima I, Duclos O, Kuduk S D, et al. Synthesis and biological activity of 3'-alkyl- and 3'-alkenyl-3'-dephenyldocetaxels. Bioorg Med Chem Lett 1994;4:2631–2634.
29b. Georg GI, Harriman GCB, Hepperle M, et al. Heteroaromatic taxol analogs: the chemistry and biological activities of 3'-furyl and 3'-pyridyl substituted taxanes. Bioorg Med Chem Lett 1994;4:1381–1384.

29c. Maring CJ, Grampovnik DJ, Yeung CM, et al. C-3'-N-acyl analogs of 9(R)-dihydrotaxol: synthesis and structure-activity relationships. Bioorg Med Chem Lett 1994;4:1429–1432.
29d. Baloglu E, Hoch JM, Chatterjee SK, Ravindra R, Bane S, Kingston DGI. Synthesis and biological evaluation of C-3'NH/C-10 and C-2/C-10 modified paclitaxel analgues. Bioorg Med Chem 2003;11:1557–1568.
29e. Nicolaou KC, Renaud J, Nantermet PG, Couladouros EA, Guy RK, Wrasidlo W. Chemical synthesis and biological evaluation of C-2 taxoids. J Am Chem Soc 1995;117:2409–2420.
29f. Chen SH, Xue M, Huang S, et al. SAR study at the 3'-N position of paclitaxel. Part 1: synthesis and biological evaluation of the 3' (t)-butylaminocarbonyloxy bearing paclitaxel analogs. Bioorg Med Chem Lett 1997;7:3057–3062.
29g. Xue M, Long BH, Fairchild CA, et al. SAR study at the 3'-N position of paclitaxel. Part 2: synthesis and biological evaluation of 3'-N-thiourea- and 3'-N-thiocarbamate-bearing paclitaxel analogues. Bioorg Med Chem Lett 2000;10:1327–1331.
30. Rose WC Fairchild CA, Lee F. Preclinical antitumor activity of two novel taxanes. Canc Chemother Pharmacol 2001;47:97.
31. Elmarakby SA, Duffel MW, Rosazza PN. In vitro metabolic transformations of vinblastine: oxidations catalyzed by human ceruloplasmin. J Med Chem 1989;32:2158–2162.
32. Noble RL. The discovery of the vinca alkaloids-chemotherapeutic agents against cancer. Biochemistry and cell biology 1990;68:1344–1351.
33. Neuss N. Therapeutic use of bisindole alkaloids from *Catharanthus*. In: Brossi A, Suffness M (eds), *The Alkaloids*, Vol. 37. Academic: 1990;229–240.
34. Pearce HL. Medicinal chemistry of bisindole alkaloids from *Catharanthus*. In: Brossi A, Suffness M (eds), *The Alkaloids*, Vol. 37. Academic: 1990; 145–204.
35. Johnson IS, Cullinan GJ, Boder GB, Grindey CB, Laguzza BC. Structural modification of the vinca alkaloids. Cancer Treat Rev 1987;14:407–410.
36. Miller JC, Gutowski GE, Poore GA, Boder GB. Alkaloids of vinca rosea L. (Catharanthus roseus G. Don). 38. 4'-Dehydrated derivatives. J Med Chem 1977;20:409–413.
37. Mangeney P, Andriamialisoa RZ, Lallemand JY, Langlois N, Langlois Y, Potier P. 5'-Nor anhydrovinblastine. Tetrahedron 1979;35:2175–2179.
38. Potier P. Search and discovery of new antitumor compounds. Chem Soc Rev 1992;113–119.
39. Gueritte F, Pouilhes A, Mangeney P, et al. Composes antitumoraux du groupe de la vinblastine: derives de la nor-5' anhydrovinblastine. Eur J Med Chem 1983;18:419–424.
40. Fahy J, Duflos A, Ribet JP, et al. Vinca alkaloids in superacidic media: a method for creating a new family of antitumor derivatives. J Am Chem Soc 1997;119:8576–8577.
41. Duflos A, Kruczynski A, Barret J-M. Noval aspects of natural and modified vinca alkaloids. Curr Med Chem-Anti-Cancer Agents 2002;2;55–70.
42. Fahy J, du Boullay VT, Bigg DCH. New method of synthesis of vinca alkaloid derivatives. Bioorg Med Chem Lett 2002;12;505–507.
43. Thompson GL, Boder GB, Bromer WW, Grindey GB, Poore GA. LY 119863, A novel potent vinca analog with unique biological properties. Proc Am Assoc Cancer Res 1982;73:792.
44. Boder GB, Bromer WW, Poore GA, Thompson GL, Williams DC. Comparative cellular responses to semisynthetic and natural vinca alkaloids. Proc Am Assoc Cancer Res 1982;73;793.
45. Conrad RA. Method of preparing vincristine. US 4375432 (Eli Lilly Co., 1983).
46. Szantay C, Szabo L, Honty K, et al. Leurosine-type alkaloids. HU 24149 (Richter, 1982).
47. Palyi I. Survival responses to new cytostatic hexitols of P388 mouse and K562 leukemia cells in vitro. Cancer Treat Rep 1986;70:279–284.
48. Szantay C, Szabo L, Honty K, et al. Bis-indole derivatives, their preparation and pharmaceutical compositions. EP 205169 (Richter, 1986).

49. Barnett CJ, Cullinan GJ, Gerzon K, et al. Structure-activity relationships of dimeric *Catharanthus* alkaloids. 1. Deacetylvinblastine amide (vindesine) sulfate. J Med Chem 1978;21;88–96.
50. Conrad RA, Cullinan GJ, Gerzon K, Poore GA. Structure-activity relationships of dimeric *Catharanthus* alkaloids. 2. Experimental antitumor activities of *N*-substituted deacetylvinblastine amide (vindesine) sulfates. J Med Chem 1979;22:391–400.
51. Bhushana Rao KSP, Collard MPM, Dejonghe JP, Atassi G, Hannart JA, Trouet A. Vinblastin-23-oyl amino acid derivatives: chemistry, physicochemical data, toxicity, and antitumor activities against P338 and L1210 leukemias. J Med Chem 1985;28:1079–1088.
52. Bhushana Rao KSP, Collard MPM, Trouet A. Vinca-23-oyl amino acid derivatives: as new anticancer agents. Anticancer Res 1985;5;379–386.
53. Lavielle G, Hautefaye P, Schaeffer C, Boutin JA, Cudennec CA, Pierre A. New α-amino phosphonic acid derivatives of vinblastine: chemistry and antitumor activity. J Med Chem 1991;34:1998–2003.
54a. Jungheim LN, Shepherd TA, Meyer DL. Synthesis of acylhydrazido-substituted cephems. Design of cephalosporin-vinca alkaloid prodrugs: substrates for an antibody-targeted enzyme. J Org Chem 1992;57:2334–2340.
54b. Johnson IS, Spearman ME, Todd GC, Zimmerman JL, Bumol TF. Monoclonal antibody drug conjugates for site-directed cancer chemotherapy: preclinical pharmacology and toxicology studies. Cancer Treat Rev 1987;14:193–196.
55. Nasioulas G, Grammbitter K, Gerzon K, Ponstingl H. Synthesis of napavin, a new photoreactive derivative of vinblastine. Tetrahedron Lett 1989;30:5881–5882.
56. Miller JC, Gutowski GE. Vinca alkaloid derivatives. DE 2753791 (Eli Lilly Co., 1978).
57. Gerzon K, Miller JC. Oxazolidinedione sulfide compounds. EP 55602 (Eli Lilly Co., 1982).
58. De Bruyn A, Verzele M, Dejonghe J-P, et al. Modification of Catharanthus roseus alkaloids: a lactone derived from 17-deacetylvinblastine. Planta Medica 1989;55:364–366.
59. Orosz F, Comin B, Rais B, et al. New semisynthetic vinca alkaloids: chemical, biochemical and cellular studies. British J Cancer 1999;79 (9/10):1356–1365.
60. Johnson IS, Hargrove WW, Harris PN, Wright HF, Boder GB. Preclinical studies with vinglycinate, one of a series of chemically derived analogs of vinblastine. Cancer Res 1966;26:2431–2436.
61. Laguzza BC, Nicoles CL, Briggs SL, et al. New antitumor monoclonal antibody-vinca conjugates LY203725 and related compounds: design, preparation, and representative in vivo activity. J Med Chem 1989;32;548–555.
62. Brady SF, Pawluczyk JM, Lumma PK, et al. Design and synthesis of a pro-drug of vinblastine targeted at treatment of prostate cancer with enhanced efficacy and reduced systemic toxicity. J Med Chem 2002;45:4706–4715.

9
Terpenoids As Therapeutic Drugs and Pharmaceutical Agents

Guangyi Wang, Weiping Tang, and Robert R. Bidigare

Summary

Terpenoids, also referred to as terpenes, are the largest group of natural compounds. Many terpenes have biological activities and are used for the treatment of human diseases. The worldwide sales of terpene-based pharmaceuticals in 2002 were approximately US $12 billion. Among these pharmaceuticals, the anticancer drug Taxol® and the antimalarial drug Artimesinin are two of the most renowned terpene-based drugs. All terpenoids are synthesized from two five-carbon building blocks. Based on the number of the building blocks, terpenoids are commonly classified as monoterpenes (C_{10}), sesquiterpenes (C_{15}), diterpenes (C_{20}), and sesterterpenes (C_{25}). These terpenoids display a wide range of biological activities against cancer, malaria, inflammation, and a variety of infectious diseases (viral and bacterial). In last two decades, natural-product bioprospecting from the marine environment has resulted in hundreds of terpenoids with novel structures and interesting bioactivities, with more to be discovered in the future. The problem of supply is a serious obstacle to the development of most terpenoid compounds with interesting pharmaceutical properties. Although total chemical synthesis plays a less important role in the production of some terpenoid drugs, it has contributed significantly to the development of terpenoid compounds and terpene-based drugs by providing critical information on structure–activity relationships (SAR) and chiral centers as well as generating analog libraries. Semisynthesis, on the other hand, has played a major role in the development and production of terpenoid-derived drugs. Metabolic engineering as an integrated bioengineering approach has made considerable progress to produce some terpenoids in plants and fermentable hosts. Cell culture and aquaculture will provide a solution for the supply issue of some valuable terpenes from terrestrial and marine environments, respectively. Recent advances in environmental genomics and other "-omics" technologies will facilitate isolation and discovery of new terpenoids from natural environments. There is no doubt that more terpenoid-based clinical drugs will become available and will play a more significant role in human disease treatment in the near future.

Key Words: Natural products; bioactive terpenoids; biosynthesis; chemical synthesis; drug discovery and development; sustainable production.

1. Introduction

Natural products have played a significant role in human disease therapy and prevention *(1)*. More than 60% and 75% of the chemotherapeutic drugs for cancer and infectious disease, respectively, are of natural origin *(2)*. With more than 23,000 known compounds, terpenoids, also referred to as terpenes, are the largest class of natural products. Among this group, many interesting compounds are extensively applied in the industrial sector as flavors, fragrances, and spices, and are also used in perfumery

From: *Natural Products: Drug Discovery and Therapeutic Medicine*
Edited by: L. Zhang and A. L. Demain © Humana Press Inc., Totowa, NJ

and cosmetics products and food additives *(1,3,4)*. Many terpenes have biological activities and are used for medical purposes. For example, the antimalarial drug Artemisinin and the anticancer drug paclitaxel (Taxol®) are two of a few terpenes with established medical applications. Natural products continue to be one of the most important sources of lead compounds for the pharmaceutical industry. At the same time, more terpenes have been discovered as efficacious compounds in human disease therapy and prevention. In particular, marine chemists and biologists have identified many marine terpenes with promising potential for medical applications. For instance, eleutherobin and sarcodictyin A from the Australian soft coral *Eleutherobia* sp. *(5)* and the stoloniferan coral *Sarcodictyon roseum (6,7)*, respectively, exhibit potent antitumor activity against a variety of tumor cells. Therefore, terpenoids presumably will play an increasingly important role as therapeutic and preventative agents for human diseases. In this chapter, we review the status of terpenoids as potential pharmaceutical agents, with an emphasis on monoterpenes, sesquiterpenes, diterpenes, and sesterterpenes.

2. Biosynthesis of Terpenoids

Terpenoids show enormous chemical and structural diversity. However, their backbones are synthesized from only two five-carbon isomers: isopentenyl diphosphate (IPP, C_5) and its highly electrophilic isomer, dimethylallyl diphosphate (DMAPP, C_5). There are two known pathways for the biosynthesis of these two universal precursors (**Fig. 1**). Details on the progress of the elucidation of these two biosynthetic pathways are summarized in several excellent reviews *(4,8–17)*. From these two basic building blocks, a group of enzymes called prenyltransferases can synthesize linear prenyl diphosphates, which serve as the precursors for terpenoid biosynthesis. During biosynthesis, the active isoprene unit (IPP) is repetitively added to DMAPP or a prenyl diphosphate in sequential head-to-tail condensations catalyzed by the prenyltrans-ferases. The reaction starts with elimination of the diphosphate ion from an allylic diphosphate to form an allylic cation, which is attacked by the IPP molecule with stereospecific removal of a proton to form a new C-C bond and a new double bond in the product (**Fig. 2**) *(18)*. Through consecutive condensations of IPP with an allylic prenyl diphosphate, a prenyltransferase can synthesize a variety of products with fixed lengths and stereochemistry. The chain length of prenyl diphosphates ranges from geranyl diphosphate (GPP, C_{10}) to natural rubber, whose carbon chain length extends to several million *(18–20)*. All prenyltransferases require divalent metals such as Mg^{2+} or Mn^{2+} for catalysis. GPP synthase and farnesyl diphosphate (FPP, C_{15}) synthase catalyze condensation reactions with IPP and DMAPP to form GPP, a precursor to monoterpenes, and FPP, a precursor to sesquiterpenes, respectively *(21,22)*. Geranylgeranyl diphosphate (GGPP, C_{20}) synthase *(18,19)* and farnesylgeranyl diphospate (FGPP, C_{25}) *(23,24)* synthase use the same condensation reactions to synthesize GGPP, a precursor to diterpene, and FGPP, a precursor to sesterterpene, respectively (**Fig. 2**).

Terpene cyclases are responsible for the biosynthesis of the thousands of natural terpenoid compounds found in terrestrial and marine organisms. Terpenes are synthesized from linear prenyl diphosphates in the cyclization cascades mediated by terpenoid cyclases (also known as terpene synthases or isoprenoid synthases), which are known to catalyze the most complex chemical reactions known in chemistry and biology *(25–27)*. A terpenoid cyclase binds and chaperones a linear prenyl diphosphate

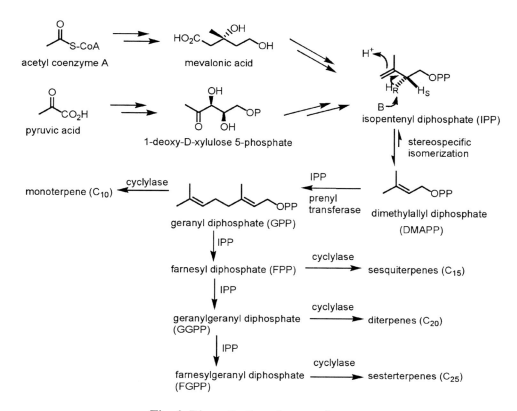

Fig. 1. Biosynthetic pathways of terpenes.

substrate through a precise and multistep cyclization cascade that is initiated by the generation and propagation of highly reactive carbocationic intermediates, which readily undergo dramatic structural rearrangements. The cyclase controls the reactions and provides a template for cyclization and rearrangement with stereochemical and regiochemical precision *(26–30)*. Over all, two-thirds of the substrate carbon atoms undergo changes in chemical bonding and/or hybridization to form a single, unique product *(27)*.

Variations in the number of isoprene unit repetitions, cyclization reactions, and rearrangements are primarily responsible for the chemical and structural diversity of terpenoids *(31)*. Many structurally distinct monocyclic and bicyclic terpenes arise from cyclization and rearrangement of GPP; the larger precursors FPP, GGPP, and FGPP give rise to an even greater variety of terpene carbon skeletons. The variety of ways in which these linear and cyclic skeletons can be arranged, results in the incredible structural diversity observed in nature. Many terpenes have carbacyclic ring systems and oxidized carbon chains such as alcohols and/or carbonyl groups. Some terpenoids have sugar moieties. The bicyclic and polycyclic ring systems with three- and four-membered rings appear commonly in terpenes *(9,13,32,33)*. Based on the number of five-carbon isoprene units in their linear precursor prenyl diphosphate, terpenoids are typically classified as C_5 hemiterpenes, C_{10} monoterpenes, C_{15} sesquiterpenes, C_{20} diterpenes, C_{25} sesterterpenes, and C_{30} triterpenes *(3,20)*.

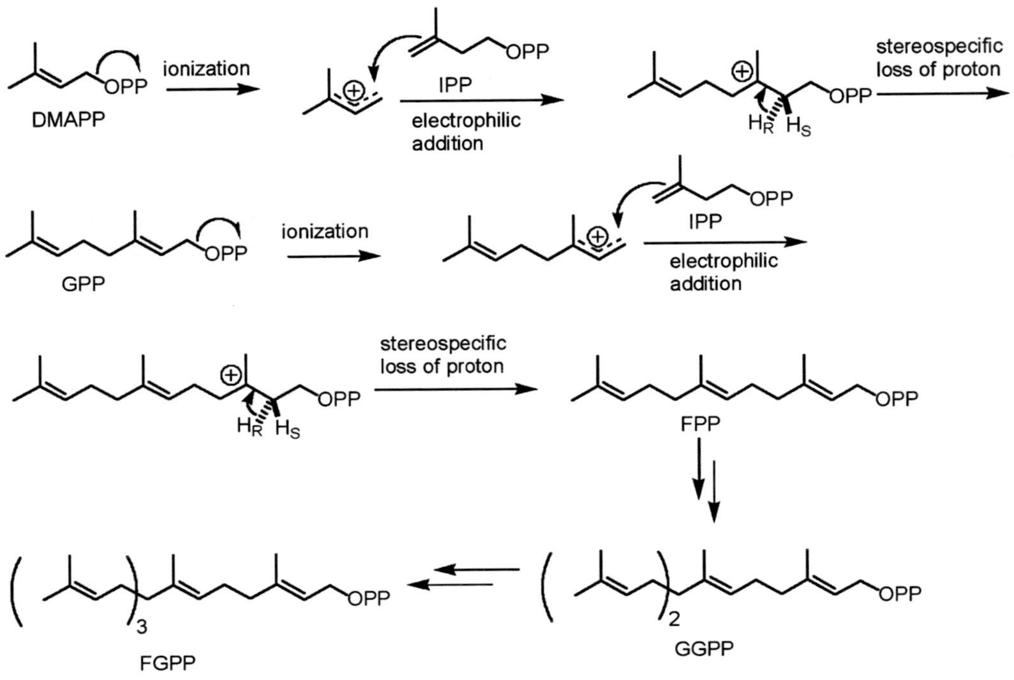

Fig. 2. Biosynthetic pathways of GPP, FPP, GGPP, and FGPP.

3. Pharmaceutical Terpenoids

Terpenoids have a very broad range of biological activities. To review all biologically active terpenoids would be a difficult task, owing to space constraints. In this review we will focus on terpenoids with activities against cancer, malaria, inflammation, and a variety of infectious diseases (viral and bacterial). Other chemotherapeutic agents for these diseases have been the subject of many excellent reviews *(34–71)*. Here, we discuss terpenoids with therapeutic and pharmaceutical functions against these diseases.

3.1. Monoterpenes

Monoterpenes are best known as constituents of the essential oils, floral scents, and the defensive resins of aromatic plants *(72,73)*. Many monoterpenes are nonnutritive dietary components found in the essential oils of citrus fruits, cherry, mint, and herbs. The formation of various monoterpenes from geranyl diphosphate is catalyzed by different terpene cyclases. The general properties of monoterpenes have been discussed in several reviews *(32,74–76)*. A number of dietary monoterpenes have antitumor activity, exhibiting not only the ability to prevent the formation or progression of cancer, but the ability to regress existing malignant tumors *(77)*.

Limonene (**Fig. 3**) is the most abundant monocyclic monoterpene found in nature, and it occurs in a variety of trees and herbs (e.g., *Mentha* spp.). It is a major constituent of peel oil from oranges, citrus, and lemons, and the essential oil of caraway. It has the same skeleton as found in a wide range of important flavor and medicinal compounds,

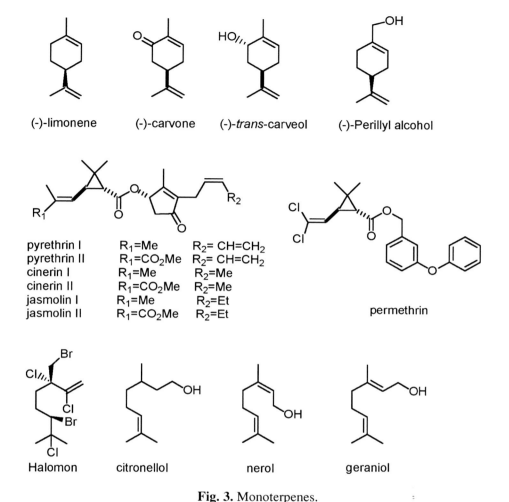

Fig. 3. Monoterpenes.

and consequently, limonene has been an interesting target molecule for chemists and biologists *(78)*. Limonene is a well-established chemopreventive and therapeutic agent against many tumor cells *(65,77,79)*. Carvone, a major monoterpene in caraway seed oil, has been shown to prevent chemically induced lung and forestomach carcinoma development *(80)*. In addition, carveol has chemopreventive activity against rat mammary cancer during the initiation phase *(81)*. Perillyl alcohol, a hydroxylated analog of limonene, exhibits chemopreventive activity against chemically induced liver cancer in rats *(82)* and tumor recurrences in animals *(83)*. Furthermore, perillyl alcohol exhibits chemotherapeutic activity against rat mammary tumors *(83)* and transplanted pancreatic cancer in hamster, with a significant portion of treated tumors being completely regressed *(84)*. Clinical trial testing of chemotherapeutic activity of limonene and perillyl alcohol is in phase I *(85)* and phase II *(86,87)*, respectively. The mechanism of action of monoterpenes against tumor cells is the induction of apoptosis and interference of the protein prenylation of key regulatory proteins *(77,79,88,89)*.

The pyrethrins represent a group of six closely related monoterpene esters and are valuable insecticidal components isolated from pyrethrum flowers, *Chrysanthemum cinerariaefolium*, and several other species in the Asteraceae family *(90,91)*. Pyrethrins are used for treatment of skin parasites such as head lice. They block sodium-channel repolarization of the arthropod neuron, leading to paralysis and death of parasites *(43,92)*. However, minor side effects such as dry and scaly patches, edema, pruritus, and erythema have been reported when pyrethrins are applied in some forms *(92)*. Permethrin, a synthetic analog of pyrethrin, is also used to treat head lice infestation. However, the emerging resistance to pyrethrin and permethrin in head lice has become a serious concern *(93)*.

Among other halogenated acyclic monoterpenes, halomin, isolated from the red alga *Portieria hornemnnii (94)*, is very effective against renal, brain, colon, and non-small cell lung cancer cell lines through a unique mode of action *(95)*. However, further elucidation of its atypical biological activity has been hampered by the limited availability of its natural source, *P. hornemnnii (96)*. Finally, several acyclic monoterpenes—citronellol, nerol, and geraniol—exhibit some activity against *Mycobacterium tuberculosis* with MIC (minimal inhibition concentration) values of 64, 128, and 64 µg/mL, respectively *(97)*.

3.2. Sesquiterpenes

In general, sesquiterpenes are less volatile than monoterpenes. Among the sesquiterpenes, the sesquiterpene lactones are widely distributed in marine and terrestrial organisms and are well known for their wide variety of biological activities *(98,99)*. The anti-inflammatory activities of some medicinal plants result from the presence of one or more sesquiterpene lactones. Feverfew has been used for at least two millennia for the treatment of fever, headaches, menstrual difficulties, and stomach aches *(100)*. Today it is widely used for the relief of arthritis, migraine, asthma, and psoriasis *(101–105)*. Parthenolide (**Fig. 4**), a sesquiterpene lactone, is responsible for the majority of the medicinal properties of this traditional herbal remedy. This sesquiterpene lactone can be found in several species (e.g., *Tanacetum parthenium*, *C. parthenium*, *Leucanthemum parthenium*, and *Pyrethrum parthenium*).

Artemisinin, another sesquiterpene lactone, contains a rare endoperoxide bridge that is essential for its antimalarial activity *(106)*. Artemisinin is derived from an ancient Chinese herbal remedy and has been isolated from *Artemisia annua* (sweet wormwood or "Qinghao"), a species of the Asteraceae family. This herbal plant has been used in Chinese herbal medicine for over 200 years. Artemisinin and its derivatives represent a very important new class of antimalarial drugs, and are used throughout the world. It is effective against both drug-resistant and cerebral malaria-causing strains of *Plasmodium falciparum*. As an antimalarial agent, artemisinin has been extensively reviewed by several researchers *(35,39,56,107,108)*. Chamazulene, α-Bisabolol, and bisabiolol oxides A and B, are terpenoids isolated from matricaria flowers (*Matricaria chamolilla*) and are commonly used in herbal medicine for the treatment of skin inflammation, and as antibacterial and antifungal agents *(13)*. Chamazulene's anti-inflammatory activity is a result of blocking of leukotriene biosynthesis *(109)*.

Many plant sesquiterpenes have also been shown to be effective against the causal agent of tuberculosis (TB), *M. tuberculosis*, which infects approx 8 million people and

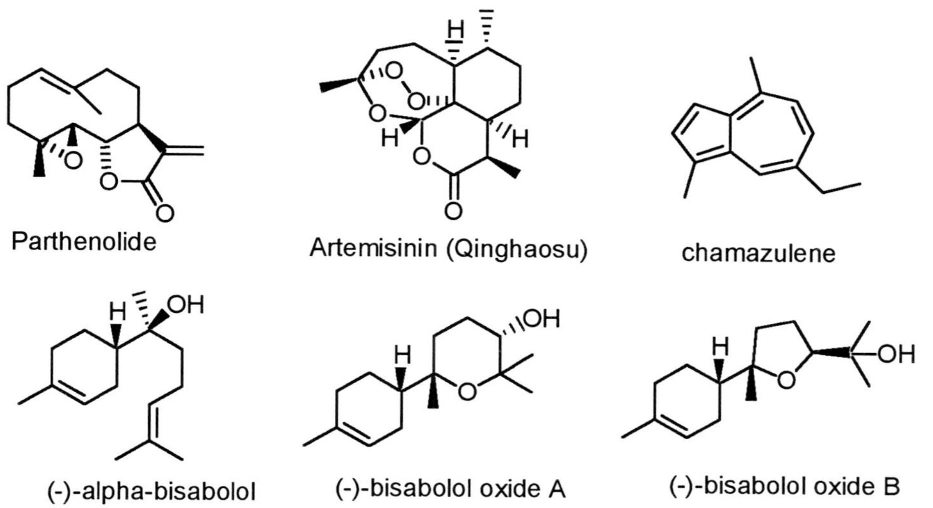

Parthenolide Artemisinin (Qinghaosu) chamazulene

(−)-alpha-bisabolol (−)-bisabolol oxide A (−)-bisabolol oxide B

Fig. 4. Sesquiterpenes I.

causes 2 million deaths each year. Tuberculosis is still the leading killer among all infectious disease, especially in developing countries. With the emergence of multidrug-resistant strains of *M. tuberculosis*, the search for new antituberculosis agents has become increasingly important *(48)*. More than 50 sesquiterpenes from plants exhibit significant antituberculosis activity. Sesquiterpene lactones of the germacranolide (**Fig. 5**), guaianolide, and eudesmanolide types are shown to have antituberculosis activity, with MICs ranging from 2 µg/mL to >128 µg/mL. Details of the antituberculosis activity of these sesquiteperne lactones are discussed thoroughly by Cantrell et al. *(45)*.

Many sesquiterpenes isolated from the marine environment also show activity against tuberculosis *(50)*. Axisonitrile-3 (**Fig. 6**), a cyanosesquiterpene isolated from the sponge *Acanthella klethra*, is a potent inhibitor of *M. tuberculosis*, with an MIC of 1.56 µg/mL *(110)*. Puupehenone, 15-cyanopuupehenone, and 15-α-cyanopuupehenol, isolated from sponges of the orders Verongida and Dictyoceratida *(111–113)*, are natural sesquiterpene-shikimate-derived metabolites. Some of them have attracted significant attention from several research groups because of their cytotoxic, antimicrobial, and immunomodulatory activities *(113)*. Puupehenone, 15-cyanopuupehenone, and 15-α-cyanopuupehenol demonstrated 99, 90, and 96% inhibition of *M. tuberculosis* (H3Rv) growth, respectively, at MICs of 12.5 µg/mL. It has been shown that the quinine-methide system in ring D of puupehenone is essential for its inhibitory activity *(50)*.

Sesquiterpene quinines and hydroxyl quoinones and their related compounds represent a prominent class of biologically active metabolites *(114)*. Their occurrence seems to be restricted to sponges of the three families Spongiidae, Thorectidae, and Dysideidae of the order Dictyoceratida, to the family Niphatidae of the order Haplosclerida, and to the algal species *Dictyopteris undulata (115)*. Their remarkable biological activities include antibacterial, cytotoxic, anti-inflammatory, anti-human

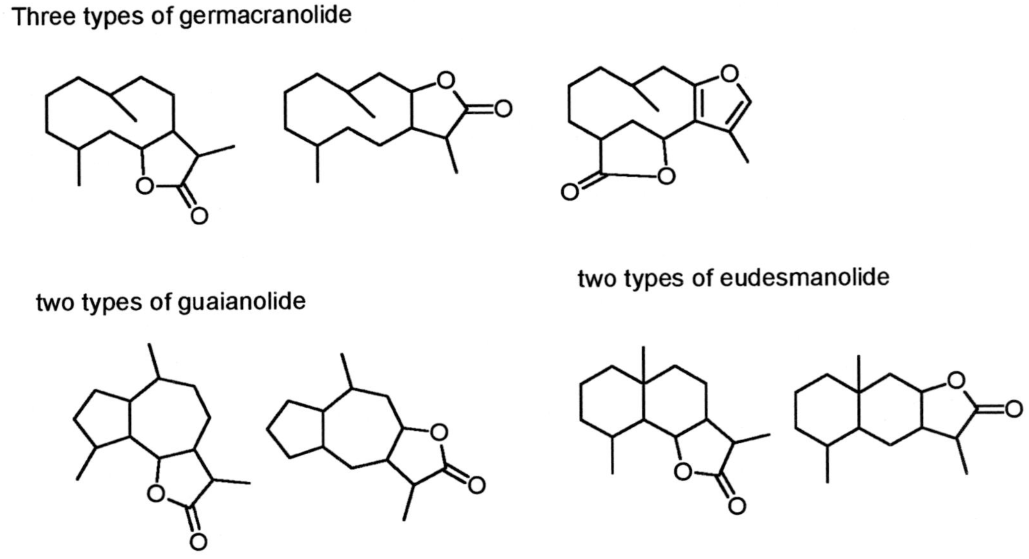

Three types of germacranolide

two types of guaianolide

two types of eudesmanolide

Fig. 5. Sesquiterpenes II.

immunodeficiency virus (HIV), and protein kinase inhibition *(116–120)*. Among these marine sesquiterpenoids, avarol and avarone, isolated from the Red Sea sponge *Dysidea cinerea* have been shown to inhibit HIV reverse transcriptase (RT) with respect to its natural substrate (dNTP) *(121)*. However, their anti-HIV activity was determined in vitro using cell cultures, so it is not yet known whether the anti-HIV activity was a result of inhibition of HIV-1 RT *(48)*. Avarol and avarone also exhibit strong anticancer and antibacterial activities *(122)*. Another quinine sesquiterpene, ilimaquinone, isolated from the Red Sea sponge *Smenospongia* sp., also has been reported to inhibit the RNase activity of the RT from human HIV type I at concentrations of 5–10 µg/mL, while being less potent against RNA-dependent DNA polymerase and DNA-dependent DNA polymerase *(123)*. Ilimaquinone also exhibits antimitotic and anti-inflammatory activities, promotes a reversible vesiculation of the Golgi apparatus, and interferes with intracellular protein trafficking *(124)*. Bolinaquinone, a sesquiterpene hydroxyl quinine, has recently been isolated from the Philippine *Dysidea* sponge *(117)*. It exhibits activity against the human colon tumor cell line HCT116 with an IC_{50} value of 1.9 µg/mL, mild inhibition of *Bacillus subtilis* at 80 µg/disk, remarkable inhibition of phospholipase A2, and anti-inflammatory activity *(117–120)*.

Illudins are a family of natural toxic sesquiterpene compounds with anti-tumor activity, isolated from the basidiomycete *Omphalotus illudens* (*O. olearius* and *Clitocybe illudens*). These compounds are believed to be responsible for the poisoning that occurs when *Omphalutus* is mistaken for an edible mushroom *(125)*. Illudins S and M are extremely cytotoxic and exhibit antitumor activity *(126,127)*. Illudins are effective against various types of tumor cells at picomolar to nanomolar concentrations. A variety of multidrug-resistant tumor cell lines remain sensitive to the illudins *(128)*. Irofulven (hydroxymethylacylfulvene), a derivative of illudin S, has been extensively investigated and is currently in phase II clinical trials. In particular, irofulven exhibits

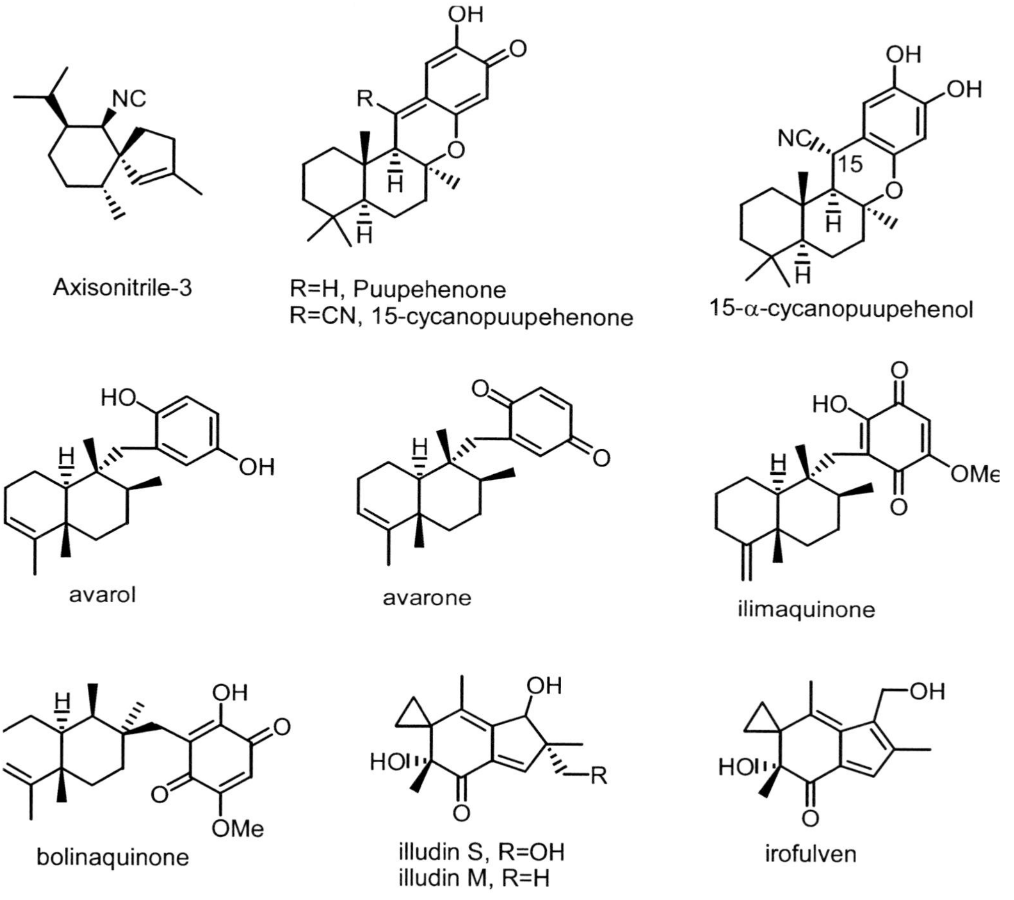

Fig. 6. Sesquiterpenes III.

efficacy against pancreatic carcinoma, a malignancy that is resistant to all other forms of chemotherapy. Irofulven rapidly enters cancer cells, where it binds to cellular macromolecules and inhibits DNA synthesis *(129,130)*. The most unique aspect of irofulven's anticancer activity seems to be its ability to act as a selective inducer of apoptosis in human cancer cell lines, and, in contrast to conventional antitumor agents, this activity of irofulven is effective against tumor cell lines regardless of their p53 or p21 expression *(131,132)*. In addition, the DNA lesion induced by illudins and irofulven is largely ignored by global repair pathways. Therefore, the irofulven and other illudins are considered a new and promising class of tumor-therapeutic agents *(133)*.

3.3. Diterpenes

The diterpenes represent a large group of terpenoids with a wide range of biological activities, isolated from a variety of organisms *(134–146)*. One of the simplest and most important acyclic diterpenes is phytol (**Fig. 7**), a reduced form of geranylgeraniol. This terpenoid is perhaps the most studied biomarker of those found in aquatic environments, and it is a side chain of chlorophyll-a *(12,147)*. (E)-phytol, isolated from

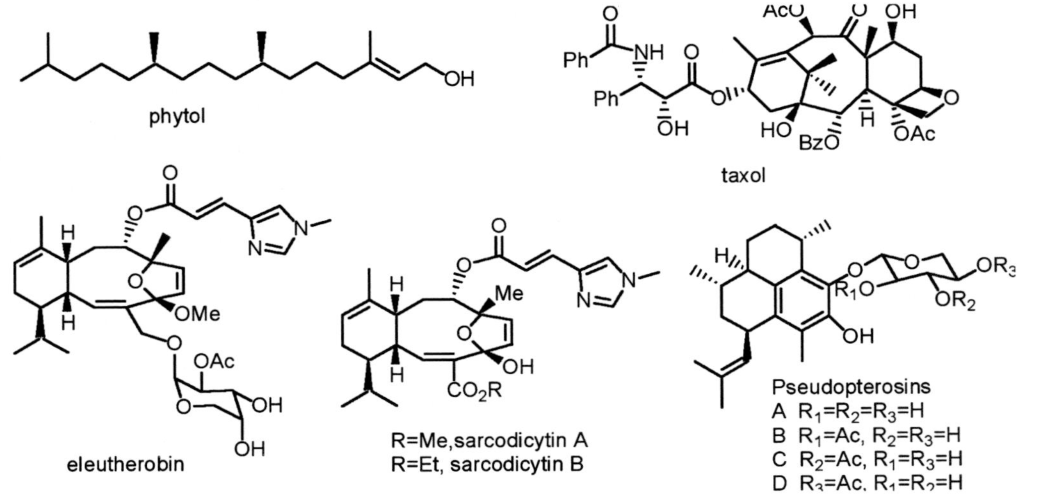

Fig. 7. Diterpene I.

Lucas volkensii, exhibits significant antituberculosis activity, with a MIC of 2 µg/mL *(97)*. Among the cyclic diterpenes, taxines isolated from the common yew (*Taxus baccata*; Taxaceae) represent an important group of compounds whose structure is based on the taxadiene skeleton. Taxines have been well studied because of their anticancer activity *(54)*.

Over 100 different taxanes have been characterized from various *Taxus* species. Paclitaxel (Taxol®) is a member of this group and possesses a four-membered oxetane ring and a complex ester side chain, both of which are essential for anticancer activity. The biosynthesis of Taxol involves complicated cyclizations and modifications (**Fig. 8**). Like epothilones and several other anticancer agents, paclitaxel is able to interact with tubulin (or microtubues) and inhibit cell proliferation by acting on the mitotic spindle through inhibition of microtubule polymerization or microtubule stabilization *(37,38,42,63,148–150)*. Paclitaxel is currently used to treat ovarian, lung, and breast cancers, head and neck carcinoma, and melanoma. It has been hailed as the "perhaps most important addition to the chemotherapeutic armamentarium against cancer over the past several decades"*(151)*.

The development of paclitaxel into an anticancer drug spans several decades. In 1966, a crude extract of bark from the Pacific yew was demonstrated to have a broad range of antitumor activities against several tumor cell lines. In 1971, paclitaxel was identified as the active constituent of the crude extract and its structure determined *(152)*. Paclitaxel was approved by the Food and Drug Administration (FDA) for the treatment of advanced ovarian cancer in 1992 and of breast cancer in 1994. Paclitaxel inhibits microtubule depolymerization, promotes the formation of unusually stable microtubles, and thereby disrupts the dynamic reorganization of the microtubule network required for mitosis and cell proliferation, and causes cellular arrest in the G_2/M phase of the cell cycle *(153,154)*. The protracted arrest of the cell cycle during the mitotic phase is considered to be an important mechanism of paclitaxel-induced cytotoxicity. Nevertheless, the precise mechanisms of cytotoxicity of paclitaxel against

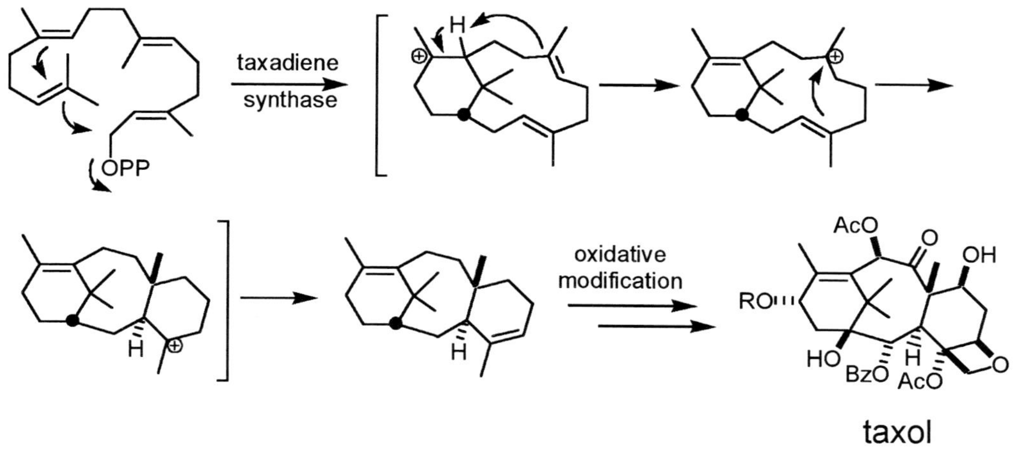

Fig. 8. Biosynthesis of Taxol®.

cancer cells is not entirely clear *(150)*. In addition, the low aqueous solubility and the development of clinical drug resistance are two problems associated with paclitaxel. Furthermore, as for many other anticancer drugs, paclitaxel has several side effects, including neutropenia, peripheral neuropathy, alopecia, and hypersensitivity reactions *(155)*.

Eleutherobin is another microtubule-stabilizing diterpenoid originally isolated from a marine soft coral species of the genus *Eleutherobia (5)*. It was shown to possess activity as an inhibitor of microtubule depolymerization with a mean cytotoxicity greater than those of taxol or the epothilones *(156)*. Eleutherobin belongs to the eleutheside family of marine diterpenoid and is extracted in extremely low yields (0.01–0.02% of the dry weight of the rare alcyonacean *Eleutherobia* sp.). Several elegant total syntheses have been reported *(157–160)*. However, neither chemical syntheses nor the original source have provided sufficient quantities of eleutherobin to permit full in vivo evaluation, and this has thwarted its further development *(161)*. Recently, *Erythropodium caribaeorum*, a relatively abundant Caribbean gorgonian, has been found to be a good source of eleuthesides *(162,163)* and can provide sufficient eleutherobin for preliminary animal studies and chemical transformation to new analogs *(164)*. Sarcodictyin A and B, analogs of eleutherobin, are marine diterpenoids isolated from the Mediterranean stoloniferan *S. roseum (6,7)* and then from the South African soft coral *E. aurea* along with two glycosylated congeners, eleuthosides A and B *(165)*. Sarcodictyin A and B demonstrated very low resistance factors against the P-glycoprotein-overexpressing human cancer cell line *(166)*, and their intrinsic antiproliferative activity against drug-sensitive cells seems significantly lower than those of eleutherobin and Taxol assayed in vitro *(38)*.

Studies involving combinatorial libraries and natural analogs of eleutherobin indicate that the loss of the sugar moiety dramatically changes the potency of eleutherobin against cancer, whereas modification of C-8 side chain has little effect on its potency *(167,168)*. However, the C-8 side chain is essential for the cytotoxicity of the sarcodyctyins; and both imidazole nitrogens are also required *(160)*.

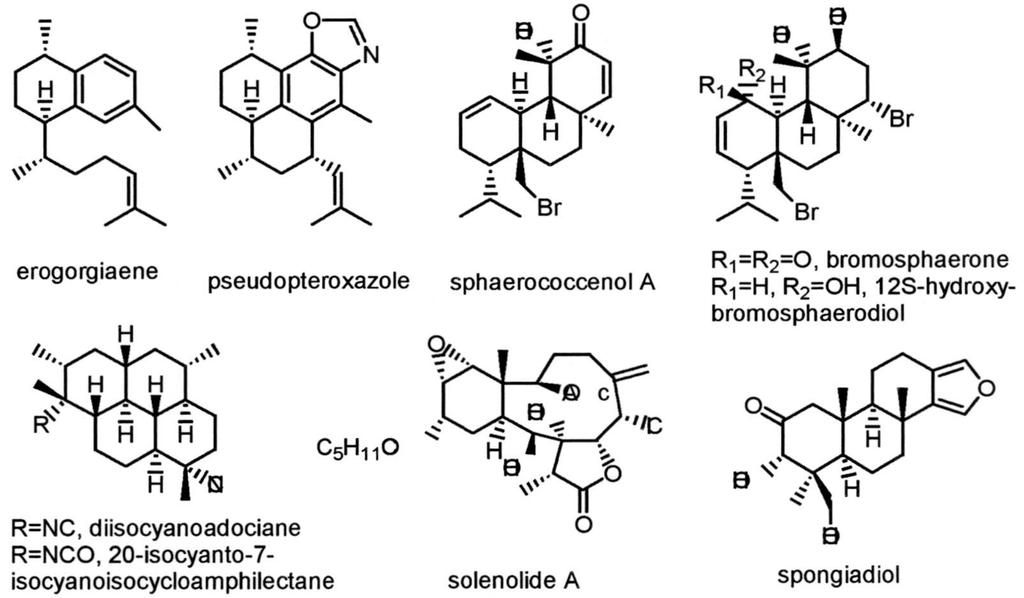

erogorgiaene   pseudopteroxazole   sphaerococcenol A $R_1=R_2=O$, bromosphaerone
$R_1=H$, $R_2=OH$, 12S-hydroxy-bromosphaerodiol

R=NC, diisocyanoadociane
R=NCO, 20-isocyanto-7-isocyanoisocycloamphilectane   solenolide A   spongiadiol

Fig. 9. Diterpene II.

Extracts of the Gorgonian *Pseudopterogorgia elisabethae* show anti-inflammatory activity and are currently used as an ingredient in cosmetic skin-care products *(70)*. The active constituents in the extracts have been identified as diterpene glycosidesm (pseudopterosins), which have analgestic properties and are used for promoting wound healing and as inhibitors of PLA_2 *(169,170)*. Methopterosin (OAS100), a semisynthetic product of pseudopterosin, also has anti-inflammatory activity and is in clinical development for the promotion of wound healing *(55,70)*. Erogorgiaene (**Fig. 9**), a serrulatane diterpene (also known as biflorane), isolated from the West Indian gorgonian *P. elisabethae*, induces 96% growth inhibition for *M. tuberculosis* H37V at a concentration of 12.5 µg/mL *(171)*. The benzoxazole diterpene alkaloid pseudopteroxazole, isolated from the same gorgonian, shows 97% growth inhibition for *M. tuberculosis* H37Rv at a concentration of 12.5 µg/mL without substantial toxic effects *(172)*.

The bromoditerpene sphaerococcenol A, isolated from the red alga *Sphaerococcus coronopifolius* collected along the Atlantic coast of Morocco, has antimalarial activity against the chloroquine-resistant *Plasmodium falsciparum* FCB1 strain, with an IC_{50} of 1 µM *(173)*. Two other bromoditerpenes, bromosphaerone and 12S-hydroxy-bromosphaerodiol, isolated from the same species, show strong antibacterial activity against *Staphylococcus aureus*, with MICs of 0.104 and 0.146 µM, respectively *(173,174)*.

Diisocyanoadociane, a tetracyclic diterpene with an isocycloamphilectane skeleton, isolated from the sponge *Cymbastela hooperi*, has been reported to have significant antimalarial activity in vitro against two clones of the malaria parasite *P. falciparum* *(175)*. The tetracyclic diterpene demonstrated significant antiplasmodial activity, with IC_{50} values of 4.7 ng/mL and 4.3 ng/mL and selectivity indices (SI) 1000 and 1100 against the two clones of *P. falciparum*. Another tricyclic diterpene, 20-isocyanto-7-isocyanoisocycloamphilectane, isolated from the same sponge, also shows the same level of antiplasmodial activity and selectivity. In addition, the potency and selectivity of the two tetracyclic diterpenes from the sponge *C. hooperi* rivals the in vitro results obtained with the currently prescribed antimalarial drugs (Artemisinin and chloroquine) *(173)*.

Finally, solenolide A, a diterpene lactone, isolated from a new Indopacific gorgonian species of the genus *Solenopodium*, has been reported to inhibit rhinovirus with an IC_{50} value of 0.39 µg/mL. This diterpene also shows inhibitory activity against poliovirus III, herpesvirus, and the Ann Arbor and Maryland viruses *(176)*. Spongiadiol, a tetracyclic furanoditerpene isolated from deep-water *Spongia* sp., shows inhibitory activity against HSV1 at a concentration of 0.5 µg/disk *(177,178)*.

3.4. Sesterterpenes

Sesterterpenes are the least common group of terpenoids. They are primarily isolated from fungi and marine organisms, and encompass relatively few structural types *(12,13)*. Although many examples of natural sesterterpenes are known, studies of their biosynthesis are rare *(12)*.

Many sesterterpenes inhibit the activity of the human secreted type IIA phospholipase A_2 (PLA_2). LPA_2 is involved in the pathogenesis of a variety of inflammatory diseases via the production of arachidonic acid, the precursor of prostaglandins and leukotrienes. Therefore, secreted PLA_2 has been considered to be a primary target for the development of anti-inflammatory drugs *(119)*. The well-known PLA_2 inhibitors among sponge sesterterpenes are the scalaranes, which were named after scalaradial (**Fig. 10**), isolated from *Cacospongia mollior* *(179,180)*. Marine sesterterpenes containing the γ-hydroxybutenolide moiety have been studied for their potent anti-inflammatory activity. Manoalide, which is the first marine PLA_2 inhibitor of sesterterpene and a reference compound for this class of natural products, is a potent analgesic and anti-inflammatory agent isolated from the pacific sponge *Luffariella variabilis* *(181)*. The anti-inflammatory activity of manoalide is ascribed to its irreversible inhibition of PLA_2 *(181–183)*, resulting in inhibition of the formation of pro-inflammatory mediators such as leukotrienes and prostaglandins *(184,185)*. This compound was formerly in phase I clinical trials, but its development was later discontinued *(55)*. Thereafter, many other related molecules have been isolated, such as seco-manoalide, luffariellolide, luffariellins, luffolide, the cacospongionolides, and the petrosaspongiolides M-R *(181,186–192)*, all of which are irreversible inhibitors of PLA_2. Among these compounds, petrosaspongiolide M has been investigated for its pharmacological activity in vivo and in vitro. This compound has significant inhibitory activity against PLA_2 (IC_{50} of 0.6 µM) *(186)* and reduces the level of prostaglandin E2, tumor necrosis factor α, and leukotriene B4 in a dose-dependent manner *(193–195)*. Moreover, it has no significant side effects. Consequently, there is considerable interest in the development of this molecule and its analogs for the treatment of acute and/or chronic inflammation *(194,196)*.

The novel sesterterpene salmahyrtisol A (**Fig. 11**) and three other new scalaranes-type sesterterpenes (3-acetyl sesterstatin 1, 19-acetyl sesterstatin 3, and salmahyrtisol) isolated from the Red Sea sponge *Hyrtios erecta*, show significant cytotoxicity

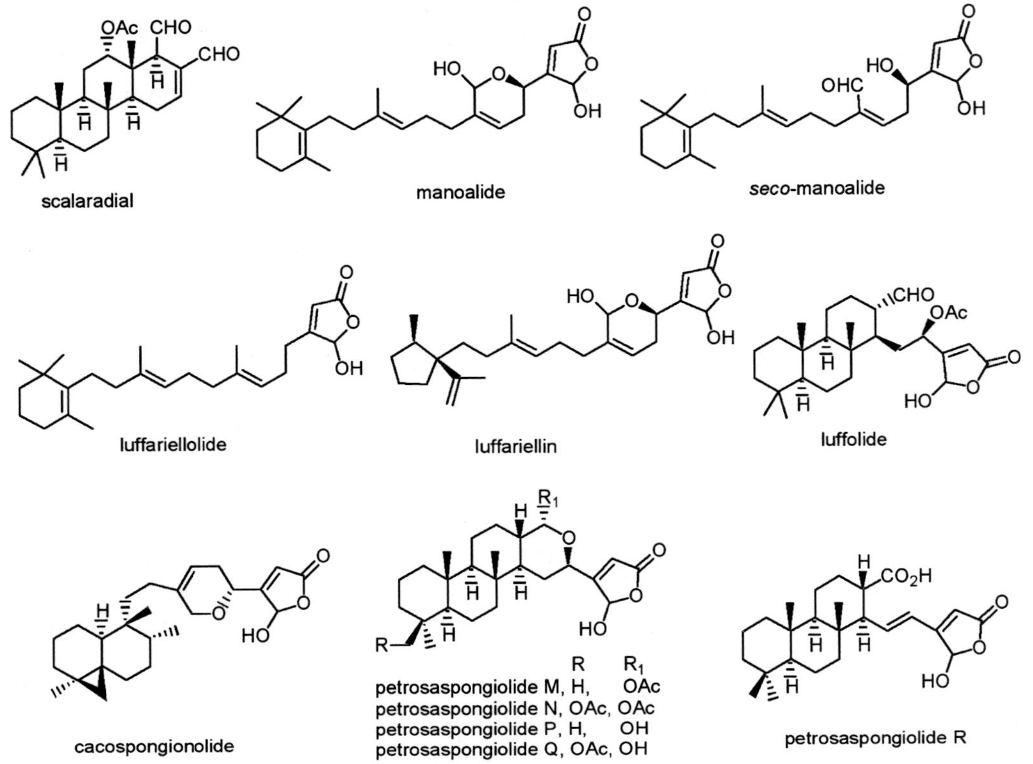

Fig. 10. Sesterterpene I.

against murine leukemia, human lung carcinoma, and human colon carcinoma *(197)*. Halorosellinic acid, an ophiobolane sesterterpene isolated from the marine fungus *Halorosellinia oceanica*, exhibits antimalarial activity, with an IC_{50} of 13 µg/mL, and weak antimycobacterial activity, with a MIC of 200 µg/mL *(198)*. Mangicols A-G are a group of sesterterpenes possessing novel spirotricyclic natural compounds, isolated from the marine fungus tentatively identified as *Fusarium heterosporum (199)*. Among these compounds, mangicols A and B show significant anti-inflammatory activity in phorbol myristate acetate-induced mouse ear edema assay.

4. Chemical Synthesis in Terpenoid Drug Production and Development

During the first part of the 20th century, total synthesis played a central role in identification and structure confirmation of the active principle in a crude extract from natural sources *(200,201)*. However, with the development of modern analytical and purification techniques and spectroscopic methods, structures of most natural products can now be determined unambiguously. Nevertheless, total synthesis is still the ultimate proof for the structure of complex natural products, especially the absolute stereochemistry or remote relative stereochemistry. Many terpenoids with promising biological activity are structurally too complex to be readily synthesized in a cost-effective way via total synthesis from commercially available materials. Semisynthesis,

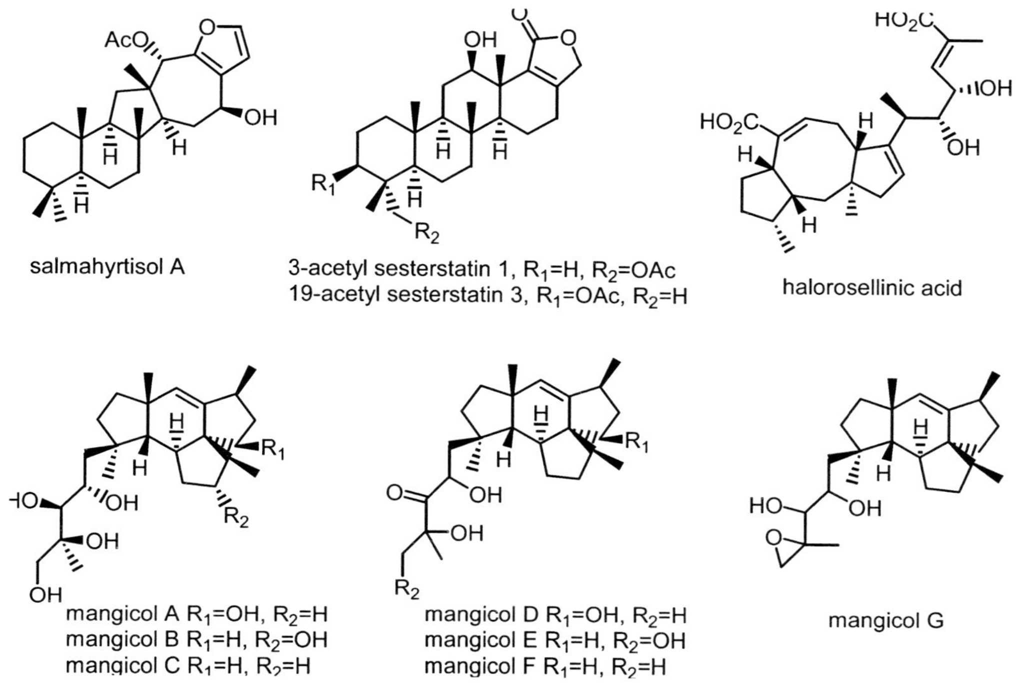

salmahyrtisol A

3-acetyl sesterstatin 1, R$_1$=H, R$_2$=OAc
19-acetyl sesterstatin 3, R$_1$=OAc, R$_2$=H

halorosellinic acid

mangicol A R$_1$=OH, R$_2$=H
mangicol B R$_1$=H, R$_2$=OH
mangicol C R$_1$=H, R$_2$=H

mangicol D R$_1$=OH, R$_2$=H
mangicol E R$_1$=H, R$_2$=OH
mangicol F R$_1$=H, R$_2$=H

mangicol G

Fig. 11. Sesterterpene II.

on the other hand, has played a major role in the development and production of terpenoid-derived drugs.

The isolation of paclitaxel from nature did not produce enough material for clinical trials *(202)*. In addition, killing yew trees for compound isolation was a big environmental concern. The four-step semisynthesis (**Fig. 12**) of paclitaxel from 10-deacetylbaccatin III greatly facilitated the development of Taxol. 10-deacetylbaccatin III was found in the needles of the common European yew tree *T. beccata* as well as a yew tree species found in India. By harvesting and extracting the needles, baccatin III or 10-deactyl baccatin III can be provided in large quantities as precursors of paclitaxel without substantially injuring the tree populations. In addition to paclitaxel, analogs such as docetaxel (Taxotere®) were synthesized in sufficient yield by semisynthesis. Compared with paclitaxel, semisynthetic docetaxel has improved water solubility and is being used clinically against breast and ovarian cancer *(151)*. Six groups completed total synthesis of paclitaxel between 1994 and 2002 *(203–216)*. However, none of their methods are comparable to semi-synthesis in terms of the cost for large-scale production.

Artemisinin (Qinghaosu) has also been produced by semisynthesis. Some phenotypes of *A. annula* have been found to produce as much as 1% Artemisinin, but the yield is normally very much less, typically 0.05–0.2%. The more abundant sesquiterpene in the plant is artemisinic acid (qinghao acid, typically 0.2–0.8%). Fortunately, artemisinic acid can be converted chemically into Artemisinin by a relatively simple

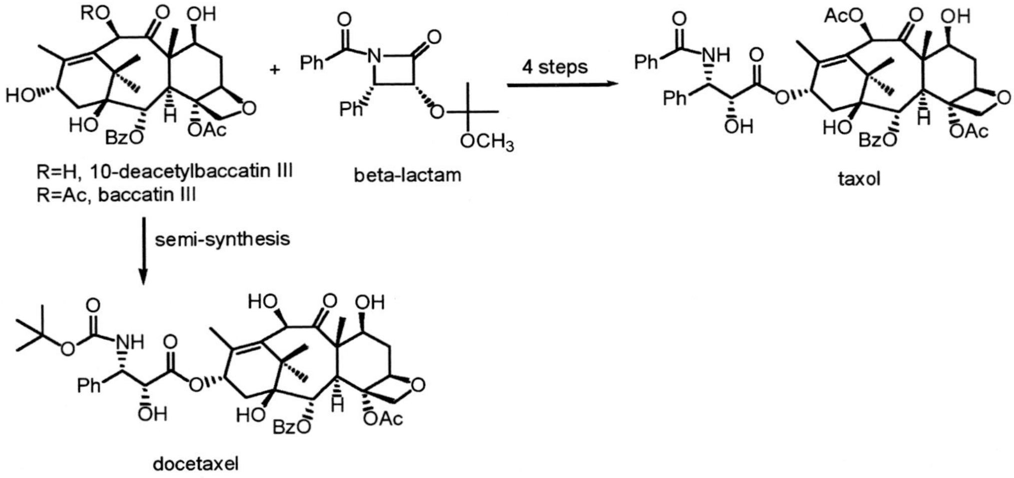

Fig. 12. Semisynthesis of Taxol and its analog.

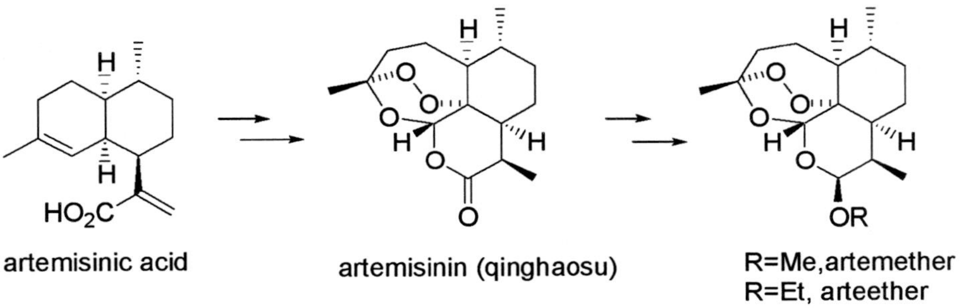

Fig. 13. Semisynthesis of Artemisinin and its analogs.

and efficient process (**Fig. 13**) *(217–219)*. At the same time, Artemisinin can be reduced to the lactol (hemiacetal), and this can be used for the semi-synthesis of a range of analogs, of which artemether, arteether, and the water-soluble sodium salts of artelinic acid and artesunic acid are promising second-generation antimalarial drugs *(220)*. The total synthesis of Artemisinin was completed in 1983, and the peroxy bridge was confirmed to be its unique feature conferring antimalarial activity *(221)*.

Although total synthesis plays a less important role in production of terpenoid drugs, it has been shown to be an important tool in the development of compounds derived from terpenoids and other natural compounds. First, total synthesis is an essential tool for studying structure–activity relationships (SAR). For example, the binding of manoalide to PLA_2 was demonstrated by using the two-masked aldehyde present in the molecule. Extensive SAR studies using chemical synthesis of various analogs have revealed that the minimum structural requirement for activity is the presence in the inhibitor of functional groups able to bind to the amino groups of PLA_2 lysine residues. Many manoalide analogs have thus been synthesized, and many of them share PLA_2 inhibitory properties *(222)*. A similar approach has also revealed the active moieties of paclitaxel, eleutherobin, sarcodictyin A, and Artemisinin *(106,160,163, 167,168,223)*.

Second, it is often difficult to correlate the relative stereochemistry of remote chiral centers by spectroscopic methods. Total synthesis is the ultimate tool for solving this problem. For example, the chemical structure of eleutherobin was elucidated by extensive two-dimensional nuclear magnetic resonance spectroscopy and mass spectrometry *(5)*. L-arabinose, the sugar moiety of eleutherobin, appeared to be favored in nature. This was proved to be wrong by total synthesis. The relative stereochemistry of the diterpenoid 4,7-oxaeunicellane skeleton and the arabinose of eleutherobin were assigned by the Danishefsky and Nicolaou groups through total synthesis *(224–227)*. It is remarkable that they could unambiguously identify the sugar unit of eleutherobin as D-arabinose despite the fact that L-arabinose is the natural abundant enantiomer.

Finally, libraries of structurally diverse natural-product-like molecules have contributed to the understanding of biological processes in small molecule-based systematic approaches *(228)*. Efficient access to diversity can be achieved in diversity-oriented synthesis using pairs of complexity-generating reactions, where the product of one is the substrate for another *(229)*. Diversity can also be achieved by chemical modification of natural products. For example, the potential of macrocyclic diterpenoids to afford natural-product-like polycyclic compounds was demonstrated by the conversion of two lathyrane *Euphorbia* factors into a series of densely functionalized diterpenoids of unnatural skeletal type (**Fig. 14**) *(230)*.

5. The Sustainable Production of Terpenoids

Like many other natural products, terpenoids are usually extracted from the source organism in extremely low yields. Producing an adequate and sustainable supply of these compounds has been one of the major challenges for developing terrestrial and marine organism-derived natural terpenoid compounds into clinically useful entities *(164)*. Harvesting source organisms from the environment for the target terpenoid is not a feasible strategy because of environmental and ecological concerns, especially for terpenoids derived from marine environments. For example, eleutherobin, a structurally complex diterpene glycoside, constitutes 0.01–0.02% of the dry weight of the rare alcyonacean coral *Eleutherobin* sp. *(5)*. The limited supply of paclitaxel from its primary source once hampered the development of paclitaxol into a clinically used drug *(150)*. The limited supply of eleutherobin has restricted the evaluation of this promising marine diterpene *(162)*. Consequently, a serious obstacle to the commercial development of most promising terpenoids is their availability from the natural environment. To provide enough material for preclinical evaluation and clinical trials, several approaches have been used to produce terpenoids with promising biological activities, and their implementation should greatly facilitate the development of novel terpenoid-related drugs.

5.1. Metabolic Engineering

The redirection of metabolic pathways for enhanced production of existing natural products or for production of "unnatural" natural products has been an active area of research. Metabolic engineering, which integrates engineering design and systematic and quantitative analysis of pathways using molecular biology, modern analytical tech-

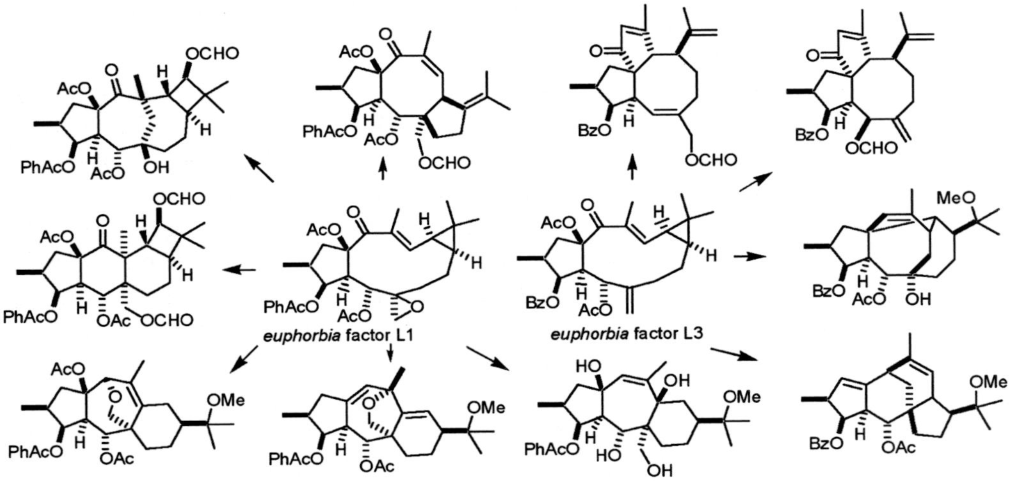

Fig. 14. Diversity via natural-product derivatization.

niques, genomic approaches, and "-omics" technologies into production of natural products, will provide alternative approaches to produce terpenoids with pharmaceutical value. Indeed, enhancing monoterpene yields and changing the metabolite composition of essential oils in plants has been shown to be feasible. Using the transgenic plants to increase the production of monoterpenes for scent and flavor has been demonstrated in flowers *(231,232)* and fruits *(233)*. Moreover, the pathway engineering approach has been used to successfully produce several interesting sesquiterpenes, including amorphadiene, the precursor of the antimalarial drug artemisinin, in *E. coli* *(234,235)*. As a powerful tool for the production of both natural products and the library of their analogs, the engineered fermentable microbial host will provide renewable and sustainable resources to generate terpenoids with interesting biological activities.

5.2. Cell Culture and Others

Advances in cell culture have also provided a promising means for the production of valuable secondary metabolites from plants. Production of paclitaxel and other taxanes using cell-culture techniques has been well established. The level of taxoid production in bioreactors has been reported to reach 612 mg/L in *T. chinensis* cell cultures *(236)*. Strategies for enhancing paclitaxel and its precursors have been reviewed extensively by Zhong *(237)*. At the same time, the production of artemisinin has also been pursued in callus, cell suspension, and shoot and hairy root cultures of *A. annua (238–242)*. Undifferentiated callus and cell suspension cultures produce extremely low yields of artemisinin. However, differentiated shoot cultures and hairy roots show promising potential for artemisinin production *(237)*. Compared with the paclitaxel production level, the production yield of artemisinin in cell culture is still not significant enough to be used for commercial purposes.

The bioprocess engineering of marine macroalgae for the production of halogenated monoterpenes has been carried out in small-scale tissue-culture bioreactors *(243–245)*. In the established system, the photosynthetic tissue culture was developed for the marine red algae *Ochtodes secondiramea* and *P. hornemannii* using callus-induction and shoot-regeneration techniques. Rates of monoterpene production have been shown to be affected by temperature, light intensity, and nitrate concentration in the growth medium *(246,247)*. Metabolic flux analysis indicated that the halogenated monoterpene production was not limited by its precursor (myrcene) and that chlorination is the target for increasing the production *(248)*.

Recently, aquaculture systems designed and engineered for the production of large quantities of biomass of two species of marine invertebrates (*Bugula neritina* and *Ecteinascidia turbinate*) have been established in order to harvest their natural chemical constituents *(249)*. Aquaculture will also provide an alternative approach for producing valuable terpenoids from marine invertebrates in a reliable, renewable, and cost-effective way. However, additional research is still needed in order to optimize controlled environment culture systems and to explore the feasibility of such systems for other marine invertebrates. Finally, the Caribbean gorgonian *E. caribaeorum* has been cultured in shallow running seawater tanks located in a greenhouse under ambient sunlight illumination *(164)*. The cultured *E. caribaeorum* produce eleutherobin and the briarane diterpenoids erythrolides A and B in yields comparable to those reported from wild-harvest reef animals. Therefore, the aquaculture could enhance large-scale marine terpenoid production in the near future.

6. Concluding Remarks

Terpenoid-derived drugs have contributed significantly to human disease therapy and prevention. Some terpenoid drugs have provided tremendous benefits for patients and for the pharmaceutical industry. Artemisinin and its derivatives comprise a multi-million-dollar market worldwide. Taxol alone is estimated to have annual sales of over $1.8 billion. Terpenoids indisputably continue to be important compounds for drug discovery. In last two decades in particular, many terpenoids with promising biological activities have been isolated from diverse marine environments. These marine terpenoids exhibit an impressive array of novel structural motifs, many of which are considered to be derived from biosynthetic pathways that are exclusive to marine organisms. Moreover, most of these marine terpenoids possess remarkable biological activities whose potential benefits extend beyond the marine ecosystem and embody the development of new antifungal, anticancer, anti-inflammatory, and antiviral drugs *(250)*. Although marine natural-product bioprospecting has begun only relatively recently, it has already yielded over a thousand novel molecules. Marine natural-compound bioprospecting will continue to provide more promising terpenoids and other natural products for drug development. In addition, marine biodiversity is estimated to be much greater than that on land *(55)*. Many marine animals, plants, fungi, and algae are rich sources of novel terpenoids. With the continuing exploration of the oceans and affordable technology for the exploration of the deep oceans, more novel terpenoids with promising biological activities are expected to be discovered from marine environments in the near future.

In addition, chemical synthesis (e.g., derivatization of natural products) and combinatorial synthesis of natural product analogs can also play a significant role in providing SAR data for a particular target. Accurate knowledge of the structural features required for activity in each compound class will give the predictive power of pharmacophore models. For example, several pharmacophore models have attempted to reconcile the SAR data for taxoids and other compound types in order to generate a sufficiently detailed understanding of tubulin-binding requirements. Information from this approach will allow rational design of new classes of microtubule-stabilizing drugs. The terpenoids encompass a wealth of significantly diverse compounds, providing chemists with great opportunities to synthesize not only new, valuable terpenoid drugs, but also novel terpenoid compounds that can be used as tools for understanding biochemical pathways.

Finally, along with the information derived from the human genome and metabolic pathways, high-throughput technologies such as proteomics, transcriptomics, and metabolomics will greatly shorten the time required for assaying natural compounds in vivo and in vitro. In parallel, new analytical instruments and synthetic approaches will further facilitate both the identification of terpenoids from natural sources and better understanding the SAR. In particular, marine biodiversity is far beyond what we have dreamed. Isolation and identification of marine terpenoids should draw more attention from drug discovery and development. Engineered new assays and high-throughput tools should facilitate identification and isolation of new terpenoids. There is no doubt that more terpenoid-based clinical drugs will become available and play a more significant role in human disease therapy.

References

1. Newman DJ, Cragg GM, Snader KM. The influence of natural products upon drug discovery. Nat Prod Rep 2000;17(3):215–234.
2. Newman DJ, Cragg GM, Snader KM. Natural products as sources of new drugs over the period 1981–2002. J Nat Prod 2003;66(7):1022–1037.
3. Hill RA. Terpenoids. In: Thomson RH (ed), The Chemistry of Natural Products. Chapman & Hall, Blackie Academic & Professional, New York:;, 1993.
4. Eisenreich W, Schwarz M, Cartayrade A, Arigoni D, Zenk MH, Bacher A. The deoxyxylulose phosphate pathway of terpenoid biosynthesis in plants and microorganisms. Chem Biol 1998;5(9):R221–R233.
5. Lindel T, Jensen PR, Fenical W, et al. Eleutherobin, a new cytotoxin that mimics paclitaxel (taxol) by stabilizing microtubules. J Am Chem Soc 1997;119(37):8744–8745.
6. D'Ambrosio M, Guerriero A, Pietra F. Sarcodictyin a and sarcodictyin b novel diterpenoidic alcohols esterified by e-n1 methylurocanic acid isolation from the mediterranean stolonifer sarcodictyon-roseum. Helv Chim Acta 1987;70(8):2019–2027.
7. D'Ambrosio M, Guerriero A, Pietra F. Isolation from the mediterranean stoloniferan coral sarcodictyon-roseum of sarcodictyin c d e and f novel diterpenoidic alcohols esterified by e or z-n1 methylurocanic acid failure of the carbon-skeleton type as a classification criterion. Helv Chim Acta 1988;71(5):964–976.
8. Dubey VS, Bhalla R, Luthra R. An overview of the non-mevalonate pathway for terpenoid biosynthesis in plants. J Biosci 2003;28(5):637–646.
9. Kuzuyama T, Seto H. Diversity of the biosynthesis of the isoprene units. Nat Prod Rep 2003;20(2):171–183.
10. Buchanan SG. Structural genomics: bridging functional genomics and structure-based drug design. Curr Opin Drug Discov Dev 2002;5(3):367–381.
11. Kuzuyama T. Mevalonate and nonmevalonate pathways for the biosynthesis of isoprene units. Biosci Biotechnol Biochem 2002;66(8):1619–1627.

12. Dewick PM. The biosynthesis of C-5–C-25 terpenoid compounds. Nat Prod Rep 2002;19(2): 181–222.
13. Dewick PM. The mevalonate and deoxyxylulose phosphate pathways: terpenoids and steroids. In: Dewick PM (ed), Medicinal Natural Products. John Wiley & Sons, Ltd., 2002.
14. Rohdich F, Kis K, Bacher A, Eisenreich W. The non-mevalonate pathway of isoprenoids: genes, enzymes and intermediates. Curr Opin Chem Biol 2001;5(5):535–540.
15. Wanke M, Skorupinska-Tudek K, Swiezewska E. Isoprenoid biosynthesis via 1-deoxy-D-xylulose 5-phosphate/2-C-methyl-D-erythritol 4-phosphate (DOXP/MEP) pathway. Acta Biochim Pol 2001;48(3):663–672.
16. Moore BS. Biosynthesis of marine natural products: microorganisms and macroalgae. Nat Prod Rep 1999;16(6):653–674.
17. Lichtenthaler HK. The 1-deoxy-D-xylulose-5-phosphate pathway of isoprenoid biosynthesis in plants. Annu Rev Plant Physiol Plant Molec Biol 1999;50:47–65.
18. Ogura K, Koyama T. Enzymatic aspects of isoprenoid chain elongation. Chem Rev 1998; 98(N4):1263–1276.
19. Wang KC, Ohnuma S-i. Isoprenyl diphosphate synthases. Biochim Biophys Acta 2000;1529(1–3):33–48.
20. Poulter CD, and Rilling, H. C. In: Spurgeon SL, and Port JW (eds), Biosynthesis of Isoprenoid Compounds. John Wiley & Sons, New York: 1981; pp. 161–224.
21. Heide L, Berger U. Partial purification and properties of geranyl pyrophosphate synthase from Lithospermum erythrorhizon cell cultures. Arch Biochem Biophys 1989;273(2):331–338.
22. Ohnuma S-I, Hirooka K, Tsuruoka N, et al. A pathway where polyprenyl diphosphate elongates in prenyltransferase: Insight into a common mechanism of chain length determination of prenyltransferases. J Biol Chem 1998;273(41):26,705–26,713.
23. Tachibana A, Yano Y, Otani S, Nomura N, Sako Y, Taniguchi M. Novel prenyltransferase gene encoding farnesylgeranyl diphosphate synthase from a hyperthermophilic archaeon *Aeropyrum pernix*—molecular evolution with alteration in product specificity. Eur J Biochem 2000;267(2): 321–328.
24. Tachibana A. A novel prenyltransferase, farnesylgeranyl diphosphate synthase, from the haloalkaliphilic archaeon, *Natronobacterium-pharaonis*. FEBS Lett 1994;341(2–3):291–294.
25. Caruthers JM, Kang I, Rynkiewicz MJ, Cane DE, Christianson DW. Crystal structure determination of aristolochene synthase from the blue cheese mold, Penicillium roqueforti. J Biol Chem 2000;275(33):25,533–25,539.
26. Lesburg CA, Zhai GZ, Cane DE, Christianson DW. Crystal structure of pentalenene synthase—mechanistic insights on terpenoid cyclization reactions in biology. Science 1997;277(5333): 1820–1824.
27. Lesburg CA, Caruthers JM, Paschall CM, Christianson DW. Managing and manipulating carbocations in biology: terpenoid cyclase structure and mechanism. Curr Opin Struct Biol 1998;8(6):695–703.
28. Abe I RM, Prestwich G. D. Enzymatic cyclization of squalene and squalene oxide to sterols and triterpenes. Chem Rev 1993;93:2189–2206.
29. Cane DE. Enzymatic formation of sesquiterpenes. Chem Rev 1990;90:1089–1103.
30. Rising KA, Starks CM, Noel JP, Chappell J. Demonstration of germacrene A as an intermediate in 5-epi-aristolochene synthase catalysis. J Am Chem Soc 2000;122(9):1861–1866.
31. Rohmer M. The discovery of a mevalonate-independent pathway for isoprenoid biosynthesis in bacteria, algae and higher plants. Nat Prod Rep 1999;16(5):565–574.
32. Wagner KH, Elmadfa I. Biological relevance of terpenoids—overview focusing on mono-, di- and tetraterpenes. Ann Nutr Metab 2003;47(3–4):95–106.

33. Kuroda C, Suzuki H. Synthesis of odd-membered rings by the reaction of beta-carbonylallylsilane or its derivative as a carbon 1,3-dipole. Curr Org Chem 2003;7(2):115–131.
34. Ridley RG. Medical need, scientific opportunity and the drive for antimalarial drugs. Nature 2002;415(6872):686–693.
35. Abdin MZ, Israr M, Rehman RU, Jain SK. Artemisinin, a novel antimalarial drug: Biochemical and molecular approaches for enhanced production. Planta Med 2003;69(4):289–299.
36. Abebe W. Herbal medication: potential for adverse interactions with analgesic drugs. J Clin Pharm Ther 2002;27(6):391–401.
37. Altaha R, Fojo T, Reed E, Abraham J. Epothilones: a novel class of non-taxane microtubule-stabilizing agents. Curr Pharm Design 2002;8(19):1707–1712.
38. Altmann KH. Microtubule-stabilizing agents: a growing class of important anticancer drugs. Curr Opin Chem Biol 2001;5(4):424–431.
39. Bez G, Kalita B, Sarmah P, Barua NC, Dutta DK. Recent developments with 1,2,4-trioxane-type artemisinin analogues. Curr Org Chem 2003;7(12):1231–1255.
40. Blunt JW, Copp BR, Munro MHG, Northcote PT, Prinsep MR. Marine natural products. Nat Prod Rep 2003;20(1):1–48.
41. Bongiorni L, Pietra F. Marine natural products for industrial applications. Chem Indus 1996(2):54–58.
42. Borzilleri RM, Vite GD. Epothilones: new tubulin polymerization agents in preclinical and clinical development. Drug Future 2002;27(12):1149–1163.
43. Burkhart CG, Burkhart CN, Burkhart KM. An assessment of topical and oral prescription and over-the-counter treatments for head lice. J Am Acad Derm 1998;38(6 Part 1):979–982.
44. Caniato R, Puricelli L. Review: natural antimalarial agents (1995–2001). Crit Rev Plant Sci 2003;22(1):79–105.
45. Cantrell CL, Franzblau SG, Fischer NH. Antimycobacterial plant terpenoids. Planta Med 2001;67(8):685–694.
46. Carte BK. Biomedical potential of marine natural products. Bioscience 1996;46(4):271–86.
47. Chakraborty Tushar K, Das S. Chemistry of potent anti-cancer compounds, amphidinolides. Curr Med Chem Anticancer Agents 2001;1(2):131–149.
48. De Clercq E. Current lead natural products for the chemotherapy of human immunodeficiency virus (HIV) infection. Med Res Rev 2000;20(5):323–349.
49. Donia M, Hamann MT. Marine natural products and their potential applications as anti-infective agents. Lancet Infect Dis 2003;3(6):338–348.
50. El Sayed KA, Bartyzel P, Shen XY, Perry TL, Kjawiony JK, Hamann MT. Marine natural products as antituberculosis agents. Tetrahedron 2000;56(7):949–953.
51. Frederich M, Dogne JM, Angenot L, De Mol P. New trends in anti-malarial agents. Curr Med Chem 2002;9(15):1435–1456.
52. Fujiki H, Suganuma M, Yatsunami J, et al. Significant marine natural products in cancer research. Gaz Chim Ital 1993;123(6):309–316.
53. Gochfeld DJ, El Sayed KA, Yousaf M, et al. Marine natural products as lead anti-HIV agents. Minni Rev Med Chem 2003;3(5):401–424.
54. Gogas H, Fountzilas G. The role of taxanes as a component of neoadjuvant chemotherapy for breast cancer. Ann Oncol 2003;14(5):667–674.
55. Haefner B. Drugs from the deep: marine natural products as drug candidates. Drug Discov Today 2003;8(12):536–544.
56. Haynes RK. Artemisinin and derivatives: the future for malaria treatment? Curr Opin Infect Dis 2001;14(6):719–726.
57. Heras BDL, Rodriguez B, Bosca L, Villar AM. Terpenoids: sources, structure elucidation and therapeutic potential in inflammation. Curr Topics Med Chem 2003;3:171–185.

58. Heym B, Cole ST. Multidrug resistance in *Mycobacterium tuberculosis*. Int J Antimicrob Agents 1997;8(1):61–70.
59. Ireland Chris M. Mining the world's oceans for medicinals. Cancer Epidem Biomark Prevention 2002;11(10 Part 2).
60. Itokawa H, Takeya K, Hitotsuyanagi Y, Morita H. Antitumor compounds isolated from higher plants. Yakugaku Zasshi 1999;119(8):529–583.
61. Jefford CW. Why artemisinin and certain synthetic peroxides are potent antimalarials. Implications for the mode of action. Curr Med Chem 2001;8(15):1803–1826.
62. Jung M, Lee S, Kim H, Kim H. Recent studies on natural products as anti-HIV agents. Curr Med Chem 2000;7(6):649–661.
63. Kavallaris M, Verrills NM, Hill BT. Anticancer therapy with novel tubulin-interacting drugs. Drug Resist Update 2001;4(6):392–401.
64. Koenig Gabriele M, Wright Anthony D, Franzblau Scott G. Assessment of antimycobacterial activity of a series of mainly marine derived natural products. Planta Med 2000;66(4):337–342.
65. Kris-Etherton PM, Hecker KD, Bonanome A, et al. Bioactive compounds in foods: their role in the prevention of cardiovascular disease and cancer. Am J Med 2002;113(Suppl 9B):71–88.
66. Lloyd AW. Marine natural products as therapeutic agents. Drug Discov Today 2000;5(1):34.
67. Mann J. Natural products in cancer chemotherapy: past, present and future. Nat Revs Cancer 2002;2(2):143–148.
68. Mayer Alejandro MS, Gustafson Kirk R. Marine pharmacology in 2000: Antitumor and cytotoxic compounds. Int J Cancer 2003;105(3):291–299.
69. Mayer Alejandro MS, Lehmann Virginia KB. Marine pharmacology in 1999: antitumor and cytotoxic compounds. Anticancer Res 2001;21(4A):2489–2500.
70. Proksch P, Edrada RA, Ebel R. Drugs from the seas—current status and microbiological implications. Appl Microbiol Biotechnol 2002;59(2–3):125–134.
71. Wylie Bryan L, Ernst Nadia B, Grace Krista JS, Jacobs Robert S. Marine natural products as phospholipase A-2 inhibitors. Progress in Surgery Uhl, W 1997;24:146–152.
72. Loza-Tavera H. Monoterpenes in essential oils-biosynthesis and properties. Adv Exp Med Biol 1999;464:49–62.
73. Little DB, Croteau R. Biochemistry of essential oil plants: a thirty year overview. In: Teranishi R, Wick EL, Hornstein I (eds), Flavor Chemistry: Thirty years of Progress: Kluwer Academic/Plenum, 1999.
74. Niinemets U, Hauff K, Bertin N, Tenhunen JD, Steinbrecher R, Seufert G. Monoterpene emissions in relation to foliar photosynthetic and structural variables in Mediterranean evergreen *Quercus* species. New Phytol 2002;153(2):243–256.
75. Sharkey TD, Yeh SS. Isoprene emission from plants. Annu Rev Plant Physiol Plant Molec Biol 2001;52:407–436.
76. Davis EM, Croteau R. Cyclization enzymes in the biosynthesis of monoterpenes, sesquiterpenes, and diterpenes. Biosynth Arom Polyket Isopren Alkal 2000;209:53–95.
77. Crowell PL. Prevention and therapy of cancer by dietary monoterpenes. J Nutr 1999;129(3):775S–778S.
78. Duetz WA, Fjallman AHM, Ren SY, Jourdat C, Witholt B. Biotransformation of D-limonene to (+) trans-carveol by toluene-grown *Rhodococcus opacus* PWD4 cells. Appl Environ Microbiol 2001;67(6):2829–2832.
79. Fabian CJ. Breast cancer chemoprevention: beyond tamoxifen. Breast Cancer Res 2001;3(2):99–103.
80. Wattenberg LW, Sparnins VL, Barany G. Inhibition of N nitrosodiethylamine carcinogenesis in mice by naturally occurring organosulfur compounds and monoterpenes. Cancer Res 1989;49(10):2689–2692.

81. Crowell PL, Kennan WS, Vedejs E, Gould MN. Chemoprevention of mammary carcinogenesis by hydroxylated metabolites of limonene. In: 81st Annual Meeting of the American Association for Cancer Research, Washington, DC, USA, May 23–26, 1990. Proc Am Assoc Cancer Res Annu Meet 1990.
82. Mills JJ, Chari RS, Boyer IJ, Gould MN, Jirtle RL. Induction of apoptosis in liver tumors by the monoterpene perillyl alcohol. Cancer Res 1995;55(5):979–983.
83. Haag JD, Gould MN. Mammary carcinoma regression induced by perillyl alcohol, a hydroxylated analog of limonene. Cancer Chem Pharm 1994;34(6):477–483.
84. Stark MJ, Burke YD, McKinzie JH, Ayoubi AS, Crowell PL. Chemotherapy of pancreatic cancer with the monoterpene perillyl alcohol. Cancer Lett 1995;96(1):15–21.
85. McNamee D. Limonene trial in cancer. Lancet 1993;342(8874):801.
86. Phillips LR, Malspeis L, Supko JG. Pharmacokinetics of active drug metabolites after oral administration of perillyl alcohol, an investigational antineoplastic agent, to the dog. Drug Metab Disposi 1995;23(7):676–680.
87. Liu G, Oettel K, Bailey H, et al. Phase II trial of perillyl alcohol (NSC 641066) administered daily in patients with metastatic androgen independent prostate cancer. Invest New Drugs 2003;21(3):367–372.
88. Ariazi EA, Satomi Y, Ellis MJ, et al. Activation of the transforming growth factor beta signaling pathway and induction of cytostasis and apoptosis in mammary carcinomas treated with the anticancer agent perillyl alcohol. Cancer Res 1999;59(8):1917–1928.
89. Crowell PL, Chang RR, Ren Z, Elson CE, Gould MN. Selective inhibition of isoprenylation of 21-26-Kda proteins by the anticarcinogen D-limonene and its metabolites. J Biol Chem 1991;266(26):17,679–17,685.
90. Hitmi A, Coudret A, Barthomeuf C. The production of pyrethrins by plant cell and tissue cultures of *Chrysanthemum cinerariaefolium* and *Tagetes* species. Crit Rev Biochem Mol Biol 2000;35(5):317–337.
91. George J, Bais HP, Ravishankar GA. Biotechnological production of plant-based insecticides. Crit Rev Biotechnol 2000;20(1):49–77.
92. Jones KN, English JC. Review of common therapeutic options in the United States for the treatment of *Pediculosis capitis*. Clin Infect Dis 2003;36(11):1355–1361.
93. Pollack RJ, Kiszewski A, Armstrong P, et al. Differential permethrin susceptibility of head lice sampled in the United States and Borneo. Arch Pediatr Adolesc Med 1999;153(9):969–73.
94. Fuller RW, Cardellina JH, Jurek J, et al. Isolation and structure activity features of halomon-related antitumor monoterpenes from the red alga *Portieria hornemannii*. J Med Chem 1994;37(25):4407–4411.
95. Egorin MJ, Rosen DM, Benjamin SE, Callery PS, Sentz DL, Eiseman JL. In vitro metabolism by mouse and human liver preparations of halomon, an antitumor halogenated monoterpene. Cancer Chem Pharm 1997;41(1):9–14.
96. Sotokawa T, Noda T, Pi S, Hirama M. A three-step synthesis of halomon. Angew Chem 2000;39(19):3430–3432.
97. Rajab MS, Cantrell CL, Franzblau SG, Fischer NH. Antimycobacterial activity of (e)-phytol and derivatives—a preliminary structure-activity study. Planta Med 1998;64(1):2–4.
98. Abraham WR. Bioactive sesquiterpenes produced by fungi: are they useful for humans as well? Curr Med Chem 2001;8(6):583–606.
99. Asakawa Y, Toyota M, Nagashima F, Hashimoto T, El Hassane L. Sesquiterpene lactones and acetogenin lactones from the Hepaticae and chemosystematics of the liverworts *Frullania*, *Plagiochila* and *Porella*. Heterocycles 2001;54(2):1057–+.
100. Tayler VE. The honest herbal—a sensible guide to the use of herbs and related remedies. Third ed. New York: The Haworth Press, 1993.

101. Smolinski AT, Pestka JJ. Modulation of lipopolysaccharide-induced proinflammatory cytokine production in vitro and in vivo by the herbal constituents apigenin (chamomile), ginsenoside Rb-1 (ginseng) and parthenolide (feverfew). Food Chem Toxicol 2003;41(10): 1381–1390.
102. Schinella GR, Giner RM, Recio MD, De Buschiazzo PM, Rios JL, Manez S. Anti-inflammatory effects of South American *Tanacetum vulgare*. J Pharm Pharmacol 1998;50(9):1069–1074.
103. Jain NK, Kulkarni SK. Antinociceptive and anti-inflammatory effects of *Tanacetum parthenium* L. extract in mice and rats. J Ethnopharmacol 1999;68(1–3):251–259.
104. Williams CA, Harborne JB, Geiger H, Robin J, Hoult S. The flavonoids of *Tanacetum parthenium* and *T. vulgare* and their anti-inflammatory properties. Phytochemistry 1999;51(3): 417–423.
105. Sheehan M, Wong HR, Hake PW, Zingarelli B. Parthenolide improves systemic hemodynamics and decreases tissue leukosequestration in rats with polymicrobial sepsis. Crit Care Med 2003;31(9):2263–2270.
106. Brossi A, Venugopalan B, Gerpe LD, et al. Arteether, a new antimalarial drug: synthesis and antimalarial properties. J Med Chem 1988;31(3):645–650.
107. Robert A, Dechy-Cabaret O, Cazelles J, Meunier B. From mechanistic studies on artemisinin derivatives to new modular antimalarial drugs. Accounts Chem Res 2002;35(3):167–174.
108. Delabays N, Simonnet X, Gaudin M. The genetics of artemisinin content in *Artemisia annua* L. and the breeding of high yielding cultivars. Curr Med Chem 2001;8(15):1795–1801.
109. Safayhi H, Sabieraj J, Sailer ER, Ammon HPT. Chamazulene—an antioxidant-type inhibitor of leukotriene B-4 formation. Planta Med 1994;60(5):410–413.
110. Konig GM, Wright AD, Franzblau SG. Assessment of antimycobacterial activity of a series of mainly marine derived natural products. Planta Med 2000;66(4):337–342.
111. Hamann MT, Scheuer PJ, Kellyborges M. Biogenetically diverse, bioactive constituents of a sponge, order verongida—bromotyramines and sesquiterpene-shikimate derived metabolites. J Org Chem 1993;58(24):6565–6569.
112. Nasu SS, Yeung BKS, Hamann MT, Scheuer PJ, Kellyborges M, Goins K. Puupehenone-related metabolites from two hawaiian sponges Hyrtios spp. J Org Chem 1995;60(22):7290–7292.
113. Zjawiony JK, Bartyzel P, Hamann MT. Chemistry of puupehenone: 1,6-conjugate addition to its quinone-methide system. J Nat Prod 1998;61(12):1502–1508.
114. Capon RJ. Marine sesquiterpene/quinone. In: Atta-ur-Rahaman (ed), Studies in Natural Products. Elsevier, New York: 1995; pp. 289–326.
115. Rodriguez J, Quinoa E, Riguera R, Peters BM, Abrell LM, Crews P. The structures and stereochemistry of cytotoxic sesquiterpene quinones from *Dactylospongia elegans*. Tetrahedron 1992;48(32):6667–6680.
116. Alvi KA, Diaz Maria C, Crews P, Slate DL, Lee RH, Moretti R. Evaluation of new sesquiterpene quinones from two *Dysidea* sponge species as inhibitors of protein tyrosine kinase. J Org Chem 1992;57(24):6604–6607.
117. Deguzman FS, Copp BR, Mayne CL, et al. Bolinaquinone—a novel cytotoxic sesquiterpene hydroxyquinone from a Philippine *Dysidea* sponge. J Org Chem 1998;63(22):8042–8044.
118. Giannini C, Debitus C, Posadas I, Paya M, D'Auria MV. Dysidotronic acid, a new and selective human phospholipase A(2) inhibitor from the sponge *Dysidea* sp. Tetrahedron Lett 2000;41(17):3257–3260.
119. Giannini C, Debitus C, Lucas R, et al. New sesquiterpene derivatives from the sponge *Dysidea* species with a selective inhibitor profile against human phospholipase A(2) and other leukocyte functions. J Nat Prod 2001;64(5):612–615.
120. Lucas R, Giannini C, D'Auria MV, Paya M. Modulatory effect of bolinaquinone, a marine sesquiterpenoid, on acute and chronic inflammatory processes. J Pharmacol Exp Ther 2003;304(3):1172–1180.

121. Loya S, Hizi A. The inhibition of human immunodeficiency virus type-1 reverse transcriptase by avarol and avarone derivatives. FEBS Lett 1990;269(1):131–134.
122. Ling TT, Xiang AX, Theodorakis EA. Enantioselective total synthesis of avarol and avarone. Angew Chem 1999;38(20):3089–3091.
123. Loya S, Tal R, Kashman Y, Hizi A. Illimaquinone a selective inhibitor of the RNAase H activity of human immunodeficiency virus type 1 reverse transcriptase. Antimicrob Agents Chemotherapy 1990;34(10):2009–2012.
124. Ling T, Poupon E, Rueden EJ, Theodorakis EA. Synthesis of (-)-ilimaquinone via a radical decarboxylation and quinone addition reaction. Org Lett 2002;4(5):819–822.
125. McMorris TC, Yu J, Lira R, et al. Structure-activity studies of antitumor agent irofulven (hydroxymethylacylfulvene) and analogues. J Org Chem 2001;66(18):6158–6163.
126. McMorris TC, Kashinatham A, Lira R, et al. Sesquiterpenes from *Omphalotus illudens*. Phytochemistry 2002;61(4):395–398.
127. McMorris TC, Lira R, Gantzel PK, Kelner MJ, Dawe R. Sesquiterpenes from the basidiomycete *Omphalotus illudens*. J Nat Prod 2000;63(11):1557–1559.
128. Kelner MJ, McMorris TC, Montoya MA, et al. Characterization of cellular accumulation and toxicity of illudin S in sensitive and nonsensitive tumor cells. Cancer Chem Pharm 1997;40(1):65–71.
129. Herzig MCS, Arnett B, MacDonald JR, Woynarowski JM. Drug uptake and cellular targets of hydroxymethylacylfulvene (HMAF). Biochem Pharmcol 1999;58(2):217–225.
130. Woynarowski JM, Napier C, Koester SK, et al. Effects on dna integrity and apoptosis induction by a novel antitumor sesquiterpene drug, 6-hydroxymethylacylfulvene (hmaf, mgi 114). Biochem Pharmcol 1997;54(11):1181–1193.
131. Izbicka E, Davidson K, Lawrence R, Cote R, MacDonald JR, Von Hoff DD. Cytotoxic effects of MGI 114 are independent of tumor p53 or p21 expression. Anticancer Res 1999;19(2A):1299–1307.
132. Woynarowska BA, Woynarowski JM, Herzig MCS, Roberts K, Higdon AL, MacDonald JR. Differential cytotoxicity and induction of apoptosis in tumor and normal cells by hydroxymethylacylfulvene (HMAF). Biochem Pharmcol 2000;59(10):1217–1226.
133. Jaspers NGJ, Raams A, Kelner MJ, et al. Anti-tumour compounds illudin S and Irofulven induce DNA lesions ignored by global repair and exclusively processed by transcription- and replication-coupled repair pathways. DNA Repair 2002;1(12):1027–1038.
134. Hanson JR. Diterpenoids. Nat Prod Rep 2003;20(1):70–78.
135. Sung PJ, Chen MC. The heterocyclic natural products of gorgonian corals of genus Briareum exclusive of briarane-type diterpenoids. Heterocycles 2002;57(9):1705–1715.
136. Bruno M, Piozzi F, Rosselli S. Natural and hemisynthetic neoclerodane diterpenoids from Scutellaria and their antifeedant activity. Nat Prod Rep 2002;19(3):357–378.
137. Hanson JR. Diterpenoids. Nat Prod Rep 2002;19(2):125–132.
138. Sung PJ, Sheu JH, Xu JP. Survey of briarane-type diterpenoids of marine origin. Heterocycles 2002;57(3):535–579.
139. Hanson JR. Diterpenoids. Nat Prod Rep 2001;18(1):88–94.
140. Hanson JR. Diterpenoids. Nat Prod Rep 2000;17(2):165–174.
141. Baloglu E, Kingston DGI. The taxane diterpenoids. J Nat Prod 1999;62(10):1448–1472.
142. Hanson JR. Diterpenoids. Nat Prod Rep 1999;16(2):209–219.
143. Hanson JR. Diterpenoids. Nat Prod Rep 1998;15(1):93–106.
144. Hanson JR. Diterpenoids. Nat Prod Rep 1997;14(3):245–258.
145. Hanson JR. Diterpenoids. Nat Prod Rep 1996;13(1):59–71.
146. Hanson JR. Diterpenoids. Nat Prod Rep 1995;12(2):207–218.
147. Rontani JF, Volkman JK. Phytol degradation products as biogeochemical tracers in aquatic environments. Org Geochem 2003;34(1):1–35.

148. Broker LE, Glaccone G. The role of new agents in the treatment of non-small cell lung cancer. Eur J Cancer 2002;38(18):2347–2361.
149. Myles DC. Emerging microtubule stabilizing agents for cancer chemotherapy. Ann Rep Med Chem 2002;37(37):125–132.
150. Mollinedo F, Gajate C. Microtubules, microtubule-interfering agents and apoptosis. Apoptosis 2003;8(5):413–450.
151. Rowinsky EK. The development and clinical utility of the taxane class of antimicrotubule chemotherapy agents. Ann Rev Med 1997;48:353–374.
152. Wani MC, Taylor HL, Wall ME, Coggon P, McPhail AT. Plant anti-tumor agents part 6: the isolation and structure of taxol a novel anti leukemic and anti tumor agent from *Taxus brevifolia*. J Am Chem Soc 1971;93(9):2325–2327.
153. Schiff PB, Horwitz SB. Taxol stabilizes micro tubules in mouse fibroblast cells. Proc Natl Acad Sci USA 1980;77(3):1561–1565.
154. Wang TH, Popp DM, Wang HS, et al. Microtubule dysfunction induced by paclitaxel initiates apoptosis through both c-Jun N-terminal kinase (JNK)-dependent and -independent pathways in ovarian cancer cells. J Biol Chem 1999;274(12):8208–8216.
155. Rowinsky EK, Eisenhauer EA, Chaudhry V, Arbuck SG, Donehower RC. Clinical toxicities encountered with paclitaxel. Semin Oncol 1993;20(4 Suppl 3):1–15.
156. Hofle GH, Bedorf N, Steinmetz H, Schomburg D, Gerth K, Reichenbach H. Epothilone A and B - novel 16-membered macrolides with cytotoxic activity—isolation, crystal structure, and conformation in solution. Angew Chem 1996;35(13–14):1567–1569.
157. Winkler JD, Quinn KJ, MacKinnon CH, Hiscock SD, McLaughlin EC. Tandem Diels-Alder fragmentation approach to the synthesis of eleutherobin. Org Lett 2003;5(10):1805–1808.
158. Scalabrino G, Sun XW, Mann J, Baron A. A convergent approach to the marine natural product eleutherobin: synthesis of key intermediates and attempts to produce the basic skeleton. Org Biomol Chem 2003;1(2):318–327.
159. Chen XT, Bhattacharya SK, Zhou BS, Gutteridge CE, Pettus TRR, Danishefsky SJ. The total synthesis of eleutherobin. J Am Chem Soc 1999;121(28):6563–6579.
160. Nicolaou KC, Xu JY, Kim S, et al. Total synthesis of sarcodictyins A and B. J Am Chem Soc 1998;120(34):8661–8673.
161. Britton R, de Silva ED, Bigg CM, McHardy LM, Roberge M, Andersen RJ. Synthetic transformations of eleutherobin reveal new features of its microtubule-stabilizing pharmacophore. J Am Chem Soc 2001;123(35):8632–8633.
162. Britton R, Roberge M, Berisch H, Andersen RJ. Antimitotic diterpenoids from *Erythropodium caribaeorum*: isolation artifacts and putative biosynthetic intermediates. Tetrahedron Lett 2001;42(16):2953–2956.
163. Cinel B, Patrick BO, Roberge M, Andersen RJ. Solid-state and solution conformations of eleutherobin obtained from X-ray diffraction analysis and solution NOE data. Tetrahedron Lett 2000;41(16):2811–2815.
164. Taglialatela-Scafati O, Deo-Jangra U, Campbell M, Roberge M, Andersen RJ. Diterpenolds from cultured *Erythropodium caribaeorum*. Org Lett 2002;4(23):4085–4088.
165. Ketzinel S, Rudi A, Schleyer M, Benayahu Y, Kashman Y. Sarcodictyin a and two novel diterpenoid glycosides, eleuthosides a and b, from the soft coral *Eleutherobia aurea*. J Nat Prod 1996;59(9):873–875.
166. Ciomei M, Albanese C, Pastori W, et al. Sarcodictyins: a new class of marine derivatives with mode of action similar to taxol. Proc Am Assoc Cancer Res Annu Meet 1997;38(0):5.
167. Cinel B, Roberge M, Behrisch H, van Ofwegen L, Castro CB, Andersen RJ. Antimitotic diterpenes from *Erythropodium caribaeorum* test pharmacophore models for microtubule stabilization. Org Lett 2000;2(3):257–260.

168. Nicolaou KC, Kim SH, Pfefferkorn J, et al. Synthesis and biological activity of sarcodictyins. Angew Chem 1998;37(10):1418–1421.
169. Ata A, Kerr RG, Moya CE, Jacobs RS. Identification of anti-inflammatory diterpenes from the marine gorgonian *Pseudopterogorgia elisabethae*. Tetrahedron 2003;59(23):4215–4222.
170. Haimes HB, Jimenez PA. Use of pseudopterosins for promoting wound healing. Official Gazette of the United States Patent & Trademark Office Patents 1997;1194(4):2589.
171. Rodriguez AD, Ramirez C. Serrulatane diterpenes with antimycobacterial activity isolated from the West Indian sea whip *Pseudopterogorgia elisabethae*. J Nat Prod 2001;64(1):100–102.
172. Rodriguez AD, Ramirez C, Rodriguez, II, Gonzalez E. Novel antimycobacterial benzoxazole alkaloids, from the West Indian sea whip *Pseudopterogorgia elisabethae*. Org Lett 1999;1(3):527–530.
173. Etahiri S, Bultel-Ponce V, Caux C, Guyot M. New bromoditerpenes from the red alga *Sphaerococcus coronopifolius*. J Nat Prod 2001;64(8):1024–1027.
174. Cafieri F, De Napoli L, Fattorusso E, Santacroce C. Diterpenes from the red alga *Sphaerococcus coronopifolius*. Phytochemistry 1987;26(2):471–474.
175. Konig GM, Wright AD, Angerhofer CK. Novel potent antimalarial diterpene isocyanates, isothiocyanates, and isonitriles from the tropical marine sponge *Cymbastela hooperi*. J Org Chem 1996;61(10):3259–3267.
176. Groweiss A, Look SA, Fenical W. Solenolides new antiinflammatory and antiviral diterpenoids from a marine octocoral of the genus Solenopodium. J Org Chem 1988;53(11):2401–2406.
177. Sakemi S, Higa T, Jefford CW, Bernardinelli G. Venustatriol a new anti-viral triterpene tetracyclic ether from *Laurencia venusta*. Tetrahedron Lett 1986;27(36):4287–4290.
178. Komoto S, McConnell OJ, Cross SS. Antiviral Furanoditerpenoids. US Patent-4801607. January 31 1989. Official Gazette of the United States Patent & Trademark Office Patents 1989;1098(5):2452.
179. Potts BCM, Faulkner DJ, De Carvalho MS, Jacobs RS. Chemical mechanism of inactivation of bee venom phospholipase A-2 by the marine natural products manoalide luffariellolide and scalaradial. J Am Chem Soc 1992;114(13):5093–5100.
180. Puliti R, De Rosa S, Mattia CA, Mazzarella L. Structure and stereochemistry of an acetate derivative of cacospongionolide a new antitumoral sesterterpenoid from marine sponge *Cacospongia mollior*. Acta Crystallog C Crystal Struct Comm 1990;46(8):1533–1536.
181. De Silva ED, Scheuer PJ. Manoalide an antibiotic sester terpenoid from the marine sponge *Luffariella variabilis*. Tetrahedron Lett 1980;21(17):1611–1614.
182. Pommier A, Stepanenko V, Jarowicki K, Kocienski PJ. Synthesis of (+)-manoalide via a copper(I)-mediated 1,2-metalate rearrangement. J Org Chem 2003;68(10):4008–4013.
183. Soriente A, De Rosa M, Scettri A, et al. Manoalide. Curr Med Chem 1999;6(5):415–431.
184. Glaser KB, De Carvalho MS, Jacobs RS, Kernan MR, Faulkner DJ. Manoalide structure-activity studies and definition of the pharmacophore for phospholipase A(2) inactivation. Mol Pharmacol 1989;36(5):782–788.
185. Reynolds LJ, Morgan BP, Hite GA, Mihelich ED, Dennis EA. Phospholipase A-2 inhibition and modification by manoalogue. J Am Chem Soc 1988;110(15):5172–5177.
186. Randazzo A, Debitus C, Minale L, et al. Petrosaspongiolides M-R—new potent and selective phospholipase A(2) inhibitors from the new caledonian marine sponge *Petrosaspongia nigra*. J Nat Prod 1998;61(5):571–575.
187. Kernan MR, Faulkner DJ, Jacobs RS. The luffariellins novel antiinflammatory sesterterpenes of chemotaxonomic importance from the marine sponge *Luffariella variabilis*. J Org Chem 1987;52(14):3081–3083.
188. Kernan MR, Faulkner DJ, Parkanyi L, Clardy J, De Carvalho MS, Jacobs RS. Luffolide, a novel anti-inflammatory terpene from the sponge *Luffariella* sp. Experientia 1989;45(4):388–390.

189. De Rosa S, De Stefano S, Zavodnik N. Cacospongionolide: a new antitumoral sesterterpene from the marine sponge *Cacospongia mollior*. J Org Chem 1988;53(21):5020–5023.
190. De Rosa S, Puliti R, Crispino A, De Giulio A, Mattia CA, Mazzarella L. A new scalarane sesterterpenoid from the marine sponge *Cacospongia mollior*. J Nat Prod 1994;57(2):256–262.
191. De Silva ED, Scheuer PJ. 3 New sester terpenoid antibiotics from the marine sponge *Luffariella variabilis*. Tetrahedron Lett 1981;22(33):3147–3150.
192. Albizati KF, Holman T, Faulkner DJ, Glaser KB, Jacobs RS. Luffariellolide: an anti-inflammatory sesterterpene from the marine sponge *Luffariella* sp. Experientia 1987;43(8):949–950.
193. Posadas I, De Rosa S, Terencio MC, Paya M, Alcaraz MJ. Cacospongionolide B suppresses the expression of inflammatory enzymes and tumour necrosis factor-alpha by inhibiting nuclear factor-kappa B activation. Br J Pharmacol 2003;138(8):1571–1579.
194. Posadas I, Terencio MC, Randazzo A, Gomez-Paloma L, Paya M, Alcaraz MJ. Inhibition of the NF-kappa B signaling pathway mediates the anti-inflammatory effects of petrosaspongiolide M. Biochem Pharmcol 2003;65(5):887–895.
195. Garcia-Pastor P, Randazzo A, Gomez-Paloma L, Alcaraz MJ, Paya M. Effects of petrosaspongiolide M, a novel phospholipase A(2) inhibitor, on acute and chronic inflammation. J Pharmacol Exper Ther 1999;289(1):166–172.
196. Dal Piaz F, Casapullo A, Randazzo A, et al. Molecular basis of phospholipase A(2) inhibition by petrosaspongiolide M. Chembiochem 2002;3(7):664–671.
197. Youssef DTA, Yamaki RK, Kelly M, Scheuer PJ. Salmahyrtisol A, a novel cytotoxic sesterterpene from the Red Sea sponge *Hyrtios erecta*. J Nat Prod 2002;65(1):2–6.
198. Chinworrungsee M, Kittakoop P, Isaka M, Rungrod A, Tanticharoen M, Thebtaranonth Y. Antimalarial halorosellinic acid from the marine fungus *Halorosellinia oceanica*. Bioorg Med Chem Lett 2001;11(15):1965–1969.
199. Renner MK, Jensen PR, Fenical W. Mangicols: structures and biosynthesis of a new class of sesterterpene polyols from a marine fungus of the genus Fusarium. J Org Chem 2000;65(16): 4843–4852.
200. Nicolaou KC, Sorensen EJ (eds), Classics in Total Synthesis. Weinheim, Germany: VCH, 1996.
201. Nicolaou KC, Vourloumis D, Winssinger N, Baran PS. The art and science of total synthesis at the dawn of the twenty-first century. Angew Chem Int Ed 2000;39(1):44–122.
202. Kingston DGI, Jagtap PG, Yuan H, Samala L. The chemistry of Taxol and related taxoids. Prog Chem Org Nat Prod 2002;84:53–225.
203. Holton RA, Kim HB, Somoza C, et al. First total synthesis of Taxol. 2. Completion of the C and D rings. J Am Chem Soc 1994;116(4):1599–1600.
204. Holton RA, Somoza C, Kim HB, et al. First total synthesis of Taxol. 1. Functionalization of the B ring. J Am Chem Soc 1994;116(4):1597–1598.
205. Nicolaou KC, Yang Z, Liu JJ, et al. Total synthesis of Taxol. Nature 1994;367(6464):630–634.
206. Masters JJ, Link JT, Snyder LB, Young WB, Danishefsky SJ. A total synthesis of Taxol. Angew Chem Int Ed 1995;34(16):1723–1726.
207. Nicolaou KC, Ueno H, Liu JJ, et al. Total synthesis of Taxol. 4. The final stages and completion of the synthesis. J Am Chem Soc 1995;117(2):653–659.
208. Nicolaou KC, Nantermet PG, Ueno H, Guy RK, Couladouros EA, Sorensen EJ. Total Synthesis of Taxol. 1. Retrosynthesis, degradation, and reconstitution. J Am Chem Soc 1995;117(2): 624–633.
209. Danishefsky SJ, Masters JJ, Young WB, et al. Total synthesis of baccatin III and Taxol. J Am Chem Soc 1996;118(12):2843–2859.
210. Wender PA, Badham NF, Conway SP, et al. The pinene path to taxanes. 5. Stereocontrolled synthesis of a versatile taxane precursor. J Am Chem Soc 1997;119(11):2755–2756.
211. Wender PA, Badham NF, Conway SP, et al. The pinene path to taxanes. 6. A concise stereocontrolled synthesis of Taxol. J Am Chem Soc 1997;119(11):2757–2758.

212. Morihira K, Hara R, Kawahara S, et al. Enantioselective total synthesis of Taxol. J Am Chem Soc 1998;120(49):12,980–12,981.
213. Nicolaou KC, Ohshima T, Hosokawa S, et al. Total synthesis of eleutherobin and eleuthosides A and B. J Am Chem Soc 1998;120(34):8674–880.
214. Shiina I, Saitoh K, Frechard-Ortuno I, Mukaiyama T. Total asymmetric synthesis of Taxol by dehydration condensation between 7-TES baccatin III and protected N-benzoylphenylisoserines prepared by enantioselective aldol reaction. Chem Lett 1998(1):3–4.
215. Mukaiyama T, Shiina I, Iwadare H, et al. Asymmetric total synthesis of Taxol. Chem- Euro J 1999;5(1):121–161.
216. Kusama H, Hara R, Kawahara S, et al. Enantioselective total synthesis of (-)-Taxol. J Am Chem Soc 2000;122(16):3811–3820.
217. Jung M, ElSohly HN, Croom EM, McPhail AT, McPhail DR. Practical conversion of artemisinic acid in desoxyartemisinin. J Org Chem 1986;51(26):5417–5419.
218. Roth RJ, Acton N. A simple conversion of artemisinic acid into artemisinin. J Nat Prod 1989;52(5):1183–1185.
219. Ye B, Wu YL. An efficient synthesis of qinghaosu and deoxoqinghaosu from arteannuic acid. J Chem Soc Chem Comm 1990(10):726–727.
220. Luo XD, Shen CC. The chemistry, pharmacology, and clinical-applications of qinghaosu (artemisinin) and its derivatives. Med Res Rev 1987;7(1):29–52.
221. Schmid G, Hofheinz W. Total synthesis of qinghaosu. J Am Chem Soc 1983;105:624–625.
222. Soriente A, De Rosa M, Scettri A, et al. Manoalide. Curr Med Chem 1999;6(5):415–431.
223. Gueritte F. General and recent aspects of the chemistry and structure-activity relationships of taxoids. Curr Pharm Design 2001;7(13):1229–1249.
224. Chen X-T, Bhattacharya SK, Zhou B, Gutteridge CE, Pettus TRR, Danishefsky SJ. The total synthesis of eleutherobin. J Am Chem Soc 1999;121(28):6563–6579.
225. Chen X-T, Zhou B, Bhattacharya SK, Gutteridge CE, Pettus TRR, Danishefsky SJ. The total synthesis of eleutherobin: a surprise ending. Angew Chem Int Ed 1998;37(6):789–792.
226. Nicolaou KC, van Delft F, Ohshima T, et al. Total synthesis of eleutherobin. Angew Chem Int Ed 1997;36(22):2520–2524.
227. Nicolaou KC, Winssinger N, Vourloumis D, et al. Solid and solution phase synthesis and biological evaluation of combinatorial sarcodictyin libraries. J Am Chem Soc 1998;120(42):10,814–10,826.
228. Schreiber SL. Target-oriented and diversity-oriented organic synthesis in drug discovery. Science 2000;287(5460):1964–1969.
229. Burke MD, Berger EM, Schreiber SL. Generating diverse skeletons of small molecules combinatorially. Science 2003;302(5645):613–618.
230. Appendino G, Tron GC, Jarevng T, Sterner O. Unnatural natural products from the transannular cyclization of lathyrane diterpenes. Org Lett 2001;3(11):1609–1612.
231. Lucker J, Bouwmeester HJ, Schwab W, Blaas J, van der Plas LHW, Verhoeven HA. Expression of *Clarkia* S-linalool synthase in transgenic petunia plants results in the accumulation of S-linalyl-beta-D-glucopyranoside. Plant J 2001;27(4):315–324.
232. Lavy M, Zuker A, Lewinsohn E, et al. Linalool and linalool oxide production in transgenic carnation flowers expressing the *Clarkia breweri* linalool synthase gene. Mol Breed 2002;9(2):103–111.
233. Lewinsohn E, Schalechet F, Wilkinson J, et al. Enhanced levels of the aroma and flavor compound S-linalool by metabolic engineering of the terpenoid pathway in tomato fruits. Plant Physiol 2001;127(3):1256–1265.
234. Martin VJJ, Yoshikuni Y, Keasling JD. The in vivo synthesis of plant sesquiterpenes by *Escherichia coli*. Biotech Bioeng 2001;75(5):497–503.

235. Martin VJ, Pitera DJ, Withers ST, Newman JD, Keasling JD. Engineering a mevalonate pathway in *Escherichia coli* for production of terpenoids. Nature Biotechnol 2003.
236. Wang ZY, Zhong JJ. Repeated elicitation enhances taxane production in suspension cultures of *Taxus chinensis* in bioreactors. Biotechnol Lett 2002;24(6):445–448.
237. Zhong JJ. Plant cell culture for production of paclitaxel and other taxanes. J Biosci Bioeng 2002;94(6):591–599.
238. Liu CZ, Guo C, Wang YC, Ouyang F. Comparison of various bioreactors on growth and artemisinin biosynthesis of *Artemisia annua* L. shoot cultures. Process Biochem 2003;39(1):45–49.
239. Liu CZ, Guo C, Wang YC, Fan OY. Factors influencing artemisinin production from shoot cultures of *Artemisia annua* L. World J Microbiol Biotechnol 2003;19(5):535–538.
240. Souret FF, Kim Y, Wysiouzil BE, Wobbe KK, Weathers PJ. Scale-up of *Artemisia annua* L. hairy root cultures produces complex patterns of terpenoid gene expression. Biotech Bioeng 2003;83(6):653–667.
241. Wang JW, Tan RX. Artemisinin production in *Artemisia annua* hairy root cultures with improved growth by altering the nitrogen source in the medium. Biotechnol Lett 2002;24(14):1153–1156.
242. Xie DY, Zou ZR, Ye HC, Li GF, Guo ZC. Selection of hairy root clones of *Artemisia annua* L. for artemisinin production. Isr J Plant Sci 2001;49(2):129–134.
243. Polzin JP, Rorrer GL. Halogenated monoterpene production by microplantlets of the marine red alga *Ochtodes secundiramea* within an airlift photobioreactor under nutrient medium perfusion. Biotech Bioeng 2003;82(4):415–428.
244. Barahona LF, Rorrer GL. Isolation of halogenated monoterpenes from bioreactor-cultured microplantlets of the macrophytic red algae *Ochtodes secundiramea* and *Portieria hornemannii*. J Nat Prod 2003;66(6):743–751.
245. Huang YM, Rorrer GL. Cultivation of microplantlets derived from the marine red alga *Agardhiella subulata* in a stirred tank photobioreactor. Biotechnol Prog 2003;19(2):418–427.
246. Huang YM, Rorrer GL. Dynamics of oxygen evolution and biomass production during cultivation of *Agardhiella subulata* microplantlets in a bubble-column photobioreactor under medium perfusion. Biotechnol Prog 2002;18(1):62–71.
247. Maliakal S, Cheney DP, Rorrer GL. Halogenated monoterpene production in regenerated plantlet cultures of *Ochtodes secundiramea* (Rhodophyta, Cryptonemiales). J Phycol 2001;37(6):1010–1019.
248. Polzin JJ, Rorrer GL, Cheney DP. Metabolic flux analysis of halogenated monoterpene biosynthesis in microplantlets of the macrophytic red alga *Ochtodes secundiramea*. Biomol Eng 2003;20(4–6 Special Issue SI):205–215.
249. Mendola D. Aquaculture of three phyla of marine invertebrates to yield bioactive metabolites: process developments and economics. Biomol Eng 2003;20(4–6 Special Issue SI):441–458.
250. Ling T, Poupon E, Rueden EJ, Kim SH, Theodorakis EA. Unified synthesis of quinone sesquiterpenes based on a radical decarboxylation and quinone addition reaction. J Am Chem Soc 2002;124(41):12,261–12,267.

10

Challenges and Opportunities in the Chinese Herbal Drug Industry

Wei Jia and Lixin Zhang

Summary

Tremendous achievements have been made in Western medicine in the past that provide fast relief of symptoms at the disease sites, particularly under critical conditions. However, some are either ineffective or produce undesirable adverse effects, or are too costly in some complex diseases, especially chronic diseases. On the other hand, traditional medicines strive to focus on the balance of the body in a holistic manner. Traditional medicines are increasing popular in the Western world, as reflected in the name changes from "alternative medicines" to "complementary medicines" or even "integrative medicines." Traditional Chinese medicines (TCM) have been used in China for more than 2000 yr and have always followed the philosophy and principle of restoration, i.e., the yin and yang (balance of the body). As an important part of the pharmaceutical sector, the Chinese herbal drug industry has made rapid progress over the past decades. This chapter provides a review of its current status, including challenges and opportunities, specifically with regard to modernization and globalization of TCM.

Key Words: Complementary and alternative medicine (CAM); integration; traditional Chinese medicine (TCM); modernization; globalization; preventive therapy.

1. Introduction

Instead of the hit-or-miss technology of the past, current biological research and much of drug discovery is being driven by the search for new molecules targeting disease-relevant proteins (1–3). In this approach, a specific protein is studied in vitro, in cells and in whole organisms, and evaluated as a drug target for a specific therapeutic indication (3–5). The historical paradigm "one drug, one target" has resulted in the identification of many effective chemical molecules that affect specific proteins, providing valuable reagents for both biology and medicine. One example is the recently US Food and Drug Administration (FDA)-approved drug Avastin, a recombinant humanized antibody designed to bind to and inhibit vascular endothelial growth factor (VEGF) in tumor angiogenesis (6). Benefiting from the advancement of biology and chemistry, the world market of drug products has increased tremendously over the past decade, from $180 billion USD in 1990 to $430 billion in 2001 (7). IMS Health, a well-known information company of the pharmaceutical industry, anticipated that the

From: *Natural Products: Drug Discovery and Therapeutic Medicine*
Edited by: L. Zhang and A. L. Demain © Humana Press Inc., Totowa, NJ

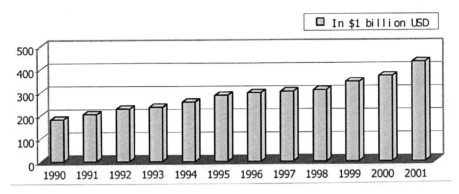

Fig. 1. Annual sales of the world drug industry.

annual sales of drug products would reach $587.9 billion USD by the year 2005, with an average growth rate of 7.8%, much higher than the average growth rate of the global economy (2.4%) (**Fig. 1**) *(7)*.

On the other hand, the expensive new technologies, innovation, and intellectual properties, combining factors such as regulatory issues and high failure rate, really cause health care costs to skyrocket. It costs about $1.7 billion to bring a new chemical entity (NCE) drug to market, a 55% increase over the average cost during 1995 to 2000 *(8–12)*. How can we increase the efficiency of drug development while reducing the failure rate of compounds in clinical trials? Pharmaceutical companies are increasingly responding with mergers, partnerships, and intensified promotion. They also shift focus to diseases of richer and older people, and away from antimicrobials, vaccines, and the like *(13)*. Ironically, even though the investment on research and development is steadily increasing, the pharmaceutical discovery pipeline has declined substantially over the past decade. One of the major limiting factors is the prevalent "one drug, one target" dogma in the biotechnology and pharmaceutical industries. Pharmaceuticals have been traditionally designed to target individual factors in a disease system but have limited success in some complex diseases, especially chronic diseases. The reason is that, in real physiology, diseases are multigenic. Therefore, to cure diseases, multiple stages along the disease pathway may need to be manipulated simultaneously for an effective treatment *(1)*. Systems and integrative biology have revealed that human cells and tissues are composed of complex, networking systems with redundant, convergent and divergent signaling pathways *(14–21)*. On the other hand, drugs are targeting human beings, and each individual is different; this leads to the popularity of pharmacogenomics and personalized medicine *(22)*.

Traditional medicines focus on the balance of the body in a holistic manner and consider the difference between individuals. Traditional medicine is increasingly popular in the Western world, as reflected in the name changes from "alternative medicines" to "complementary medicines" or even "integrative medicines" *(23–29,119)*. Name changes reflect a dissatisfaction with conventional Western medicine as well as a cultural rebellion against the biomedical community, and a process to bring traditional medicine into alliance with and integrate it into a pivotal part of mainstream health care. As many as 42% of individuals in the United States are adopting complementary and alternative medicine (CAM) approaches to help meet their personal health needs *(30,31)*. The popularity of CAM has spread to the whole world *(32,33)*. There probably will be a paradigm shift to bridge conventional Western medicines with TCM and enable the drug discovery pipeline to become more productive with respect to safer and better drugs.

In 1998, responding to public demand for better guidance regarding the myriad of CAM options, the United States Congress authorized the creation of the National Center for Complementary and Alternative Medicine (NCCAM) at the National Institutes of Health (NIH). The establishment of the center represented an expansion in scope and authority of the Office of Alternative Medicine (OAM), which was first established in 1992 *(34)*. Patients who choose CAM approaches are seeking ways to improve their health and well-being, and to relieve symptoms associated with chronic or even terminal illnesses, or with the side effects of conventional treatments. The effectiveness, scientific rationale, side effects, and cost-effectiveness of different kinds of CAM (including dietary modification; supplementation with antioxidant vitamins, soy, herbs; acupuncture; massage; exercise; psychological and mind–body interventions) have been studied and compared in parallel on cancer patients *(35)*. TCM emphasizes the proper balance or disturbances of Qi—or vital energy—in health and disease, respectively. TCM consists of techniques and methods such as acupuncture; microbial, plant, and animal products; physical exercises and calisthenics with, or without, meditation; Qi Gong; massage; and other forms. In this chapter, we will focus on the status of herbal remedies (typical natural products) either as single-chemical entity drugs or complexed botanical supplements.

More than 60% of small-molecule drugs are either natural products or their derivatives *(36,37)*. The total number of natural products produced by plants has been estimated to be over 500,000 *(38)*. About 160,000 natural products have been identified *(39)*, a number growing by 10,000 per year *(40–43)*. Many important drugs are derived directly or indirectly from the active ingredients of herbal remedies, such as the predecessor of aspirin from willow tree bark *(44)*, reserpine from the herb *Rauwolfia serpentina*, Taxol from the Pacific yew tree, Qinghaosu (artemisinin and senna) from *Artemisia annua* and *Cassia angustifolia*, quinine from cinchona bark, digitalis from foxglove leaf, morphine from the poppy herb, and vincristine from rosy periwinkle *(45–49)*.

In the past decade, the American government issued or proposed several new regulations in favor of developing Chinese medical herbal products in the United States. Following the Dietary Supplement Health and Education Act (DSHBA), passed by the US Congress in 1994 *(86)*, the FDA issued the Final Rule of Dietary Supplements on February 7, 2000. In accordance with current FDA regulations, many TCM products may be promoted and marketed in the United States as dietary supplements. The FDA also drafted "Guidance for Industry: Botanical Drug Products," which encourages medical societies and pharmaceutical companies to develop botanical materials for curing human diseases *(87)*. These developments provide a great opportunity to the TCM pharmaceutical industry.

In Europe, the same substance may be considered a dietary supplement in one country while being sold as an over-the-counter (OTC) drug in another country *(50,51)*. This situation will change if the European Union develops a uniform set of laws to be followed by its member states. There exists a long tradition in Asian countries dating back thousands of years of identifying and using herbal ingredients to successfully treat various diseases. Usage rates range from 29% to 53% among various patient populations in Korea *(26)*. It is very popular in China, and China's herbal drug industry has contributed significantly to this rich reservoir of both herbal drugs and dietary supplements *(53–61)*.

2. The Development of the Chinese Herbal Drug Industry

TCM has been used in Chinese medical practice for more than 2000 years. Based on available literature, TCM products were safe and effective for the treatment of many human diseases before Western medicines were introduced and marketed in China. TCM has been an integral part of China's health care system, along with conventional Western medicine *(59–61)*. Patients generally enjoy the benefit of combining the power of traditional and conventional Western medicine.

Although herbal medicine has a long history and pharmacological foundation in China, a TCM industry was only established a century ago by adopting some modern technologies. A system for the production and circulation of herbal drug products has been formed in China with the production of Chinese medicinal materials as its foundation, Chinese herbal manufacturing industry as its main body, and a Chinese herbal drug-marketing network as its linkage.

China is rich in medicinal resources, with more than 11,100 species of medicinal herbs and 2000 prescription recipes well documented. There are more than 600 plantation bases for the production of TCM raw materials. About 5 million mus (1 mu = 0.165 acre) of land are used to produce Chinese medicinal materials every year, with an output of about 400,000 tons. Currently, there are more than 1000 herbal pharmaceutical factories in China, producing more than 8000 herbal drug products in more than 40 different kinds of dosage forms. There are more than 30,000 wholesale and retail shops for herbal medicines in China *(62–64)*. With the back-to-nature trend and the fact that China recently became a member of the World Trade Organization (WTO), an increase in the use of Chinese herbal medicine is expected, with a projected $400 billion worldwide market by 2010 *(59)*.

The seemingly overwhelmed herbal market will facilitate new herbal product development. Since the late 1970s, the Chinese pharmaceutical industry has achieved a very high growth rate, averaging 17.7% annually, a number much higher than the average growth rate of its national GDP (8%). The consumer market of pharmaceutical products increased from annual sales of $1.83 billion USD in 1990 to $18 billion USD in 2002, nearly a 10-fold increase, and has become one of the fastest-growing sectors in China (**Fig. 2**). At the end of 1999, 28 herbal drug products from 107 Chinese manufacturers reached individual annual sales beyond $12 million USD *(65)*.

The herbal drug industry is more profitable than most other industries in China. The role of government in managing the TCM industry has changed fundamentally as a result of China's economic reform and opening policies, especially the in-depth reform

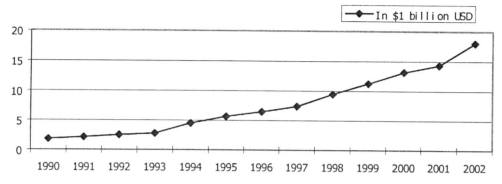

Fig. 2. Fast-growing traditional Chinese medicine industry in China.

of the social security, health care, and medical insurance systems. According to the latest data from the Chinese National Economy and Trading Commission, the annual growth rate of the TCM industry has averaged 20%, with a profit rate reaching 24% annually in the past decade *(62,63)*. From the latest survey of Chinese industries by the National Bureau of Statistics, the TCM industry is among the best (as judged by eight economic indices), in comparison with the 41 other industries, including petroleum and rubber. Its profit margin ranked number two, next to the tobacco industry. The Strategic Development Center of the State Council in China has recently completed a project entitled "The Strategic Development of Traditional Chinese Medicine," under the ninth Five-Year Strategic Plan, in which the general competitiveness of about 10 industries, including herbal drugs, food manufacturing, pharmaceutical manufacturing, and electronic and communication equipment, were evaluated based on 12 indices. From the report, the TCM industry was ranked number four and was believed to have great potential, with strong development capability, creativity, and economy-driving impact *(66)*. Government funding of TCM research and development has been and will be continually increasing.

3. Modernization of TCM

Medicines are products that are claimed to treat, cure, mitigate, or prevent a disease, and are regulated by national authorities. Since the establishment of the US FDA in 1938, all drugs have had to be proven safe for their intended use to gain FDA premarket approval (unless they had been "grandfathered" as old drugs) *(84)*. Since 1962, approval has also required that efficacy should be shown in adequate and well-controlled studies. Furthermore, all drug manufacturers must follow current good manufacturing practices (CGMPs) to assure quality and standardization of their drugs, and must list their facilities and products with the FDA *(85–88)*. The Chinese government established their own standard for TCM drugs and dietary supplements by comparing different criteria from the United States, European Union, and Japanese systems.

Systematic government agencies are also established to monitor and regulate affairs in the health care and pharmaceutical industry: the State Administration of Industry & Commerce (SAIC) is in charge of ethical promotion and business transition; the State

Table 1
Resources of Chinese Materia Medica

Resource	Family	Genus	Species	Percentage (%)
Medicinal plants	383	2309	11,146	87.0
Medicinal animals	395	862	1581	12.3
Medicinal mines	–	–	80	0.63
Total			12,807	

Development and Planning Commission (SDPC) controls prices; the State Food and Drug Administration (SFDA) focuses on drug regulatory issues; the Ministry of Labor and Social Security (MLSS) deals with issues of medical insurance, reimbursement, and co-payment; the Ministry of Health (MOH) is responsible for health service and hospitals. Relatively detailed documents are written on a complete system of quality control, supervision, and standardization of TCM, such as current good agriculture practice (CGAP), to protect the medicinal plant resource for sustainable development; CGMP for manufacturing; current good laboratory practice (CGLP) for consistent bookkeeping; and current good clinical practice (CGCP) for TCM development. Other standard operation protocols, such as current good quality control practice (CGQP) and current good extraction practice (CGEP) are proposed and will be implemented soon.

3.1. The Status of Medicinal Plant Resources in China

More than 11% of the world plant species are found in China, including 240 rare genera. A national survey indicated that China had 12,807 species of medicinal materials *(67,68)* (**Table 1**), in which 11,146 species (9933 taxonomic species and 1213 taxonomic units under species) are medicinal plants, including 10,687 species of seed plants, bryophytes, or pteridophytes and 459 species of algae, bacteria, fungi, or lichens (**Table 2**).

As a result of over-exploitation, the reserves and production of wild medicinal plants gradually decreased. For example, *Radix Glycyrrhiza uralensis* was originally produced in the province of Inner Mongolia, but its reserve and production has decreased very quickly. Recently, the annual production of *Radix Glycyrrhiza* was reduced by 40% compared to the 1950s, and the main harvest place of this plant shifted to Xinjiang Province *(69)*. The wild plant output of another common medicinal herb, *Radix Astragalus membranaceus* and *Radix Astragalus mongolicus*, was more than 2000 tons in the 1960s, but has decreased to less than 100 tons in recent years *(70)*. Many other medicinal plant species are in danger of extinction, such as *Acanthopanax senticosus*; *Atractylodes lancea*; *Anemarrhena asphodeloides*; *Asarum sieboldii*; *Cistanche salsa*; *Cynomorium songaricum*; *Dichroa febrifuga*; *Ephedra sinica*; *Gastrodia elata*; *Gentiana macrophylla*; *Gentiana scabra*; *Glycyrrhiza uralensis*; *Lithospermum erythrorhizon*; *Notopterygium incisum*; *Paris polyphylla*; *Phellodendron amurense*; *Pinellia ternate*; *Rheum officinale*; *Saposhnikovia divaricata*; *Scutellaria baicalensis*; *Stellaria gypsophiloides*; *Tripterygium wilfordii*; *Uncaria rhynchophylla*; *Vitex trifolia*; and *Ziziphus jujuba*.

Table 2
Medicinal Plant Resources in China

Resource	Families	Genera	Species
Algae	42	56	115
Bacteria	40	117	292
Fungi	9	15	52
Lichens	21	33	43
Bryophytes	49	116	456
Pteridophytes	222	1972	10,188
Seed plants	383	2309	11,146

The Chinese government has recognized the situation and has begun to implement CGAP to protect the medicinal plant resources for sustainable development. Thus far, around 600 CGAP cultivation bases have been established nationwide, and the total cultivation area of medicinal plants has reached 6 million mus *(71–74)*. The research institutes, herbal drug manufacturing plants, and local farmers are the three important and indispensable participants in this effort. The CGAP cultivation base for each drug material is usually built in its genuine or original production place(s) to achieve the geo-authenticity of the plants. **Table 3** lists some significant CGAP cultivation bases.

3.2. Status of Chinese Prepared Medicinal Herbs

In the process of the TCM industry, there is one sector called "prepared medicinal herbs," situated between the cultivation of medicinal plants and extraction/manufacturing of herbal drug products. The manufacturers specializing in prepared medicinal herbs collect, clean, cut, and sometimes process the herbs through boiling or steaming procedures. Although many researchers in the industry suggest that a good processing practice (GPP) should be put in place, much less research and process development work has been done to improve the processing of Chinese prepared medicinal herbs, as compared to CGAP in the cultivation of medicinal materials and CGMP in the production of herbal drug products.

By the end of 2001, there were 1175 Chinese herbal drug factories, equivalent to 18% of the total number of the medical and pharmaceutical manufacturing plants in China. Currently, there are 48 manufacturing plants of prepared medicinal herbs, 14 of which are operating at a deficit. Sales of prepared medicinal herbs in 2001 reached $60 million USD, accounting for 1.08% of the total market of herbal drugs. **Table 4** lists the top 18 large manufacturers of prepared medicinal herbs in China *(61,75)*.

The quality control methods and standards of Chinese prepared herbs are being established with modern analytical technologies, including microscopic analysis and chromatographic fingerprints. In addition, regulatory and market control laws are being strengthened to eliminate false and low-quality Chinese prepared herbs.

To meet the growing demand for consistency in herbal material handling and quality control, a new form of prepared medicinal herb, the granule of herbal extracts, has recently been developed and marketed in the major TCM hospitals in China. The two

Table 3
Some Current Good Agriculture Practice Bases of Chinese Materia Medica in China

Site of bases	Cultivated medicinal plants
Anhui province	*Paeonia suffruticosa*
Chongqing	*Pinellia ternata*
Gansu province	*Angelica sinensis*
Guangxi province	*Momordica grosvenorii*
GuiZhou province	*Dendrobium candidum; Eucommia ulmoides; Ginkgo biloba*
Hebei province	*Angelica dahurica; Scutellaria baicalensis*
Heilongjiang province	
Henan province	*Panax ginseng*
Hubei province	*Rehmannia glutinosa*
Hunan province	*Lespedeza cyrtobotrya*
Inner Mongolia	*Eucommia umloides*
Jiangsu province	*Glycyrrhiza uralensis*
Jilin province	*Chrysanthemum morifolium*
Liaoning province	*Panax ginseng; Panax quinquefolium*
Ningxia province	*Panax ginseng*
Shandong province	*Lycium chinensis*
Shanghai	*Lonicera japonica*
Shanxi province	*Crocus sativus*
Sichuan province	*Salvia miltiorrhiza*
Tibet province	*Crocus sativus; Ligusticum chuanxiong*
Yunnan province	*Rhodiola rosea*
	Dracaena draco; Panax notoginseng

major suppliers are Guangdong E-fong Pharmaceuticals and Jiangyin Tianjiang Pharmaceutical Company. The prepared herbal granule is made by a modern technique of processing single medicinal herbs, in which herbs are boiled until a thick syrup emerges and then are dried by a combined spray drying and fluidized bed drying technique. Granules are considered to have the highest effectiveness among all the preparations, because they maintain the most active ingredients through such a process and retain potency for a long time in storage. Nearly 500 kinds of prepared herbal granules have been made available on the market, and many TCM practitioners and consumers prefer to use such materials for combination formula preparations because granules are easier to keep and handle, and require lower amounts per volume than liquid extracts or raw materials. At the same time, this new form of prepared medicinal herbs is being challenged by many conventional herbal drug manufacturers and researchers. One of the major critiques is that the use of such a modern preparation in traditional combination formulas omitted an important step, i.e., decoction of mixed raw herbal materials, where synergistic chemical interactions occur to enhance activity and reduce toxicity; this does not comply with TCM philosophy. Additionally, the direct use of granule preparations in hospitals is not regulated or controlled.

Table 4
Top 18 Manufacturers of Chinese Prepared Medicinal Herbs in 2001

No.	Name of the manufacturing plants	Sales ($ 1,000 USD)
1	Xinjiang Tefeng Pharmaceutical Inc.	8605
2	Shenzhen Jinchun Pharmaceutical Co. Ltd.	7321
3	Shanghai Tong-han-chun-tang Manufacturing Plant of Prepared Medicinal Herbs	6343
4	Tianjin Manufacturing Plant of Prepared Medicinal Herbs	4841
5	Shanghai Lei-yun-shang Plant of Prepared Medicinal Herbs	4711
6	Shanghai Xu-chong-dao Manufacturing Plant of Prepared Medicinal Herbs	3865
7	Shanghai Baoshan Manufacturing Plant of Prepared Medicinal Herbs	2131
8	Shanghai Xuhui Manufacturing Plant of Prepared Medicinal Herbs	1929
9	Beijing Renwei Manufacturing Plant of Prepared Medicinal Herbs	1818
10	Qijia Prepared Herbals Co. Ltd. of Heibei Province	1811
11	The Prepared Herb Manufacturing Plant of Shanghai Jiabo Pharmacy	1810
12	Hebei Meizhu Company of Chinese Materia Medica	1594
13	The Sichuan Tibet Plateau Pharmaceuticals Co. Ltd.	1529
14	The Beiqi Pharmaceuticals of the Greater Xing'an Mountains	1504
15	Anguo Guangming Manufacturing Plant of Prepared Herbs	1384
16	Shuangqiao Yanjing Manufacturing Plant of Prepared Medicinal Herbs	1196
17	Shanghai Yangpu Manufacturing Plant of Prepared Medicinal Herbs	1131
18	Liuzhou Shennong Manufacturing Plant of Prepared Medicinal Herbs	945

3.3. Status of TCM Manufacturing Enterprises

In the 1950s, the professional factories making herbal drug products were situated in herbal pharmacies, with simple and crude equipment and poor preparation technology. Through technology innovation and transformation, production by these herbal drug manufacturers has increased tremendously. By the end of 1995, there were 1020 herbal drug enterprises in China (of which 158 were large and medium-sized) with a total output value of $2 billion USD (of which the large and medium-sized enterprises contributed about 60%) *(76)*. In this industry, the conventional mode of production was gradually replaced by modern facilities and technology, and administrative and technical standards have been greatly enhanced. With the reform of the TCM industry, the working environment, the level of technology and equipment, and the total quality control of products have been upgraded to CGMP standards in about two-thirds of the TCM manufacturers. Thus far, approx 8000 herbal drug products, prepared medicinal herbs, and natural health food products of different dosage forms, including pills, pallets, capsules, granules, syrups, injectables, topical creams and ointments, patches, aerosols, and so on have been supplied to the market, with consistently good quality, reasonably high curative effect, and few side effects. By the end of 2001, the total output value of the herbal drug industry reached $10–13 billion USD, and a number of pharmaceutical companies specializing in herbal drug product manufacturing and marketing have evolved (see **Tables 5** and **6**), with advanced processing technology, strong R&D, and effective marketing capabilities *(62,77)*.

Table 5
The Top 20 Chinese Herbal Drug Manufacturers in 2001

No.	Name and place	Sales ($Million USD)
1	Taiji Group Ltd., Chongqing	317
2	Tianjin Zhong-xin Pharmaceuticals Ltd., Tianjin	262
3	Shanghai Lei Yun Shang Pharmaceutical Co. Ltd., Shanghai	190
4	Hui Ren Group Co. Ltd, Jiangxi Province	188
5	Tianjin Tasly Group, Tianjin	164
6	China Beijing Tong-ren-tang (Group) Co. Ltd., Beijing	137
7	Xiu Zheng Pharmaceutical Co. Ltd., Jilin Province	125
8	Chengdu Diao Group, Sichuan Province	106
9	Chitai Qing Chun Bao Pharmaceutical Co., Zhejiang Province	94
10	Zhenzhou Hua Xia Medical and Health Product Co. Ltd., Henan Province	92
11	Shandong Lu Nan Pharmaceutical Inc., Shandong Province	83
12	Shanghai Traditional Chinese Materia Medica Co., Shanghai	81
13	Guilin San Jin Pharmaceuticals Co. Ltd., Guangxi Province	73
14	Dong-E E-geltin Group Ltd., Shandong	65
15	Tong Ren Tang Science and Technology Development Co. Ltd., Beijing	62
16	Guangxi Jin Shang Zi Co. Ltd., Guangxi Province	60
17	The Guangzhou First TCM Pharmaceutical Plant, Guangdong Province	59
18	Shi Jia Zhuang Shenwei Pharmaceuticals Co. Ltd., Hebei Province	57
19	Jiu Zhi Tang Inc., Hunan Province	48
20	Qingdao Guofeng Pharmaceuticals Inc., Shandong Province	47

Compared with 1999, annual sales in 2001 increased by 41.3%, whereas total profit value increased by 49.0%. There were eight herbal drug companies having a net income of more than 100 million Yuan (**Table 7**). There were 284 enterprises running deficits, accounting for 25.7% of the total number *(62,76,77)*.

One of the obvious problems with the herbal drug industry is that there are many small businesses (more than 70% of the total number of the TCM manufacturers) whose product pipelines and technology are out of date, and competitiveness and financial status are relatively poor. These small enterprises are becoming a main source of economic loss and employee layoffs in the TCM industry.

3.4. Status of Plant Extracts

China has recently become one of the world's largest medicinal plant exporters and importers. It is reported that China exported nearly 144,000 tons of medicinal plants annually from 1993 to 1998 *(78)*. The medicinal plants were sold to more than 90 foreign countries, such as the United States, the Republic of Korea, and Japan. Meanwhile, China imported about 9200 tons of medicinal plants each year over the same period from more than 30 foreign countries. The demand for medicinal plants around the world has increased in recent years, as more and more people preferred natural therapies. Because the United States is the largest consumer of plant extracts (the market of diet supplements based on plant extracts reached $337 million in 2001, and the sales of the top 20 plants amounted to 96% of the market, *see* **Table 8**), the majority of

Table 6
Comparison of Economic Indices for Herbal Drug Manufacturers in 1999–2001

Year	No. of enterprises	Fixed asset ($million USD)	Output value ($million USD)	Sales ($million USD)	Total profit ($million USD)
1999	1033	2434	4807	4379	459
2000	1072	2657	5982	5537	608
2001	1104	2841	6621	6189	684

extract products are sold to the United States. The quantity of plant extracts supplied from China accounts for 7% of the world market (78).

Currently there are approx 200 companies manufacturing and supplying plant extract products; most are small businesses, and the largest has annual sales of less than $10 million USD. There are only seven or eight suppliers capable of exporting over $4 million USD products each year. Main products are extracts from various plants, including grape seed, pine bark, apple, astragalus, cat's claw, celery seed, chrysanthemum, corn silk, epimedium, ginkgo leaf, ginseng, giant knotweed, hawthorn leaf, red yeast rice, rhodiola, serrate clubmoss, shiitake mushroom, soy bean, tobacco, uniflower swisscentaury, wolfberry and so on (78).

New drug formulations and delivery systems are also emerging. For example, nanoparticles (NPs) are composed of solid colloidal particles ranging in size from 1 to 1000 nm (97), and their primary advantages are stability both in the body and in stock, site-specific targeting, and slow release. The application of NPs in herbal medicine should increase the effectiveness of herbs.

3.5. Status of Preclinical and Clinical Research

The experience of TCM needs to be substantiated and advanced by the available toolbox of science and technology. Quantitative and qualitative analysis, biological activity, bioavailability, absorption, metabolism, elimination, toxicity, and mode of action studies have been employed using either cell or animal models (113–115). Randomized controlled clinical trials and double-blind experiments have been done to address the question of efficacy (116–118). A detailed warning label should be issued to notify consumers and medical practitioners about potential harm and assure that the claimed health benefits were really there (119).

4. The Internationalization or Globalization of TCM

TCM constitutes a multi-billion-dollar industry worldwide, and more than 1500 herbals are sold as dietary supplements or ethnic traditional medicines. **Table 9** lists the exports of herbal drug products and Chinese material medica, including plant extracts, from 1990 to 2001 (62). However, the great potential of TCM has not been reached. There are many challenges in the course of herbal product internationalization, and the entire industry is making a great effort to advance into the international market.

Table 7
Most Profitable Herbal Drug Manufacturers in 2001

No.	Name of the manufacturers	Profit ($million USD)
1	Chengdu Diao Group	29.6
2	Tianjin Tasly Group	28.8
3	Xiu Zheng Pharmaceuticals Group Ltd.	28.3
4	China Beijing Tong Ren Tang (Group) Co. Ltd.	27.7
5	Chitai Qing Chun Bao Pharmaceuticals	22.6
6	Gong-E E-gelatin Group Ltd.	17.1
7	Guilin San Jin Pharmaceuticals Co. Ltd.	14.4
8	Jilin Au Dong Group Ltd.	13.8
9	Tong Ren Tang Sci. & Tech. Development Co. Ltd.	11.9
10	Tianjin Zhong-xin Pharmaceuticals Ltd.	11.3
11	Jiang Zhong Pharmaceutical Co. Ltd.	10.5
12	Tonghua Jinma Pharmaceutical Co. Ltd.	10.3
13	Shandong Lu Nan Pharmaceutical Inc.	10.1
14	Zhangzhou Pian Zai Huang Pharmaceutical Co. Ltd.	10.1
15	Tibet Qi Zheng Pharmaceutical Co.	9.4
16	Guangxi Jin Shang Zi Co. Ltd.	9.4
17	Yunnan Bai Yao Pharmaceuticals Group Ltd.	9.3
18	Tianjin Guang Xia Group Ltd.	9.3
19	Zhejiang Kang Eng Bei Group Ltd.	9.2
20	The Guangzhou First TCM Pharmaceutical Plant	9.0

4.1. Open Dialog and Mutual Understanding Among All Parties Involved in Health Care System

Western medicine is built on reproducible experiments and statistical analysis, whereas CAM and TCM are built on clinical experience. Compared with conventional Western medicine, TCM is poorly researched. Many studies in TCM therapies have flaws, such as insufficient statistical power, poor controls, inconsistent treatment, and lack of comparisons. TCM is one of the oldest evidence-based alternative and complementary therapies, and its formulations are often not subjected to premarket toxicity testing. With the booming market of TCM worldwide, a new strategy must be formulated for the assessment of drug efficacy and toxicity to optimize the therapeutic and preventive potential of Chinese herbal medicine.

The validation of TCM theory is perhaps more important than the commercialization of herbal drugs in the international market. The international standardization of traditional medical terms and practices will facilitate worldwide acceptance of TCM products. The safe and effective application of TCM has much to do with the skills of traditional medical practitioners as well as the basic understanding of the mechanism of TCM. As an example, *Ginkgo biloba* is a dioecious tree with a long history of TCM. The standardized extracts of its leaves have been used widely as a phytomedicine in Europe and as a dietary supplement in the United States. The primary active constituents of its leaves include flavonoid glycosides and unique diterpenes known as ginkgolides; the latter are potent inhibitors of platelet activating factor *(94)*. Clinical

Table 8
Sales of 20 Herbal Extracts in International Mmarket in 2002

Rank	Botanical name	Active ingredients	Sales ($million USD)
1	Gingko	Flavones/Lactones	5.8
2	Echinacea	Total phenolics	5.0
3	Garlic	Alicins	4.4
4	Ginseng	Saponins	3.9
5	Soy bean	Isoflavones	3.5
6	Saw Palmetto	Fatty acids	3.1
7	St. John's wort	Hypericins	3.0
8	Valerian	Valeric acids	1.5
9	Bilberry	Anthocyanidins	1.3
10	Black Cohosh	Triterpenes	1.2
11	Kava Kava	Lactones	1.2
12	Milk Fruit	Silymarins	0.9
13	Evening primrose	Oils	0.7
14	Grape seed	Proanthocyanidins	0.5
15	Yohimbin	Alkaloids	0.3
16	Green tea	Polyphenols	0.2
17	Pycnogenol Pine	Polyphenols	0.2
18	Rhodiola rosea	Salidrosides, rosavins	0.2
19	Ginger	Gingerols	0.2
20	Pyrethrum	Total flavones	0.1
Total			41.9

studies have shown that ginkgo extracts exhibit therapeutic effects in a variety of conditions, including Alzheimer's disease, memory loss, age-related dementia, cerebral and ocular blood flow occlusion, premenstrual problems, and altitude sickness *(95)*. As a result of its potent antioxidant properties and ability to enhance peripheral and cerebral circulation, ginkgo shows prospective value in the treatment of cerebrovascular dysfunctions and peripheral vascular disorders *(96)*.

4.2. Deciphering the Preventive Nature of TCM

A very important philosophy for TCM is to cure disease before it happens—the prototype of preventive medicine *(98–101)*. Three functional levels of preventive medicine have been mentioned in the past: preventing the occurrence of disease and injury; early detection and intervention, by reversing, halting, or retarding the progression of a disease; minimizing the effects of disease and disability by surveillance and maintenance to prevent complications.

As an example, cardiovascular disease is a complex and multifactorial disease, characterized by such factors as high cholesterol, hypertension, reduced fibrinolysis, increase in blood-clotting time, and increased platelet aggregation. Evidence from numerous studies points to the fact that garlic can bring about the normalization of plasma lipids, enhancement of fibrinolytic activity, inhibition of platelet aggregation, and reduction of blood pressure and glucose *(120)*. Drug developers in the oncology

**Table 9
Annual Export of Herbal Products From 1990 to 2001**

Year	Chinese material medica (in $1000 USD)	Herbal drug products (in $1000 USD)	Nation's total exports (in $million USD)
1990	304,150	112,870	62,091
1991	241,250	126,680	71,843
1992	241,340	135,280	84,940
1993	329,190	120,910	91,744
1994	537,210	130,260	121,006
1995	537,010	135,350	148,780
1996	443,570	126,000	151,048
1997	424,650	107,520	182,792
1998	332,130	84,520	183,712
1999	295,170	73,870	194,931
2000	348,990	90,940	249,212
2001	354,000	102,000	266,154

arena have much to learn from their cardiovascular counterparts. The quest for primary and secondary cancer prevention has been ongoing for some years. One of the most debated is the anti-inflammatory drug COX-2 for colorectal carcinoma and other types of cancer. Other examples include hormonal therapy for prevention of prostate and breast cancer. Tamoxifen, a selective estrogen receptor modulator (SERM), was actually approved for cancer prevention for women with high risk of breast cancer in 1998, thus being the first cancer prevention drug ever approved by the FDA *(121–127)*.

The further validation of preventive medicine is awaiting the advancement of biostatistics, epidemiology, gene biomarker identification, and other diagnostic tools in disease initiation and progression. One promising example of cancer-preventive effects that are not specific to any organ is *Panax ginseng*, an herb with a long medicinal history. The genus name of ginseng, *Panax*, is derived from the Greek *pan* (all) *akos* (cure), meaning "cure-all." No single herb can be considered a panacea, but ginseng comes close to it. Ginseng is a tonic herb, or an adaptogen that helps to improve overall health and restore the body to balance, and helps the body to heal by itself. Its protective influence against cancer has been shown by extensive preclinical and epidemiological studies *(89,90)*. Ginseng is a slow-growing perennial herb, reaching about 2 ft in height. The older the root, the greater is the concentration of ginsenosides, the active chemical compounds; thus the ginseng becomes more potent with time. More than 28 ginsenosides have been extracted from ginseng, and might be associated with a wide range of therapeutic actions in the central nervous system (CNS) and cardiovascular and endocrine systems *(91)*. Indeed, ginseng promotes immune function and metabolism, and possesses antistress and anti-aging activities. Several ginsenosides were proven to be non-organ-specific tumor suppressors and to improve learning and memory in patients with Alzheimer's disease *(91–93)*.

4.3. Respect and Protect Intellectual Property

The Dietary Supplement Act of 1994 opened the door for a surge of products into the nutraceutical marketplace *(86)*. However, maintaining exclusive market rights for any natural product remains challenging. Many people in the natural-products industry perceive few opportunities to obtain defensible patents for natural products that have been used in the public domain for many years. Thus, with little chance to achieve market exclusivity and without a requirement for premarket approval, companies have little incentive to conduct expensive clinical trials of dietary supplements. However, opportunities do exist with respect to identifying new active ingredients, improving methodologies of secondary metabolites production, applying modern technology to discover new modes of action of TCM *(102)*, and fingerprinting the new natural product remedy. Therefore, it is wise to seek multiple facets of intellectual property protection. The most common form for a new natural-product remedy is patent protection. A patentable invention needs to be novel, useful, and nonobvious (inventive). In some instances, these remedies may qualify for protection as trade secrets. With the existence of many manufacturers of the same type of product, creating a brand name that can assure high quality may be the ultimate challenge.

4.4. Safety Issues of TCM Need To Be Seriously Evaluated

Widespread favor of TCM also brings serious concern about its safety, regulation, efficacy, and mode of action *(103,104)*. In some cases, the use of TCM has been connected to many undesirable side effects, resulting in nephropathy, acute hepatitis, coma, and so on *(105–108)*. Cases of poisoning have been linked to variations in the chemical composition of different brands of the same herb. Such differences may arise as the result of inadequate processing (processes normally involve soaking and boiling the raw material), resulting in toxins being retained or adulteration with cheaper substitutes *(109–111)*. Cases of poisoning may also be attributed to contamination by heavy metals *(112)*.

TCM is guided and supported by medical theory. There will be serious consequences if the drugs are used but the traditional medical theory is discarded. People in many countries are using Chinese herbal medicines as daily diet supplements without having a basic understanding of Chinese medical theory or the rationale behind their use. One of the most serious examples is the case of xiao-chai-hu-tang (decoction of bupleuri for regulating *Shaoyang*), observed in Japan. This ancient formula was used to treat febrile diseases in *Shaoyang* meridian with symptoms of alternate attacks of chills and fever, fullness in the chest, discomfort, dizziness, dry throat, vomiting, and so on *(80)*. Based on modern pharmacological findings, some TCM practitioners used such a formula to treat hepatitis specifically showing the above symptoms and obtained reasonably good results. However, the formula was widely prescribed in Japan for the long-term treatment of all types of hepatitis, and unfortunately, resulted in severe adverse effects, including a number of deaths *(81)*.

It was recently reported that a specific type of nephropathy occurred due to ingestion of certain Chinese herbs such as *Aristolochiae manshuriensis*. The case highlighted the role of aristolochic acid and its metabolite, aristololactam I, in causing this nephropathy, which was first observed in a Belgian cohort *(82)*. The phenomenon, now called

"Chinese herbal nephropathy," led to a public misconception of the toxic nature of TCM. While further research on aristolochic acid-containing herbs is being conducted in China *(83)*, we would like to make the following points: In ancient times, the herbs Mu-tong (*Akebia quinata*), San-ye-mu-tong (*Akebia trifoliate*), and Bai-mu-tong (*Akebia quinata var. australis*) were used in combination formulas rather than the herb Guan-mu-tong (*Aristolochiae manshuriensis*). The three herbs used in ancient formulas have no aristolochic acid and have been replaced nowadays with Guan-mu-tong due to much decreased resources. As a result, they were no longer collected in the Chinese Pharmacopoeia 2000. Such a replacement greatly altered the toxicology of these Mu-tong-containing formulas. Secondly, the preparation of many TCMs nowadays has been greatly modified for convenience, i.e., the drug decoction process is replaced with "instant" water-soluble dosage forms of herbal extracts. This has minimized chemical interactions including drug detoxification among herbal medicines. Additionally, the use of a single herbal medicine, or refined fractions in large quantity, may exhibit certain therapeutic activities, but are often more toxic and less efficacious than combination formulas. Any use of herbal products in high dosage or for long-term medication is not advisable unless patients see a practitioner for a diagnosis and holistic approach to their conditions and obtain a customized and personalized prescription, taking the various manifestations of their symptoms and perceived causes into full account. In fact, a carefully prepared combination formula works rather differently, because it contains synergistic and balanced elements that may interact in different ways and neutralize the negative effects of some toxic constituents that the plants might contain. This is the hypothesis, but the public is entitled to have medicines that are proven to work by rigorous tests and that are safe and cost-effective. The standards and the criteria for judging the safety and the effectiveness of treatment and diagnostic interventions must be formulated by critical scientific evaluation from practitioners with different epistemology. Satisfaction from patients using both Western medicine and TCM has been scored and compared in Korea *(26)*.

4.5. The Small Scale and Limited Capability of Herbal Drug Enterprises

Most Chinese pharmaceutical companies specialized in herbal drug manufacturing and marketing are currently very weak in terms of their product quality, proprietary technology, international sales, and overall competitiveness. Those TCM enterprises are also weak in terms of sales, profit, and equity capital in the international pharmaceutical industry. China's largest herbal drug company, Taiji Group (**Table 5**), has annual sales of only $300 million USD. Such herbal drug companies will have financial and technical difficulties in the development of international business. The R&D investment by these companies supported by their revenues will be limited. This vicious circle will prevent them from experiencing rapid and sustainable growth with a well-structured product pipeline *(79)*. At present, the majority of manufacturers primarily produce products in traditional dosage forms, such as large spherical pills consisting of powders of raw materials which are not subjected to sophisticated extraction processes and are inconsistent in quality.

There are more than 1000 companies in China making TCM, but large and medium-sized enterprises account for less than 20% of the total business players in the Chinese market. There are too many companies scrambling into the same business with almost the same strategies, leading to a glut in production that reduces their profit. A limited high-tech content in the production of TCM has also prevented China from competing effectively on the world stage *(79)*. Merger and acquisition should take place to create firms large enough to be cost-effective.

5. Closing Remarks

Historical precedent predicts the potential of TCM to expand the health-care repertoire, either as single-chemical entities or as complex botanical drugs. With an increase in the aging population and changes in the epidemiology of health problems (such as chronic and degenerative diseases) that have frustrated Western medicine, TCM and CAM will gain more popularity in the new millennium. Most survey results in Western countries showed there is considerable interest in TCM and CAM among primary-care professionals, and many are already referring or suggesting referrals. Such referrals are driven mainly by patient demand and by dissatisfaction with the results of conventional medicine. The trend of integrating TCM and CAM into mainstream primary care is unstoppable. There is an urgent need to further educate and inform primary-care health professionals about TCM and CAM, which bring unprecedented opportunities and challenges to TCM specifically. Modernization and globalization of TCM is the right way to go. We strongly believe that the future of TCM as a healing art is promising, with its success in research and education and, very importantly, in commercialization in the international market. With the further development of systems biology, pharmacogenomics, synergistic medicine, and personalized medicine, the real time for integration of TCM and Western medicine will come. It is hopeful that some day patients will benefit from an integrated medicine—wisdom combining the strengths of both the East and the West.

References

1. Borisy AA, Elliott PJ, Hurst NW, et al. Systematic discovery of multicomponent therapeutics. Proc Natl Acad Sci USA 2003;100(13):7977–7982.
2. Shawver LK, Slamon D, Ullrich A. Smart drugs: tyrosine kinase inhibitors in cancer therapy. Cancer Cell 2002;1, 117–123.
3. Gibbs JB. Mechanism-based target identification and drug discovery in cancer research. Science 2000;287, 1969–1973.
4. Lenz GR, Nash HM, Jindal S. Chemical ligands, genomics and drug discovery. Drug Discovery Today 2000;5, 145–156.
5. Kalgutkar AS, Crews BC, Rowlinson SW., et al. Biochemically based design of cyclooxygenase-2 (COX-2) inhibitors: facile conversion of nonsteroidal antiinflammatory drugs to potent and highly selective COX-2 inhibitors. Proc Natl Acad Sci USA 2000;97:925–930.
6. Nanda A, St. Croix B. Tumor endothelial markers: new targets for cancer therapy. Curr Opin Oncol 2004;16:44–49.
7. Review of Global Pharmaceutical Industry in 2002. Institute of the South Economic Research of SFDA, 2003.
8. Service RF. Surviving the blockbuster syndrome. Science 2004;303(5665):1796–1799.
9. Lathers CM. Challenges and opportunities in animal drug development: a regulatory perspective. Nat Rev Drug Discov 2003;2(11):915–918.
10. Szuromi P, Vinson V, Marshall E. Rethinking drug discovery. Science 2004:1795.
11. Knight V, Sanglier JJ, DiTullio D, et al. Diversifying microbial natural products for drug discovery. Appl Microbiol Biotech 2003;62:446–458.

12. Landers P. Cost of developing a new drug increases to about $1.7 billion. The Wall Street Journal 12/8/2003.
13. Kennedy D. Drug discovery. Science 2004;303(5665):1729.
14. Kitano H. Systems biology: a brief overview. Science. 2002;295, 1662–1664.
15. Kanehisa M, Goto S, Kawashima S, Nakaya A. The KEGG databases at GenomeNet. Nucleic Acids Res 2002;30:42–46.
16. Blume-Jensen P, Hunter T. Oncogenic kinase signalling. Nature 2001;411:355–365.
17. Brent R. Genomic biology. Cell 2000;100:169–183.
18. Kitano H. Computational systems biology. Nature 2002;420(6912):206–210.
19. Jorgensen WL. The many roles of computation in drug discovery. Science 2004;303(5665): 1813–1818.
20. Shaheen RM, Tseng WW, Davis DW, et al. Tyrosine kinase inhibition of multiple angiogenic growth factor receptors improves survival in mice bearing colon cancer liver metastases by inhibition of endothelial cell survival mechanisms. Cancer Res 2001;61:1464–1468.
21. Zhang P. **Cell** cycle control and development: redundant roles of **cell** cycle regulators. Curr Opin Cell Biol 1999;11:655–662.
22. Chan-Hui PY. From PGx to molecular diagnostics and personalized medicine. Drug Discov Today 2003;8(18):829–831.
23. Van Haselen RA, Reiber U, Nickel I, Jakob A, Fisher PA. Providing complementary and alternative medicine in primary care: the primary care workers' perspective. Complement Ther Med 2004;12(1):6–16.
24. Giordano J, Boatwright D, Stapleton S, Huff L. Blending the boundaries: steps toward an integration of complementary and alternative medicine into mainstream practice. J Altern Complement Med 2002;8(6):897–906.
25. McClure MW. An overview of holistic medicine and complementary and alternative medicine for the prevention and treatment of BPH, prostatitis, and prostate cancer. World J Urol 2002;20(5):273–284.
26. Hong CD. Complementary and alternative medicine in Korea: current status and future prospects. J Altern Complement Med 2001;7:S33–S40.
27. Dossey BM. Holisitic nursing: taking your practice to the next level. Nurs Clin North Am 2001;36(1):1–22.
28. Larson L. Natural selection. Trustee 2001;54(4):6–12.
29. Norred CL, Zamudio S, Palmer SK. Use of complementary and alternative medicines by surgical patients. AANA J 2000;68(1):13–18.
30. Eisenberg DM, Davis RB, Ettner SL, et al. Trends in alternative medicine use in the United States, 1990–1997. J Am Med Assoc 1998;280:1569–1575 .
31. Kessler RC, Davis RB, Foster DF, et al. Long-term trends in the use of complementary and alternative medical therapies in the United States. Ann Intern Med 2001;135:262–268.
32. Goldbeck-Wood S, Dorozynski A, Lie LG, et al. Complementary medicine is booming worldwide. Br Med J 1996;313:131–133.
33. Carlsson M, Arman M, Backman M, Flatters U, Hatschek T, Hamrin E. Evaluation of quality of life/life satisfaction in women with breast cancer in complementary and conventional care. Acta Oncol 2004;43(1):27–34.
34. Engel LW, Straus SE. Development of therapeutics: opportunities within complementary and alternative medicine. Nat Rev Drug Discov 2002;1(3):229–237.
35. Weiger WA, Smith M, Boon H, Richardson MA, Kaptchuk TJ, Eisenberg DM. Advising patients who seek complementary and alternative medical therapies for cancer. Ann Intern Med 2002;137(11):889–903.
36. Cragg GM, Newman DJ, Snader KM. Natural products in drug discovery and development. J Nat Prods 1997;60:52–60.

37. Newman DJ, Cragg GM, Snader KM. Natural products as sources of new drugs over the period 1981–2002. J Nat Prod 2003;66(7):1022–1037.
38. Mendelson R, Balick MJ. The value of undiscovered pharmaceuticals in tropical forests. Econ Bot 1995;49:223–228.
39. Dictionary of Natural Products. Chapman and Hall/CRC, London: 2001.
40. Henkel T, Brunne RM, Müller H, Reichel F. Statistical investigation into the structural complementarity of natural products and synthetic compounds. Angew Chem Int Ed Engl 1999;38:643–647.
41. Berdy J. Are actinomycetes exhausted as a source of secondary metabolites? Proc. 9th Internat Symp Biol Actinomycetes, Part 1. Allerton, New York: 1995; pp. 3–23.
42. Roessner CA, Scott AI. Genetically engineered synthesis of natural products: from alkaloids to corrins. Ann Rev Microbiol 1996;50:467–490.
43. Fenical W, Jensen PR. Marine microorganisms: a new biomedical resource. In: Attaway DH, Zaborsky OR (eds), Marine Biotechnology I: Pharmaceutical and Bioactive Natural Products. Plenum, New York: 1993; pp. 419–475.
44. Kiefer DM. A century of pain relief. Todays Chem Work 1997;6(12):38–42.
45. Clark AM. Natural products as a resource for new drugs. Pharm Res 1996;13:1133–1141.
46. Tyler VE, Brady LR, Robbers JE. Pharmacognosy, 9th edn. Lea & Febiger, Philadelphia: 1988.
47. Schultz V, Hansel R, Tyler VE. Rational Phytotherapy: A Physician's Guide to Herbal Medicine, 4th edn. Springer, Berlin: 2001.
48. Klayman DL. Qinghaosu (artemisinin): an antimalarial drug from China. Science 1985;228:1049–1055.
49. Horwitz SB. Personal recollections on the early development of taxol. J Nat Prod 2004;67(2):136–138.
50. Clark J. Regulation of natural health products challenged. CMAJ. 2004;170(8):1217.
51. Blumenthal M. In: Blumenthal M et al (eds), The Complete German Commission E Monographs: Therapeutic Guide to Herbal Medicines 17. Austin, Texas: American Botanical Council, 1998.
52. Unschuld PU. The past 1000 years of Chinese medicine. Lancet 1999;354 Suppl:SIV9.
53. Opara EI. The efficacy and safety of Chinese herbal medicines. Br J Nutr 2004;91(2):171–173.
54. Kam PC, Liew S. Traditional Chinese herbal medicine and anaesthesia. Anaesthesia 2002;57(11):1083–1089.
55. Ernst E. Adulteration of Chinese herbal medicines with synthetic drugs: a systematic review. J Intern Med 2002;252(2):107–113.
56. Normile D. Asian medicine. The new face of traditional Chinese medicine. Science 2003;299(5604):188–190.
57. Lampert N, Xu Y. Chinese herbal nephropathy. Lancet 2002;359(9308):796–797.
58. Hesketh T, Zhu WX. Health in China. Traditional Chinese medicine: one country, two systems. BMJ 1997;315(7100):115–117.
59. Wang ZG, Ren J. Current status and future direction of Chinese herbal medicine. Trends Pharmacol Sci 2002;23(8):347–348.
60. Bodeker G. Lessons on integration from the developing world's experience. BMJ 2001;322(7279):164–167.
61. Scheid V. The globalization of Chinese medicine. Lancet 1999;354 Suppl:SIV10.
62. Research Report of Chinese Pharmaceutical Market Analysis of Chinese Herbal Drug Industry, WAN Fang Database Co. Ltd., 2003.
63. Summary of Chinese Herbal Pharmaceutical Technology and Process. Information Center of State Drug Administration, 2002.
64. Liu ZH, Li ZJ. The Industrial Developing Strategy of Chinese Herbal Drug Modernization. Traditional Chinese Medicine, Beijing: 2003.

65. Chinese Medicines and Health Products Report. Ming Jing Market Research & Consultation Co., Ltd., 2002.
66. Developing and Future Trend of National and International Pharmaceutical Industry in 2002. Institute of Economic Research of NDRC, 2003.
67. Huang LQ, Cui GH, Chen ML, et al. Study on complex system of Chinese material medica GAP fulfilling-the situation, problems, and prospects of Chinese material medica germplasm. China J Chin Mate Med 2002;27:481–483.
68. Huang LQ, Cui GH, Dai RW. Study on complex system of Chinese Materia Medica GAP fulfilling. China J Chin Mater Med 2002;27:1–3.
69. Yuan CQ, Wang NH, Lu Y. Conservation of endangered medicinal plants in China. In: Zhang ED, Zheng HC (eds), Conservation of Endangered Medicinal Wildlife Resources in China. Second Military Medical University Press, Shanghai: 2000; pp. 25–32.
70. Wang LX, Xiao LL. Significance of medicinal botany to the conservation of endangered species. In: Zhang ED, Zheng HC (eds), Conservation of Endangered Medicinal Wildlife Resources in China. Second Military Medical University Press, Shanghai: 2000; pp. 83–86.
71. Liu HG, Zhan YH, Chen JC. The cultivation and GAP of Chinese Materia Medica in Hubei province in China. J Hubei College TCM 2001;3:45–46.
72. Shang MF, Zhu Y. The situation of cultivation base of Chinese materia medica in China. Res Inform Trad Chin Med 2000;2(10):23–27.
73. Wang WQ, Liu CS, Sun ZR, et al. The study of GAP and its application in Chinese materia medica. Res Inform Trad Chin Med 2001;3(10):14–16.
74. Wang JM, Liu HM, Jiang CZ, et al. The establishment of GAP cultivation base. Chin Pharm 2002;16:32–34.
75. Yao ZG, Fu YQ. Recent questions and improved measures of Chinese Prepared Medicinal Herbs. Lishizhen Mater Med Res 2002;13(9):552–553.
76. Economic Operating Analytic Report of Pharmaceutical Industry in 2001. Bureau of Economic Operation of SETC, 2002.
77. Review and Prospect of the Pharmaceutical Companies on the Stock Market. Bureau of Economic Operation of SETC, 2003.
78. Shao YD, Gao WY, Jia W, et al. The quality control of plan extract. China J Chin Mater Med 2003;28 (10):899–903.
79. Jia W, Gao W Y, Xiao PG. The discussion of internationalization of Chinese herbal drugs. China J Chin Mater Med 2002;27(9):645–648.
80. Fan QL. Science of Prescription. Shanghai TCM University Press, Shanghai: 2002.
81. Sasaki H. International Senior Forum of Traditional Chinese Medicines & Botanical Products. Speech Collection: 38–41, Hangzhou, China, 2001.
82. Lord GM, Cook T, Arlt VM, et al. Urothelial malignant disease and Chinese herbal nephropathy. Lancet 2001;358:1515–1516.
83. Ma HM, Zhang BL, Xu ZP, et al. Experimental research on the renal toxicity of Guan-mu-tong (*Aristolochiae manshuriensis*). Clin Pharmacol Tradit Chin Med 2001;12:404–409.
84. Pub L no. 75–717, 52 Stat 1040 (1938) 21 USC 9 (Food, Drug and Cosmetic Act 1938).
85. Pub L no. 87–781, § 102(c), 76 Stat 780 (1962) (codified at 21 USC § 355(d) (5)) (Kefauver Harris Amendment of 1962).
86. Pub L no. 103–417, 108 Stat 4325 (1994) 21 USC 231 (The Dietary Supplement Health and Education Act of 1994).
87. Bass IS, Raubicheck CJ. Marketing Dietary Supplements. Washington DC: Food and Drug Law Institute, 2000.
88. Federal Register. House Government Reform Committee. Six Years After the Enactment of DSHEA: The Status of National and International Dietary Supplement Regulation (March 20, 2001).

89. Qi G. Protective effect of gypenosides on DNA and RNA of rat. Acta Pharmacol Sin 2000;21:1193–1196.
90. Yun TK. Panax ginseng—a non-organ-specific cancer preventive? Lancet Oncol 2001;2:49–55.
91. Shibata S. Chemistry and cancer preventing activities of ginseng saponins and some related triterpenoid compounds. J Korean Med Sci 2001;16:S28–S37.
92. Yun TK. Update from Asia. Asian studies on cancer chemoprevention. Ann NY Acad Sci 1999;889:157–192.
93. Zhao X, et al. Effects of ginsenoside of stem and leaf in combination with choline on improving learning and memory of Alzheimer disease. Chin Pharmacol Bull 2000;16:544–547.
94. McKenna DJ, Jones K, Hughes K. Efficacy, safety, and use of ginkgo biloba in clinical and preclinical applicagtions. Altern Ther Health Med 2001;7:70–90.
95. Gao Y, et al. Effect of allicin on the regulation of VEGF mRNA expression in human hepatocellular carcinomal cells. Chin Pharmacol Bull 2001;17:531–536.
96. Xu JP, Rui YC, Li TJ. Antagonistic effects of ginkgo biloba extract on adhesion of monocytes and neutrophils to cultured cerebral microvascular endothelial cells. Zhongguo Yaoli Xue Bao 1999;20:423–425.
97. Deng R. General research situation of long-circulation nanoparticles. Chin Pharm J 2001;36:511–513.
98. Nie QH, Luo XD, Zhang JZ, Su Q. Current status of severe acute respiratory syndrome in China. World J Gastroenterol 2003;9(8):1635–1645.
99. Fugh-Berman A. Herbs and dietary supplements in the prevention and treatment of cardiovascular disease. Prev Cardiol 2000;3(1):24–32.
100. Koo LC. Concepts of disease causation, treatment and prevention among Hong Kong Chinese: diversity and eclecticism. Soc Sci Med 1987;25(4):405–417.
101. Bibeau G. From China to Africa: the same impossible synthesis between traditional and western medicines. Soc Sci Med 1985;21(8):937–943.
102. Wang WK, Wang YY. Biomedical engineering basis of traditional Chinese medicine. Med Prog Technol 1992;18(3):191–197.
103. Ernst E. Complementary medicine: evidence base, competence to practise and regulation. Clin Med 2003;3(5):481–482.
104. Ernst E. Herbal medicines put into context. BMJ 2003;327(7420):881–882.
105. Hohmann N, Koffler K. Risk of adverse reactions from contaminants in Chinese herbal medicines can be minimized by using quality products and qualified practitioners. Int J Environ Health Res 2002;12(1):99–100.
106. Chan TY, Critchley JA. Usage and adverse effects of Chinese herbal medicines. Hum Exp Toxicol 1996;15(1):5–12.
107. Chan TY. Monitoring the safety of herbal medicines. Drug Saf 1997;17(4):209–215.
108. Matsumoto K, Mikoshiba H, Saida T. Nonpigmenting solitary fixed drug eruption caused by a Chinese traditional herbal medicine, ma huang (Ephedra Hebra), mainly containing pseudoephedrine and ephedrine. J Am Acad Dermatol 2003;48(4):628–630.
109. Li XM, Sampson HA. Novel approaches for the treatment of food allergy. Curr Opin Allergy Clin Immunol 2002;2(3):273–278.
110. Fugh-Berman A. Herb-drug interactions. Lancet 2000;355(9198):134–138.
111. Pan CX, Morrison RS, Ness J, Fugh-Berman A, Leipzig RM. Complementary and alternative medicine in the management of pain, dyspnea, and nausea and vomiting near the end of life. A systematic review. J Pain Symptom Manage 2000;20(5):374–387.
112. Ernst E. Toxic heavy metals and undeclared drugs in Asian herbal medicines. Trends Pharmacol Sci 2002;23:136–139.
113. Xu QF, Fang XL, Chen DF. Pharmacokinetics and bioavailability of ginsenoside Rb1 and Rg1 from Panax notoginseng in rats. J Ethnopharmacol 2003;84(2–3):187–192.

114. Zhang BL, Wang YY, Chen RX. Clinical randomized double-blinded study on treatment of vascular dementia by jiannao yizhi granule. Zhongguo Zhong Xi Yi Jie He Za Zhi 2002;22(8): 577–580.
115. Chen LC, Chen YF, Chou MH, Lin MF, Yang LL, Yen KY. Pharmacokinetic interactions between carbamazepine and the traditional Chinese medicine Paeoniae Radix. Biol Pharm Bull 2002;25(4):532–535.
116. Fugh-Berman A, Kronenberg F. Complementary and alternative medicine (CAM) in reproductive-age women: a review of randomized controlled trials. Reprod Toxicol 2003;17(2): 137–152.
117. Kronenberg F, Fugh-Berman A. Complementary and alternative medicine for menopausal symptoms: a review of randomized, controlled trials. Ann Intern Med 2002;137(10):805–813.
118. Tang JL, Zhan SY, Ernst E. Review of randomised controlled trials of traditional Chinese medicine. BMJ 1999;319(7203):160–161.
119. Koop CE. The future of medicine. Science 2002;295(5553):233.
120. Rahman K. Historical perspective on garlic and cardiovascular disease. J Nutr 2001;131(3s): 977S–979S.
121. Freedman AN, Graubard BI, Rao SR, McCaskill-Stevens W, Ballard-Barbash R, Gail MH. Estimates of the number of US women who could benefit from tamoxifen for breast cancer chemoprevention. J Natl Cancer Inst 2003;95(7):526–532.
122. Pappas SG, Jordan VC. Raloxifene for the treatment and prevention of breast cancer? Expert Rev Anticancer Ther 2001;1(3):334–340.
123. Honig SF. Tamoxifen for the reduction in the incidence of breast cancer in women at high risk for breast cancer. Ann NY Acad Sci 2001;949:345–348.
124. Goldstein SR. The effect of SERMs on the endometrium. Ann NY Acad Sci 2001;949:237–242.
125. Prout MN. Breast cancer risk reduction: what do we know and where should we go? Medscape Womens Health 2000;5(5):E4.
126. Osborne MP. Chemoprevention of breast cancer. Surg Clin North Am 1999;79(5):1207–1221.
127. Goldstein SR. Selective estrogen receptor modulators: a new category of compounds to extend postmenopausal women's health. Int J Fertil Womens Med 1999;44(5):221–226.

11

Arsenic Trioxide and Leukemia

From Bedside to Bench

Guo-Qiang Chen, Qiong Wang, Hua Yan, and Zhu Chen

Summary

Cumulative evidence indicates that traditional Chinese medicine (TCM) is an important resource for discoveries of drugs against cancer. Arsenic, a common, naturally existing substance, is rarely found in its pure elemental state in nature. In addition to the organic arsenicals, there are three major inorganic arsenic forms: red arsenic (As_2S_2), yellow arsenic (As_2S_3), and white arsenic (As_2O_3, ATO). Based on the basic TCM theory of "using poison against poison," a group from Harbin Medical University in the northeastern region of China in the early 1970s introduced intravenous infusion of "Ailing-1 (anticancer-1)," a solution of crude ATO and herbal extracts, into cancer therapy. After a lengthy study in more than 1000 patients with various kinds of cancers, acute promyelocytic leukemia (APL), a specific subtype of acute myeloid leukemia characterized by the failure of differentiation/maturation towards granulocytic cells at the promyelocytic stage, was determined to be an excellent target for "Ailing-1" therapy. At the same time, there were reports of using the formula called "niu huang jie du pian" (containing xiong huang) or just using xiong huang (realgar) by itself to treat APL. A formula of Qing Dai (indigo) containing xiong huang as the main Chinese herb was reported to result in 98.3% total remission.

Thus, some cautious clinical trials with pure ATO and state-of-the-art bench studies were conducted worldwide. The results revealed that ATO brings the majority of relapsed APL patients to a second complete morphologic, cytogenetic, and even molecular remission. Successful experiences that used ATO together with the differentiation-inducing agent *all-trans* retinoic acid (ATRA) and chemotherapeutic drugs make APL the most curable subtype of acute myeloid leukemia in adults, an unprecedented achievement in the field of hematologic malignancies. The clinical data from our group showing striking efficacy and safety of ATO in patients with APL led to clinical trials in the United States and subsequent approval by US Food and Drug Administration (FDA) in September 2000. Inspired by such a bedside discovery, many investigators are researching pharmacological mechanisms of ATO, and finding that ATO exerts wide-spectrum cellular and molecular activities on APL cells and other cancer cells, such as growth arrest, apoptosis induction, differentiation induction, and anti-angiogenesis, although its exact mechanisms remain to be determined. More importantly, the clinical potential of ATO in other cancers beyond APL is also under study. In this chapter, we discuss recent clinical practice and the mechanism of ATO action on APL and other cancer cells.

Key Words: Arsenic trioxide; acute promyelocytic leukemia (APL); clinical trials; differentiation; apoptosis; leukemia; cancer.

1. Introduction

Microbial and plant secondary metabolites helped to double our life span during the twentieth century, reduced pain and suffering, and revolutionized medicine. For instance,

From: *Natural Products: Drug Discovery and Therapeutic Medicine*
Edited by: L. Zhang and A. L. Demain © Humana Press Inc., Totowa, NJ

all-trans retinoic acid (ATRA), a vitamin A derivative, has been successfully used for acute promyelocytic leukemia (APL) by inducing terminal differentiation of leukemic cells towards granulocytes, the only well-established example of cancer differentiation therapy so far *(1,2)*. In the past 20 yr, a therapeutic strategy with ATRA and anthracycline-based chemotherapy for induction, anthracycline-based consolidation, and maintenance with ATRA and/or low-dose chemotherapy has kept more than 70% of APL patients alive and disease-free for 5 yr, an unprecedented achievement in the field of oncology *(3)*.

Increasing evidence demonstrated that herbs-based traditional Chinese medicine (TCM) could be used for treatment of many diseases, including cancer. As documented, some Chinese herbs, such as Qing Dai (*Indigo naturalis*), Ya Dai Zi (*Brucea javanica*), She Xiang (*Moschus moschiferus*), and E Shu (rhizoma *Curcuma phaeocaulis*), can directly kill cancer cells. Also, other herbs can inhibit cancer cell growth, such as Jiang Huang (*Curcuma aromatica*), Chang Chun Teng (*Hedera helix*), Tian Hua Feng (*Trichosanthes kirilowii*), Tian Nan Xing (*Arisaemi japonicum*), and Shan Zi Gu (*Cremastra variabilis*) *(4,5)*. In addition, up to 100 kinds of Chinese herbs can stimulate the patient's immune system to attack cancer cells *(6)*. Taken together, Chinese herbs are becoming important resources for anticancer drug discovery *(7)*.

Chinese herbs are often mixed into formulae for clinical application. For example, the formula "Qing Huang San" (one of the popular formulae that grinds two herbs, Qing Dai and Xiong Huang, at a ratio of 9:1 into fine powder, which is loaded into capsules) was shown to effectively treat chronic myeloid leukemia (CML) *(8)*. In 1995, Huang et al. from Da Lian city, China, used a formula combining these two herbs as main ingredients to treat 60 APL patients, which resulted in 98.3% experiencing total remission *(9)*. Before that, there were reports of using a formula called "niu huang jie du pian" (containing Xiong Huang) or just using Xiong Huang (realgar) by itself to treat APL *(8)*.

APL, an unique subtype of acute myeloid leukemia (AML), is characterized by the blockage of granulocytic differentiation at the promyelocytic stage and specific reciprocal chromosomal translocation t(15;17)(q22;q21). The latter leads to the expression of the fusion protein PML-RARα (promyelocytic leukemia—retinoic acid receptor-α), whose leukemogenic role has been demonstrated using transgenic mouse models *(10)*. Although conventional chemotherapy such as anthracyclines and cytosine arabinoside help about 70% of newly diagnosed APL patients to reach clinically complete remission (CR), a high frequency of early death (mainly due to exacerbation of bleeding syndrome) and a lower 5-yr disease-free survival (DFS) rate have emphasized a need for new drugs. Inspired by the Chinese philosophy that it is better to transform and reverse a bad element than to simply get rid of it, and by important discoveries in the 1970s and early 1980s that leukemic cells undergo phenotypic reversion in vitro when treated with some agents, the Shanghai Institute of Hematology (SIH) started to screen a large number of compounds. SIH identified ATRA as a strong differentiation-inducing agent for APL cells, and improved clinical efficacy as mentioned above. However, there is still a 30% relapse rate, especially in APL patients having higher white blood cells (WBC) counts. In addition, relapsed patients after ATRA treatment are resistant to this drug, which becomes the main reason for failure to be cured *(11,12)*. Fortunately, arsenic trioxide (As_2O_3, ATO, commercially named Trisenox™ [Cell

Therapeutics, Inc., Seattle, WA]) made the majority of relapsed and newly diagnostic APL patients achieve complete morphologic, cytogenetic, and even molecular remission (13). Such an exciting ingredient, which was rapidly approved by the State Drug Administration (SDA) in China (13) and the Food and Drug Administration (FDA) of the United States (14), encouraged many investigators to focus on mechanisms of action of ATO on APL and to explore the potentials of ATO in treatment of cancers beyond APL. In this chapter, we shall concentrate on recent clinical practice and the mechanism of action of ATO on APL and other cancer cells.

2. Historic Overview of Medicinal Arsenicals

Arsenic, a common, naturally occurring substance, is rarely found in its pure elemental state in nature. In addition to the organic arsenicals formed by covalent linking of a trivalent or pentavalent arsenic atom to a carbon atom, there are three major inorganic arsenic forms: red arsenic (As_2S_2, also named realgar or sandaraca), yellow arsenic (As_2S_3, referred to as arsenikon, aurum pigmentum, or orpiment), and white arsenic (ATO). That arsenic is a poison that has been known for many centuries. Its odorless and tasteless properties and acute/chronic poisoning has resulted in symptoms confused with a variety of other natural disorders, such as hemorrhagic gastroenteritis, cardiac arrhythmias, and psychiatric disease, which has made arsenic a common homicide or suicide tool since the Middle Ages. For example, Napoleon might have been poisoned by arsenic-tainted wine that was served to him in exile (15). At present, long-term exposure to arsenic is still a serious public health problem in some parts of the world, especially in developing countries. It has been estimated that tens of millions of people are at high risk of being exposed to excessive levels of arsenic from both contaminated water and arsenic-bearing coal from natural sources (16). Apart from the poisoning, arsenic exposure has long been known to be correlated with human skin, lung, liver, kidney and urinary bladder cancers, although arsenic itself has not been shown to be carcinogenic in animal models to date.

Paradoxically, arsenic usage has been medicinally practiced for 24 centuries, providing a history of utility, dishonor, and redemption (for details, see ref. 10). Since the days of the ancient Greek and Roman civilizations, arsenic has been an active ingredient in folk remedies in central and southern Asia. In the well-known Ben Cao Gang Mu (Compendium of Materia Medica) compiled by Herbalist Li Shizhen in the Ming Dynasty of China, arsenic was recorded as a treatment for several diseases. In the eighteenth century, Dr. William Withering (1741–1799) found that one of his patients with dropsy (congestive heart failure) improved remarkably after taking a traditional herbal remedy. He discovered that the active ingredient was digitalis, which was contained in the leaves of foxglove. His comment, "Poisons in small doses are the best medicines; and the best medicines in too large doses are poisonous," made him a strong proponent of arsenic-based therapies (17). Many arsenic preparations were used therapeutically in the 18th century. At that time, Thomas Fowler compounded a potassium bicarbonate-based solution of ATO named Fowler's solution to treat a variety of diseases, including asthma, chorea, eczema, and psoriasis. The solution became a standard remedy to treat anemia, Hodgkin's disease, and leukemia until the early 20th century (17). Arsenic's reputation as a therapeutic agent was further explored in 1910 when Nobel laureate and the founder of chemotherapy Paul Ehrlich developed salvarsan, an organic arsenical, to treat syphilis, which is still used for trypanosomiasis today *(17)*. Taken together, arsenic was one of the mainstays of the materia medica in the 19th century.

Arsenic's antileukemic activity was first reported in the late 1800s. In 1878, a report from Boston City Hospital described the effect of Fowler's solution on the reduction of WBC counts in two healthy persons and one patient with "leukocythemia" *(18)*. After discontinuation of treatment, the leukocyte count progressively increased until arsenic therapy was reinstituted. Arsenic then experienced a brief resurgence in popularity following a report in 1931 of nine CML patients who responded well to ATO therapy. They experienced a reduction in total WBC counts and in the size of enlarged livers and spleens, a return to apparently normal hematopoiesis in bone marrow (BM) biopsy specimens, and a sense of well-being *(18)*. However, Kandel et al. *(19)* reported the development of chronic arsenic poisoning in five of six CML patients and recommended careful monitoring with its application. Thereafter, the utility of Fowler's solution progressively declined, and it was supplanted by radiotherapy and cytotoxic chemotherapy.

Nevertheless, a group from Harbin Medical University, in the northeastern region of China, has been challenging its disuse. They deeply believed in the basic theory of TCM, i.e., "using poison against poison," and determined that the possible adverse effects should not prevent physicians from applying arsenic to treat patients with life-threatening diseases. Actually, most chemotherapeutic agents in use today are genotoxic and carcinogenic. In the early 1970s, they re-introduced intravenous infusion of Ailing-1 (anticancer-1), a solution of crude ATO and herbal extracts, into cancer therapy. After a lengthy study in more than 1000 patients with various kinds of cancer, APL was determined to be an excellent target for Ailing-1 therapy. In 1992, they revealed their preliminary clinical trials on APL patients, showing clinically CR was achieved in 21 out of 32 (65.6%) patients, with 5- and 10-yr survival rates of 50.0% and 28.2%, respectively *(20)*. Since 1994, a careful clinical trial with pure ATO and state-of-the-art bench studies conducted at SIH in collaboration with Harbin's group made the remarkable achievement well known in the world *(21,22)*. Since then, these results have been confirmed in randomized clinical trials in the United States and other countries *(23–26)*, resulting in a new era of medicinal arsenic.

3. Clinical Experiences With Arsenic in the Treatment of APL

3.1. Pharmacokinetics

In order to understand pharmacological mechanisms and potential toxicity of ATO in the treatment of APL, pharmacokinetic analysis in eight relapsed APL patients receiving intravenous infusions of the present standard dose (0.16 mg/kg/d) of ATO were performed in our group *(27)*. The results revealed that the plasma arsenic concentration rapidly reached the peak level of 6.85 (range, 5.54 to 7.30) µmol/L of the mean maximal plasma concentration (C_{pmax}). In one Japanese APL patient who developed BM necrosis during ATO treatment, the C_{pmax} and half-life time was 6.9 µmol/L and 3.2 h, respectively *(28)*. The mean C_{pmax} in six APL patients given a low dose of ATO (0.08 mg/kg/d) fell to 2.63 (range, 1.54 to 3.42) µmol/L, which was nearly half the level of a

Table 1
CR Rate in APL Treated With ATO in Some Clinical Trials

Reference	Status of the disease	N	CR rate (%)
Sun et al. (20)	Primary + relapsed	32	65.6
Zhang et al. (32)	Primary	30	73.3
	Relapsed	42	52.4
Shen et al. (27)	Relapsed	15	93.8
Soignet et al. (23)	Relapsed + refractory	12	91.7
Niu et al. (34)	Primary	11	72.4
	Relapsed	47	85.1
Soignet et al. (24)	Relapsed + refractory	40	85.0
Zhang et al. (33)	Primary	124	87.9
	Relapsed + refractory	118	61.0
Shen et al. (29)	Relapsed	20	80.0
Lazo et al. (26)	Relapsed	12	100

CR, complete remission.

standard dose (29). Furthermore, continuous administration of ATO over 30 d did not change its pharmacokinetic behavior. It was very different from the observations with ATRA, the C_{pmax} of which decreased progressively with continuous usage (30). However, most of the time during treatment with either a standard or a low dose, the plasma concentration was maintained in the range of 0.1 to 0.5 µmol/L. Of note was the report that the C_{pmax} of ATO (median 0.46 µmol/L) was slowly achieved at 8–15 d with a standard dose in a phase II trial of an ATO/interferon (IFN)-α combination in seven patients with relapsed/refractory adult T-cell leukemia/lymphoma (ATL) (31). In both standard and low-dose groups, moreover, urinary arsenic content was slightly increased during drug administration, and the total amount of arsenic excreted daily in the urine accounted for less than one-tenth of the total daily dose. Arsenic accumulated gradually in the hair and nails, and the peak level at CR stage was five to seven times higher than prior to treatment. However, the amount of arsenic in hair and nails tended to decrease after withdrawal of the drug (27,29).

3.2. Higher Complete Remission Rate

In 1997, we reported that 93.8% (14/15) of relapsed APL patients obtained CR with an intravenous drip of pure ATO at 0.16 mg/kg/d, which was diluted in 250 mL to 500 mL of a 5% glucose-normal saline solution (27). Later, a series of clinical trials were carried out in various countries and regions, and a similar CR rate was obtained (**Table 1**). For instance, the report from Memorial Sloan Kettering Cancer Center in New York revealed that out of 12 relapsed APL patients receiving ATO, 11 got CR. It is worthwhile to note that treating 20 cases of relapsed APL with low-dose ATO (0.08 mg/kg/d) yielded a similar CR rate (80%) (29).

It should be mentioned that some groups, including ours (25,34), tried to use ATO to treat newly diagnosed APL patients and obtained similar CR rates. For example, 14 newly diagnosed patients were recruited to an ongoing trial of Mathews's group, during which ATO was administered at a dose of 10 mg/d until CR was achieved. Afterwards, a consolidation course and a maintenance schedule consisting of ATO as a single agent were administered over 6 mo. Exclusive of three early deaths (two on d 3 and one on d 4) related to intracerebral hemorrhage and one patient who died on d 21 secondary to uncontrolled sepsis, 10 of the remaining 11 patients (91%) attained CR with the average time to CR of 52.3 d (range: 34–70 d). This included one unfortunate patient who developed an isolated central nervous system (CNS) relapse and subsequently went into a second CR following therapy with triple intrathecal chemotherapy, cranial irradiation, and an additional 4-wk course of systemic ATO *(25)*.

3.3. Correction of Coagulopathy

APL cells contain several distinct procoagulant factors, including tissue factor (TF) and cancer procoagulants as well as plasminogen activators (t-PA, u-PA). These factors induce generation of thrombin and activation of plasmin, resulting in disseminated intravascular coagulation and hyperfibrinolysis respectively, which are aggravated by chemotherapeutic drugs *(35)*. Similar to ATRA treatment, ATO induction can improve the bleeding diathesis and coagulation-test abnormalities. Those are manifested by decreased soluble fibrin monomer complex, thrombin-antithrombin complex, fibrinogen/fibrin degradation products, and D-dimer, and increased fibrinogen and plasminogen in the plasma *(36,37)*. Recent studies have confirmed that ATO inhibited the aberrant expression of TF at the transcriptional level, in parallel to the improvement in hemorrhagic syndrome during ATO treatment *(36,38)*.

3.4. Low Toxicity

A few patients developed BM depression and significant cross-resistance to the currently used drugs for APL. But upon treatment with ATO, about 70% of the patients presented with increased WBC counts, and hyperleucocytosis with signs mimicking retinoic acid syndrome was reported in one-third to one-half of the patients *(27,32, 33,39,40)*. This new "arsenic syndrome" responded to dexamethasone treatment. More surprisingly, although arsenic is known as a poison, most patients could tolerate this treatment. ATO-related adverse effects, most of which were mild, included skin reactions (dryness, itching, erythematous lesion and pigmentation), gastrointestinal disturbance with loss of appetite, nausea, vomiting and diarrhea, peripheral neuropathy with dysesthesia, joint and muscle pain, facial edema, ECG changes with low-flat T wave, sinusal tachycardia, or grade 1 atrioventricular blockage. Thirty percent or so developed impaired liver function with elevation of AST, ALT, alkaline phosphatase, GGT, or total bilirubin *(14,20,23,24,33,34,41–43)*. Other less frequent adverse effects were enlargement of the salivary gland, enlarged thyroid gland without hyperthyroidism, oral ulcer, toothache, and gingival or nose bleeding. However, one should be cautious in interpreting the unexplained BM necrosis after a daily dose of 0.15 mg/kg ATO therapy in a 60-yr-old Japanese male with refractory APL *(28)*. ATO may also suppress cell-mediated immunity by inducing apoptosis of T-helper lymphocytes and exert the immunosuppressive effect of ATO in vivo. That was activated by an occurrence of herpes zoster during ATO treatment in two patients who were already in remission and received ATO as a consolidation treatment *(44)*. These mild side effects can be prevented or managed successfully with careful patient monitoring during treatment. Tanvetyanon and Nand have proposed that early antiviral treatment might help avoid such complications as postherpetic neuralgia *(44)*. It is worth noting that hepatotoxicity seemed to be higher in newly diagnosed cases in our experience *(34)*.

3.5. Survival Duration and Postremission Treatment

In the standard-dose group, among 33 relapsed patients followed up for 7 to 48 mo after CR in our institute *(34)*, the estimated disease-free survival (DFS) rate and overall survival (OS) rate at 2 yr were 49.11 ± 15.09% and 61.55 ± 15.79%, respectively. These results showed no statistically significant superiority to those obtained among 14 relapsed patients followed up for 7 to 33 mo on the low dose *(29)*. Hence, low-dose ATO treatment seems to be equally effective in APL patients, although a larger-scale randomized study should be launched to reach a definitive conclusion. Soignet et al. *(23)* reported that after two courses of ATO therapy, 27 out of 37 patients obtained molecular remission, i.e., APL-related PML-RARα fusion gene transcripts were negative by RT-PCR assay. Similar findings were also reported in studies at M. D. Anderson *(26)*, and other small-range studies *(45)* suggest that molecular remission may be achieved at the time of CR in some APL patients with durable DFS.

3.6. Combination of ATO With ATRA on APL

Although some reports on the in vitro joint effects of ATRA and ATO are controversial, several lines of evidence support a strong synergistic effect between ATRA and ATO, supporting the feasibility of their combination in clinical application for newly diagnosed APL. In a randomized study of ATO alone vs a combination of ATO and ATRA in 20 relapsed APL patients, all previously treated with ATRA-containing chemotherapy between 1998 and 2001, Raffoux et al. *(46)* found that the CR rate (80%), the hematologic and molecular responses, the time necessary to reach CR, and the outcome were comparable in both treatment groups. This led to a conclusion that ATRA did not significantly improve the response to ATO in relapsed patients. In our recent study *(47)*, 61 newly diagnosed APL subjects were randomized into three treatment groups—ATRA, ATO, and the combination of the two drugs. The results revealed that all three groups reached a high CR rate of >90%, among which the combination group experienced the shortest time to CR and a significantly decreased tumor burden as reflected by change of PML-RARα transcripts at CR. All 20 cases in the combination group remained in CR, whereas 7 of 37 cases treated with mono-therapy relapsed after a follow-up of 8–30 mo (median: 18 mo) (**Fig. 1**).

3.7. Other Arsenic Agents in Leukemia

Lu et al. *(48)* used oral tetra-arsenic tetra-sulfide (As_4S_4) to treat 129 APL patients, including 19 newly diagnosed, 7 first relapse, and 103 hematologic complete remission (HCR) patients. The results revealed that CR was achieved in all patients with newly diagnosed APL and with hematologic relapse. Like ATO, As_4S_4 was well tolerated by APL patients with moderate side effects, including asymptomatic prolongation of corrected QT interval, transient elevation in liver enzyme levels, rash, and mild gastrointestinal discomfort, without any evidence of myelosuppression or appreciable long-term side effects. More interestingly, 14 out of 16 newly diagnosed patients, 5 of the 7 patients

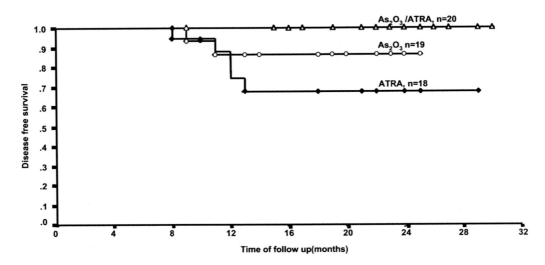

Fig. 1. Kaplan-Meier disease-free survival curve. The successive cases of newly diagnosed acute promyelocytic leukemia patients were randomized into three groups for induction therapy. The relapsed rate of the *all-trans* retinoic acid-alone treatment group is significantly higher than that of combined group ($p = 0.0202$, Fisher's exact test). When patients of two monotherapy groups are put together, the relapse rate is also statistically higher than the combined group ($p = 0.038$) (cited from ref. *47*).

with hematologic relapse, and 35 of 44 patients in the HCR group had cytogenetic and molecular CR. Obviously, oral As_4S_4 treatment is highly effective and safe enough in both remission induction and maintenance therapy for APL patients.

4. In Vivo Action of ATO on APL

The clinical effectiveness of ATO on APL has stimulated research aimed at understanding its mechanism of action. Our initial clinical observations showed that during daily continuous intravenous infusion of ATO on APL, with the gradual reduction of leukemic promyelocytes, there were a large percentage of myelocyte-like cells carrying the PML-RARα gene. This was revealed by fluorescence *in situ* hybridization (FISH) analysis, indicating that ATO could induce in vivo partial differentiation of APL cells *(49)*. The presence of leukocytosis and even a retinoic acid syndrome-like syndrome during ATO treatment supported this suggestion *(49)*. Moreover, during the ATO treatment, cells with condensed nuclei hallmarking the apoptosis process appeared in both BM and peripheral blood, which could be positively labeled by *in situ* terminal deoxynucleotidyl transferase *(49)*. Similar observations were described in other clinical reports *(23,24,32,33,35)*. With a deeper in vivo analysis, for example, Soignet et al. *(23)* reported that ATO treatment induced a progressive decrease in the proportion of cells expressing CD33 (an antigen typically associated with primitive myeloid cells) along with an increase in the proportion of cells expressing CD11b (an antigen associated with mature myeloid elements). ATO also increased the number of cells that simultaneously expressed both antigens and displayed a hybrid signal of PML and RARα genes on FISH analysis. Meanwhile, serial Western blot analysis of BM mononuclear cells revealed the precursor forms of caspase 2 and caspase 3 were upregulated in response to ATO treatment. In addition, ATO-induced apoptosis and modest differentiation of APL cells in vivo could also be seen in an APL model using syngenic grafts of leukemic blasts from PML-RARα transgenic mice *(50)*. The same was observed in subcutaneous tumors that were formed by implantation of ATRA-resistant UF-1 APL cells into hGM-CSF-producing transgenic SCID mice *(51)*.

5. In Vitro Action of ATO
5.1. Apoptosis Induction
5.1.1. Apoptosis Induction in a Relatively Wide Spectrum of Tumors

Since we reported that ATO induced the NB4 cell line (a commonly used APL cell line with t(15;17) and sensitized to ATRA-induced differentiation *(52)*) to undergo apoptosis with downregulation of the expression of anti-apoptotic gene Bcl-2 and modulation of PML-RARα/PML proteins *(21)*, the apoptosis-inducing and growth-inhibitory effects of ATO have been widely confirmed in many cancer cells. These include myeloid leukemia cell lines and/or fresh cells from CML, acute megakaryocytic leukemia, lymphoproliferative diseases, multiple myeloma, human T-cell leukemia virus type I (HTLV-I)-infected T-cell lines and adult T-cell leukemia cells, as well as some solid tumor cells such as neuroblastoma cell lines, esophageal carcinoma cell lines, gastric cancer cells, ovarian cancer cells, and prostate cancer cells (for details, *see* ref. *53*).

5.1.2. Potential Mechanisms of Apoptosis Induction of ATO
5.1.2.1. MITOCHONDRIA, CASPASES, AND ATO-INDUCED APOPTOSIS

A large body of studies have shown that mitochondrial modulation, especially the mitochondrial transmembrane potentials ($\Delta\Psi m$) collapse, is a pivotal event in the apoptotic process. One of the important factors leading to $\Delta\Psi m$ collapse is the opening of the mitochondrial permeability transition pore (MPT), which results in the release of pro-apoptotic factors such as cytochrome c and the apoptosis-inducing factor into the cytosol. These substances, in turn, activate some members of the caspase family *(54)*. ATO has been shown to induce $\Delta\Psi m$ disruption in APL cells, malignant lymphocytes, and myeloma cell lines, and in a cell-free system *(55–57)*. Arsenite has also been shown to open the purified, reconstituted MPT in vitro in a cyclosporin A- and Bcl-2-inhibitable fashion *(58)*. Furthermore, ATO-induced apoptosis in cultured myeloid leukemia cells was carried out by caspases from the distal, PARP-cleaving part of the activation cascade; PKC activation had no effect on apoptosis induced by either ATO or VP-16 in these cells *(59–61)*. In addition to caspase-3, caspase-8 and Bid were activated by ATO in a GSH concentration-dependent manner in NB4. An inhibitor of caspase-8 blocked not only the activation of caspase-3 but also the loss of $\Delta\Psi m$ *(62)*. However, Liu et al. *(63)* suggested that caspase-8 and caspase-3 were primarily involved in ATO-induced apoptosis in myeloma cells with mutated p53, while primarily caspase-9 and caspase-3 were involved in cells expressing wild-type (wt) p53. It is worth noting that ATO did not increase the p53 level or activity in most cell types tested. ATO also did not induce p53-dependent transactivation, and there was no difference in apoptosis induction between cells with p53$^{+/+}$ or p53$^{-/-}$ *(64)*. However, a marked increase in p53 protein levels was seen during apoptosis in a human gastric cancer cell line induced by ATO, while co-incubation with p53 antisense oligonucleotide suppressed ATO-induced intracellular p53 overexpression and apoptosis failed *(65)*. Taken together, depolarization of mitochondria and activation of caspases are common features involved in ATO-induced apoptosis *(66)*.

5.1.2.2. CELLULAR GLUTATHIONE, REACTIVE OXYGEN SPECIES, AND ATO-INDUCED APOPTOSIS

It has long been known that the toxic effects of trivalent arsenicals are mediated through interaction with active tissue thiols. Although one report *(67)* showed that dithiothreitol (DTT) unexpectedly enhanced ATO-induced apoptosis in leukemic cells, most groups have observed that 0.2 mM DTT could block ATO-induced $\Delta\Psi$m collapse, caspase activation, and apoptosis in NB4 cells and malignant lymphocytes. On the other hand, buthionine sulfoximine (BSO), a depleter of cellular glutathione (GSH), substantially enhanced sensitivity. Simultaneous treatment with BSO and ATO could increase the sensitivity of HL60 and U937 cells to ATO. Furthermore, GSH depletion overcame resistance to ATO in arsenic-resistant cell lines *(68)*. We and others *(56,69–71)* have found that intracellular GSH levels and/or the activity of enzymes including glutathione peroxidase (GPx), glutathione-S-transferase π (GSTπ), and catalase (which regulate cellular H_2O_2 content) are important factors in determining cell sensitivity to ATO. According to these data, we proposed that the ATO-induced $\Delta\Psi$m collapse and apoptosis are associated with thiols of some important proteins, especially those constituting the MPT complex. Of note, ATO did not cause the oxidation of a critical cysteine residue (Cys 56) of purified adenine nucleotide translocator *(72)*, an important vicinal SH-group-containing component-constituted MPT complex.

The amount of cellular thiol groups is regulated by the redox system, while reactive oxygen species (ROS) are important factors to induce MPT opening. The generation of ROS in apoptosis induction by ATO, partly inhibiting mitochondrial respiratory function *(73)*, has been widely confirmed, although the apoptotic activity of ATO in prostate and ovarian cancer cells is not mediated by superoxide generation *(74)*. Of note, ATO at clinically achievable therapeutic concentrations also induced apoptosis in peripheral blood T-cells by enhancing oxidative stress *(75)*. Different compounds, such as ascorbic acid *(69,76)* and emodin *(77)*, a natural anthraquinone derivative, sensitized cells to ATO via generation of ROS. More intriguingly, ATO-induced oxidative stress promoted telomere attrition, chromosome end-to-end fusions, and apoptotic cell death, which were effectively prevented by the antioxidant N-acetylcysteine. Furthermore, embryos with shortened telomeres from late generation telomerase-deficient mice exhibited increased susceptibility to arsenic-induced oxidative damage. Unexpectedly, arsenite did not cause chromosome end-to-end fusions in telomerase RNA knockout mouse embryos, despite progressively damaged telomeres and disrupted embryo viability *(78)*. The authors suggested that ATO could initiate oxidative stress and telomere erosion, leading to apoptosis and antitumor therapy on the one hand and chromosome instability and carcinogenesis on the other. In addition, the generation of ROS was associated with enhanced apoptosis induced by tetra-arsenic oxide (As_4O_6, 2,4,6,8,9,10-Hexaoxa-1,3,5,7-tetraarsatricyclo [3.3.1.1^{3,7}] decane) in ATO-resistant U937 leukemic cells *(79)*.

5.1.2.3. c-Jun NH$_2$-Terminal Kinases (JNKs) and ATO-Induced Apoptosis

ATO activated JNK at a dose range similar to that for induction of apoptosis. Activation was almost totally blocked by expression of a dominant-negative mutant of JNK1 in JB6 cells *(64)* and in ATO-sensitive but not in ATO-resistant NB4-derived subclones *(80)*. The JNK inhibitor dicumarol significantly increased cell growth and prolonged survival in response to ATO. Inhibition of nuclear factor (NF)-κB kinase (IKKβ) gene knockout (Ikkβ$^{-/-}$) resulted in prolongation of ATO-induced JNK activation, which was closely associated with oxidative stress response, as indicated by elevated expression of heme oxygenase-1 and the accumulation of H$_2$O$_2$ *(81)*. Furthermore, activation of JNK in NKM-1, which was established from a patient with acute myeloid leukemia (M2), was sustained from 6 to 24 h after 1 µmol/L ATO treatment, and preceded changes in cellular H$_2$O$_2$, Δϕm, and caspase-3 activation *(82)*. These data indicated an essential role of JNK signaling in the induction of growth inhibition and apoptosis induction by ATO.

5.1.2.4. Mitotic Arrest and Apoptosis Induction

Apoptosis induction by ATO in normal hematopoietic progenitors and leukemic cells might be dependent on their cell-cycle status *(83)*. According to several recent reports *(84–86)*, ATO stimulated G$_2$/M arrest in promonocytic leukemic cells U937 *(85)*, and in other human tumor cell lines such as HeLa S3, which was inhibited by pretreatment with the DNA polymerase inhibitor aphidicolin and was related to tubulin polymerization without affecting GTP binding to β-tubulin. Recently, we *(87)* showed that APL cells are arrested at early mitotic phase before the collapse of Δϕm and apoptosis after treatment with pharmacological concentrations of ATO. On the contrary, myeloma cells with wild-type (wt) p53 were relatively resistant to ATO, with maximal apoptosis of about 40% concomitant with partial arrest of cells in G$_1$ and up-regulation of p21 *(63)*. These results lead to our proposal that mitotic arrest is one of the common mechanisms for ATO-induced apoptosis in cancer cells. ATO induced rapid and extensive (more than 90%) apoptosis in a time- and dose-dependent manner concomitant with arrest of cells in G$_2$/M phase of the cell cycle.

5.2. Induction of Differentiation

As described above, plasma concentrations of ATO were relatively low (0.25–0.1 µ*M*) most of the time during remission induction with 0.16 mg/kg/d iv of ATO. Thus, we investigated effects of such low-dose ATO on APL cells. The results revealed that a relatively longer use of low-dose ATO (0.1–0.5 µmol/L) induced partial differentiation. Furthermore, fresh cells from APL cases were resistant to 1.0 µmol/L ATO-induced ΔΨm collapse/apoptosis, whereas they were sensitive to 0.1 µmol/L ATO-induced differentiation. The arsenic-resistant NB4 subclone (NB4-AsR) did not undergo apoptosis but maintained the partial differentiation response to this drug *(88)*, concomitant with the finding on AP-1060, a newly established APL cell line from a multiple-relapse patient *(89)*. Obviously, different pathways might be involved in differentiation and apoptosis as induced by ATO. However, the mechanism of ATO-induced differentiation remains elusive. Recently, the possibility of a modulation of histone acetylation *(90)* and potent inhibition of the interaction of SMRT *(91)*, which participates in transcriptional repression by a diverse array of vertebrate transcription factors, with its transcription factor partners by ATO has also been proposed.

The in vitro differentiation-inducing ability of ATO did not appear to be better than its in vivo activity. Thus, it is reasonable to speculate that some factors in BM microenvironments modulate the in vivo activity of ATO. These factors could be some cytokines and other molecules. Indeed, APL cells in vitro exposed to ATO at 1 µmol/L showed a significant increase of IL-1β and G-CSF ($P < 0.05$) production but a significant decrease of IL-6 and IL-8 *(92)*. More recently, Chelbi-alix et al. *(93)* suggested that IFN-α or IFN-γ combined with 0.1 µmol/L ATO led to an increased maturation effect on NB4 cells and on two ATRA-resistant NB4-derived cell lines, NB4-R1 and NB4-R2. A strong synergy was discovered between 0.25 µmol/L ATO and the cyclic adenosine monophosphate (cAMP) analog in the full induction of differentiation in NB4, NB4-R1, and fresh APL cells *(94)*.

Although leukemic cells do not form a well-circumscribed "mass" in BM like that in solid tumors, oxygen levels of BM in AML patients may be decreased as a result of rapid growth of leukemic cells. The lack of oxygen in BM is possibly further aggravated by the anemia that often accompanies newly diagnosed AML patients. Leukemic cells are cultured in vitro at ambient oxygen (21%) in most circumstances, while in vivo cells are physiologically exposed to much lower oxygen levels, ranging from 16% in pulmonary alveoli to less than 6% in most peripheral organs of the body. Based on these facts, we tested the effect of ATO on APL cells under hypoxia. Unexpectedly, we found that cobalt chloride ($CoCl_2$)/ desferrioxamine (DFO)-mimicked hypoxia or moderate hypoxia (2% and 3% O_2) triggered and enhanced ATO-induced leukemic cell differentiation to various extents in an AML subtype-independent manner (*[95]* and data not shown).

5.3. ATO and PML/PML-RARa

About 95% APL expresses PML-RARα chimeric proteins, which can block cell differentiation and apoptosis through dominant negative inhibition of biologic functions of wild-type PML, RARα, as well as other proteins (for details, *see* ref. *1*). PML, a tumor suppressor involved in many complex functions through gene transcription regulation (e.g., growth arrest and apoptosis) is normally localized in a nuclear domain named the PML nuclear body (NB), PML oncogenic domain (POD), or nuclear domain 10 (ND-10). In APL cells, due to the heterodimerization of PML-RARα with wild-type PML, PML-NBs are disrupted into a micro-speckle nuclear pattern, resulting in the loss of functions of PML and/or NB. Modulation and/or degradation of the chimeric proteins and reorganization of NB components have been widely accepted as important molecular mechanisms based on ATRA-induced APL cell differentiation. Although arsenic exerts its effect on many cell types, APL seems to be the most striking clinical target for ATO. This encouraged exploration of possible effects of ATO on APL-specific oncogenic mechanisms, such as PML-RARα fusion proteins. Interestingly, ATO can rapidly modulate the subcellular localization of PML and PML-RARα and induce the degradation of these proteins in APL cells and cells expressing exogenous PML-RARα *(21,49,70,71,88,96–98)*. The latter are associated with the SUMO-1-polymodified forms of PML.

Sternsdorf et al. *(98)* reported that ATO induces apoptosis only in U937 cells expressing PML-RARα. Of note, trivalent antimonial-induced apoptosis in NB4 cells accompanies degradation of PML-RARα and the reorganization of PML NBs *(99)*. Comparative studies between cells expressing PML-RARα and PLZF-RARα proteins, the latter being produced by another APL-related chromosome translocation t(11;17), showed that ATO can induce apoptosis on U937 and APL cells expressing the former, but not the latter, fusion protein*(100,101)*. It thus seems that ATO induces apoptosis of APL cells by targeting PML-RARα. However, the wide spectrum of ATO-induced apoptosis, as well as the degradation of PML/PML-RARα in a subline of NB4 cells (NB4/As) resistant to ATO-induced apoptosis *(62)*, show that PML/RARα degradation is not the only mediator for arsenic sensitivity.

Puccetti et al. *(102)* reported that sensitivity to ATO-induced apoptosis in U937 cells can be increased either by overexpression of PML or by conditional expression of activated RAS, which upregulates PML. Moreover, ATO treatment induced phosphorylation of the PML protein through a mitogen-activated protein (MAP) kinase pathway. Increased PML phosphorylation is associated with increased sumoylation of PML and increased PML-mediated apoptosis. Conversely, MAP kinase cascade inhibitors or the introduction of phosphorylation- or sumoylation-defective mutations of PML impair apoptosis, indicating that phosphorylation by MAP kinase cascades potentiates the antiproliferative functions of PML and helps mediate the proapoptotic effects of ATO *(103)*.

Interestingly, among those nonresponder-to-ATO patients described in the SIH clinical trial series *(27,29)*, two relapsed cases, one from a standard-dose group and the other from a low-dose group, were PML-RARα negative at the time of treatment in spite of the fact that fusion gene transcripts were positive at initial disease presentation. Another patient from the standard-dose group developed a new malignant clone with the AML1-ETO fusion gene, a hallmark of AML-M2, in addition to PML-RARα *(34)*. Therefore, the in vivo sensitivity to ATO appears to require the expression of PML-RARα.

5.4. Inhibition of Angiogenesis

Angiogenesis plays a critical role in the growth of solid tumors and may be important for the expansion of leukemic cell populations. An interesting finding by Roboz et al. *(104)* was that treatment of proliferating layers of human umbilical vein endothelial cells (HUVECs) with ATO resulted in a reproducible dose- and time-dependent alteration including activation of endothelial cells, upregulation of endothelial cell adhesion molecules, apoptosis of endothelial cells, and inhibition of vascular endothelial growth factor production. Incubation of HUVECs with ATO prevented capillary tubule and branch formation, as shown by in vivo and in vitro endothelial cell differentiation assays. In experimental solid tumors, a single administration of ATO produced preferential vascular shutdown in the tumor tissue with resultant hemorrhagic necrosis *(105)*. This phenomenon was repeatable, and no apparent toxic effects were observed on normal skin, muscle, or kidneys of the experimental animals.

5.5. Effects of Metabolites of ATO

Biomethylation is the major metabolic pathway for inorganic arsenic (iAs) in humans and in many animal species, within which iAs undergoes metabolic conversion that includes reduction of iAs^V to iAs^{III}, with subsequent methylation yielding mono (MM)- and dimethylated (DM) metabolites. The postulated scheme is as follows:

$iAs^V \rightarrow iAs^{III} \rightarrow MAs^V \rightarrow MAs^{III} \rightarrow DMAs^V \rightarrow DMAs^{III} \rightarrow MAs^{III}$ and $DMAs^{III}$. NB4 cells did not methylate iAs^{III}, but mono- and dimethylated metabolites formed from iAs^{III} in HepG2 cells were released into the medium *(106)*. Indeed, MMA^{III}, DMA^{III}, MMA^V, and DMA^V could be detected in urine samples collected consecutively for 48 h in four tested cases of APL patients receiving ATO treatment at standard dose; As^{III}, MMA^V, and DMA^V accounted for >95% of the total arsenic excreted *(107)*. Therefore, it is possible that methylated ATO metabolites that form in vivo may contribute to the therapeutic effect of ATO in APL. Ochi et al. *(108)* concluded that DMA^{III} was the most potent in terms of the ability to cause apoptosis in HL-60 cells; arsenite was less potent. Our in vitro studies *(106)* indicated that MMA^{III}, and to a lesser extent DMA^{III}, were more potent growth inhibitors and apoptotic inducers than iAs^{III} in leukemia and lymphoma cells, but not in human BM progenitor cells. H_2O_2 accumulation and GPx inhibition, but not degradation of PML-RARα, correlated with greater $MAs^{III}O$-induced apoptosis of NB4 cells.

6. Clinical Trials of ATO in Other Cancers

Preclinical investigations as described above indicated that the biological targets of ATO, including induction of apoptosis, nonterminal differentiation, and suppression of proliferation and angiogenesis, extend to a variety of malignancies other than APL. Thus, many clinical physicians and scientists have tried to enumerate the potential benefits of ATO in cancers other than APL *(109)*. The National Cancer Institute is working cooperatively with research centers across the United States to evaluate clinical activity of ATO in hematologic malignancies such as AML, acute lymphocytic leukemia, CML, non-Hodgkin's lymphoma, Hodgkin's disease, chronic lymphocytic leukemia, myelodysplastic syndrome, and multiple myeloma. Similar research was also carried out on solid tumors, such as advanced hormone-refractory prostate cancer and renal cell cancer, and in cervical cancer and refractory transitional cell carcinoma of the bladder. The safety and pharmacokinetics of ATO are being evaluated in pediatric patients with refractory leukemia and lymphoma. The results of these ongoing studies should provide important insights into the clinical utility of ATO in these diseases *(110,111)*. We now summarize results of a clinical trial on adult T-cell leukemia/lymphoma (ATL). ATL is a severe chemotherapy-resistant malignancy associated with prolonged infection by the human T-cell-lymphotropic virus 1 (HTLV-1), a retrovirus. Substantial data revealed that ATO combined with IFN-α could induce cell-cycle arrest and apoptosis of ATL cells both *ex vivo* and in vitro *(112–114)*. Moreover, ATO was shown to rapidly and selectively block the transcription of NF-κB-dependent genes in HTLV-1-infected cells only via dramatic stabilization of IκB-α and IκB-β *(115–116)*. Inspired by these in vitro observations, Hermine *(31)* initiated a phase II trial of an ATO/IFN combination in seven patients with relapsed/refractory ATL (four acute and three lymphoma). Four patients exhibited a clear initial response (one complete remission and three partial remissions). Yet, the treatment was discontinued after a median of 22 d because of toxicity (three patients) or subsequent progression (four patients). Six patients eventually died from progressive disease (five patients) or infection (one patient), but the remaining patient is still alive and disease-free at 32 mo. In conclusion, arsenic/IFN treatment is feasible and exhibits an antileukemic effect in very poor prognostic ATL patients despite significant toxicity.

Finally, 14 patients with metastatic renal cell carcinoma (RCC) with bidimensionally measurable disease, a Karnofsky performance status of at least 70%, life expectancy of greater than 3 mo, and no evidence of brain metastases, were treated in a phase II trial with ATO given intravenously at a dose of 0.3 mg/kg/d for five consecutive days every 4 wks. The most common forms of toxicity observed were grade II elevation in liver function tests (36%), anemia (21%), renal insufficiency (14%), rash (7%), and diarrhea (7%). Best response was stable disease in three patients, with one patient remaining in the study at 8+ mo. At the dose and schedule used in this trial, ATO did not achieve a complete or partial response in metastatic renal cell carcinoma *(117)*.

7. Conclusions

Although treatment with ATRA has greatly improved patient survival, 30% or so of patients still relapse following combination therapy with ATRA and cytotoxic chemotherapy. From recent studies that have demonstrated the efficacy and safety of ATO in treating APL patients, ATO is now considered the standard of care in relapsed/refractory APL, which has excited Western and Chinese oncologists *(22)*. Unlike patients treated with ATRA alone, more intriguingly, patients treated with ATO alone have a high rate of molecular remission, i.e., molecular conversion to PML-RARα negativity. On the other hand, the multiple mechanisms by which ATO can induce cell differentiation, apoptosis, and anti-angiogenesis suggest its potential for clinical application alone or combined with other agents to manage a variety of malignancies. The experience with ATO leads us to predict that a merger between TCM and modern molecular biology will bring benefits for cancer therapy. It also emphasizes the importance and feasibility of natural products in discovery of drugs against ailments including cancer.

There are many questions remaining to be addressed in the future, which will help to uncover how ATO functions within the cell, to explore new drugs and new clinical applications of ATO. For example, does ATO achieve its clinical effectiveness by apoptosis and/or differentiation induction, and does PML-RARα involve clinical action of ATO in APL patients? Answering such questions will provide important clues for its possible clinical use in leukemia and other cancers and improve its application. If it is true that differentiation induction and PML-RARα degradation are critical for the in vivo effects of ATO in APL, whereas apoptosis induction is relevant to its toxicity, we can speculate on its ineffectiveness in cancers beyond APL. Second, it is very important to determine by which mechanism(s) ATO induces APL cells to differentiate. The answer will not only advance our understanding of mechanisms of leukemic cell differentiation, but will also promote identification of differentiation-related drug target(s), and could lead to some new differentiation-inducing drug(s). Third, many drugs are synthesized to achieve better efficacy and safety profiles according to the chemical structure of natural products. For example, development of Lipitor and Atorvastatin were clearly inspired by natural-product HMG-CoA reductase inhibitors such as lovastatin and simvastatin. In this sense, it is very meaningful and significant to synthesize new compounds similar to ATO or to chemically modify ATO so as to reduce its toxicity and to keep or even enhance its clinical effectiveness. Based on the in vivo synergistic effects of ATO and ATRA, it has been proposed that ATO be integrated into the backbone of ATRA *(47)*. Finally, the long-term effect of ATO in treatment of APL remains to be illustrated, in spite of higher molecular remission. It is also essential to investigate new combinations of arsenic agents with other chemotherapy agents and ATRA in large clinic trials.

Acknowledgments

This work is supported in part by the National Key Program (973) for Basic Research of China (NO2002CB512806, NO2002CB512805), National Natural Science Foundation of China (30370592), and International Collaborative Items of Ministry of Science and Technology of China (2003 DF000038). Grants from the Science and Technology Committee of Shanghai and the 100-Talent Program of the Chinese Academy of Sciences are gratefully acknowledged.

References

1. Huang ME, Ye YC, Chen SR, et al. Use of all-trans retinoic acid in the treatment of acute promyelocytic leukemia. Blood 1988;72:567–572.
2. Wang ZY, Chen Z. Differentiation and apoptosis induction therapy in acute promyelocytic leukaemia. Lancet Oncol 2002;1:101–106.
3. Tallmann MS. Curative therapeutic approaches to APL. Ann Hematol 2004;83:S81–S82.
4. Wicke RW, Cheung CS. Principles for applying traditional Chinese medicine to cases of cancer. Integr Cancer Ther 2002;1:175–178.
5. Chu DT, Wong WL, Mavligit GM. Immunotherapy with Chinese medicinal herbs. I. Immune restoration of local xenogeneic graft-versus-host reaction in cancer patients by fractionated Astragalus membranaceus in vitro. J Clin Lab Immunol 1998;25:119–123.
6. Cohen I, Tagliaferri M, Tripathy D. Traditional Chinese medicine in the treatment of breast cancer. Semin Oncol 2002;29:563–574.
7. Tang W, Hemm I, Bertram B. Recent development of antitumor agents from Chinese herbal medicines. Part II. High molecular compounds. Planta Med 2003;69:193–201.
8. Wang ZY. Arsenic compounds as anticancer agents. Cancer Chemother Pharmacol 2001;48: S72–S76.
9. Huang SL, Guo EX, Xiang Y. Qing Dai tablet treats acute promyelocytic leukemia: clinical reports. Chinese J Hematology 1995;16:26–28.
10. Melnick A, Licht JD. Deconstructing a disease: RARalpha, its fusion partners, and their roles in the pathogenesis of acute promyelocytic leukemia. Blood 1999;93:3167–2151.
11. Freemantle SJ, Spinella MJ, Dmitrovsky E. Retinoids in cancer therapy and chemoprevention: promise meets resistance. Oncogene 2003;22:7305–7315.
12. Wang ZY, Chen Z. Differentiation and apoptosis induction therapy in acute promyelocytic leukaemia. Lancet Oncol 2000;1:101–106.
13. Chen GQ and Cai X. Clinical and experimental studies on arsenicals in treatment of hematopoietic malignancies. In: Niho Y (ed). Molecular Target for Hematological Malignancies and Cancer. Kyushu University Press, 2000; pp. 85–93.
14. Cohen MH, Hirschfeld S, Flamm Honig S, et al. Drug approval summaries: arsenic trioxide, tamoxifen citrate, anastrazole, paclitaxel, bexarotene. Oncologist 2001;6:4–11.
15. Jones DEH and Ledingham KWL Arsenic in Napoleon's wallpaper. Nature 1982;299:626–627.
16. Ng JC, Wang J, Shraim A. A global health problem caused by arsenic from natural sources. Chemosphere 2003;52:1353–1359.
17. Waxman S, Anderson KC. History of the development of arsenic derivatives in cancer therapy. Oncologist 2001;6:S2–S10.
18. Forkner CE, Scott TFM. Arsenic as a therapeutic agent in chronic myelogenous leukemia. JAMA 1931;97:3–5.

19. Kandel EV, LeRoy GV. Chronic arsenical poisoning during the treatment of chronic myeloid leukemia. Arch Intern Med 1937;60:846–866.
20. Sun HD, Ma L, Hu XC, Zhang TD. Treatment of acute promyelocytic leukemia by A1-1 therapy. Chin J Integrat Chin Trad Med Western Med 1992;12:170–117.
21. Chen GQ, Zhu J, Shi XG, et al. In vitro studies on cellular and molecular mechanisms of arsenic trioxide (ATO) in the treatment of acute promyelocytic leukemia: ATO induces NB4 cell apoptosis with downregulation of Bcl-2 expression and modulation of PML-RARa/PML proteins. Blood 1996;88:1052–1061.
22. Mervis J. Ancient remedy performs new tricks. Science 1996;273:578.
23. Soignet SL, Maslak P, Wang ZG, et al. Complete remission after treatment of acute promyelocytic leukemia with arsenic trioxide. N Engl J Med 1998;339:1341–1348.
24. Soignet SL, Frankel S, Tallman M. US multicenter trial of arsenic trioxide (AT) in acute promyelocytic leukemia (APL). Blood 1999;94:698a.
25. Mathews V, Balasubramanian P, Shaji RV, George B, Chandy M, Srivastava A. Arsenic trioxide in the treatment of newly diagnosed acute promyelocytic leukemia: a single center experience. Am J Hematol 2002;70:292–299.
26. Lazo G, Kantarjian H, Estey E, Thomas D, O'Brien S, Cortes J. Use of arsenic trioxide (As_2O_3) in the treatment of patients with acute promyelocytic leukemia: the M. D. Anderson experience. Cancer 2003;97:2218–2224.
27. Shen ZX, Chen GQ, Ni JH, et al. Use of arsenic trioxide (As_2O_3) in the treatment of acute promyelocytic leukemia (APL): II clinical efficacy and pharmacokinetics in relapsed patients. Blood 1997;89:3354–3360.
28. Ishitsuka K, Shirahashi A, Iwao Y, et al. Bone marrow necrosis in a patient with acute promyelocytic leukemia during re-induction therapy with arsenic trioxide. Eur J Haematol 2004;72:280–284.
29. Shen Y, Shen ZX, Yan H, et al. Studies on the clinical efficacy and pharmacokinetics of low-dose arsenic trioxide in the treatment of relapsed acute promyelocytic leukemia: a comparison with conventional dosage. Leukemia 2001;15:735–741.
30. Chen GQ, Shen ZX, Wu F, et al. Pharmacokinetics and efficacy of low-dose all-trans retinoic acid in the treatment of acute promyelocytic leukemia. Leukemia 1996;10:825–828.
31. Hermine O, Dombret H, Poupon J, et al. Phase II trial of arsenic trioxide and alpha interferon in patients with relapsed/refractory adult T-cell leukemia/lymphoma. Hematol J 2004;5:130–134.
32. Zhang P, Wang SY, Hu XH. Arsenic trioxide-treated 72 cases of acute promyelocytic leukemia. Chin J Hematol 1996;17:58–60.
33. Zhang P, Wang SY, Hu LH, et al. Seven years' summary report on the treatment of acute promyelocytic leukemia with arsenic trioxide: an analysis of 242 cases. Chin J Hematol 2000;21:67–70.
34. Niu C, Yan H, Yu T, et al. Studies on treatment of acute promyelocytic leukemia with arsenic trioxide: remission induction, follow-up and molecular monitoring in 11 newly diagnosed and 47 relapsed acute promyelocytic leukemia. Blood 1999;94: 3315–3324.
35. Barbui T, Finazzi G, Falanga A. The impact of all-trans-retinoic acid on the coagulopathy of acute promyelocytic leukemia. Blood 1998;91: 3093–3102.
36. Zhu J, Guo WM, Yao YY, et al. Tissue factors on acute promyelocytic leukemia and endothelial cells are differently regulated by retinoic acid, arsenic trioxide and chemotherapeutic agents. Leukemia 1999;13:1062–1070.
37. Agis H, Weltermann A, Mitterbauer G, et al. Successful treatment with arsenic trioxide of a patient with ATRA-resistant relapse of acute promyelocytic leukemia. Ann Hematol 1999; 78(7):329–332.

38. Ohsawa M, Koyama T, Shibakura M, Kamei S, Hirosawa S. Arsenic trioxide (As_2O_3) gradually downregulates tissue factor expression without affecting thrombomodulin expression in acute promyelocytic leukemia cells. Leukemia 2000;14:941–943.
39. Camacho LH, Soignet SL, Chanel S, et al. Leukocytosis and the retinoic acid syndrome in patients with acute promyelocytic leukemia treated with arsenic trioxide. J Clin Oncol 2000;18:2620–2625.
40. Lin CP, Huang MJ, Chang IY, Lin WY, Sheu YT. Retinoic acid syndrome induced by arsenic trioxide in treating recurrent all-trans retinoic acid resistant acute promyelocytic leukemia. Leuk Lymphoma 2000;38:195–198.
41. Ohnishi K, Yoshida H, Shigeno K, et al. Prolongation of the qt interval and ventricular tachycardia in patients treated with arsenic trioxide for acute promyelocytic leukemia. Ann Intern Med 2000;133:881–885.
42. Galm O, Fabry U, Osieka R. Pseudotumor cerebri after treatment of relapsed acute promyelocytic leukemia with arsenic trioxide. Leukemia 2000;14:343–344.
43. Hu J, Shen ZX, Sun GL, Chen SJ, Wang ZY, Chen Z. Long-term survival and prognostic study in acute promyelocytic leukemia treated with all-trans-retinoic acid, chemotherapy, and As_2O_3: an experience of 120 patients at a single institution. Int J Hematol 1999;70:248–260.
44. Tanvetyanon T, Nand S. Herpes zoster during treatment with arsenic trioxide. Ann Hematol 2004;83:198–200.
45. Zhang T, Westervelt P, Hess JL. Pathologic, cytogenetic and molecular assessment of acute promyelocytic leukemia patients treated with arsenic trioxide (As_2O_3). Mod Pathol 2000;13: 954–961.
46. Raffoux E, Rousselot P, Poupon J, et al. Combined treatment with arsenic trioxide and all-trans-retinoic acid in patients with relapsed acute promyelocytic leukemia. J Clin Oncol 2003;21:2326–2334.
47. Shen ZX, Shi ZZ, Fang J, et al. All-trans retinoic acid/ATO combination yields a high quality remission and survival in newly diagnosed acute promyelocytic leukemia. Proc Natl Acad Sci USA 2004;101:5328–5335.
48. Lu DP, Qiu JY, Jiang B, et al. Tetra-arsenic tetra-sulfide for the treatment of acute promyelocytic leukemia: a pilot report. Blood 2002;99:3136–3143.
49. Chen GQ, Shi XG, Tang W, et al. Use of arsenic trioxide (As2O3) in the treatment of acute promyelocytic leukemia (APL): I.As_2O_3 exerts dose-dependent dual effects on APL cells. Blood 1997;89:3345–3353.
50. Lallemand-Breitenbach V, Guillemin MC, Janin A, et al. Retinoic acid and arsenic synergize to eradicate leukemic cells in a mouse model of acute promyelocytic leukemia. J Exp Med 1999;189:1043–1052.
51. Kinjo K, Kizaki M, Muto A, et al. Arsenic trioxide (As_2O_3)-induced apoptosis and differentiation in retinoic acid-resistant acute promyelocytic leukemia model in hGM-CSF-producing transgenic SCID mice. Leukemia 2000;14:431–438.
52. Lanotte M, Martin-Thouvenin V, Najman S, Balerini P, Valensi F, Berger R. NB4, a maturation inducible cell line with t(15;17) marker isolated from a human acute promyelocytic leukemia (M3). Blood 1991;77:1080–1086.
53. Zhang TD, Chen GQ, Wang ZG, Wang ZY, Chen SJ, Chen Z. Arsenic trioxide, a therapeutic agent for APL. Oncogene 2001;20:7146–7153.
54. Wang X. The expanding role of mitochondria in apoptosis. Genes Dev 2001;15:2922–2933.
55. Zhu XH, Shen YL, Jing YK, et al. Apoptosis and growth inhibition in malignant lymphocytes after treatment with arsenic trioxide at clinically achievable concentrations. J Natl Cancer Inst 1999;91:772–778.

56. Cai X, Shen YL, Zhu Q, et al. Arsenic trioxide-induced apoptosis and differentiation are associated respectively with mitochondrial transmembrane potential collapse and retinoic acid signaling pathways in acute promyelocytic leukemia. Leukemia 2000;14:262–270.
57. Larochette N, Decaudin D, Jacotot E, et al. Arsenite induces apoptosis via a direct effect on the mitochondrial permeability transition pore. Exp Cell Res 1999;249:413–421.
58. Kroemer G, de The H. Arsenic trioxide, a novel mitochondriotoxic anticancer agent? J Natl Cancer Inst 1999;91:743–745.
59. Liu P, Han ZC. Treatment of acute promyelocytic leukemia and other hematologic malignancies with arsenic trioxide: review of clinical and basic studies. Int J Hematol. 2003;78:32–39.
60. Akao Y, Yamada H, Nakagawa Y. Arsenic-induced apoptosis in malignant cells in vitro. Leuk Lymphoma 2000;37:53–63.
61. Huang XJ, Wiernik PH, Klein RS, Gallagher RE. Arsenic trioxide induces apoptosis of myeloid leukemia cells by activation of caspases. Med Oncol 1999;16:58–64.
62. Kitamura K, Minami Y, Yamamoto K, et al. Involvement of CD95-independent caspase 8 activation in arsenic trioxide-induced apoptosis. Leukemia 2000;14:1743–1750.
63. Liu Q, Hilsenbeck S, Gazitt Y. Arsenic trioxide-induced apoptosis in myeloma cells: p53-dependent G1 or G2/M cell cycle arrest, activation of caspase-8 or caspase-9, and synergy with APO2/TRAIL. Blood 2003;101:4078–4087.
64. Huang C, Ma WY, Li J, Dong Z. Arsenic induces apoptosis through a c-Jun NH_2-terminal kinase-dependent, p53-independent pathway. Cancer Res 1999;59:3053–3058.
65. Jiang XH, Chun-Yu Wong B, et al. Arsenic trioxide induces apoptosis in human gastric cancer cells through up-regulation of p53 and activation of caspase-3. Int J Cancer 2001;91:173–179.
66. Rojewski MT, Korper S, Thiel E, Schrezenmeier H. Depolarization of mitochondria and activation of caspases are common features of arsenic(III)-induced apoptosis in myelogenic and lymphatic cell lines. Chem Res Toxicol 2004;17:119–128.
67. Gurr JR, Bau DT, Liu F, Lynn S, Jan KY. Dithiothreitol enhances arsenic trioxide-induced apoptosis in NB4 cells. Mol Pharmacol 1999;56:102–109.
68. Davison K, Cote S, Mader S, Miller WH. Glutathione depletion overcomes resistance to arsenic trioxide in arsenic-resistant cell lines. Leukemia 2003;17:931–940.
69. Dai J, Weinberg RS, Waxman S, Jing Y. Malignant cells can be sensitized to undergo growth inhibition and apoptosis by arsenic trioxide through modulation of the glutathione redox system. Blood 1999;93:268–277.
70. Quignon F, Chen Z, de The H. Retinoic acid and arsenic: towards oncogene-targeted treatments of acute promyelocytic leukaemia. Biochim Biophys Acta 1997;1333:M53–M61.
71. Andre C, Guillemin MC, Zhu J, et al. The PML and PML/RARalpha domains: from autoimmunity to molecular oncology and from retinoic acid to arsenic. Exp Cell Res 1996;229:253–260.
72. Costantini P, Belzacq AS, Vieira HL, et al. Oxidation of a critical thiol residue of the adenine nucleotide translocator enforces Bcl-2-independent permeability transition pore opening and apoptosis. Oncogene 2000;19:307–314.
73. Pelicano H, Feng L, Zhou Y, et al. Inhibition of mitochondrial respiration: a novel strategy to enhance drug-induced apoptosis in human leukemia cells by a reactive oxygen species-mediated mechanism. J Biol Chem 2003;278:37,832–37,839.
74. Uslu R, Sanli UA, Sezgin C, et al. Arsenic trioxide-mediated cytotoxicity and apoptosis in prostate and ovarian carcinoma cell lines. Clin Cancer Res 2000;6:4957–4964.
75. Gupta S, Yel L, Kim D, Kim C, Chiplunkar S, Gollapudi S. Arsenic trioxide induces apoptosis in peripheral blood T lymphocyte subsets by inducing oxidative stress: a role of Bcl-2. Mol Cancer Ther 2003;2:711–719.
76. Grad JM, Bahlis NJ, Reis I, Oshiro MM, Dalton WS, Boise LH. Ascorbic acid enhances arsenic trioxide-induced cytotoxicity in multiple myeloma cells. Blood 2001;98:805–813.

77. Yi J, Yang J, He R, et al. Emodin enhances arsenic trioxide-induced apoptosis via generation of reactiveoxygen species and inhibition of survival signaling. Cancer Res 2004;64:108–116.
78. Liu L, Trimarchi JR, Navarro P, Blasco MA, Keefe DL. Oxidative stress contributes to arsenic-induced telomere attrition, chromosome instability, and apoptosis. J Biol Chem 2003;278: 31,998–32,004.
79. Park IC, Park MJ, Woo SH, et al. Tetraarsenic oxide induces apoptosis in U937 leukemic cells through a reactive oxygen species-dependent pathway. Int J Oncol 2003;23:943–948.
80. Davison K, Mann KK, Waxman S, Miller WH Jr. JNK activation is a mediator of arsenic trioxide-induced apoptosis in acute promyelocytic leukemia cells. Blood 2004;103:3496–3502.
81. Chen F, Castranova V, Li Z, Karin M, Shi X. Inhibitor of nuclear factor kappaB kinase deficiency enhances oxidative stress and prolongs c-Jun NH2-terminal kinase activation induced by arsenic. Cancer Res 2003;63:7689–7693.
82. Kajiguchi T, Yamamoto K, Hossain K, et al. Sustained activation of c-jun-terminal kinase (JNK) is closely related to arsenic trioxide-induced apoptosis in an acute myeloid leukemia (M2)-derived cell line, NKM-1. Leukemia 2003;17:2189–2195.
83. Huang MJ, Hsieh RK, Lin CP, Chang IY, Liu HJ. The cytotoxicity of arsenic trioxide to normal hematopoietic progenitors and leukemic cells is dependent on their cell-cycle status. Leuk Lymphoma 2002;43:2191–2199.
84. Huang S, Huang CF, Lee T. Induction of mitosis-mediated apoptosis by sodium arsenite in HeLa S3 cells. Biochem Pharmacol 2000;60:771–780.
85. Park JW, Choi YJ, Jang MA, et al. Arsenic trioxide induces G2/M growth arrest and apoptosis after caspase-3 activation and bcl-2 phosphorylation in promonocytic U937 cells. Biochem Biophys Res Commun 2001;286:726–734.
86. Ling YH, Jiang JD, Holland JF, Perez-Soler R. Arsenic trioxide produces polymerization of Mtubules and mitotic arrest before apoptosis in human tumor cell lines. Mol Pharmacol 2002;62:529–538.
87. Cai X, Yu Y, Huang Y, et al. Arsenic trioxide-induced mitotic arrest and apoptosis in acute promyelocytic leukemia cells. Leukemia 2003;17:1333–1337.
88. Jing Y. The PML-RARalpha fusion protein and targeted therapy for acute promyelocytic leukemia. Leuk Lymphoma 2004;45(4):639–648.
89. Sun Y, Kim SH, Zhou DC, et al. Acute promyelocytic leukemia cell line AP-1060 established as a cytokine-dependent culture from a patient clinically resistant to all-trans retinoic acid and arsenic trioxide. Leukemia 2004;18(7):1258–1269.
90. Chen Z, Chen GQ, Shen ZX, Chen SJ, Wang ZY. Treatment of acute promyelocytic leukemia with arsenic compounds: In vitro and in vivo studies. Semin Hemat 2001;38:26–36.
91. Hong SH, Yang Z, Privalsky ML. Arsenic trioxide is a potent inhibitor of the interaction of SMRT corepressor with Its transcription factor partners, including the PML-retinoic acid receptor alpha oncoprotein found in human acute promyelocytic leukemia. Mol Cell Biol 2001;21:7172–7182.
92. Jiang G, Bi K, Tang T, et al. Effect of arsenic trioxide on cytokine expression by acute promyelocytic leukemia cells. Chin Med J (Engl) 2003;116:1639–1643.
93. Chelbi-alix MK, Bobe P, Benoit G, Canova A, Pine R. Arsenic enhances the activation of Stat1 by interferon gamma leading to synergistic expression of IRF-1.Oncogene 2003;22:9121–9130.
94. Zhu Q, Zhang JW, Zhu HQ, et al. Synergic effects of arsenic trioxide and cAMP during acute promyelocytic leukemia cell maturation subtends a novel signaling cross-talk. Blood 2002;99:1014–1022.
95. Huang Y, Du KM, Xue ZH, et al. Cobalt chloride and low oxygen tension trigger differentiation of acute myeloid leukemic cells: possible mediation of hypoxia-inducible factor-1alpha. Leukemia 2003;17:2065–2073.

96. Zhu J, Koken MH, Quignon F, et al. Arsenic-induced PML targeting onto nuclear bodies: implications for the treatment of acute promyelocytic leukemia. Proc Natl Acad Sci USA 1997;94:3978–3983.
97. Muller S, Matunis MJ, Dejean A. Conjugation with the ubiquitin-related modifier SUMO-1 regulates the partitioning of PML within the nucleus. EMBO J 1998;17:61–70.
98. Sternsdorf T, Puccetti E, Jensen K, et al. PIC-1/SUMO-1-modified PML-retinoic acid receptor alpha mediates arsenic trioxide-induced apoptosis in acute promyelocytic leukemia. Mol Cell Biol 1999;19:5170–5178.
99. Muller S, Miller WH, Dejean A. Trivalent antimonials induce degradation of the PML-RAR oncoprotein and reorganization of the promyelocytic leukemia nuclear bodies in acute promyelocytic leukemia NB4 cells. Blood 1998;92:4308–4316.
100. Kitamura K, Hoshi S, Koike M, Kiyoi H, Saito H, Naoe T. Histone deacetylase inhibitor but not arsenic trioxide differentiates acute promyelocytic leukaemia cells with t(11;17) in combination with all-trans retinoic acid. Br J Haematol 2000;108:696–702.
101. Koken MH, Daniel MT, Gianni M, et al. Retinoic acid, but not arsenic trioxide, degrades the PLZF/RARalpha fusion protein, without inducing terminal differentiation or apoptosis, in a RA-therapy resistant t(11;17)(q23;q21) APL patient. Oncogene 1999;18:1113–1118.
102. Puccetti E, Beissert T, Guller S, et al. Leukemia-associated translocation products able to activate RAS modify PML and render cells sensitive to arsenic-induced apoptosis. Oncogene 2003;22:6900–6908.
103. Hayakawa F, Privalsky ML. Phosphorylation of PML by mitogen-activated protein kinases plays a key role in arsenic trioxide-mediated apoptosis. Cancer Cell 2004;5:389–401.
104. Roboz GJ, Dias S, Lam G, et al. Arsenic trioxide induces dose- and time-dependent apoptosis of endothelium and may exert an antileukemic effect via inhibition of angiogenesis. Blood 2000;96:1525–1530.
105. Lew YS, Brown SL, Griffin RJ, Song CW, Kim JH. Arsenic trioxide causes selective necrosis in solid murine tumors by vascular shutdown. Cancer Res 1999;59:6033–6037.
106. Chen GQ, Zhou L, Styblo M, et al. Methylated metabolites of arsenic trioxide are more potent than arsenic trioxide as apoptotic but not differentiation inducers in leukemia and lymphoma cells. Cancer Res 2003;63:1853–1859.
107. Wang Z, Zhou J, Lu X, Gong Z, Le XC. Arsenic speciation in urine from acute promyelocytic leukemia patients undergoing arsenic trioxide treatment. Chem Res Toxicol 2004;17:95–103.
108. Ochi T, Nakajima F, Sakurai T, et al. Dimethylarsinic acid causes apoptosis in HL-60 cells via interaction with glutathione. Arch Toxicol 1996;70:815–821.
109. Hussein MA. Trials of arsenic trioxide in multiple myeloma. Cancer Control 2003;10:370–374.
110. Bazarbachi A, Hermine O. Treatment of adult T-cell leukaemia/lymphoma: current strategy and future perspectives. Virus Res 2001;78:79–92.
111. Murgo AJ. Clinical trials of arsenic trioxide in hematologic and solid tumors: overview of the National Cancer Institute Cooperative Research and Development Studies. Oncologist 2001;6:22–28.
112. Ishitsuka K, Hanada S, Suzuki S, et al. Arsenic trioxide inhibits growth of human T-cell leukaemia virus type I infected T-cell lines more effectively than retinoic acids. Br J Haematol 1998;103:721–728.
113. Bazarbachi A, El-Sabban ME, Nasr R, et al. Arsenic trioxide and interferon-alpha synergize to induce cell cycle arrest and apoptosis in human T-cell lymphotropic virus type I-transformed cells. Blood 1999;93:278–283.
114. El-Sabban ME, Nasr R, Dbaibo G, et al. Arsenic-interferon-alpha-triggered apoptosis in HTLV-I transformed cells is associated with tax down-regulation and reversal of NF-kappa B activation. Blood 2000;96:2849–2855.

115. El-Sabban ME, Nasr R, Dbaibo G, et al. Arsenic-interferon-alpha-triggered apoptosis in HTLV-I transformed cells is associated with tax down-regulation and reversal of NF-kappaB activation. Blood 2000;96:2849–2855.
116. Nasr R, Rosenwald A, El-Sabban ME, et al. Arsenic/interferon specifically reverses 2 distinct gene networks critical for the survival of HTLV-1-infected leukemic cells. Blood 2003;101: 4576–4582.
117. Vuky J, Yu R, Schwartz L, Motzer RJ. Phase II trial of arsenic trioxide in patients with metastatic renal cell carcinoma. Invest New Drugs 2002;20:327–330.

PART IV
MICROBIAL DIVERSITY

12

New Methods to Access Microbial Diversity for Small Molecule Discovery

Karsten Zengler, Ashish Paradkar, and Martin Keller

Summary

Natural-product-derived drugs are a major portion of the total number of approved drugs in the antibacterial area. The majority of bacteria and fungi in the environment is only known by molecular fingerprints and has resisted cultivation. Therefore, new methods have been developed to access this tremendous microbial diversity for the discovery of novel small molecules. These culture-dependent and -independent methods include a novel high-throughput cultivation technology as well as a recombinant approach to discover and express novel natural products.

Key Words: Microbial diversity; high-throughput cultivation; recombinant approach; natural products; genomic libraries; high-throughput screening; cultivation-dependent and -independent methods.

1. Introduction

Natural products with high therapeutic potential derived from bacteria and fungi have played a tremendous role in drug discovery since the discovery of the first antibiotic—penicillin—by Alexander Fleming in 1929 *(1)*. Since then, almost 20,000 metabolites of microbial origin have been described *(2)*. Around 80% of all microbial-derived secondary metabolites are produced by members of the order *Actinomycetales* (phylum *Actinobacteria*) with the genus *Streptomyces* accounting for approx 50% *(3–5)*. The remaining 20% of secondary metabolites are produced by members of the phyla *Bacteroidetes* (e.g., *Chryseobacterium*), *Cyanobacteria* (e.g., *Nostoc*), *Firmicutes* (e.g., *Bacillus*), and *Proteobacteria* (e.g., *Myxococcus*) *(6–9)*. Despite attempts to isolate and culture new microorganisms by traditional cultivation methods, the discovery rate of new structural classes of antimicrobial molecules has declined *(10,11)*. Isolated culture extracts have yielded numerous previously described metabolites *(12)*, and the rate of rediscovery of known antibiotics is approaching 99.9% *(13)*. In addition, pharmaceutical companies shifted from traditional natural-product discovery to more combinatorial chemistry-based discovery. The generation of large chemical-compound libraries pushed the development of high-throughput screens based on molecular targets. Combinatorial chemistry, and more recently high-throughput parallel syntheses, has been implemented to generate these diverse chemical libraries. Consequently, many pharmaceutical companies have de-emphasized research on natural products in favor of

Table 1
Number of Approved Antibacterial Drugs (15)

Year	Number of antibacterial drugs	Natural product-derived	Totally synthetic
1990–1995	24	17	7
1996–2000	11	6	5
2001–2002	2	2	0

mass-produced combinatorial libraries. However, the expected surge in productivity has not materialized, and the number of new active substances introduced into the market by the pharmaceutical industry has hit a 20-yr low of 37 substances in 2001, and is still declining *(14)*. The US Food and Drug administration (FDA) received 16 new drug applications in 2001, down from 24 the previous year *(14)*. The number of antibacterial approved drugs is declining rapidly (**Table 1**).

Natural-product-derived drugs are still a major portion of the total number of approved drugs in this therapeutic area (**Table 1**). The generation of truly novel diversity from natural-product sources, combined with total and combinatorial synthetic methodologies, including the manipulation of biosynthetic pathways, provides the best solution to the current productivity crisis faced by the scientific community engaged in drug discovery and development *(14–16)*. Although almost 20,000 microbial metabolites and approximately 100,000 plant products have been described so far, secondary metabolites still appear to be an inexhaustible source of lead structures for new antibacterial, antiviral, antitumor, and agricultural agents *(2)*. But how can we generate novel diversity from natural-product sources? Are there still novel organisms to be discovered and isolated that harbor novel chemical structures, and how can we access such novel microorganisms?

When assessing microbial diversity, Torsvik and co-workers estimated that a pasture soil sample contained about 3500 to 8800 genome equivalents *(17)*. This could result in around 10,000 different species of equivalent abundances *(18)*. Other estimates of species diversity within a single soil sample vary from 467 species *(19)* up to 500,000 species *(20)*. It is estimated that only 1% or less of the bacteria that are known to exist in the soil environment have been cultured. Therefore, the soil environment harbors a vast number of uncharacterized microorganisms *(18,21–27)*.

In addition, studies of different environments such as open sea water and marine sediment, and even extreme environments such as hot springs, demonstrate that the diversity of microorganisms is far larger than ever anticipated *(28–30)*. Analysis of microbial diversity within the past decades has resulted in a tremendous increase of new phylotypes. When Carl Woese and colleagues defined the major bacterial phyla almost 20 yr ago *(31)*, they recognized 11 phyla. After two decades of environmental 16S rRNA gene sequencing from almost every habitat on Earth, 41 additional phyla have been described *(32)*, bringing the number to a total of 53 bacterial phyla (the former Gram-positive bacteria are now two separate phyla, the *Firmicutes* and *Actinobacteria*). Only 27 out of 53 bacterial phyla contain previously cultivated microorganisms (**Fig. 1A**). Half of the bacterial phyla recognized to date consist entirely of so far uncultured bacteria, and have been described solely by their 16S rRNA gene sequences *(32,33)*. In addition, only a small fraction of organisms that can be cultured

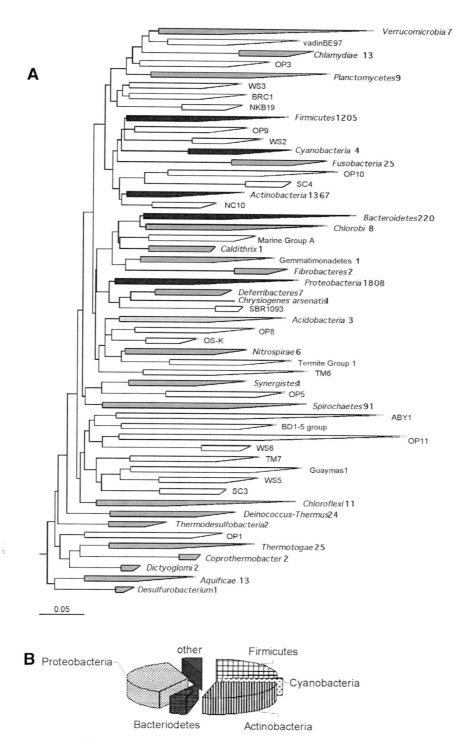

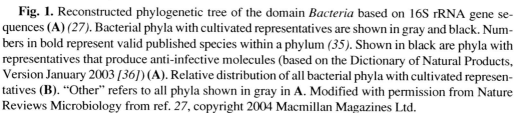

Fig. 1. Reconstructed phylogenetic tree of the domain *Bacteria* based on 16S rRNA gene sequences (**A**) *(27)*. Bacterial phyla with cultivated representatives are shown in gray and black. Numbers in bold represent valid published species within a phylum *(35)*. Shown in black are phyla with representatives that produce anti-infective molecules (based on the Dictionary of Natural Products, Version January 2003 *[36]*) (**A**). Relative distribution of all bacterial phyla with cultivated representatives (**B**). "Other" refers to all phyla shown in gray in **A**. Modified with permission from Nature Reviews Microbiology from ref. *27*, copyright 2004 Macmillan Magazines Ltd.

have been at least partially characterized *(34)*. Surprisingly, so far only five phyla include isolates producing bioactive molecules (**Fig. 1A**). These five phyla (*Actinobacteria*, *Bacteroidetes*, *Cyanobacteria*, *Firmicutes*, and *Proteobacteria*) on the other hand represent 95% of all cultivated and validly published species (**Fig. 1B**) *(27)*. The rest of the cultivated phyla (22 phyla) represent only 5% of all species *(27)*. Within these cultivated phyla, certain bacterial groups are represented to a greater extent than others. For example, representatives of the order *Actinomycetales* (phylum *Actinobacteria*) account for 27% of all available and recognized bacterial species, and members of the genus *Streptomyces* (order *Actinomycetales*) represent 10% of published bacterial species *(35)*. On the other hand, many phyla are represented by only very few isolates; some contain only one described species. This untapped microbial diversity represents a tremendous potential for discovery of novel bioactive small molecules. But how can we get access to this incredible microbial biodiversity?

In order to develop novel drugs derived from natural products, three key issues must be addressed.

1. Access to truly novel diversity.
2. Faster chemical dereplication methods to determine whether the active compound is novel or known.
3. New methods to address production and chemical modification of the novel chemical entities.

Recently, advances in microbiology, molecular biology, and analytical chemistry have started to address many of these issues. Within this chapter we shall discuss new methods to access microbial biodiversity with culture-dependent and independent methods.

2. Access to Microbial Diversity Through New Cultivation Approaches

Environmental microbiologists are facing a major challenge today. In order to gain a comprehensive understanding of microbial physiology or to fully access metabolic pathways containing genes dispersed throughout the genome *(37,38)*, cultivation of microorganisms will be required. The enrichment and cultivation techniques developed by Pasteur, Koch, Beijerinck, and Winogradsky *(39–42)* facilitated new discoveries in microbiology for over 100 yr. Such conventional cultivation of microorganisms, however, is selective and biased towards growth of specific microorganisms *(43,44)*. The majority of cells obtained from the environment and visualized by microscopy are viable, but they do not generally form visible colonies on plates. Visible colonies on plate count media require at least 10^5 cells to be identified by the naked eye, and traditional media select strongly for microbes that grow rapidly to high density, are resistant to high concentrations of nutrients, and are able to grow in isolation.

It could be argued that these traditional culturing strategies use conditions that are completely different from the normal growth habit of many microbes, and are a major contributing factor to the failure to cultivate most microorganisms in pure culture *(18,21,44)*. It has often been claimed that plate count methodologies are not suitable for cultivation of bacteria and that the members of groups without cultivated representatives are somehow "nonculturable" *(45,46)*. Janssen and co-workers addressed the issue of high concentration of nutrients in cultivation media and investigated the culturability of soil microorganisms using diluted nutrient broth, assuming that many of the so far uncultivated microorganisms are inhibited by high concentrations of nutrients *(47)*. Extended incubation times for at least 10 wk were required to allow maximum colony development. It was demonstrated that at least some of the bacteria detected by molecular ecological methods were not unculturable and that their isolation in pure culture did not require elaborate or expensive cultivation strategies *(47–50)*.

Other studies dealt with the effect of signal compounds on the cultivation efficiency of aquatic bacteria *(51–55)*. Pure cultures of bacteria, typical for these aquatic environments, could be recovered from natural planktonic assemblages. Bruns et al. *(54,55)* demonstrated that addition of cAMP to complex medium effectively increased the cultivation success, and up to 10% of the total bacterial counts determined by epifluorescence microscopy after staining with 4',6-diamidino-2-phenylindole (DAPI) could be cultivated after the addition of cAMP.

Button et al. developed a dilution cultivation method for growing bacteria from oligotrophic aquatic environments *(56)*. Populations of marine microorganisms were measured, diluted to a small and known number of cells, inoculated into plain sterilized seawater, and examined for the presence of 10^4 or more cells per mL over a 9-wk time interval. Doubling times observed were in the range of 1 d to 1 wk. Button speculated based on his experiments that most marine bacteria (60%) are viable and that low estimates of viability by traditional techniques resulted from the fact that most marine bacteria reach stationary phase before attaining visible turbidity *(56)*.

Others also reasoned that previously uncultivable microorganisms might grow in pure culture if provided with the chemical components of their natural environment *(57)*. To allow access to these components, they placed marine microorganisms in diffusion chambers and incubated the chambers in an aquarium that simulated the natural setting for these organisms. The membranes allowed exchange of chemicals between the chamber and the environment but restricted movement of cells *(57)*. Although the isolated cultures did not represent new phylotypes, large numbers of colonies of varying morphologies were obtained after 1 wk of incubation in the chambers.

In addition to cultivating novel microorganisms, the speed of isolation can be a significant factor for the discovery of novel small molecules. Development of automated isolation methods that generate thousands of different microorganisms *(58)* would be a significant advance in the discovery process. Connon and Giovannoni described the development of a high-throughput cultivation method enabling the identification of a large number of extinction cultures *(59)*. Over a time course of 3 yr, many new microbial strains were isolated, including members of previously uncultured groups that are abundant in seawater *(60)*. The use of microtiter dishes in combination with an automated cell array and imaging process enabled them to increase sensitivity for the detection of cells and therefore shorten incubation times for cells with low growth rates relative to those in previous studies that employed the concept of extinction culturing in natural media *(56)*. The percentage of cells that could be cultured by this approach was several orders of magnitude higher than that obtained by cultivation on agar plates *(59)*.

A universally applicable method that allows access to the immense reservoir of untapped microbial diversity is an automated and extremely high-throughput cultivation approach developed by Zengler et al. *(58)*. This technique combines encapsulation of single cells in microcapsules for massively parallel microbial cultivation under low

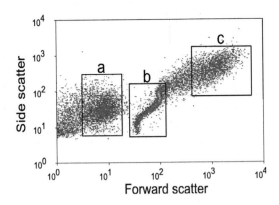

Fig. 2. Discrimination among (**A**) free-living cells, (**B**) singly occupied or empty microcapsules, and (**C**) microcapsules containing microcolonies was accomplished by flow cytometry in forward and side light-scatter mode.

nutrient flux conditions, followed by flow cytometry to detect microcapsules containing microcolonies (**Fig. 2**).

The microbial community, represented in single encapsulated cells, is grown in columns under very low nutrient flux conditions using mineral medium supplemented with low concentrations of nutrients extracted from the sampling site (**Fig. 3**). Over time, each cell capable of growth under the conditions in the column forms a microcolony within its enclosure (**Figs. 3, 4**). The high-throughput nature of this technology is based on the use of high-speed flow cytometry and cell sorting to identify microcapsules containing microcolonies of >20 cells (**Fig. 2**), and then sort each positive microcapsule into a well of a microtiter plate for further cultivation and analysis. The change in forward and side scatter is proportional to the colony size within the microcapsule. This allows the separation of microcapsules containing microcolonies from those without growth (**Fig. 2**).

Using this technology, it was demonstrated that novel, previously uncultured organisms can form microcolonies within the microcapsules and can generate a pure culture once separated *(58)*. During further incubation, microorganisms are able to divide and outgrow the microcapsules. The cultivation of microorganisms within the microcapsules allows the reconstituted microbial community to be simultaneously cultivated "together" and "apart" because each "caged microcolony" can later be separated and analyzed. The ability to reconstitute the community in the column of microcapsules allows for diffusive cross-feeding of metabolites and other molecules (e.g., regulatory and signaling molecules) between members of the community. This feature also simulates the natural environment, and thus preserves some of the community interactions and other specific requirements that may be needed for successful cultivation.

In addition, microbes are grown in an open flow system that simulates natural environments (**Fig. 3**), where microbes are exposed continually to a low concentration of nutrients. This is in contrast to a closed batch system where microbes receive a high concentration of nutrients at one point in time and metabolic byproducts can build up

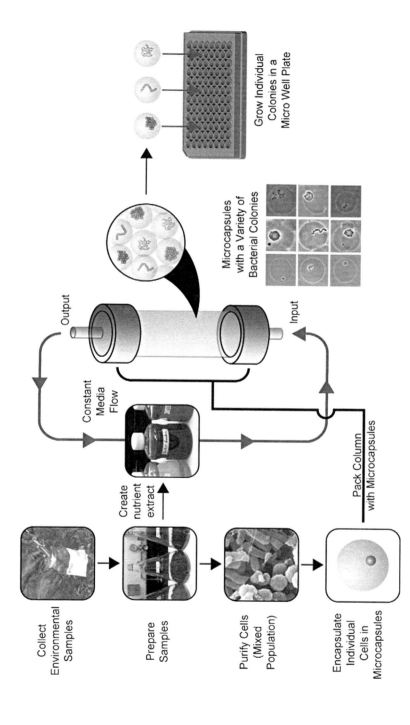

Fig. 3. Flow diagram of Diversa's high throughput cultivation approach based on encapsulation of single cells in microcapsules for massively parallel microbial cultivation. Modified with permission from Nature Reviews Microbiology from ref. 27, copyright 2004, Macmillan Magazines Ltd.

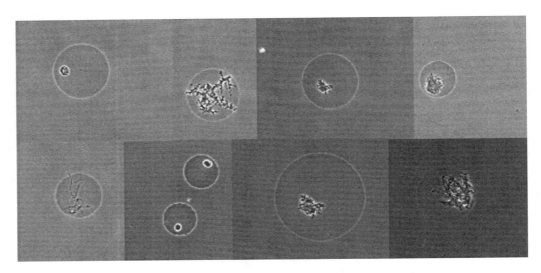

Fig. 4. Photomicrographs of microcapsules containing microcolonies of microorganisms. The size of a microcapsule is between 50 and 80 µm.

to unnaturally high and inhibitory concentrations. The low concentration of nutrients also minimizes overgrowth by fast-growing organisms, thereby allowing propagation of microorganisms with extremely slow growth rates and/or that only grow to low cell densities. The high-throughput production, screening, and sorting of microcapsules is automated, and easily and economically scaleable.

This high-throughput cultivation technology *(58)* is capable of isolating massive numbers of cultures and is therefore sufficient to supply increased diversity for modern high-throughput screening systems in drug discovery. High-throughput cultivation can provide tens of thousands of bacterial and fungal isolates per environmental sample. However, this increase in throughput requires advanced methods to characterize unique strains rapidly. To deal with this tremendous number of strains, a fully automated high-throughput method to determine the uniqueness of the bacterial and fungal isolates has been developed. Using Fourier transform infrared spectroscopy (FT-IR) *(61,62)* in combination with novel spectra-comparison algorithms, microtiter plate cultures can be analyzed and unique strains grown at larger scale (for example, 4-mL cultures) for organic extract preparation (Diversa Corporation, unpublished data).

To prove that this technology is suitable to quickly isolate cultures with biological activity, 10,000 unique bacterial and 1500 fungal strains were isolated during a 4-mo period, and the organic extracts prepared were tested against a panel of Gram-positive and Gram-negative bacteria. Several hundred cultures with chemically relevant antibacterial activities have been identified (Diversa Corporation, unpublished data) (**Fig. 5**).

Most of the recently described cultivation technologies try to simulate and mimic the natural environment. This has led to the use of extremely dilute culture media supporting only minimal cell growth, resulting in prolonged incubation times. Accordingly, technology has to be developed to work with minute amounts of cells, for example microcolonies on plates or within microcapsules. Some of the newly isolated microorganisms will not reach high cell densities, such as routine laboratory strains of

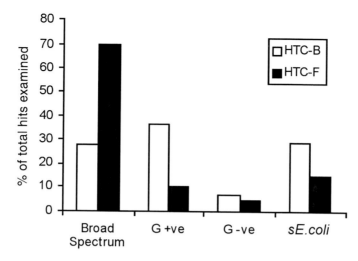

Fig. 5. Distribution of hits with regard to spectrum of activity. Percentage of hits against broad spectrum, Gram-positive (G +ve), Gram-negative (G -ve), and a supersensitive *Escherichia coli* strain (s*E.coli*) are shown. Hits from bacterial high-throughput cultures (HTC-B) are shown in white; hits from fungal high throughput cultures (HTC-F) are shown in black.

Bacillus or *Escherichia* when transferred into organic rich medium. Many of these new isolates can be a challenge for culture collections and will therefore require new methods for preservation, storage, and handling. In addition, these cultures will require the further refinement of molecular and analytical tools in order to study them. For natural-product discovery, the scale-up of some of these cultures for large-scale production could also be a potential issue. New analytical tools such as liquid chromatography (LC)-tandem mass spectrometry (MS-MS) or micro-coil nuclear magnetic resonance (NMR) allow structure elucidation of interesting molecules from very small amounts of material (Diversa Corporation, unpublished data). However, it can still be a challenge to isolate enough material for further biological studies or medicinal chemistry. A promising approach to overcome these issues lies within the recombinant natural products (RNP) platform.

3. Recombinant Approaches to Discover or Express Novel Natural Products

Recent advances in molecular biology led to the development of a recombinant approach for natural-product discovery *(24,63–66)*. The genetic information for the biosynthesis of natural small molecules is encoded in compact gene clusters that can be cloned from the producing organism and expressed in a heterologous tractable host. This approach aims to access pathways of so far uncultured and intractable microorganisms. Genomic DNA is extracted directly from an environmental sample without the need to grow the associated microbes. Alternatively, genomic DNA is isolated from cultures, and cloned into suitable vectors. The libraries thus made are introduced into an engineered surrogate host for expression of small-molecule pathways *(63,64,67–72)*. The pathways typically contain biosynthetic genes, resistance genes involved in self-

Table 2
Illustrative Examples of Inter-Species and Inter-Genus Heterologous Expression of Natural-Product Pathways

Donor organism	Vector system	Heterologous host	Pathway, natural product	References
Streptomyces	Plasmid, cosmid	*Streptomyces lividans, Streptomyces parvulus, Streptomyces coelicolor*	Polyketides, nonpolyketides	80,83–86
Rare Actinomycetes, e.g *Saccharothrix aerocolonigenes"*	Plasmid, cosmid	*Streptomyces albus*	Rebeccamycin	87
Myxobacteria, e.g., *Sorangium cellulosum*	Plasmid	*Streptomyces coelicolor*	Epothilone	88
	Integration into chromosome	*Myxococcus xanthus*	Epothilone	89
Bacillus licheniformis	Integration into chromosome	*Bacillus subtilis*	Bacitracin	90
Lysobacter lactamgenus	Plasmid	*Pseudomonas putida*	Cephalosporin	91
Pseudomonas fluorescens	Plasmid, cosmid	*Pseudomonas putida*	Pseudomonine	92
Serratia sp.	Cosmid	*Erwinia carotovora*	Prodigiosin	93
Soil DNA library	Cosmid	*Escherichia coli*	*Violacein*	94
Soil DNA library	Cosmid	*Escherichia coli*	long chain *N*-acyl-tyrosine	95,96
Soil DNA library	BAC	*Escherichia coli*	Indigo blue, indirubin	71
Soil DNA library	BAC	*Escherichia coli*	Turbomycin	97

protection, regulatory genes, and genes for the export of the product, all tightly linked on the chromosome *(73,74)*. The genes within these pathways are either expressed individually or expressed coordinately, along with the adjacent genes, forming operons. Pathways vary in size from as small as 5 to 8 kb (phosphomycin *[75]* and pyrrolnitrin *[76]*), to over 100 kb (candicidin *[77]* and rifamycin *[78]*). The majority of the pathways range from 20 to 200 kb in size *(79)*. The cloning of the pigmented antibiotic pathway for actinorhodin from *Streptomyces coelicolor* and its expression in *Streptomyces parvulus* was the first example of heterologous expression of a pathway in a representative of the *Actinomycetales (80)*. Since then, many examples of heterologous expression of small-molecule pathways in *Actinomycetales* as well as other Gram-positive and Gram-negative bacteria have been described (**Table 2**).

So far, several genomes from different members of the *Actinomycetales* have been sequenced *(79)*—for example, *Thermonospora fusca* (Integrated Genomics), *Saccharopolyspora erythraea* (Biotica), *Saccharopolyspora spinosa* (Dow AgroSciences), *Streptomyces avermitilis (81)*, *Streptomyces coelicolor (82)*, and *Streptomyces diversa*™ (Diversa Corporation). The published genome sequences of two *Streptomy-*

ces species—*S. avermitilis (81)* and *S. coelicolor (82)*—revealed that *Streptomyces* contains many more as yet unexplored secondary metabolic pathways than previously thought. The genomes of *S. avermitilis* and *S. coelicolor* encode more than 20 known and putative secondary metabolic gene clusters each. These apparently cryptic pathways consist of type I, type II, and type III polyketide synthases (PKS), nonribosomal peptide synthetase (NRPS), as well as hybrid PKS-NRPS pathways *(81,82)*. In addition, genomic partial DNA sequencing suggested that several actinomycete genera have the capability to produce many as yet undiscovered secondary metabolites *(98)*. Out of 50 strains examined, 8 strains contained core sequences for the biosynthesis of the enediyne anticancer compounds *(98)*. These compounds had not been detected in these strains before. Furthermore, 15% of all examined actinomycetes revealed sequences for the enediyne pathway *(98)*. These data suggest that microorganisms such as members of the *Actinomycetales* may contain many more novel natural products than previously observed. Therefore new technologies have been developed to access these natural products *(63)*.

4. Genomic Libraries From Single Isolates

Natural-product pathways have been cloned and characterized using plasmids, cosmids, and fosmids, with a general capacity to hold about 40 to 50 kb of DNA. To clone larger DNA fragments, vector systems such as PAC and BAC vectors were developed. The PAC vectors are based on the *Escherichia coli*-transducing phage P1, and can carry inserts >100 kb *(99,100)*. The BAC vector was initially developed to construct human genomic libraries *(101)*. It is based on the *E. coli* F'-plasmid and can accommodate inserts of up to 300 kb. These vectors can be shuttled between *E. coli* and the host of interest by adding appropriate sequences for replication and maintenance. Several single genome fosmid and BAC libraries have been constructed and published for bacterial isolates such as *Aquifex (102)*, *Mycobacterium (103)*, *Enterococcus (104)*, *Bacillus (68)*, *Corynebacterium (105)*, and *Streptomyces (100,106)*. The average insert size varied between 30 kb and 70 kb. Inserts larger than 130 kb for *Actinomycetales* libraries *(106)* and a *Bacillus* library *(68)* have also been reported.

5. Genomic Libraries From the Environment

The construction of DNA libraries from the environment represents a bigger challenge than making libraries from single isolates. One of the reasons is the presence of different types of contaminants within the isolated DNA, which can inhibit effective cloning. Many of these contaminants such as humic acids can be difficult to eliminate without shearing of the DNA. Several methods for preparing DNA suitable for cloning recently have been published *(107–111)*. Environmental small DNA insert expression libraries (5 kb or less) were pioneered by Short and colleagues from different environments to study microbial complexity and to screen for bioactive molecules such as enzymes and small pathways *(23,24,26,63,65,112,113)*. However, there have been only a few published examples of large *(64,114,115)* insert libraries (40 kb and above) constructed from environmental samples. Recently, soil DNA libraries have been constructed in either a cosmid vector *(116)* or BAC vector *(68,71)* with an average insert size ranging between 27 and 45 kb, although inserts as large as 120 kb have been captured in the BAC libraries *(71)*. These libraries contained genes from previously unidentified phylotypes, as revealed by 16S rRNA gene sequencing (Diversa Corporation, unpublished data; *[68,71,117]*). The size of the library (number of clones) to cover the microbial diversity in an environmental sample is dependent on the number of species present in the sample, the size of their genomes, the size of the pathway, and the insert size. To identify clones carrying intact pathways assuming a random distribution of genomic fragments in the library, several hundreds to thousands of clones per genome need to be screened.

6. Screening for Novel Natural-Product Pathways

Environmental libraries can be screened for novel natural products by three general methods—a sequence-based approach, a hybridization-based approach, and an expression-based approach.

Sequence-based approach. In the early 1990s, Diversa Corporation pioneered the use of environmental clone libraries for the sequence-based discovery of novel enzymes and small-molecule pathways *(118)*. In this sequence-based approach, the libraries can be screened for novel pathways such as those encoding type I and type II polyketide synthases (PKS) as well as nonribosomal peptide synthetase (NRPS). The compounds in these families possess diverse chemical structures with a variety of biological activities, but are conserved in their core biosynthetic gene sequences *(119–121)*. Similarly, there are biosynthetic genes involved in the modifications of natural products, e.g., glycosylation, which are highly conserved *(122)*. These conserved secondary metabolic sequences have been used successfully as probes to isolate corresponding genes and the associated gene clusters from single-isolate genomic libraries *(123,124)*. This principle has been extended to environmental genomic libraries. For example, using conserved sequences within type II PKS genes as polymerase chain reaction (PCR) primers, novel type II PKS genes were amplified from soil genomic DNA *(125)*. Such amplified PKS genes can be used to identify clones carrying the entire pathway in environmental libraries. Conserved type I PKS primers were used to identify and isolate several novel type I PKS pathways *(116)*. One of these pathways derived from a soil genomic library contained PKS/NRPS hybrid genes. Introduction of the PKS-containing cosmids into *S. lividans* resulted in the expression of novel fatty dienic alcohol isomers.

Hybridizaton-based approach. Alternatively, identification of novel natural-product pathways is achieved through a high-throughput genome-hybridization method to discover metabolic loci independent of their expression *(98)*. Genes that are likely to be involved in the biosynthesis of natural products are identified by sequence comparisons to a database of microbial gene clusters known to be involved in natural-product biosynthesis. This information is used to design screening probes to identify cloned subgenomic fragments (for example, cosmids or BACs) containing the gene(s) of interest as well as neighboring genes that together may constitute a biosynthetic gene cluster *(98,126)*.

Expression-based approach. In this approach, libraries consisting of large DNA inserts in a surrogate host are screened for novel activities. This approach has been very successful for the discovery of novel enzymes from small-insert environmental libraries *(24,26,113,117)*. The screening of large-insert environmental soil libraries expressed in *E. coli* or other expression hosts such as *Pseudomonas* or *Streptomyces* for antimicrobial activities or for colony pigmentation, led to the isolation of known molecules, such as turbomycins *(97)*, violacein *(94)*, indigo blue and indirubin *(71)*, and many other known and novel molecules (Diversa Corporation, unpublished data). Therefore, the groundwork described for the discovery of novel small molecules derived from environmental genomic libraries encourages further exploration.

7. Outlook

It has been demonstrated that natural-product pathways can be captured from different genomes on large-insert vectors and expressed in heterologous systems. The simplified construction of environmental libraries with an average insert size larger than 100 kb is still challenging. To cover microbial diversity within recombinant libraries requires screening of multiple tens of thousands of clones per environmental library. Hence, novel high-throughput screening approaches need to be developed to screen this large number of recombinant clones without organic extract preparation. The expression level of recombinant small molecules is host dependent and can be low. Therefore, expression hosts have to be developed with elevated levels of heterologous expressed molecules *(127)*. Different types of hosts should be explored to maximize gene expression levels. The knowledge gained from the growing number of bacterial genome sequences will help to engineer surrogate hosts for optimum expression *(93,128)*.

New technologies in microbiology and molecular biology described in this chapter will continue to advance the transformation of natural-product discovery, with the focus on new culture-dependent and -independent methods to access microbial diversity. Future investigations will demonstrate whether microorganisms belonging to phyla that have no cultivated representatives so far, produce new bioactive molecules. However, in addition to accessing novel biodiversity, faster chemical dereplication methods as well as methods to address production and chemical modification of novel chemical entities are desired.

Acknowledgments

We thank J. M. Short, G. Woodnutt, E. J. Mathur, and M. Simon for support.

References

1. Fleming A. On the antibacterial action of cultures of a *Penicillium*, with a special reference to their use in the isolation of *B. influenze*. Brit J Exp Path 1929;10:226–236.
2. Vicente MF, Basilio A, Cabello A, Peláez F. Microbial natural products as a source of antifungals. Clin Microbiol Infect 2003;9:15–32.
3. Strohl WR. Industrial antibiotics: today and the future. In: Strohl WR (ed), Biotechnology of Antibiotics. Marcel Dekker, New York: 1997; pp. 1–47.
4. Demain AL, Lancini G. Bacterial pharmaceutical products. In: Dworkin Mea (ed), The Prokaryotes: An Evolving Electronic Resource for the Microbiological Community. New-York: Springer-Verlag, 2001; http://link.springer-ny.com/link/service/books/10125/.
5. Lazzarini A, Cavaletti L, Toppo G, Marinelli F. Rare genera of actinomycetes as potential producers of new antibiotics. Antonie van Leeuwenhoek 2001;79:399–405.
6. Bellais S, Poirel L, Leotard S, Naas T, Nordmann P. Genetic diversity of carbapenem-hydrolyzing metallo-beta-lactamases from *Chryseobacterium* (*Flavobacterium*) *indologenes*. Antimicrob Agents Chemother 2000;44:3028–3034.

7. Golakoti T, Yoshida WY, Chaganty S, Moore RE. Isolation and structure determination of nostocyclopeptides A1 and A2 from the terrestrial cyanobacterium *Nostoc* sp. ATCC53789. J Nat Prod 2001;64:54–59.
8. Kawulka K, Sprules T, McKay RT, et al. Structure of subtilosin A, an antimicrobial peptide from *Bacillus subtilis* with unusual posttranslational modifications linking cysteine sulfurs to alpha-carbons of phenylalanine and threonine. J Am Chem Soc 2003;125:4726–4727.
9. Reichenbach H. Myxobacteria, producers of novel bioactive substances. J Ind Microbiol Biotechnol 2001;27:149–156.
10. Silver L, Bostian K. Screening of natural products for antimicrobial agents. Eur J Clin Microbiol Infect Dis 1990;9:455–461.
11. Strohl WR. The role of natural products in a modern drug discovery program. Drug Discov Today 2000;5:39–41.
12. Silva CJ, Brian P, Peterson T. Screening of combinatorial biology libraries for natural products discovery. In: Seethala R, Fernandes PB (eds), Handbook of Drug Screening. New York: Marcel Dekker, 2001; pp. 357–382.
13. Zähner H, Fiedler H-P. The need for new antibiotics: possible ways forward. In: Russell NJ (ed), Fifty Years of Antimicrobials: Past Perspectives and Future Trends. Cambridge University Press, Cambridge, England: 1995; pp. 67–84.
14. Class S. Pharma Overview. Chem Eng News 2002;80:39–49.
15. Newman DJ, Cragg GM, Snader KM. Natural products as sources of new drugs over the period 1981–2002. J Nat Prod 2003;66:1022–1037.
16. Walsh C. Where will new antibiotics come from? Nat Rev Microbiol 2003;1:65–70.
17. Torsvik V, Salte K, Sørheim R, Goksøyr J. Comparison of phenotypic diversity and DNA heterogeneity in a population of soil bacteria. Appl Environ Microbiol 1990;56:776–781.
18. Torsvik V, Øvreås L, Thingstad TF. Prokaryotic diversity—magnitude, dynamics, and controlling factors. Science 2002;296:1064–1066.
19. Hughes JB, Hellmann JJ, Ricketts TH, Bohannan BJ. Counting the uncountable: statistical approaches to estimating microbial diversity. Appl Environ Microbiol 2001;67:4399–4406.
20. Dykhuizen DE. Santa Rosalia revisited: why are there so many species of bacteria? Antonie van Leeuwenhoek 1998;73:25–33.
21. Amann RI, Ludwig W, Schleifer K-H. Phylogenetic identification and in situ detection of individual microbial cells without cultivation. Microbiol Rev 1995;59:143–169.
22. Whitman WB, Coleman DC, Wiebe WJ. Prokaryotes: the unseen majority. Proc Natl Acad Sci USA 1998;95:6578–6583.
23. Gray KA, Richardson TH, Kretz K, et al. Rapid evolution of reversible denaturation and elevated melting temperature in a microbial haloalkane dehalogenase. Adv Synth Catal 2001;343:607–617.
24. DeSantis G, Zhu Z, Greenberg WA, et al. An enzyme library approach to biocatalysis: development of nitrilases for enantioselective production of carboxylic acid derivatives. J Am Chem Soc 2002;124:9024–9025.
25. Dunbar J, Barns SM, Ticknor LO, Kuske CR. Empirical and theoretical bacterial diversity in four Arizona soils. Appl Environ Microbiol 2002;68:3035–3045.
26. Richardson TH, Tan X, Frey G, et al. A novel, high performance enzyme for starch liquefaction. Discovery and optimization of a low pH, thermostable alpha-amylase. J Biol Chem 2002;277: 26,501–26,507.
27. Keller M, Zengler K. Tapping into microbial diversity. Nat Rev Microbiol 2004;2:141–150.
28. Hugenholtz P, Pitulle C, Hershberger KL, Pace NR. Novel division level bacterial diversity in a Yellowstone hot spring. J Bacteriol 1998;180:366–376.
29. Curtis TP, Sloan WT, Scannell JW. Estimating prokaryotic diversity and its limits. Proc Natl Acad Sci USA 2002;99:10,494–10,499.

30. Venter JC, Remington K, Heidelberg JF, et al. Environmental genome shotgun sequencing of the Sargasso Sea. Science 2004;304:66–74.
31. Woese CR, Stackebrandt E, Macke TJ, Fox GE. A phylogenetic definition of the major eubacterial taxa. Syst Appl Microbiol 1985;6:143–151.
32. Rappé MS, Giovannoni SJ. The uncultured microbial majority. Annu Rev Microbiol 2003;57: 369–394.
33. Dojka MA, Harris JK, Pace NR. Expanding the known diversity and environmental distribution of an uncultured phylogenetic division of bacteria. Appl Environ Microbiol 2000;66:1617–1621.
34. Rosselló-Mora R, Amann R. The species concept for prokaryotes. FEMS Microbiol Rev 2001;25:39–67.
35. DSMZ. http://www.dsmz.de/bactnom/bactname.htm. Bacterial nomenclature up-to-date, 2003.
36. Dictionary of Natural Products New York: Chapman & Hall/CRC, 2003.
37. Keller NP, Hohn TM. Metabolic pathway gene clusters in filamentous fungi. Fungal Genet Biol 1997;21:17–29.
38. Yu TW, Bai L, Clade D, et al. The biosynthetic gene cluster of the maytansinoid antitumor agent ansamitocin from *Actinosynnema pretiosum*. Proc Natl Acad Sci USA 2002;99:7968–7973.
39. Pasteur L. Animalcules infusoires vivant sans gaz oxygène libre et déterminant des fermentations. CR Acad Sci 1861;52:344–347.
40. Koch R. Untersuchungen über Bakterien VI. Verfahren zur Untersuchung, zum Conservieren und Photographieren. Beitr Biol Pflanz 1877;2:399–434.
41. Beijerinck WM. Ueber *Spirillum desulfuricans* als Ursache von Sulfatreduktion. Zentralblatt Bakteriol 1895;1:1–9, 49–59, 104–114.
42. Winogradsky S. Ueber Schwefelbacterien. Botanische Zeitung 1887;45:489–507, 513–523, 529–539, 545–559, 569–576, 585–594 and 606–610.
43. Ferguson RL, Buckley EN, Palumbo AV. Response of marine bacterioplankton to differential filtration and confinement. Appl Environ Microbiol 1984;47:49–55.
44. Eilers H, Pernthaler J, Glöckner FO, Amann R. Culturability and *in situ* abundance of pelagic bacteria from the North Sea. Appl Environ Microbiol 2000;66:3044–3051.
45. Staley JT, Konopka A. Measurement of in situ activities of nonphotosynthetic microorganisms in aquatic and terrestrial habitats. Annu Rev Microbiol 1985;39:321–346.
46. Xu HS, Roberts N, Singleton FL, Attwell RW, Grimes DJ, Colwell RR. Survival and viability of nonculturable *Escherichia coli* and *Vibrio cholerae* in the estuarine and marine environment. Microb Ecol 1982;8:313–323.
47. Janssen PH, Yates PS, Grinton BE, Taylor PM, Sait M. Improved culturability of soil bacteria and isolation in pure culture of novel members of the divisions *Acidobacteria*, *Actinobacteria*, *Proteobacteria*, and *Verrucomicrobia*. Appl Environ Microbiol 2002;68:2391–2396.
48. Chin KJ, Hahn D, Hengstmann U, Liesack W, Janssen PH. Characterization and identification of numerically abundant culturable bacteria from the anoxic bulk soil of rice paddy microcosms. Appl Environ Microbiol 1999;65:5042–5049.
49. Sait M, Hugenholtz P, Janssen PH. Cultivation of globally distributed soil bacteria from phylogenetic lineages previously only detected in cultivation-independent surveys. Environ Microbiol 2002;4:654–666.
50. Joseph SJ, Hugenholtz P, Sangwan P, Osborne CA, Janssen PH. Laboratory cultivation of widespread and previously uncultured soil bacteria. Appl Environ Microbiol 2003;69:7210–7215.
51. Guan LL, Onuki H, Kamino K. Bacterial growth stimulation with exogenous siderophore and synthetic *N*-acyl homoserine lactone autoinducers under iron-limited and low-nutrient conditions. Appl Environ Microbiol 2000;66:2797–2803.

52. Guan LL, Kamino K. Bacterial response to siderophore and quorum-sensing chemical signals in the seawater microbial community. BMC Microbiol 2001;1:27.
53. Bussmann I, Philipp B, Schink B. Factors influencing the cultivability of lake water bacteria. J Microbiol Methods 2001;47:41–50.
54. Bruns A, Cypionka H, Overmann J. Cyclic AMP and acyl homoserine lactones increase the cultivation efficiency of heterotrophic bacteria from the Central Baltic Sea. Appl Environ Microbiol 2002;68:3978–3987.
55. Bruns A, Nübel U, Cypionka H, Overmann J. Effect of signal compounds and incubation conditions on the culturability of freshwater bacterioplankton. Appl Environ Microbiol 2003;69:1980–1989.
56. Button DK, Schut F, Quang P, Martin R, Roberston BR. Viability and isolation of marine bacteria by dilution culture: theory, procedures, and initial results. Appl Environ Microbiol 1993;59:881–891.
57. Kaeberlein T, Lewis K, Epstein SS. Isolating "uncultivable" microorganisms in pure culture in a simulated natural environment. Science 2002;296:1127–1129.
58. Zengler K, Toledo G, Rappe M, et al. Cultivating the uncultured. Proc Natl Acad Sci USA 2002;99:15,681–15,686.
59. Connon SA, Giovannoni SJ. High-throughput methods for culturing microorganisms in very-low-nutrient media yield diverse new marine isolates. Appl Environ Microbiol 2002;68:3878–3885.
60. Morris RM, Rappe MS, Connon SA, et al. SAR11 clade dominates ocean surface bacterioplankton communities. Nature 2002;420:806–810.
61. Orsini F, Ami D, Villa AM, Sala G, Bellotti MG, Doglia SM. FT-IR microspectroscopy for microbiological studies. J Microbiol Methods 2000;42:17–27.
62. Wenning M, Seiler H, Scherer S. Fourier-transform infrared microspectroscopy, a novel and rapid tool for identification of yeasts. Appl Environ Microbiol 2002;68:4717–4721.
63. Short JM. Recombinant approaches for accessing biodiversity. Nat Biotechnol 1997;15:1322–1323.
64. Short JM et al. Patents US 5,763,239; US 5,958,672; US 6,001,574; US 6,004,788; US 6,030,779; US 6,054,267; US 6,057,103; US 6,168,919; US 6,174,673; US 6,368,798; US 6,444,426; US 6,455,254; AU718,573; AU 720,334; AU 756,201 (1201–1AU).
65. Greenberg WA, Varvak A, Hanson SR, et al. Development of an efficient, scalable, aldolase-catalyzed process for enantioselective synthesis of statin intermediates. Proc Natl Acad Sci USA 2004;101:5788–5793.
66. Robertson DE, Steer BA. Recent progress in biocatalyst discovery and optimization. Curr Opin Chem Biol 2004;8:141–149.
67. Schloss PD, Handelsman J. Biotechnological prospects from metagenomics. Curr Opin Biotechnol 2003;14:303–310.
68. Rondon MR, Raffel SJ, Goodman RM, Handelsman J. Toward functional genomics in bacteria: analysis of gene expression in *Escherichia coli* from a bacterial artificial chromosome library of *Bacillus cereus*. Proc Natl Acad Sci USA 1999;96:6451–6455.
69. Rondon MR, August PR, Bettermann AD, et al. Cloning the soil metagenome: a strategy for accessing the genetic and functional diversity of uncultured microorganisms. Appl Environ Microbiol 2000;66:2541–2547.
70. Wang GY, Graziani E, Waters B, et al. Novel natural products from soil DNA libraries in a streptomycete host. Org Lett 2000;2:2401–2404.
71. MacNeil IA, Tiong CL, Minor C, et al. Expression and isolation of antimicrobial small molecules from soil DNA libraries. J Mol Microbiol Biotechnol 2001;3:301–308.
72. Piel J. A polyketide synthase-peptide synthetase gene cluster from an uncultured bacterial symbiont of *Paederus* beetles. Proc Natl Acad Sci USA 2002;99:14,002–14,007.

73. Hopwood DA. *Streptomyces* genes: from Waksman to Sanger. J Ind Microbiol Biotechnol 2003;30:468–471.
74. Martin JF, Liras P. Organization and expression of genes involved in the biosynthesis of antibiotics and other secondary metaboites. Annu Rev Microbiol 1989;43:173–206.
75. Hidaka T, Goda M, Kuzuyama T, Takei N, Hidaka M, Seto H. Cloning and nucleotide sequence of fosfomycin biosynthetic genes of *Streptomyces wedmorensis*. Mol Gen Genet 1995;249:274–280.
76. Hammer PE, Hill DS, Lam ST, Van Pée KH, Ligon JM. Four genes from *Pseudomonas fluorescens* that encode the biosynthesis of pyrrolnitrin. Appl Environ Microbiol 1997;63:2147–2154.
77. Aparicio JF, Caffrey P, Gil JA, Zotchev SB. Polyene antibiotic biosynthesis gene clusters. Appl Microbiol Biotechnol 2003;61:179–188.
78. August PR, Tang L, Yoon YJ, et al. Biosynthesis of the ansamycin antibiotic rifamycin: deductions from the molecular analysis of the rif biosynthetic gene cluster of *Amycolatopsis mediterranei* S699. Chem Biol 1998;5:69–79.
79. Paradkar A, Trefzer A, Chakraburtty R, Stassi D. Streptomyces genetics: a genomic perspective. Crit Rev Biotechnol 2003;23:1–27.
80. Malpartida F, Hopwood DA. Molecular cloning of the whole biosynthetic pathway of a *Streptomyces* antibiotic and its expression in a heterologous host. Nature 1984;309:462–464.
81. Omura S, Ikeda H, Ishikawa J, et al. Genome sequence of an industrial microorganism *Streptomyces avermitilis*: deducing the ability of producing secondary metabolites. Proc Natl Acad Sci USA 2001;98:12,215–12,220.
82. Bentley SD, Chater KF, Cerdeño-Tárraga AM, et al. Complete genome sequence of the model actinomycete *Streptomyces coelicolor* A3(2). Nature 2002;417:141–147.
83. Motamedi H, Hutchinson CR. Cloning and heterologous expression of a gene cluster for the biosynthesis of tetracenomycin C, the anthracycline antitumor antibiotic of *Streptomyces glaucescens*. Proc Natl Acad Sci USA 1987;84:4445–4449.
84. Gould SJ, Hong ST, Carney JR. Cloning and heterologous expression of genes from the kinamycin biosynthetic pathway of *Streptomyces murayamaensis*. J Antibiot 1998;51:50–57.
85. Bormann C, Mohrle V, Bruntner C. Cloning and heterologous expression of the entire set of structural genes for nikkomycin synthesis from *Streptomyces tendae* Tu901 in *Streptomyces lividans*. J Bacteriol 1996;178:1216–1218.
86. Lacalle RA, Tercero JA, Jiménez A. Cloning of the complete biosynthetic gene cluster for an aminonucleoside antibiotic, puromycin, and its regulated expression in heterologous hosts. EMBO J 1992;11:785–792.
87. Sanchéz C, Butovich IA, Braña AF, Rohr J, Méndez C, Salas JA. The biosynthetic gene cluster for the antitumor rebeccamycin: characterization and generation of indolocarbazole derivatives. Chem Biol 2002;9:519–531.
88. Tang L, Shah S, Chung L, et al. Cloning and heterologous expression of the epothilone gene cluster. Science 2000;287:640–642.
89. Julien B, Shah S. Heterologous expression of epothilone biosynthetic genes in *Myxococcus xanthus*. Antimicrob Agents Chemother 2002;46:2772–2778.
90. Eppelmann K, Doekel S, Marahiel MA. Engineered biosynthesis of the peptide antibiotic bacitracin in the surrogate host *Bacillus subtilis*. J Biol Chem 2001;276:34,824–34,831.
91. Kimura H, Miyashita H, Sumino Y. Organization and expression in *Pseudomonas putida* of the gene cluster involved in cephalosporin biosynthesis from *Lysobacter lactamgenus* YK90. Appl Microbiol Biotechnol 1996;45:490–501.
92. Mercado-Blanco J, van der Drift KM, Olsson PE, Thomas-Oates JE, van Loon LC, Bakker PA. Analysis of the *pmsCEAB* gene cluster involved in biosynthesis of salicylic acid and the siderophore pseudomonine in the biocontrol strain *Pseudomonas fluorescens* WCS374. J Bacteriol 2001;183:1909–1920.

93. Thomson NR, Crow MA, McGowan SJ, Cox A, Salmond GP. Biosynthesis of carbapenem antibiotic and prodigiosin pigment in *Serratia* is under quorum sensing control. Mol Microbiol 2000;36:539–556.
94. Brady SF, Chao CJ, Handelsman J, Clardy J. Cloning and heterologous expression of a natural product biosynthetic gene cluster from eDNA. Org Lett 2001;3:1981–1984.
95. Brady SF, Clardy J. Long-chain *N*-acyl amino acid antibiotics isolated from heterologously expressed environmental DNA. J Am Chem Soc 2000;122:12,903–12,904.
96. Brady SF, Chao CJ, Clardy J. New natural product families from an environmental DNA (eDNA) gene cluster. J Am Chem Soc 2002;124:9968–9969.
97. Gillespie DE, Brady SF, Bettermann AD, et al. Isolation of antibiotics turbomycin a and B from a metagenomic library of soil microbial DNA. Appl Environ Microbiol 2002;68:4301–4306.
98. Zazopoulos E, Huang K, Staffa A, et al. A genomics-guided approach for discovering and expressing cryptic metabolic pathways. Nature Biotechnol 2003;21:187–190.
99. Ioannou PA, Amemiya CT, Garnes J, et al. A new bacteriophage P1-derived vector for the propagation of large human DNA fragments. Nature Genet 1994;6:84–89.
100. Sosio M, Giusino F, Cappellano C, Bossi E, Puglia AM, Donadio S. Artificial chromosomes for antibiotic-producing actinomycetes. Nature Biotechnol 2000;18:343–345.
101. Shizuya H, Birren B, Kim UJ, et al. Cloning and stable maintenance of 300-kilobase-pair fragments of human DNA in *Escherichia coli* using an F-factor-based vector. Proc Natl Acad Sci USA 1992;89:8794–8797.
102. Deckert G, Warren PV, Gaasterland T, et al. The complete genome of the hyperthermophilic bacterium *Aquifex aeolicus*. Nature 1998;392:353–358.
103. Brosch R, Gordon SV, Billault A, et al. Use of a *Mycobacterium tuberculosis* H37Rv bacterial artificial chromosome library for genome mapping, sequencing, and comparative genomics. Infect Immun 1998;66:2221–2229.
104. Xu Y, Jiang L, Murray BE, Weinstock GM. *Enterococcus faecalis* antigens in human infections. Infect Immun 1997;65:4207–4215.
105. Tauch A, Homann I, Mormann S, et al. Strategy to sequence the genome of *Corynebacterium glutamicum* ATCC 13032: use of a cosmid and a bacterial artificial chromosome library. J Biotechnol 2002;95:25–38.
106. Alduina R, De Grazia S, Dolce L, et al. Artificial chromosome libraries of *Streptomyces coelicolor* A3(2) and *Planobispora rosea*. FEMS Microbiol Lett 2003;218:181–186.
107. Zhou J, Bruns MA, Tiedje JM. DNA recovery from soils of diverse composition. Appl Environ Microbiol 1996;62:316–322.
108. Stein JL, Marsh TL, Wu KY, Shizuya H, DeLong EF. Characterization of uncultivated prokaryotes: isolation and analysis of a 40-kilobase-pair genome fragment from a planktonic marine archaeon. J Bacteriol 1996;178:591–599.
109. Han J, Craighead HG. Separation of long DNA molecules in a microfabricated entropic trap array. Science 2000;288:1026–1029.
110. Martin-Laurent F, Philippot L, Hallet S, et al. DNA extraction from soils: old bias for new microbial diversity analysis methods. Appl Environ Microbiol 2001;67:2354–2359.
111. Short JM et al. U.S. Patents: US 5,763239; US 5,763,239; US 6,001,574; US 6,057,103; US 6,174,673; US 6,368,798.
112. Palackal N, Brennan Y, Callen WN, et al. An evolutionary route to xylanase process fitness. Protein Sci 2004;13:494–503.
113. Robertson DE, Chaplin JA, DeSantis G, et al. Exploring nitrilase sequence space for enantioselective catalysis. Appl Environ Microbiol 2004;70:2429–2436.
114. Béjà O, Suzuki MT, Koonin EV, et al. Construction and analysis of bacterial artificial chromosome libraries from a marine microbial assemblage. Environ Microbiol 2000;2:516–529.

115. Berry AE, Chiocchini C, Selby T, Sosio M, Wellington EMH. Isolation of high molecular weight DNA from soil for cloning into BAC vectors. FEMS Microbiol Lett 2003;223:15–20.
116. Courtois S, Cappellano CM, Ball M, et al. Recombinant environmental libraries provide access to microbial diversity for drug discovery from natural products. Appl Environ Microbiol 2003;69:49–55.
117. Gray KA, Richardson TH, Robertson DE, Swanson PE, Subramanian MV. Soil-based gene discovery: a new technology to accelerate and broaden biocatalytic applications. Adv Appl Microbiol 2003;52:1–27.
118. Short JM. US 6,455,254.
119. Leadlay PF. Combinatorial approaches to polyketide biosynthesis. Curr Opin Chem Biol 1997;1:162–168.
120. Hutchinson CR. Antibiotics from genetically engineered microorganisms. In: Strohl WR (ed), Biotechnology of Antibiotics. Vol. 82. Marcel Dekker, New York: 1997; pp. 683–702.
121. Zuber P, Marahiel MA. Structure, function, and regulation of genes encoding multidomain peptide synthetases. In: Strohl WR (ed), Biotechnology of Antibiotics. Vol. 82. Marcel Dekker, New York: 1997; pp. 187–216.
122. Strohl WR, Dickens ML, Rajgarhia VB, Woo AJ, Priestley ND. Anthracyclines. In: Strohl WR (ed), Biotechnology of Antibiotics. Vol. 82. Marcel Dekker, New York: 1997; pp. 577–657.
123. Ichinose K, Ozawa M, Itou K, Kunieda K, Ebizuka Y. Cloning, sequencing and heterologous expression of the medermycin biosynthetic gene cluster of *Streptomyces* sp. AM-7161: towards comparative analysis of the benzoisochromanequinone gene clusters. Microbiology 2003;149: 1633–1645.
124. Izumikawa M, Murata M, Tachibana K, Ebizuka Y, Fujii I. Cloning of modular type I polyketide synthase genes from salinomycin producing strain of *Streptomyces albus*. Bioorg Med Chem 2003;11:3401–3405.
125. Seow KT, Meurer G, Gerlitz M, Wendt-Pienkowski E, Hutchinson CR, Davies J. A study of iterative type II polyketide synthases, using bacterial genes cloned from soil DNA: a means to access and use genes from uncultured microorganisms. J Bacteriol 1997;179:7360–7368.
126. Short JM et al. U.S. Patents: US 6,030,779; US 6,344,288 B1; US 6,368,798 B1; US 6,455,254 B1.
127. Short JM. WO0196551.
128. Donadio S, Monciardini P, Alduina R, et al. Microbial technologies for the discovery of novel bioactive metabolites. J Biotechnol 2002;99:187–198.

13

Accessing the Genomes of Uncultivated Microbes for Novel Natural Products

Asuncion Martinez, Joern Hopke, Ian A. MacNeil, and Marcia S. Osburne

Summary

Recent findings suggest that only 1% or less of the total number of soil microbial species can be easily cultivated. The fact that uncultured species represent spectacular microbial diversity has sparked great interest in these microorganisms as a potentially prolific source of untapped genetic diversity encoding novel natural products. Multiple approaches are being developed to access this diversity, such as methods to improve the ability to cultivate some of these organisms, most of which are not easily grown under standard laboratory conditions. Here we discuss an alternative approach, aimed at developing technologies for gaining access to the genomes of uncultivated microbes by creating environmental DNA libraries. This method involves isolating large DNA fragments (100–300 kb) from soil microorganisms (or from microorganisms derived from other environments), and inserting these fragments into a bacterial vector, thus generating recombinant DNA libraries. Such libraries are then used to identify novel natural products by various means, including expression of the DNA in a heterologous host strain and screening for activities, or by directly analyzing the DNA for genes of interest. The recombinant approach thus obviates the need for culturing diverse microorganisms and provides a relatively unbiased sampling of the vast untapped genetic diversity present in various microenvironments. As an additional advantage, the genes encoding a product of interest are already isolated and can be analyzed using the tools of bioinformatics, thus providing a potential boost to the efforts of analytical chemists to identify the product. Furthermore, the possibility of regulating the expression of isolated environmental gene clusters or combining them with genes for other pathways to obtain new compounds could furnish a further advantage over traditional natural-product discovery methodologies.

Over the last few years, several laboratories have focused their research efforts on a proof of concept for the feasibility of this recombinant approach. Areas that have been addressed include (1) procedures to generate environmental libraries, (2) assessing the degree of phylogenetic diversity encoded in environmental libraries, (3) assessing the feasibility of expressing heterologous DNA in multiple expression host strains, and (4) high-throughput methods to screen environmental libraries successfully in a reasonable time frame. Here we review the current status of our work in developing environmental DNA cloning technologies with respect to these major areas, including compounds we and others have found by screening environmental libraries. We also discuss our vision of how to transform this exciting new technology from interesting ideas and exciting laboratory results into a realistic, productive, high-throughput industrial method for discovering novel natural products from environmental microbes.

Key Words: Natural products; uncultivated microorganisms; diversity; biodiversity; high-throughput screening; environmental DNA libraries; heterologous expression; prescreens; reporter assays; soil microbes.

1. Introduction

Historically, natural products have played a major role as chemotherapeutic agents. The majority of greater than 5000 known anti-infective compounds are derived from natural products, with over 100 of these used clinically *(1,2)*. In addition to anti-infectives, natural products are strongly represented in the areas of cancer chemotherapeutics *(3)* and immunomodulation *(4,5)*. Methodologies for discovering useful natural products in the 20th century evolved from the accidental discovery on a Petri dish of the antibacterial effect of penicillin, to more sophisticated discovery technologies, initiated by Selman Waksman in 1939. Waksman developed a rational antibiotic screening program that resulted in the discovery of streptomycin, thus leading to effective treatments for a number of bacterial illnesses. His drug discovery approach provided a foundation for the increasingly sophisticated technologies to follow, which included the isolation of thousands of microorganisms from soil and other environments, derived from areas ranging from New Jersey to exotic Pacific islands, and the development of sophisticated high-throughput screening methods to capture whatever novel and useful activities were produced by these microorganisms. Although the discovery of new structures was initially quite prolific, the discovery rate of novel classes of antimicrobial molecules from natural products has declined in the latter part of the 20th century through the present time *(6,7)*, leading industrial research laboratories to reduce or abandon their natural-product efforts on the assumption that little remained to be discovered. Yet even in the past few years, promising leads for antibacterial cell-wall inhibitors derived from natural products have been discovered, and the theory has been put forward that compounds that can enter cells and avoid efflux (a problem for many compounds derived from rational drug design strategies) have been better designed by natural selection *(8)*. Additionally, the "limits for life" continue to be extended, with the microorganism that grows at temperatures ranging from 85°C to 121°C as a case in point *(9)*. These and other intriguing findings discussed below strongly suggest that vast numbers of microorganisms, and hence natural products, remain to be discovered, and that there is a strong argument to be made for continuing and extending natural-products discovery programs. Recent research using 16S rRNA gene analysis has uncovered the exciting finding that the vast majority of bacteria in environmental samples are still unknown *(10–14)*. Such analyses *(15,16)* have shown that these previously unknown bacteria belong either to known families or to apparently novel groups. Many of these microorganisms are "uncultureable" under standard laboratory conditions, accounting for their prior lack of detection *(17)*. Because the number of microbial species currently cultivatable from soil is thought to represent only 1% or less of the total population, these uncultured species have the potential to provide a large pool of novel natural products *(18–20)*, if methodologies for tapping into this pool could be found. Multiple approaches are being developed to access this diversity for discovery-related efforts. One approach is to improve the ability to cultivate some of these previously uncultivated organisms, many of which are difficult to grow. Another approach that we and others have taken is to create recombinant environmental DNA libraries from the genomes of uncultivated microbes, thus providing access to their encoded products. This approach allows us to circumvent the difficulty in culturing and screening most soil microorganisms. The recombinant approach involves extracting microbial

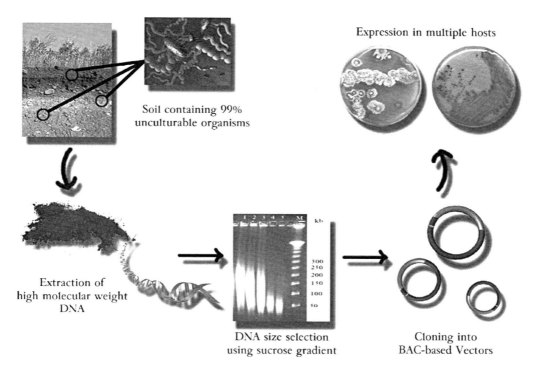

Fig. 1. Extraction and cloning of environmental DNA. Large fragments of DNA (>50 kb) obtained from uncultured soil microrganisms are ligated into a BAC vector, then transferred into various expression hosts.

DNA from the soil matrix or other microenvironment, and inserting large fragments of this DNA into vectors propagated in bacterial strains, such as *Escherichia coli*, that are easy to grow and manipulate genetically. DNA libraries are then screened in various ways to detect novel activities, including direct expression of the heterologous genes in *E. coli* and a variety of other, perhaps more suitable, expression hosts, and also by screening the recombinant DNA for homologies to known genes of interest. The process for generating libraries is depicted schematically in **Fig. 1**. Theoretically, this approach provides a nonbiased mechanism for accessing biodiversity for drug discovery. In addition, it offers some distinct advantages over traditional natural-products screening methodologies, which sometimes run into difficulties when trying to reproduce an activity of interest from a particular organism. Although the recombinant approach also requires analytical chemistry to isolate and identify compounds of interest, the DNA encoding the activity is pre-isolated on a recombinant plasmid, and can be identified by genetic and other means. The relevant genes can then be sequenced and analyzed using bioinformatics tools, thus providing both valuable clues as to the identity of the encoded product, and the potential to manipulate levels of production genetically. The success of this approach, which we refer to as "molecular biodiversity" technology, depends upon many factors, including the ability to clone sufficiently large contiguous fragments of environmental DNA into appropriate vectors, and the ability to express or otherwise screen heterologous DNA effectively. Thus, appropriate technologies are required to convert this novel approach into an efficient natural-products drug discovery tool. Therefore, we have concerned ourselves with developing such technologies, including a major emphasis on proving the concept that such an innovative approach could be successful in discovering novel and useful natural products.

2. Generation of Environmental DNA Libraries Encompassing Microbial Diversity

2.1. Environmental DNA

The ability to obtain large DNA fragments is important because many biosynthetic gene clusters encoding natural products range in size from approx 30 to 100 kb *(22–24)*. Because of their inherent fragility, the isolation of large fragments of environmental DNA from complex environments such as soil is a difficult process, especially because the DNA needs to be sufficiently clean for the enzymatic reactions required for cloning. Our group has used two basic strategies—direct and indirect DNA extraction—to generate environmental DNA libraries. For the direct approach, DNA is extracted from the soil matrix and cloned into a suitable vector, as depicted schematically in **Fig. 1**. Humic acids have been found to be a major contaminant of soil samples and can inhibit molecular reactions and reduce transformation efficiency *(25)*. Simple electrophoresis and phenol extraction are insufficient for their removal. Therefore we developed a process of extracting DNA from soil using gentle hot phenol extraction followed by treatment with CTAB *(26)*. The DNA is further sized and purified using a discontinuous sucrose density gradient (**Fig. 1**). According to the indirect approach, soil microbes are first isolated by Nycodenz gradient centrifugation, which separates microbial cells from the soil matrix, followed by DNA extraction *(26)*. Studies have shown that DNA obtained by the indirect method shows no major phylogenetic bias as compared with the direct DNA extraction method *(27)*, and that Nycodenz gradients recover only bacteria, with no detectable DNA contamination from other organisms *(27,28)*. Purified and sized DNA derived by either the direct or indirect methods is then cloned into a BAC or cosmid vector following digestion with a restriction enzyme. Alternatively, to avoid digesting environmental DNA before cloning (and possibly reducing the insert size and/or introducing bias based on G-C content), terminal transferase can be used to add polynucleotide tails to the 3' ends of the insert and vector DNA *(26)*.

2.2. Vectors

We have used both cosmid and BAC vectors as the backbone of environmental libraries, and both have been developed with an awareness of the need to express environmental libraries optimally in multiple expression hosts (at minimum, *E. coli* and *Streptomyces lividans*). BAC vectors have the advantage of replicating in single copy in *E. coli* (potentially useful for expression of a product that may be toxic at higher gene dosage), and can also stably maintain heterologous DNA inserts of up to 300 kb *(13,29,30)*. We constructed a new series of BAC vectors, represented by vector pMBD14 (**Fig. 2**), that encompass the elements required for conjugation of DNA into *Streptomyces* (*ori*T) and, to enhance the stability of the DNA in this host, elements

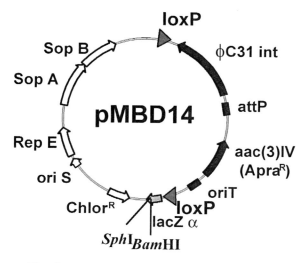

Fig. 2. Three-way BAC cloning vector pMBD14.

required for subsequent DNA integration into the *S. lividans* chromosome (i.e., the φC31 integrase and attachment site, and an apramycin-resistance marker), all flanked by *loxP* sites *(31)*. The utility of conjugation as opposed to transformation into *S. lividans* is discussed later. The *Bam*HI site in pMBD14 has been used to construct a 13,000-clone soil DNA library, with insert sizes ranging from 11.5 to at least 85 kb *(31)*. On the other hand, cosmid vectors are generally easier to work with, and are very suitable when smaller DNA inserts (40–50 kb) are appropriate. Although environmental DNA still needs to be carefully purified for cloning into cosmid vectors, cosmids are constrained to accept only 40–50 kb DNA, so that DNA sizing steps are unnecessary. Additionally, cosmids are multicopy vectors, which may be advantageous in circumstances where an increased gene dosage is desirable, e.g., in the case of a poorly expressed but nontoxic product. Cosmid shuttle vector pOS700I was constructed for generating environmental libraries. This vector is integrative in *S. lividans* via the *attP* site and *int* gene from the *Streptomyces* integrative element pSAM2 *(32)*, permitting site-specific integration of pOS700I into the chromosome of many *Streptomyces* species. It also encodes both an ampicillin-resistance gene for selection in *E. coli*, and a cassette *(33)* that confers hygromycin resistance in both *S. lividans* and *E. coli* *(26)*.

2.3. Phylogenetic Diversity of Environmental Libraries

To evaluate the microbial diversity represented by the DNA used to construct our soil libraries, we carried out phylogenetic analyses of DNA sequences encoding small subunit ribosomal RNA (SSU rRNA) *(10)*.

Typical results of these studies (**Fig. 3**) are consistent with a growing volume of work that documents the diversity found in DNA extracted from various soils *(25,27,34,35)*. Such studies consistently uncover a wide representation of sequences that could be placed within known families isolated from all over the world, but a large number are from previously unidentified bacterial families. In another study, 47 rRNA

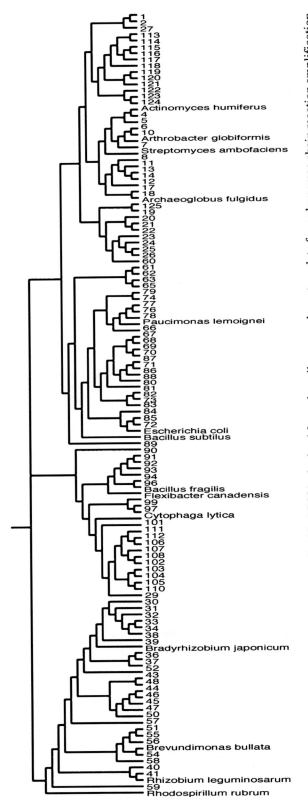

Fig. 3. Phylogenetic analysis of soil DNA. Purified DNA obtained from the soil was used as a template for polymerase chain reaction amplification. For the construction of the phylogenetic tree, alignment of the first 417 nucleotides (5') from 111 independent clones and known 16S rRNA sequences (obtained from Ribosomal Database Project [RDP] at the University of Illinois) was accomplished using ClustalX (freeware). Phylogenetic analysis was performed using MacClade (*48*). The tree was inferred by neighbor-joining analysis of 437 homologous positions.

gene sequences were amplified directly from environmental DNA library clones *(26)*. Analysis of these sequences showed that the majority were from Proteobacteria, that all 47 sequences were unique, and that they were derived from phylogenetically diverse microorganisms, most of which have never been screened or isolated. The significant finding here was that the diversity seen previously in soil DNA analyses was carried over to large pieces of DNA actually extracted from soil, and, importantly, that this diversity was then effectively captured for screening in an environmental DNA library.

3. New Streptomyces and Pseudomonas Expression Host Strains

Because different microbial strains and species have differing expression capabilities, the ability to express environmental libraries in a variety of host strains in order to improve the chances of detecting novel natural products in screening assays is very advantageous. *E. coli* was the obvious choice as a primary host, as many genetic tools are available and it is easy to manipulate. However, the fact cannot be ignored that many known (and useful) natural products have been derived from naturally occurring polyketides. Actinomycetes are a major source of useful natural products, including polyketides and nonribosomal peptides. *E. coli* does not express functional polyketide synthases and nonribosomal protein synthases efficiently, at least partially because it lacks the phosphopantetheinyl transferases required for their activation. For these reasons, and because the expression of heterologous gene clusters and polyketide genes is well established in Streptomyces *(36)*, we chose to develop *S. lividans* as an additional host strain for environmental library expression. Optimally, host expression strains should lack endogenous activities, including pigments, as their synthesis may potentially waste available metabolites and therefore interfere with the expression of heterologous compounds. Additionally, endogenous activities could obscure the detection of novel compounds that are produced from heterologous DNA. The *act* and *red* gene clusters in *S. lividans* encode the pigments actinorhodin and various undecylprodiginines, respectively. We deleted these clusters from the *S. lividans* chromosome by positive selection of unmarked allelic exchange mutants *(31)*. The resulting Δ*act* cluster deletion encompassed 24.2 Kb (nucleotides 143959 to 168217, GeneBank accession number SCO939122), and the Δ*red* cluster deletion encompassed 28.6 Kb (nucleotides 144623 to 173286, GeneBank accession number SCO939125). The deletions were verified by polymerase chain reaction (PCR), and the new *S. lividans* host strain, Δ*act*Δ*red*, no longer produced actinorhodin or undecylprodiginine. *S. lividans* Δ*act*Δ*red* was shown to grow and sporulate as well as the TK24 parent strain, and thus provides the desired background for heterologous natural-product expression and analysis *(31)*. An additional improvement to the *S. lividans* screening system encompasses an environmental DNA sequence found to confer early and enhanced expression of secondary metabolites in *S. lividans* (A. Martinez, manuscript in preparation), thus improving our ability to detect heterologous expression in this host. We developed an additional host to further extend the host range for expression of environmental libraries, the soil organism *Pseudomonas putida* KT-2440 *(37)*. Many genetic tools are available for pseudomonads, including plasmid DNA transfer by conjugation. Pseudomonads are known to colonize soil, water, and other diverse environments. They encompass rich metabolic diversity, including the ability to degrade xenobiotics and to

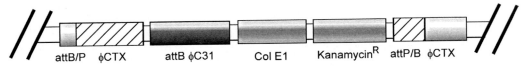

Fig. 4. Expression host *Pseudomonas putida* MBD1. The *P. putida* expression host contains the *Streptomyces* phage φC31 *attB* site, enabling integration of BAC plasmids encoding the φC31 *attP* site and *int* gene.

produce secondary metabolites, including polyketides *(37,38)*. Our *E. coli–Streptomyces* pMBD14-based shuttle BAC vectors and libraries can be transferred by conjugation into this well-characterized nonpathogenic *P. putida* strain, and we have engineered the strain to integrate the incoming vectors into the chromosome by making use of the integration properties of the φC31 (*Streptomyces* phage) system. The *attB* site for φC31 integration was inserted into the chromosome of *P. putida* by means of the site-specific integration system of *P. aeruginosa* phage φCTX *(31)*, thus giving rise to expression strain *P. putida* MBD1 (**Fig. 4**). Integration of incoming pMBD14 library plasmids into the chromosome of *P. putida* MBD1 is predicated on the ability of the φC31 integrase gene, present on the pMBD14 vector, to be expressed in the *P. putida* host strain. This was found to be the case, as discussed later. Note that crude methanol extracts of *P. putida* MBD1 did not produce detectable endogenous antibacterial or antifungal activities. Thus our two new host strains, *S. lividans* ΔactΔred and *P. putida* MBD1, provide an apparently "clean" background for production and detection of heterologous compounds.

4. Transfer and Expression of Recombinant pMBD14 Plasmids

As described previously, we developed our shuttle vector and expression host systems such that a recombinant vector library encoding environmental DNA inserts could be transferred from *E. coli* to *S. lividans* and *P. putida* via conjugation. Especially for *S. lividans*, alternative genetic transfer systems, such as protoplast transformation, are laborious and somewhat inconsistent, and the ultimate success of a molecular biodiversity screening program depends upon the ability to screen in various expression host strains in a high-throughput fashion. Conjugation can be carried out by spotting cultures on agar Petri plates, and is therefore amenable to a high-throughput robotic procedure. We found that *E. coli* donor strains efficiently transferred recombinant pMBD14 plasmids containing inserts of up to 85 kb (the largest size tested) into *S. lividans*, significantly larger than any reported previously, and recombinant plasmids successfully integrated into the *S. lividans* chromosome *(31)*. Despite the finding that some of the larger integrated inserts contained small deletions at one integration junction, most of the environmental insert DNA was present in the chromosome of the exconjugants, and provided long contiguous stretches of environmental DNA available for expression in the new host. Similar analyses showed that these pMBD14 recombinant plasmids can be transferred into *P. putida* MBD1 and maintained stably in the chromosome; conjugation resulted in stable apramycin-resistant *P. putida* MBD1 colonies, confirming that the φC31 integrase gene and the apramycin-resistance gene encoded on pMBD14 were expressed and functional in *P. putida* (31). After confirm-

Table 1
Expression[a] of Heterologous Compounds by Bacterial Expression Hosts

	Escherichia coli DH10B	*Streptomyces lividans* ΔactΔred	*Pseudomonas putida* MBD1
MG1.1	+	−	−
Granaticin	−	+	−
DAPG	−	−	+

[a]Expression of compounds was measured in cell extracts by high-performance liquid chromatography analysis and by antibacterial and antifungal assays, as described in the text.

ing the feasibility of conjugating recombinant vectors containing large environmental DNA inserts from *E. coli* to both *S. lividans* and *P. putida* and ascertaining that the recombinant vectors were integrated into the respective chromosomes, we developed a plate conjugation method, using a 96-pin array, thus enabling the high-throughput transfer of recombinant plasmids into the new expression hosts (31). This method for transferring pMBD14-based large-insert libraries from *E. coli* to *S. lividans* and *P. putida* resulted in a 95% success rate for conjugation and is the first example of BAC vectors that can be shuttled by conjugation from *E. coli* to both *Streptomyces* and *Pseudomonas*. Expression of several heterologous gene clusters present on the pMBD14 vector was tested in the various hosts using a series of constructs (pSGran, pSMG1.1, and pSDAPG [31]) encoding the synthesis of known antibiotics (granaticin, indirubin, and the antifungal compound 2,4-diacetylphloroglucinol [DAPG]). Cells containing these plasmids were grown, extracted with ethyl acetate, and the extracts analyzed. Results showed that each of the hosts expressed a different cluster, but was unable to express the other two (**Table 1**), as measured by antibacterial or antifungal assays, or by high-performance liquid chromatography (HPLC) analysis of extracts. These results call attention to the necessity of screening in multiple expression hosts, and underscore the critical importance of the three-way conjugative shuttle BAC vector. Further, our findings point out the advisability of screening in as many hosts as feasible, thereby increasing the chances of detecting the expression of molecules of interest.

5. Screening Recombinant Libraries

Recombinant environmental DNA libraries can be screened for encoded activities by a variety of means that encompass two fundamental strategies: expression of encoded activities by host strains, and detection of DNA homologies to known genes of interest. Expressed activities can be assayed in live cells (colony screens) or in cell extracts. Cell extracts can in turn be assayed for activities in targeted assays or by LC/mass spectrometry (MS) analysis, which allows for the rapid dereplication of known natural products and can detect the presence of new molecules not found in the host background. When "hits" (i.e., clones that are positive in an initial screen) are identified, they are retested at least once. The corresponding BAC plasmid is then isolated and introduced into a fresh host cell and the phenotype rechecked, in order to verify that the environmental DNA insert in the BAC plasmid is the source of the phenotype. Once confirmed, transposon-mediated insertion mutagenesis is used to disrupt the encoding environmental genes. Cells containing transposon insertions are screened for loss of the acquired phenotype *(25)*. The encoding environmental DNA can then be sequenced from the transposon ends in order to determine the sequence of the gene(s) of interest. The resulting DNA sequence can then be analyzed using bioinformatics tools, which can help identify function, novelty, and phylogeny of the genes and expressed products. Our initial screening efforts (and those reported by others) have concentrated principally on screening *E. coli* libraries. This is because our major focus was first the proof of concept for molecular biodiversity technology, followed by the development of technological improvements aimed at incorporating these first principles into a realistic and high-throughput drug discovery program. However, despite the fact that *E. coli* is not an ideal expression host for reasons already discussed, a number of interesting genes and activities have been discovered in *E. coli* environmental libraries, lending much encouragement to the enormous potential for this emerging technology.

5.1. E. coli *Colony Screens*

We, as well as other workers in the field, have screened *E. coli* colonies containing environmental DNA. Colonies were grown on agar Petri dishes and screened for activities that could be assessed easily using plate assays, including antibacterial (against *Bacillus subtilis* or *Staphylococcus aureus*) and antifungal (against *Candida albicans*) activities. Early libraries were also assayed for enzyme activities in agar plate screens, in order to provide data addressing the frequency of appearance and expression of heterologous genes *(25,26,34)*. These screens can normally be carried out robotically, using large (9" × 9") agar trays onto which colonies are transferred using a 96-pin array. Individual screens make use of assay-specific agar, which is often commercially available in the case of enzyme assays. For antibacterial screening, colonies are normally grown for several days, usually at 37°C for 1 d and at room temperature for some additional days, and overlaid with soft agar containing an indicator strain (e.g., *B. subtilis* or other Gram-positive bacterium). After further incubation, plates are scored for activity by looking for a zone of inhibition in the *B. subtilis* lawn surrounding an environmental library colony. Antifungal screening follows a similar process *(31)*. The choice of incubation temperatures for environmental libraries is an interesting issue, since *E. coli*'s optimum temperature of 37°C is not necessarily (or likely to be) optimal for expression of a particular heterologous gene derived from a soil environment. Therefore, it is beneficial to screen library colonies after growing for various amounts of time and at several temperatures.

5.2. HPLC Screens

An additional method of screening for expression of heterologous DNA in environmental libraries involves HPLC or LC/MS analysis of cultured library clones, aimed at detecting and characterizing new metabolites that are produced by the cells and appear in cell extracts. Peaks that do not match those found in the host strain bearing the vector alone potentially reveal the presence of homologous compounds not present in the host. Although cumbersome, the method can be carried out in a fairly high-throughput fashion, and has succeeded in detecting novel compounds in environmental libraries

Table 2
Some Molecular Structures Isolated From *E. Coli* (Soil) Environmental Libraries

Library Type	Number of clones	Average insert size/range	DNA origin	Type of screen[a]	Number of hits	Ref.	Structure
E. coli BAC	3648	27 kb	soil Madison, WI	antibacterial, colony screen	11 enzymes[b] 1 antibacterial	33	Amylase, lipase, DNase enzymes. Antibacterial likely to be a membrane protein
E. coli BAC	12,000	5-125kb	soil Lexington, MA	antibacterial, colony screen	4 antibacterial	24	1. family of compounds related to indirubin; 2. homogentisic acid; 3., 4. unknown structures
E. coli shuttle cosmid	5000	30-60kb	soil Isere, France	HPLC	2-6 new structures	25	$CH_3(CH_2)_x$...$(CH_2)_y CH_3$, OH; $CH_3(CH_2)_x$...$(CH_2)_y CH_3$, HO; $x+y=12$
E. coli shuttle cosmid	5000	30-60kb	soil Isere, France	PKS 1 homology	11	25	11 distinct clones, including 1 partial pathway
S. lividans	1020	c	c	HPLC	18	38	Terragines: family of 5 compounds (A-E)
E. coli BAC	24,576	44.5 kb	soil Madison, WI	Pigmented colonies	3	39	Turbomycin family of antibacterial molecules
E. coli cosmid	700,000	c	c	Antibacterial Colony screen	65	40	3 families of novel antibacterial long-chain acyl phenols

[a] Environmental library screens are described in the text.
[b] 8 amylase, 2 lipase, and 1 DNase activities were detected, all distinct and novel (33). [c] Information not available.

(**Table 2**) *(26,39)*. In one such library *(26)*, five novel structures were detected in a library of 5000 clones, each clone containing about 40 kb of insert DNA. However, none of these novel structures had a detectable activity, and therefore an estimate of the number of colonies needed to be screened by this method in order to obtain an interesting molecule cannot yet be derived.

5.3. Environmental DNA Homology Screens (Molecular Screens)

Environmental DNA encoded in library clones can also be screened for homologies to known genes of functional relevance. The advantage of this type of screen is that detectable expression of the encoding DNA is not required in a particular host strain. A disadvantage, on the other hand, is that the screen is restricted to known classes of molecules. One method of screening for DNA homologies involves amplifying specific biosynthetic gene sequences of interest from recombinant vectors. For example, polyketides, a vast group of structurally diverse natural compounds produced by a large variety of soil microorganisms, are a major class of natural products. The existence of highly conserved regions of actinomycetes type I polyketide synthases (PKS 1) DNA, flanking the active site of the keto acyl synthetase domain, provided two sets of primer sequences that could be used for the PCR amplification of homologous genes, and thus furnished a means for gauging the frequency of appearance of genes encoding polyketide biosynthetic enzymes in environmental libraries (26).

5.4. Results of Screening Environmental Libraries

Much of the work describing the potential for drug discovery from environmental DNA libraries by these and similar means has not yet been published; therefore the list of molecules that have been discovered via molecular biodiversity screening is very incomplete. However, a partial list of molecules, derived from the work of several different laboratories, is shown in **Table 2**. Again, when considering these results it is important to keep in mind that the majority of published screening results describe work only with *E. coli* as the expression host. Extensive multi-host large-scale screening of environmental libraries using our three-way shuttle BAC technologies is a promising way forward but has only just begun. Nevertheless, available screening results provide very useful data and firmly prove the concept that novel and interesting molecules can be found in environmental libraries by the means described here, even when such libraries are expressed in a less-than-ideal expression host. **Table 2** also shows that activities derived from *E. coli* libraries frequently consist of families of small molecules, rather than single discrete compounds. These compounds or families are encoded by single genes, or gene clusters ranging from 3 to 13 genes (25,41).

Although some screening has been carried out in streptomycetes, this has so far been on a small scale, as the cumbersome transformation procedure previously required to transfer plasmids into streptomycetes has precluded a genuinely high-throughput effort. The three-way shuttle BAC vector pMBD14, which permits high-throughput conjugation from *E. coli* to *S. lividans*, will greatly enhance the ability to screen libraries in the *S. lividans* host. Very encouragingly, DNA homology screens to determine the prevalence of PKS I genes in our cosmid environmental library revealed 11 distinct type I PKS gene sequences, a number much higher than expected in a random and relatively small (<250 Mb) DNA library sample (26). Additionally, the identification of a partial PKS 1 pathway with nearly six genes in one clone of this library strongly suggests that complete clusters of polyketide or other biosynthetic genes would be expected to appear in a library containing more clones and larger inserts. Our screening focus has been primarily on the detection of heterologous small molecules, consistent with our goal of augmenting chemical libraries. We have also screened for enzyme activities as a means of gauging the frequency of appearance and expression of heterologous sequences in our environmental libraries. Note that because enzymes are normally encoded by a single gene rather than by large multigene pathways, their discovery requires much smaller environmental DNA inserts.

6. Technologies for Generating Quality Environmental Libraries for High-Throughput Screening

The data in **Table 2** give us a rough idea of the frequency with which heterologous molecules might be detected in *E. coli* environmental libraries, and lend credence to the concept that this type of technology is feasible for drug discovery. Note that one *E. coli* BAC library generated a hit rate for antibacterial activities of roughly one antibacterial clone per 60 MB of soil-derived DNA *(25)*, although this number is expected to vary depending upon insert size and many other factors. Natural-product screens directed against a variety of targets usually require the preparation and screening of cell extracts, however. For libraries with hit rates similar to those described above, most of the extracts derived from library clones would be unlikely to contain novel molecules, let alone molecules active against specific targets of interest. For this reason, more recent efforts of our group have been directed toward generating quality libraries, in which every (or nearly every) clone is known to contain a heterologous product that can be screened productively in an assay. To this end, we have replaced the concept of screening static environmental libraries with a more dynamic approach that involves the continuous generation and prescreening of environmental DNA clones to select out those that are producing heterologous products. The goal is to generate an efficient high-throughput method for identifying and separating out those cells in environmental libraries that encode heterologous bioactive natural products. Prescreens thus serve as a filter to isolate producing clones. We have begun to develop reporter strains of *E. coli* for use in prescreening. The method (**Fig. 5**) utilizes a battery of *E. coli* strains that have been engineered to encode, in single chromosomal copy, chemoresponsive promoters that respond to particular classes of metabolites, fused to a reporter gene (encoding green fluorescence protein, GFP). Bacterial stress-response genes and multidrug efflux pumps are known to be activated in response to certain chemicals, and therefore a cell producing a compound expressed from heterologous DNA might be expected to be induced for a stress and/or efflux response, resulting in GFP production *(42,43)*. For example, certain classes of antibiotics are known to produce characteristic stress responses *(44)*. The GFP prescreen can be carried out with a multitude of different stress and efflux pump fusion strains, many of which we have already constructed (Ian MacNeil, unpublished results). An additional attractive feature is the very high throughput nature of this prescreen: cells producing GFP can be sorted by flow cytometry and then isolated *(45)*, as diagrammed in **Fig. 5**. Clearly this potentially powerful process enables rapid sorting of producing clones from a background of millions of nonproducing clones. Simple prescreens for *S. lividans* have also begun to be developed (e.g., overproduction of actinorhodin in response to the production of a heterologous molecule *(26)*), and should further facilitate the ability to generate quality libraries from a continuous supply of millions of environmental DNA clones. Other prescreens for all hosts comprise high-throughput (robotic) antibacterial and antifungal colony or extract screens, as already described. The availability of multiple means for prescreening millions of clones is likely to be a critical step in deriving value from the shotgun approach to natural-product discovery from environmental microbes.

In addition to developing effective prescreens for generating quality environmental libraries, it is clearly advantageous to incorporate into the process as many expression

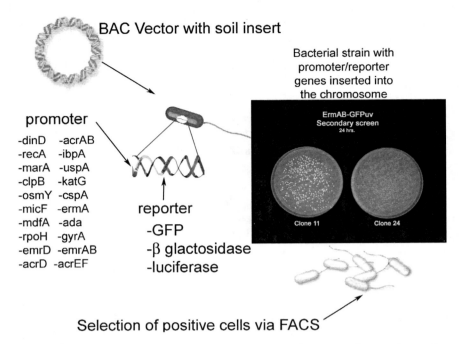

Fig. 5. *Escherichia coli* prescreening strains. A battery of prescreening strains each contains a stress or efflux promoter fused to a green fluorescence protein reporter gene, integrated into the *E. coli* chromosome.

hosts with differing expression capabilities as possible. For example, *Myxcoccus xanthus* is an excellent candidate. Myxobacteria are widely distributed in soils *(46)* and are know to synthesize many natural products. More than 80 basic structures have been isolated from myxobacteria, including many that are novel *(47)*. The genome of *M. xanthus* is nearly twice the size of that of *E. coli*. As a large percentage of its genome appears to be devoted to secondary metabolite biosynthesis, it is reasonable to expect that this host would have many of the tools required to express environmental DNA and possibly modify the expressed product in a variety of ways. Furthermore, genetic tools are available for *M. xanthus*, including transduction and conjugation *(48)*, suggesting that its genome could also be engineered to accept the three-way shuttle BAC plasmid. In addition to bacteria, lower eukaryotic hosts, such as fungi, for which genetic tools are also available, are attractive possibilities for expression hosts.

With the technologies already developed, we have set up a process for generating and prescreening environmental libraries, diagrammed in **Fig. 6**. According to this process, thousands of environmental library clones are generated weekly and transformed into *E. coli*. The clones are then prescreened in a variety of ways, including reporter stress/efflux pump and antibacterial/antifungal screens. Positive clones, once validated, are added to the high-quality library, and are also conjugated into *S. lividans* and *P. putida*. In addition, random *E. coli* clones that are not positive in the *E. coli* prescreen would also be conjugated into the other host strains for prescreening. Confirmed positives from at least one of the prescreens in one of the hosts would comprise the high-quality environmental library, and organic extracts of these clones would be

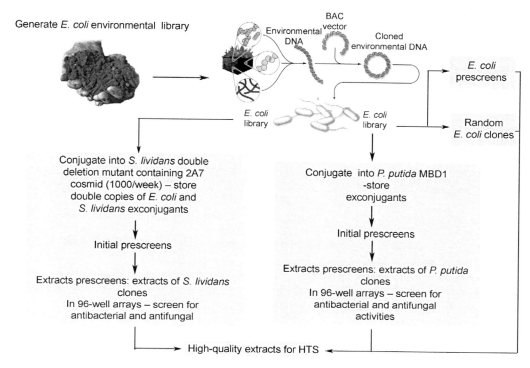

Fig. 6. High-throughput screening process for expression of environmental libraries. The goal of this process is to screen millions of environmental clones quickly and efficiently in order to identify those clones producing heterologous molecules. Producing clones are then selected to be included in a "quality" environmental library, which is then screened as cell extracts against targets of interest.

prepared for high-throughput screening against targets of interest. This screening scheme is just one of many possibilities for generating high-quality environmental extract libraries. Note that the fluid nature of molecular biodiversity technology permits its use in nearly any type of screening program, since prescreens can be specifically tailored toward targets of interest.

7. Conclusions

Screening as yet uncultured microorganisms for novel and useful natural products by means of shotgun cloning environmental DNA is an exciting new technology that has the potential to become a powerful tool for drug discovery. We and others have confirmed that genes encoding natural products can be readily captured using this strategy, and that heterologous host strains can be optimized to express environmental libraries. In addition, conjugative transfer of BAC plasmids containing environmental DNA makes feasible the transfer and screening of entire large-insert DNA libraries in multiple host expression strains. It is our view that in order to maximize the chances of discovery, environmental libraries need to be generated and screened in a high-throughput fashion, using as many different expression hosts as practical, thus greatly increasing the chances for success. At this point the technology has succeeded in the proof of concept stage, and is poised to move forward rapidly. The next important step to a productive drug-discovery program, in which this potentially vast source of natural products derived from uncultivated environmental microorganisms can be fully exploited, is critically dependent upon a firm commitment and adequate funding.

Acknowledgments

We are especially grateful to Kara Brown, Steve Kolvek, and Choi Lai Tiong-Yip for their very large contributions to this work. We also thank all of our collaborators, including K. Loiacono, B. Lynch, T. Grossman, M. Cappellano, S. Courtois, P. Jeannin, G. Helynck, P. August, J.-L. Pernodet, C. Ribard, P. Simonet, M. Rondon, J. Handelsman, and B. Goodman.

References

1. Harvey A. Strategies for discovering drugs from previously unexplored natural products. Drug Discovery Today 2000;5:294–300.
2. Dixon B. In: Power Unseen: How Microbes Rule the World. New York: W.H. Freeman, 1994.
3. Cragg GM, Newman DJ. Antineoplastic agents from natural sources: achievements and future directions. Expert Opin Investig Drugs 2000;9:2783–2797.
4. Clark AM. Natural products as a resource for new drugs. Pharm Res 1996;13:1133–1144.
5. Morris RE, Wu J, Shorthouse R. A study of the contrasting effects of cyclosporine, FK 506, and rapamycin on the suppression of allograft rejection. Transplant Proc 1990;22:1638–1641.
6. Newman DJ, Cragg GM, Snader KM. Natural products as sources of new drugs over the period 1981–2002. J Nat Prod 2003;66:1022–1037.
7. Silver L, Bostian K. Screening of natural products for antimicrobial agents. Eur J Clin Microbiol Infect Dis 1990;9:455–461.
8. Silver LL. Novel inhibitors of bacterial cell wall synthesis. Curr Opin Microbiol 2003;6:431–438.
9. Kashefi K, Lovely DR. Extending the upper temperature limit for life. Science 2003;301:934.
10. Pace NR. A molecular view of microbial diversity and the biosphere. Science 1997;276:734–740.
11. Stahl DA. The natural history of microorganisms. ASM News 1993;59:609–661.
12. Stahl DA, Lane DJ, Olsen GJ, Pace NR. Characterization of a Yellowstone hot spring microbial community by 5S rRNA sequences. Appl Environ Microbiol 1985;49:1379–1384.
13. Suzuki MT, Rappe MS, Haimberger ZW, et al. Bacterial diversity among small subunit rRNA gene clones and cellular isolates from the same seawater sample. Environ Microbiol 1997;63:983–989.
14. Ward DM, Weller R, Bateson MM. 16S rRNA sequences reveal numerous uncultured microorganisms in a natural community. Nature 1990;345:63–65.
15. Bintrim SB, Donohue TJ, Handelsman J, Roberts GP, Goodman RM. Molecular phylogeny of Archaea from soil. Proc Natl Acad Sci USA 1998;94:277–282.
16. Hugenholtz P, Goebel BM, Pace NR. Impact of culture independent studies on the emerging phylogenetic view of bacterial diversity. J Bacteriol 1998;180:4765–4774.
17. Bull AT, Ward AC, Goodfellow M. Search and discovery strategies for biotechnology: the paradigm shift. Microbiol Mol Biol Rev 2000;64:573–606.
18. Griffiths BS, Ritz K, Glover LA. Broad-scale approaches to the determination of soil microbial community structure:application of the community DNA hybridization technique. Microbial Ecol 1996;31:269–280.
19. Torsvik V, Salte K, Sørheim R, Goksøyr J. Comparison of phenotypic diversity and DNA heterogeneity in a population of soil bacteria. Appl Environ Microbiol 1990;56:776–781.

20. Whitman WB, Coleman DC, Wiebe WJ. Prokaryotes: the unseen majority. Proc Natl Acad Sci USA 1998;95:6578–6583.
21. Kaeberlein T, Lewis K, Epstein SS. Isolating "uncultivable" microorganisms in pure culture in a simulated natural environment. Science 2002;296:1127–1129.
22. Donadio S, Staver MJ, McAlpine JB, Swanson SJ, Katz L. Modular organization of genes required for complex polyketide biosynthesis. Science 1991;252:675–679.
23. Krause M, Marahiel MA. Organization of the biosynthesis genes for the peptide antibiotic gramicidin S. J. Bacteriol 1988;170:4669–4674.
24. Schwecke T, Aparicio JF, Molnár I, et al. The biosynthetic gene cluster for the polyketide immunosuppressant rapamycin. Proc Natl Acad Sci USA 1995;92:7839–7843.
25. MacNeil IA, Tiong CL, Minor C, et al. Expression and isolation of antimicrobial small molecules from soil DNA libraries. J Mol Microbiol Biotechnol 2001;3:301–308.
26. Courtois S, Cappellano CM, Ball M, et al. Recombinant environmental libraries provide access to microbial diversity for drug discovery from natural products. Appl Environ Microbiol 2003;69:49–55.
27. Courtois S, Frostegard A, Goransson P, Depret G, Jeannin P, Simonet P. Quantification of bacterial subgroups in soil: comparison of DNA extracted directly from soil or from cells previously released by density gradient centrifugation. Environ Microbiol 2001;3:431–439.
28. Bakken LR, Lindahl V. Recovery of bacterial cells from soil. In: van Elsas JD, Trevors JT (eds), Nucleic Acids in the Environment: Methods and Applications. Springer Verlag, Berlin: 1995; pp. 9–27.
29. Cai L, Taylor JF, Wing RA, Gallagher DS, Woo SS, Davis SK. Construction and characterization of a bovine bacterial artificial chromosome library. Genomics 1995;29:413–425.
30. Stone NE, Fan JB, Willour V, et al. Construction of a 750-kb bacterial clone contig and restriction map in the region of human chromosome 21 containing the progressive myoclonus epilepsy gene. Genome Res 1996;6:218–225.
31. Martinez A, Kolvek SJ, Tiong Yip CL, et al. Genetically modified bacterial strains and novel shuttle BAC vectors for constructing environmental libraries and detecting heterologous natural products in multiple expression hosts. Appl Environ Microbiol 2004;70(4):2452–2463.
32. Smokvina TP, Mazodier P, Boccard F, Thompson CJ, Guerineau M. Construction of a series of pSAM2-based integrative vectors for use in actinomycetes. Gene 1990;94:53–59.
33. Blondelet-Rouault MH, Weiser J, Lebrihi A, Branny P, Pernodet JL. Antibiotic resistance gene cassettes derived from the omega interposon for use in *E. coli* and *Streptomyces* Gene 1997;190:315–317.
34. Rondon MR, August PR, Bettermann AD, et al. Cloning the soil metagenome: a strategy for accessing the genetic and functional diversity of uncultured microorganisms. Appl Environ Microbiol 2000;66:2541–2547.
35. Rondon MR, Goodman RM, Handelsman J. The Earth's bounty: assessing and accessing soil microbial diversity. Trends Biotechnol 1999;17:403–409.
36. Kieser TMJ, Bibb M, Buttner J, Chater KF, Hopwood DA. Practical Streptomyces Genetics. John Innes Foundation, Norwich: 2000.
37. Timmis KN. *Pseudomonas putida*: a cosmopolitan opportunist par excellence. Environ Microbiol 2002;4:779–781.
38. Bender, C., V. Rangaswamy, and J. Loper. Polyketide production by plant-associated Pseudomonads. Annu Rev Phytopathol 1999;37:175–196.
39. Wang GYS, Graziani E, Waters B, et al. Novel natural products from soil DNA libraries in a streptomycete host. Organic Lett 2000;2:2401–2404.
40. Gillespie DE, Brady SF, Bettermann AD, et al. Isolation of antibiotics turbomycin A and B from a metagenomic library of soil microbial DNA. Appl Environ Microbiol 2002;68:4301–4306.

41. Brady SF, Chao CJ, Clardy J. New natural product families from an environmental DNA (eDNA) gene cluster. J Am Chem Soc 2002;124:9968–9969.
42. Putman M, van Veen HW, Konings WN. Molecular properties of bacterial multidrug transporters. Microbiol Mol Biol Rev 2000;64:672–693.
43. Nishino K, Yamaguchi A. Analysis of a complete library of putative drug transporter genes in *Escherichia coli*. J Bacteriol 2001;183:5803–5812.
44. Bianchi AA, Baneyx F. Stress responses as a tool to detect and characterize the mode of action of antibacterial agents. Appl Environ Microbiol 1999;65:5023–5027.
45. Valdivia RH, Falkow S. Bacterial genetics by flow cytometry: rapid isolation of *Salmonella typhimurium* acid-inducible promoters by differential fluorescence induction. Mol Microbiol 1996;22:367–378.
46. Dawid W. Biology and global distribution of myxobacteria in soils. FEMS Microbiol Rev 2000;24:403–427.
47. Reichenbach H. Myxobacteria, producers of novel bioactive substances. J Ind Microbiol Biotechnol 2001;27(3):149–156.
48. Kaiser D. Genetic systems in myxobacteria. Methods Enzymol 1991;204:357–372.
49. Maddison WP, Maddison DR. MacClade: Analysis of phylogeny and character evolution Version 3.08. Sunderland, Massachusetts: Sinauer Associates, 1999.

PART V
SPECIFIC SOURCES

14

New Natural-Product Diversity From Marine Actinomycetes

Paul R. Jensen and William Fenical

Summary

There is currently renewed interest in the study of microorganisms as a source of structurally unique and pharmacologically active natural products. Actinomycetes, being the single most productive source of naturally occurring antibiotics, are a logical component of these studies, and success with this group will be enhanced by the inclusion of previously unknown taxa. Recent studies of marine-derived actinomycetes have revealed the widespread distribution of unique marine taxa residing in ocean sediments. Chemical studies of these strains, focusing on members of the new genus *Salinospora*, have led to a high rate of novel secondary metabolite discovery, including molecules with potent biological activity. Given the encouraging results from preliminary studies of these newly described marine bacteria, it seems clear that marine actinomycetes represent an important future resource for small-molecule drug discovery.

Key Words: Marine actinomycetes; natural products; secondary metabolites; phylogenetic diversity.

1. Introduction

Members of the order Actinomycetales include slow-growing, filamentous bacteria that are abundant in soil, where they play important roles in the breakdown of recalcitrant organic materials *(1)*. Historically, these Gram-positive bacteria are best known for their unparalleled ability to produce structurally diverse secondary metabolites, a capacity that made actinomycetes a central component in the development of the modern pharmaceutical industry. Their role as a source of useful pharmaceuticals is highlighted by the remarkable fact that, as of 1988, actinomycetes accounted for approximately two-thirds of the naturally derived antibiotics discovered *(2)*, clearly making them the single most important source of naturally occurring medicines.

Although the positive impact of actinomycete secondary metabolites on human health is clear, there is a perception that 50 yr of intensive research by the pharmaceutical industry has exhausted the supply of compounds that can be discovered from this group of bacteria. This perception has been a driving force behind the recent shift away from natural products as a resource for the discovery of small-molecule therapeutics towards other drug-discovery platforms, including high-throughput combinatorial synthesis and rational drug design. These new technologies, however, have proven to be less productive at yielding quality leads de novo (as opposed to, e.g., combinatorial

From: *Natural Products: Drug Discovery and Therapeutic Medicine*
Edited by: L. Zhang and A. L. Demain © Humana Press Inc., Totowa, NJ

synthesis as a method for modifying a pre-existing chemical scaffold), and as a result there has been a recent groundswell of renewed interest, largely from the biotechnology sector, in returning to nature as a source of new structural motifs.

So the question can be asked, is it logical to reinvestigate actinomycetes as a source of natural products given the intensity with which these bacteria were studied in the final five decades of the twentieth century? In other words, are the returns that can be anticipated from these bacteria so skewed towards the isolation of known chemical scaffolds that the effort required to discover the few remaining new molecules cannot be justified? That is, has the chemical potential of the actinomycetes been exhausted? In the case of *Streptomyces*, the genus from which the largest number of antibiotics have been discovered *(3)*, it was recently estimated that only 3% of the compounds produced by this taxon have been characterized *(4)*, thus implying that further studies, even of this relatively well-characterized genus, are warranted. Given that improved analytical techniques now allow for the rapid dereplication of known compounds and the detection and characterization of molecules that occur at very low concentrations, the search for new natural products can be carried out with greatly enhanced efficiency. When coupled with high-throughput screening, target-specific assays, improved cultivation techniques, functional genomics, and our growing understanding of the molecular basis of secondary metabolite production, there can be little doubt that many promising actinomycete secondary metabolites await discovery. In fact, we may now be entering a new era in which these technologies are creating a "paradigm shift" in our approach to natural-product discovery *(5)*, a shift with the potential to result in a rate of pharmaceutical lead discovery that will eclipse even that which occurred during the "Golden Age" of the pharmaceutical industry.

There are many approaches that can be applied to the discovery of novel actinomycete secondary metabolites, and certainly high among these is the search for new actinomycete taxa. More than 20 yr ago, as success rates from streptomycetes dropped, efforts to explore the ability of nonstreptomycete actinomycetes, such as *Micromonospora* sp., to produce unique metabolites proved invaluable *(6)*. Today, however, *Micromonospora* is no longer underexplored, relatively speaking, and a different set of "rare" genera, such as *Actinokineospora*, are garnering attention for their chemical prolificacy *(7)*. Thus, the strategy of seeking out new actinomycete taxa remains a valid approach to drug discovery, with questions addressing where these taxa occur and what is required for their cultivation becoming key to the success of future natural-product discovery programs.

Being at an oceanographic institution, we have long been interested in the concept of the existence of marine actinomycetes and the potential of these bacteria as a source of new secondary metabolites (e.g., refs. *8,9*). Although a case for the existence of indigenous marine actinomycete populations was made more than 30 yr ago *(10)*, the occurrence of marine actinomycetes has not been widely embraced either by the marine microbiology community or industrial microbiologists, and a prevailing sentiment expressed in the literature suggests that the vast majority of marine-derived strains are of terrestrial origin (e.g., ref. *11*). Given that many soil actinomycetes produce resistant spores that are washed in large numbers from shore into the sea, where they have the potential to remain viable yet dormant for an undetermined period of time, this skepticism was not without basis. However, evidence in support of the existence of marine actinomycetes has been accumulating with the discovery that some strains display specific marine adaptations *(12)* while others appear to be metabolically active in marine sediments *(13)*. Further support came from the description of *Rhodococcus marinonascens*, the first marine actinomycete species to be characterized *(14)*. This prior evidence in now further corroborated by the recent discovery of a new marine actinomycete taxon that is widely distributed in ocean sediments *(15)*.

This chapter is not an attempt to review the approx 100 novel compounds that have been reported from marine-derived actinomycetes over the last 20 yr, as much of this information has been summarized elsewhere (e.g., refs. *16,17* and references cited therein). We will instead focus on recent advances in our understanding of marine actinomycete diversity and the secondary metabolites that have been discovered from strains that either represent new marine taxa or are clearly adapted to life in the ocean.

2. Studies of Marine Actinomycetes

As part of a program to explore marine actinomycetes as a source of novel secondary metabolites, we have been studying actinomycete populations that occur in ocean sediments. These studies have incorporated modern concepts in natural-product isolation and structure elucidation along with new cultivation methods and the application of molecular techniques to assess both culture-dependent and culture-independent microbial diversity. Although actinomycetes can readily be isolated from marine sediments if the appropriate selective techniques are applied, it is only through the use of molecular phylogenetics that we have begun to get a clear picture of the diversity, distributions, and taxonomic uniqueness of certain populations. Once recognized as unique, these populations can then be selected as the focal point of intensive natural-product studies.

Using the SSU rRNA gene to trace evolutionary relationships among actinomycetes, it has become increasingly clear that the oceans harbor phylogenetically unique actinomycete populations and that these bacteria can be readily obtained in culture and studied as a source of new natural products. More importantly, from what can only be considered preliminary studies, these bacteria have been found to produce a remarkable range of structurally unique secondary metabolites. This finding suggests that adaptations to life in the sea, coupled with genetic isolation from terrestrial strains, has led not only to the evolution of new actinomycete species that are highly adapted to life in the ocean but to the production of secondary metabolites that are unlikely to be discovered from nonmarine strains.

2.1. Salinospora

Our first glimpse into the phylogenetic diversity of marine actinomycetes came following a study of sediment populations cultured from samples collected from the Bahamas *(15)*. Many of these bacteria required seawater for growth, and, since this physiological requirement had never previously been reported for actinomycetes, we suspected that they represented a new marine taxon. However, it wasn't until we applied SSU rRNA sequencing methods that it became apparent that these seawater-requiring strains were also unique at the genetic level. These findings led us to the conclusion that highly adapted marine taxa occurred in ocean sediments and that these bacteria were significantly different from their terrestrial counterparts. Looking back at strains isolated from some of the same locations more than 10 yr earlier *(12)*, we realized that they also belonged to this new marine phylotype, and that members of this group represent a consistent component of the sediment bacterial community. These observations, coupled with the recovery of related strains from tropical marine locations worldwide, led us to believe that we had cultivated the first marine actinomycete taxon that was persistent and widespread in tropical ocean sediments. Based on their level of phylogenetic uniqueness, we proposed the generic epithet *Salinospora* to describe this group (**Fig. 1**). As of this writing, a formal description of the genus "Salinospora," including two type species (*"S. tropica"* and *"S. arenicola"*), has been submitted for publication *(18)*. If accepted, "Salinospora" will be the first marine actinomycete genus to be formally recognized, marking an important milestone in a slow march towards the recognition that actinomycetes are part of the indigenous marine microbiota.

Given our new appreciation of the phylogenetic diversity of marine actinomycetes, we began detailed chemical evaluations of members of the *Salinospora* group. Organic extracts of these actinomycetes, whose closest relatives are *Micromonospora* spp., another important group of secondary metabolite producers, yielded remarkable levels of biological activity. Bioassay-guided fractionation of the extract of one strain (CNB-392) quickly led to the discovery of an extraordinarily potent cytotoxin that we have called salinosporamide A *(19)*.

Salinosporamide A (**1, Fig. 2**) contains an unusual bicyclic γ-lactam-β-lactone ring system that forms the core of the molecule. It displayed potent in vitro cytotoxicity in assays against the human colon tumor cell line HCT-116 with an IC_{50} of 35 nM. When tested in the National Cancer Institute's (NCI's) 60-cell panel, the mean IC_{50} was <10 nM and it displayed a four-log range in activity among the least and most sensitive cell lines in both total growth inhibition (TGI) and LC_{50} (concentration required to kill 50% of the cells), indicating a high degree of cell line selectivity. Based on its potency and selectivity, salinosporamide A (**1**) was selected for the NCI's hollow fiber assay, where the score achieved was not sufficient for the compound to advance to xenograft studies, possibly a result of aqueous hydrolysis of the β-lactone, which we have observed in some cases, depending upon handling conditions.

Salinosporamide A (**1**) is most closely related to omuralide (**2**), also known as *clasto*-lactacystin β-lactone, the hydrolysis product of lactacystin *(20)* that has been described as the only known truly specific proteasome inhibitor ever discovered *(21)*. Although salinosporamide A (**1**) shares an identical bi-cyclic ring structure with omuralide (**2**), it is uniquely functionalized, possessing a methyl at the C-3 ring juncture, a chloroethyl instead of a methyl at C-2, and a cyclohexene instead of an isopropyl at the C-5 position. Like omuralide (**2**), salinosporamide A (**1**) is also a potent proteasome inhibitor, inhibiting the chymotrypsin-like proteolytic activity of the 20S proteasome with an IC_{50} value of 1.3 nM, which is approx 35 times more potent than omuralide (**2**, IC_{50} = 49 nM) when tested in the same assay (data provided by Nereus Pharmaceuticals, Inc., San Diego).

Like many bacteria that possess the genetic capacity to produce secondary metabolites, strain CNB-392 does not produce only one molecule but a series of related compounds. To date, in addition to salinosporamide A (**1**), we have isolated one additional

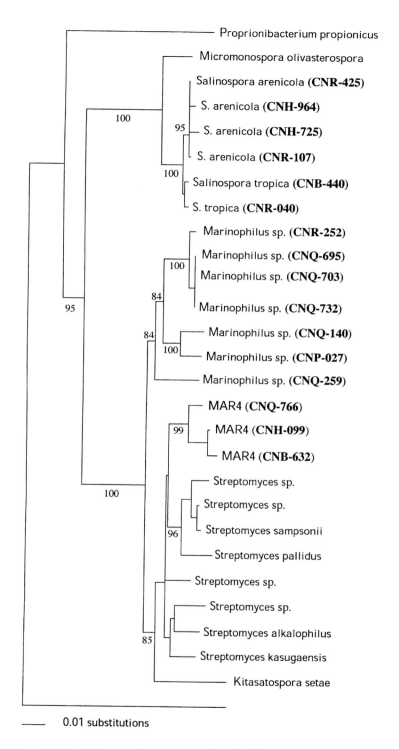

Fig. 1. Neighbor-joining phylogenetic representation of cultured marine actinomycetes and their closest National Center for Biotechnology Information (NCBI) (BLASTn) relatives based on almost complete SSU rRNA gene sequences. Phylogenetically unique marine actinomycetes include the new marine genus *"Salinospora"* along with the proposed genus *"Marinophilus"* and a potential third genus that is currently being described as MAR4. Representatives of all three new phylotypes produce secondary metabolites that possess heretofore undescribed carbon skeletons. Our strain numbers are in bold and bootstrap values (1000 re-samplings) are presented at the respective nodes.

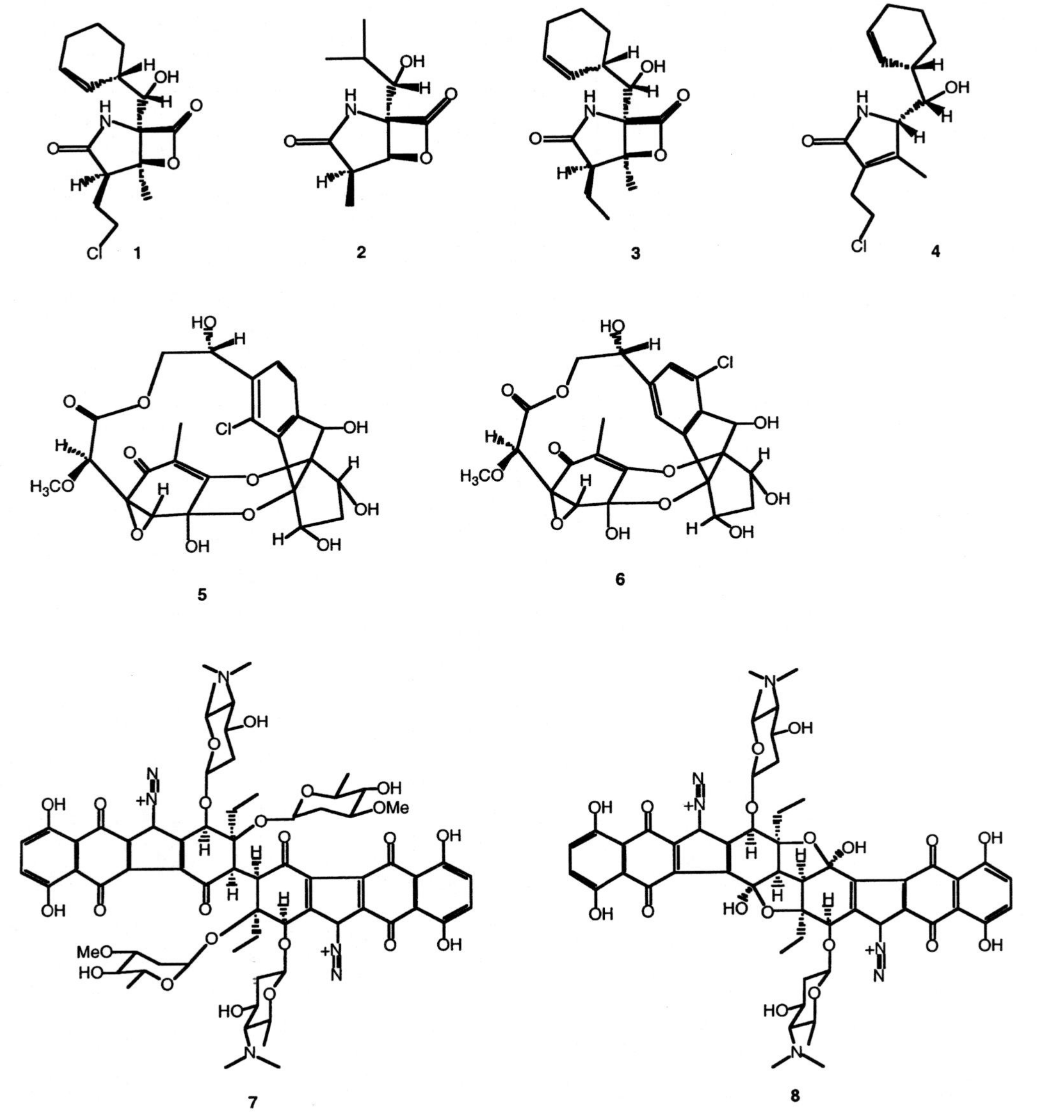

Fig. 2. Actinomycete secondary metabolites.

natural product and a series of related degradation products *(22)*, and from structure activity relationships have been able to gain insight into the features of compound **1** that are associated with its potent biological activity.

The first structural feature of salinosporamide A (**1**) that was found to be associated with its HCT-116 cytotoxicity became apparent following the isolation of salinosporamide B (**3**), the dechloro-analog of **1**. Loss of the chlorine atom at C-13 results in a greater than 500-fold loss in activity, with a reduction in the IC_{50} to 20 µM. This is not unprecedented, as chlorine, being a good leaving group, facilitates biological nucleophilic attack at the adjacent carbon by nucleophiles such as DNA or, in this case, most likely a protein. Chlorine has been shown to be required for activity in a variety of microbial metabolites including vancomycin *(23)*, the antibiotic of last choice against MRSA, and thus salinosporamide A (**1**) joins the ranks of chlorinated metabolites in which a chlorine substituent is an essential element for biological activity.

But the chlorine atom at C-13 is not the only active site in the salinosporamide A (**1**) molecule. This became clear with the isolation of the salinosporamide degradation products (**4**), a derivative in which the chloro-ethyl functionality is maintained but the β-lactone is missing. Loss of the β-lactone ring reduces the activity of compound **4** 5000-fold in comparison to compound **1**. Thus it appears that at least two sites in the salinosporamide A (**1**) molecule are associated with its biological activity. Of the remaining molecules in the series, all but salinosporamide B (**3**) are chlorinated, while only A (**1**) and B (**3**) maintain the intact β-lactone ring. As of this writing, Salinosporamide A (**1**) is proceeding through preclinical development as an anticancer agent at Nereus Pharmaceuticals, Inc.

The exquisite selectivity with which omuralide (**2**) is capable of inhibiting proteasome function, without inhibiting other proteases or lysosomal protein degradation *(21)*, has made it a powerful tool in the study of protein biochemistry and cell biology. It has also generated interest in the structural features of compound **2** that are required for proteasome inhibition. Of specific interest to our studies of salinosporamide A (**1**) was the finding that the omuralide (**2**) C-9 isopropyl is optimal for activity and that other substituents at this position, including phenyl, resulted in significantly reduced activity *(24)*. In addition, the X-ray crystal structure of the omuralide-inactivated 20S proteasome indicated that the isopropyl group fits into a lipophilic pocket. Because activity was greatly reduced by the substitution of diverse alkyl groups at C-9, it could be implied that there was a fairly snug fit of the isopropyl group into the complementary binding pocket *(24)*. Given that salinosporamide A (**1**) possesses the bulkier cyclohexene substituent at C-9, and that in omuralide a phenyl at this position resulted in a complete loss of activity, compound **1** may have a unique 20S proteasome binding site.

2.2. Additional Compounds From "Salinospora" spp.

In the course of performing time-dependant studies of salinosporamide A (**1**) production, we observed the production of two new molecules late in the fermentation period. Analysis of these compounds by liquid chromatography (LC)/mass spectrometry (MS) indicated that both molecules possessed a molecular weight of 536 and a single chlorine atom. Subsequent cultivation and extraction of a second strain (CNB-477), which was found to produce these compounds in higher yield, followed by a series of chromatography steps, led to the isolation of the heterocyclic macrolides sporolides A (**5**) and B (**6**), which differ only in the position of the chlorine substituent on the aromatic ring. These compounds possess unprecedented carbon skeletons and appear to be formed from two independently produced polyketide precursors. Both compounds were inactive in our cytotoxicity and drug resistant antimicrobial screens, and possessed insufficient activity in the NCI 60-cell panel (GI_{50} 24.5 and 49.0 µ*M*, respectively) to be considered for further action. Given the fact that they are chlorinated and possess epoxide functionalities, the lack of activity is somewhat surprising, and further testing may be warranted.

Thus far from one *"Salinospora"* strain (CNB-392), we have isolated two new series of molecules (the salinosporamides, compounds **1** and **3**, and the sporolides, compounds **5** and **6**) and are currently working on the structure elucidation of other new bioactive compounds, including a series of polyketide-derived antibiotics, from the extracts of additional *"Salinospora"* strains. In general, *"Salinospora"* extracts display an extraordinary level of biological activity, with >50% producing crude organic extracts with HCT-116 IC_{50} values of <10 µg/mL. In some cases, the observed extract activities have been ascribed to compounds related to staurosporines and rifamycin, two well-known classes of actinomycete cytotoxins. This brings up the point that it is unrealistic to expect that new actinomycete taxa will only produce secondary metabolites with previously undescribed carbon skeletons. The logic for seeking out new taxa is the belief that they will have a greater likelihood of producing new compounds, as has been the case with *"Salinospora"* species, not the unrealistic hope that all of the molecules that they produce will be new.

Other research groups worldwide have also been investigating marine-derived actinomycetes, and some, such as Okami's group at the Institute of Microbial Chemistry in Tokyo, were among the first to explore the chemical potential of these bacteria and should be recognized for their pioneering work *(25)*. As was our experience, many early efforts did not emphasize taxonomically unique marine actinomycetes or those that were specifically adapted to life in the ocean, and, as a result, many known compounds or new derivatives of known skeletons were obtained. Given our growing understanding of marine actinomycete diversity, along with improvements in our ability to culture these bacteria, we feel confident that future chemical studies, targeting specific marine species, will begin to reveal the true chemical potential of these bacteria.

Although our goal here is not to summarize the findings of other research groups, it is of particular relevance to discuss a recent discovery by the Wyeth-Ayerst group (for a review of this program, *see* ref. 26). As part of their effort to investigate the potential microbial origin of the potent antitumor agent namenamicin, originally isolated from extracts of the marine ascidian *Polysyncraton lithostrotum*, the Wyeth group cultivated what appeared to be a new *Micromonospora* sp. from tissues of *P. lithostrotum* collected off Fiji *(27)*. In culture, this halophilic strain did not produce namenamicin; however, culture extracts displayed potent DNA-damaging activity, and bioassay-guided fractionation led to the isolation of the new dimeric diazobenzofluorene glycoside antitumor antibiotics lomaivitacins A (**7**) and B (**8**). The authors were kind enough to provide us with the SSU rRNA sequence data for the producing strain, and it clearly falls within the *"Salinospora"* clade. Thus, as has been our experience, strains belonging to the *"Salinospora"* clade can be isolated from the tissues of marine invertebrates as well as from marine sediments. The Wyeth-Ayerst discovery adds a third new chemotype to the diverse repertoire of secondary metabolites produced by this new group of marine actinomycetes.

2.3. "Marinophilus"

Given the high rate of new compound production by *"Salinospora"* strains, we initiated a study to cultivate new *"Salinospora"* species as well as to determine whether additional marine actinomycete taxa could be recovered from marine samples *(28)*. Using the requirement of seawater for growth as a selective criterion, we quickly recognized a second new phylotype, for which we hope to propose the generic epithet "*Marinophilus*" (**Fig. 1**). Unlike "*Salinospora*" strains, "*Marinophilus*" are not limited to tropical waters. They are also, unfortunately, unlike "*Salinospora*" in that they cannot be recovered in large numbers from marine sediments using any of our current isolation methods. Although we have cultured relatively few "*Marinophilus*" strains thus far, they do appear to have one thing in common with "*Salinospora*" spp.—the ability to produce unique secondary metabolites.

Initial chemical studies of the "*Marinphilus*" strains in our collection quickly led to the discovery of a series of structurally unique antitumor-antibiotics that we have called marinomycins A-D (*[9–12]*, **Fig. 3**). The structure elucidation of these compounds proved difficult due to the symmetry of the molecules and their many stereo centers; however, these issues have now been resolved and the compounds will soon be submitted for publication *(29)*. Although the marinomycins are polyene-like macrolides, they do not possess the antifungal activities typically associated with many polyene metabolites. They do, however, display potent cytotoxicities in the NCI 60-cell panel. Of the three compounds thus far tested (compounds 9–11), the mean GI_{50} values were 18.6 nM, 12.6 nM, and 5.1 nM, respectively. Possibly of greater significance, all three compounds demonstrated a four-log range in LC_{50} values between the least and most sensitive cell lines, with a pronounced spike in activity against one specific melanoma. These compounds are currently being subjected to additional testing at the NCI. We hope in the future to develop improved cultivation techniques for "*Marinophilus*" spp. and to further explore the chemical potential of this new actinomycete taxon.

2.4. Marinone Producers

Although we now possess a better understanding of the taxonomic uniqueness of the marine-derived actinomycetes we are cultivating, this was not always the case, and, in the early days of our program, we routinely relied on broad chemical and biological screening in our search for new metabolites. For example, in 1989, we isolated an actinomycete (strain CNB-632) from the Torrey Pines Estuary in La Jolla, CA. Chemical studies of this strain led to the isolation of two new meroterpenoid antibiotics, marinone (compound **13**) and debromomarinone (compound **14**), examples of rare biosynthetic hybrids of mixed polyketide-terpenoid origin *(30)*. These compounds share a novel sequiterpene napthoquinone carbon skeleton and are most closely related to the actinomycete-derived monoterpene napthaquinones naphterpin *(31)* and the naphthgeranines *(32)*. They also possess antibiotic activities against Gram-positive bacteria, with MIC values ranging from 2 to 4 µM.

Four years later, while collecting sediment samples from the Batiquitos Lagoon, Carlsbad, CA, approx 50 km north of the original collection site, we isolated another actinomycete strain (CNH-099), which produced not only compounds 13 and 14 but also several new derivates (**15, 16**) in the marinone series as well as the new metabolite neomarinone (compound **17** *[33]*), the structure of which was subsequently revised *(34)*. While marinone and its analogs (compounds **13–16**) contain a nonrearranged sesquiterpene residue, neomarinone (compound **17**) features a rearranged C-15 substituent that occurs on the nonquinone side of the molecule. Feeding experiments with ^{13}C-labeled acetate by Moore's group at the University of Arizona indicated that, like in naphterpin, the naphthoquinone portions of both marinone and neomarinone were

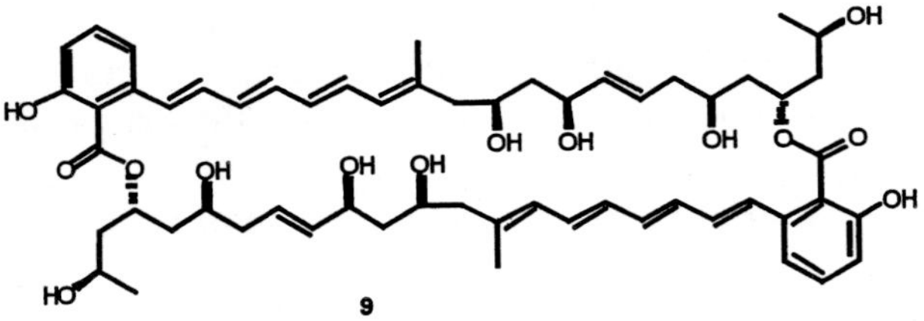

Fig. 3. Actinomycete secondary metabolites.

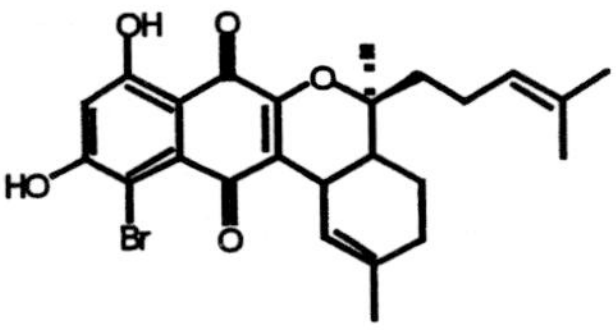

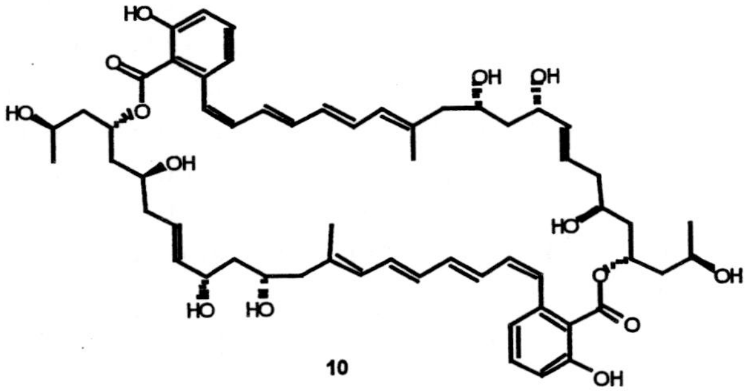

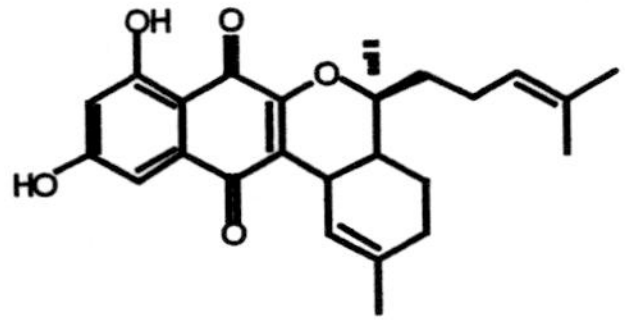

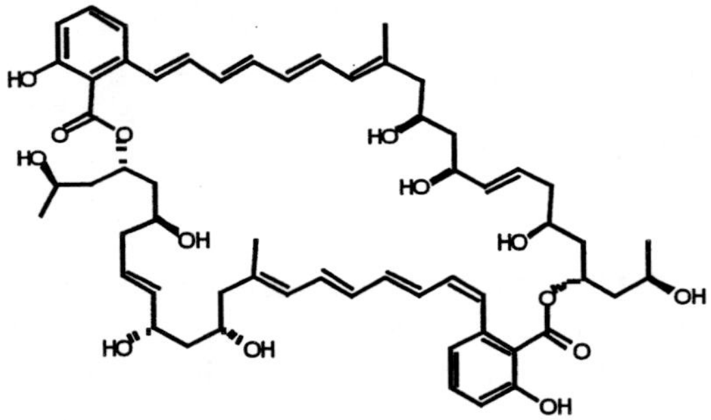

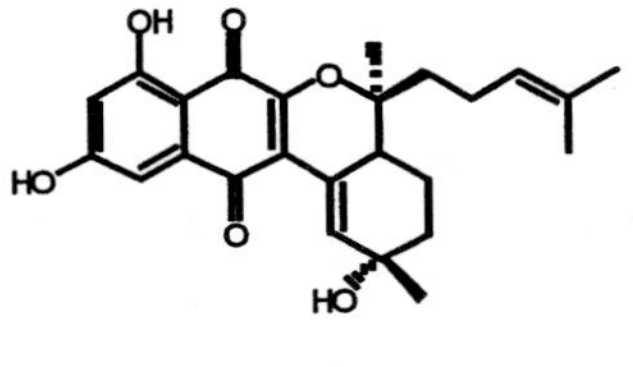

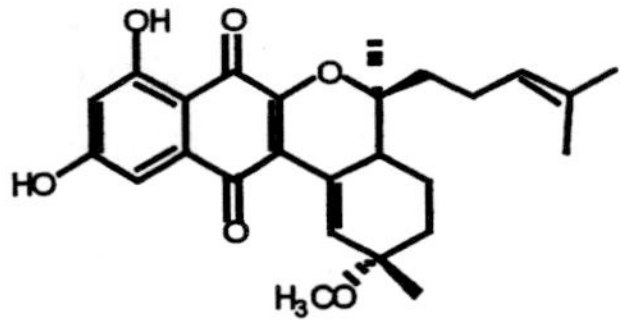

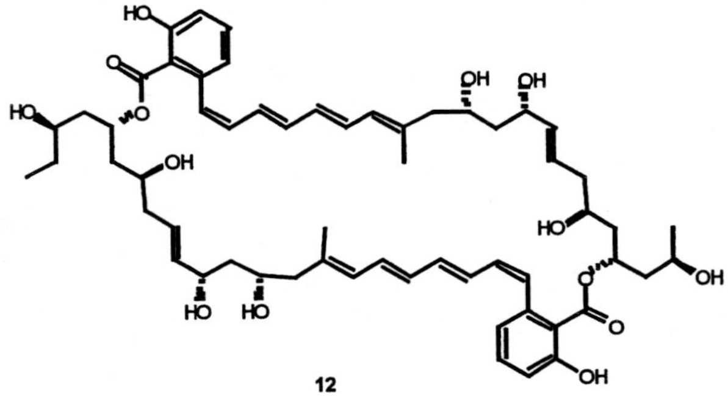

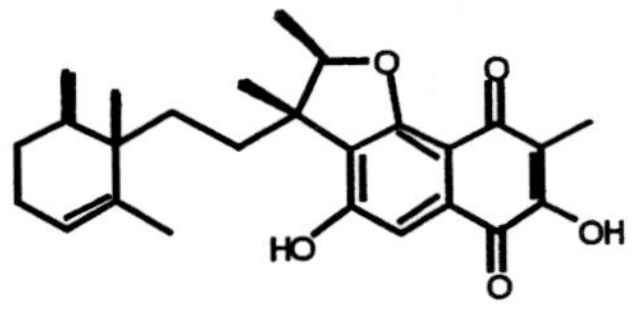

derived from a symmetrical pentaketide intermediate and thus are of polyketide origin. Unlike the monoterpene portion of naphterpin however, ^{13}C-labeled glucose feeding studies revealed that the sesquiterpene portions of compounds **14** and **17** are not derived from the mevalonate pathway but instead appear to originate from the recently described nonmevalonate terpene biosynthetic pathway in a manner similar to the related *Streptomyces* antibiotic furaquinocin *(34)*.

Subsequent phylogenetic analyses of the SSU rRNA genes of the marinone-producing strains CNB-632 and CNH-099 revealed that they are closely related to each other but are significantly different from any known members of the Streptomycetaceae (**Fig. 1**). These strains, along with our strain CNQ-766, form a new marine clade within the family and appear to represent a new marine taxon. Thus, chemical studies leading to a highly unusual class of secondary metabolites inadvertently led to the discovery of a new actinomycete taxon, supporting our contention that there is a correlation between taxonomic and chemical novelty.

Interestingly, neither CNB-632 nor CNH-099 requires seawater for growth. Although we have used this criterion reliably as a rapid method to recognize unique marine phylotypes, it is becoming increasingly clear that not all marine strains possess this trait (*see* **Subheading 2.5.**). It is also of interest that related strains producing similar compounds were isolated over a 4-yr period. This has been our experience with other marine phylotypes, e.g., *"Salinospora,"* and is inconsistent with the concept that these actinomycetes are merely random strains originating as run-off from terrestrial sources. It also raises questions about biogeographical distributions and the ecological significance of compounds whose production appears to be maintained over time.

2.5. Seawater Requirements

The observation that non-seawater-requiring marine-derived strains can be phylogenetically unique suggests that the requirement of seawater for growth, although clearly a marine adaptation, should not be considered a defining factor used to delineate marine from nonmarine actinomycetes. Marine and terrestrial environments are merely habitats from which actinomycetes can be recovered, and there is no reason to assume that these boundaries, as we perceive them, are as clearly delineated in the microbial world. As to what defines a marine actinomycete, it may not be wise to speculate on this subject, as any seemingly reasonable definition has a high probability of one day proving incomplete. It may be better instead to seek evidence that marine-derived actinomycetes are metabolically active or capable of growth in the marine environment. Other useful characteristics include the display of specific marine adaptations, the formation of associations with plants or invertebrates, and the demonstration of a role in nutrient cycling, all of which provide an indication that these bacteria are not merely present as dormant spores. When coupled with phylogenetic novelty, evidence for any of the above traits provides a strong case that the specific populations under study can be considered marine.

In the case of actinomycetes, the persistence with which the requirement of seawater for growth is maintained for any specific population remains unclear, as does the timescale within which this trait can be lost or acquired. In the case of *"Salinospora"* strains, they have thus far without exception maintained this physiological requirement even after more than 10 yr in culture. *"Marinomyces"* strains however appear to vary in this trait, as only a portion of those initially isolated required seawater for growth and it is not yet clear whether they maintain this requirement after prolonged laboratory culture. Thus, a better understanding of the role of sodium in actinomycete physiology, and the time frame within which a requirement of sodium for growth can be acquired, is warranted if we are to understand the evolutionary significance of this physiological trait.

3. Discussion

Although much excitement has been generated over the application of combinatorial biology to the heterologous expression of biosynthetic genes harvested directly from nature, these methods remain in the exploratory stage and at present are no substitution for the direct cultivation of chemically prolific taxa. Although efforts to culture marine actinomycetes are only beginning, preliminary studies have been highly productive, leading to the isolation of new marine taxa and a high rate of new-compound discovery. Old concepts that the vast majority of environmental bacteria cannot be cultured are being rapidly disproved (e.g., ref. 35), and there can be little doubt that with the continued development of new culture techniques, untold new actinomycete diversity will become apparent. Continued cultivation efforts, along with culture-independent studies, will clarify how effectively total actinomycete diversity is being accessed in the search for new microbial products.

We may now be witnessing the first glimpse of a major resurgence in natural products as a source of new pharmaceutical agents. *Chemical and Engineering News* devoted its October 2003 issue to this concept, speaking in detail about "rediscovering natural products" and the potential role of microorganisms in this process. Clearly, actinomycetes have the potential to once again play a major role in microbial drug discovery, and the recent observation that significant populations of marine-adapted actinomycetes occur in ocean sediments certainly emphasizes the potential importance of these strains in future discovery efforts. Given the advances made in analytical techniques, coupled with the potential for lead optimization using combinatorial synthesis and recombinant pathway expression, these tools now make it possible to capitalize on microbial products with an efficiency that was heretofore unimaginable.

References

1. Goodfellow M, Williams ST. Ecology of actinomycetes. Annu Rev Microbiol 1983;37:189–216.
2. Okami Y, Hotta K. Search and discovery of new antibiotics. In: Goodfellow M, Williams ST, Mordarski M (eds), Actinomycetes in Biotechnology. Academic, New York: 1988; pp. 33–67.
3. Goodfellow M, O'Donnell AG. Search and discovery of industrially significant actinomycetes. In: Bamberg S, Hunter I, Rhodes M (eds), Microbial Products. Cambridge University Press, Cambridge: 1989; pp. 343–383.
4. Watve MG, Tickoo R, Jog MM, Bhole BD. How many antibiotics are produced by the genus *Streptomyces*? Arch Microbiol 2001;176:386–390.
5. Bull AT, Ward AC, Goodfellow M. Search and discovery strategies for biotechnology: the paradigm shift. Microbiol Mol Biol Rev 2000;64:573–606.
6. Horan AC. Aerobic actinomycetes: a continuing source of novel natural products. In: Gullo V (ed), The Discovery of Natural Products with Therapeutic Potential. Butterworth-Heinemann, Boston: 1994; pp. 3–30.

7. Otoguro M, Hayakawa M, Yamazaki T, Iimura Y. An integrated method for the enrichment and selective isolation of *Actinokineospora* spp. in soil and plant liter. J Appl Microbiol 2001;91:118–130.
8. Fenical W. Chemical studies of marine bacteria: developing a new resource. Chem Rev 1993;93:1673–1683.
9. Jensen PR, Fenical W. Marine microorganisms and drug discovery: current status and future potential. In: Fusetani N (ed), Drugs from the Sea. Karger, Basel: 2000; pp. 6–29.
10. Weyland H. Actinomycetes in North Sea and Atlantic Ocean sediments. Nature 1969;223:858.
11. Goodfellow M, Haynes JA. Actinomycetes in marine sediments. In: Ortiz-Ortiz L, Bojalil LF, Yakoleff V (ed), Biological, Biochemical, and Biomedical Aspects of Actinomycetes. Academic, New York: 1984; pp. 453–472.
12. Jensen PR, Dwight R, Fenical W. Distribution of actinomycetes in near-shore tropical marine sediments. Appl Environ Microbiol 1991;57:1102–1108.
13. Moran MA, Rutherford LT, Hodson RE. Evidence for indigenous *Streptomyces* populations in a marine environment determined with a 16S rRNA probe. Appl Environ Microbiol 1995;61:3695–3700.
14. Helmke E, Weyland H. *Rhodococcus marinonascens* sp. nov., an actinomycete from the sea. Int J Syst Bacteriol 1984;34:127–138.
15. Mincer TJ, Jensen PR, Kauffman CA, Fenical W. Widespread and persistent populations of a major new marine actinomycete taxon in ocean sediments. Appl Environ Microbiol 2002;68: 5005–5011.
16. Davidson BS. New dimensions in natural products research: cultured marine microorganisms. Curr Biol 1995;6:284–291.
17. Blunt JW, Copp BR, Munro MHG, Northcote PT, Prinsep MR. Marine natural products. Nat Prod Rep 2003;20:1–48.
18. Maldonado L, Fenical W, Goodfellow M, Jensen PR, Mincer TJ, Ward AC. *Salinospora* gen. nov., a home for obligate marine actinomycetes belonging to the family *Micromonosporaceae*. In preparation.
19. Feling RH, Buchanan GO, Mincer TJ, Kauffman CA, Jensen PR, Fenical W. Salinosporamide A: A highly cytotoxic proteasome inhibitor from a novel microbial source, a marine bacterium of the new genus *Salinospora*. Angew Chem Int Ed 2003;42:355–357.
20. Dick LR, Cruikshank AA, Grenier L, Melandri FD, Nunes SL, Stein RL. Mechanistic studies on the inactivation of the proteasome by lactacystin. J Biol Chem 1996;271:7273–7276.
21. Fenteany G, Schreiber SL. Lactacystin, proteasome function, and cell fate. J Biol Chem 1998;273:8545–8548.
22. Williams PG, Buchanan GO, Feling RH, Mincer TJ, Kauffman CA, Jensen PR, Fenical W. Salinosporamide B and other products from the culture broth of the marine actinomycete "*Salinospora*" sp. In preparation.
23. Naumann K. Influence of chlorine substituents on biological activity of chemicals. J Prakt Chem 1999;341:417–435.
24. Corey EJ, Li Wei-Dong Z, Nagamitsu T, Fenteany G. The structural requirements for inhibition of proteasome function by the lactacystin-derived β-lactone and synthetic analogs. Tetrahedron 1999;55:3305–3316.
25. Okami Y. The search for bioactive metabolites from marine bacteria. J Mar Biotechnol 1993;1:59–65.
26. Bernan VS, Greenstein M, Maiese WM. Marine microorganisms as a source of new natural products. Adv Appl Microbiol 1997;43:57–90.
27. He H, Ding W-D, Bernan VS, et al. Lomaiviticins A and B, potent antitumor antibiotics from *Micromonospora lomaivitensis*. J Am Chem Soc 2001;123:5362–5363.

28. Jensen PR, Gontang E, Mafnas C, Mincer TJ, Fenical W. Culturable marine actinomycete diversity from tropical Pacific Ocean sediments. Environ Microbiol, in press.
29. Kwon HC, Jensen PR, Fenical W. Marinomycins A-D, antitumor antibiotics of a new structure class from a marine actinomycete of the genus "Marinophilus." In preparation.
30. Pathirana C, Jensen PR, Fenical W. Marinone and debromomarinone, two phenylated napthoquinones from a marine actinomycete. Tetrahedron Lett 1992;33:7663–7666.
31. Shin-Ya K, Imai S, Furihata K, et al. Isolation and structure elucidation of an antioxidative agent, naphterpin. J Antibiot. 1990;43:444–447.
32. Wessels P, Gohrt A, Zeeck A. Metabolic products of microorganisms. 260. Naphthgeranines, new naphthoquinone antibiotics from *Streptomyces* sp. J Antibiot 1991;44:1013–1018.
33. Hardt IH, Jensen PR, Fenical W. Neomarinone, and new cytotoxic marinone derivatives, produced by a marine filamentous bacterium (actinomycetales). Tetrahedron Lett 2000;41: 2073–2076.
34. Kalaitzis JA, Hamano Y, Nilsen G, Moore BS. Biosynthesis and structural revision of neomarinone. Org Lett 2003;5:4449–4452.
35. Joseph SJ, Hugenholtz P, Sangwan P, Osborne CA, Janssen PH. Laboratory cultivation of widespread and previously uncultured soil bacteria. Appl Environ Microbiol 2003;69:7210–7215.

15
Novel Natural Products From Rainforest Endophytes

Gary Strobel, Bryn Daisy, and Uvidelio Castillo

Summary

Endophytic microorganisms are found in virtually every higher plant on earth. These organisms reside in the living tissues of the host plant and do so in a variety of relationships, ranging from symbiotic to pathogenic. Endophytes may contribute to their host plant by producing a plethora of substances that provide protection and survival value to the plant. Ultimately, these compounds, once isolated and characterized, may also have potential for use in modern medicine. Novel antibiotics, antimycotics, immunosuppressants, and anticancer compounds are only a few examples of what has been found after the isolation and culturing of individual endophytes followed by purification and characterization of some of their natural products. The potential of finding new drugs that may be effective candidates for treating newly developing diseases in humans is great.

Key Words: Munumbicins; kakadumycin; taxol; volatile antibiotics; streptomycetes; anticancer agents; immunosuppressants.

1. Introduction

The need for new and useful compounds to provide assistance and relief in all aspects of the human condition is ever growing. Drug resistance in bacteria, the appearance of new life-threatening viruses, the recurrent problems of diseases in persons with organ transplants, and the tremendous increase in the incidence of fungal infections in the world's population all underscore our inadequacy to cope with these medical problems. Environmental degradation, loss of biodiversity, and spoilage of land and water also add to problems facing humanity, and each of these in turn can have health-related consequences.

Endophytes, microorganisms that reside in the tissues of living plants, are relatively unstudied as potential sources of novel natural products for exploitation in medicine. However, some of the most extensive and comprehensive work on natural products produced by endophytes has been done on the *Neotyphodium* sp. found on grasses *(1)*. Alkaloids synthesized by this fungus in its grass hosts have been implicated in fescue toxicosis in rangeland animals *(1)*. The chemistry and biology of this and other grass endophytes are reviewed elsewhere *(2)*. Unfortunately, because this work is so comprehensive, one may be led to the conclusion that endophytes produce toxic compounds only in their respective hosts and hold no promise for any medicinal applications whatsoever *(2)*. It turns out that this is simply not the case. As endophytes are examined from a plethora of sources, an overwhelming number have been found to produce natural products with promising potential for medicinal applications.

Of the approx 300,000 higher plant species that exist on the earth, each individual plant, of the billions that exist here, is host to one or more endophytes. Only a handful of these plants (grass species) have ever been completely studied relative to their endophytic biology *(2)*. Consequently, the opportunity to find new and interesting endophytic microorganisms among myriads of plants in different settings and ecosystems is very great. The intent of this review is to provide insights into their occurrence in nature, the products that they make, and indicate how some of these organisms are beginning to show some potential for human use. The majority of the report discusses rationale for study, methods used, and examples of a number of endophytes isolated and studied in the authors' laboratories over the course of many years. This review, however, also includes some specific examples that illustrate the work of others in this emerging field of bioprospecting the microbes of the world's rainforests.

2. Needs for New Natural Products

There is a general call for new antibiotics, and for chemotherapeutic agents that are highly effective and possess low toxicity. This search is driven by the development of resistance in infectious microorganisms (e.g., *Staphylococcus*, *Mycobacterium*, *Streptococcus*) to existing drugs and by the menacing presence of naturally resistant organisms. The ingress to the human population of new disease-causing agents such as acquired immunodeficiency syndrome (AIDS), Ebola, and severe acute respiratory syndrome (SARS) requires the discovery and development of new drugs to combat them. Not only do diseases such as AIDS require drugs that target them specifically, but new therapies are needed for treating ancillary infections, which are a consequence of a weakened immune system. Furthermore, others who are immunocompromised (e.g., cancer and organ transplant patients) are at risk of infection by opportunistic pathogens, such as *Aspergillus*, *Cryptococcus*, and *Candida*, which normally are not major problems in the human population. In addition, more drugs are needed to efficiently treat parasitic protozoan and nematodal infections such as malaria, leishmaniasis, trypanomiasis, and filariasis. Malaria by itself is more effective in claiming lives each year than any other single infectious agent with the exception of AIDS and tuberculosis (TB) *(3)*. However, the enteric diseases claim the most lives each year of any disease complex, and unfortunately, the victims are mostly children *(3)*.

Novel natural products and the organisms that make them offer opportunities for innovation in drug discovery. Exciting possibilities exist for those who are willing to venture into the wild and unexplored territories of the world to experience the thrill of engaging in the discovery of endophytes, their biology, and potential usefulness.

3. Endophytic Microbes

It may also be true that a reduction in interest in natural products for use in drug development has happened as a result of people growing weary of dealing with the traditional sources of bioactive compounds, including plants of the temperate zones and microbes from a plethora of soil samples gathered in different parts of the world by armies of collectors. In other words, why continue to do the same thing when robots, combinatorial chemistry, and molecular biology have arrived on the scene? Furthermore, the logic and rationale for time and effort spent on drug discovery using a target-site-directed approach has been overwhelming.

While combinatorial synthesis produces compounds at random, secondary metabolites, defined as low-molecular-weight compounds not required for growth in pure culture, are produced as an adaptation for specific functions in nature *(4)*. Shutz notes that certain microbial metabolites seem to be characteristic of certain biotopes, both on an environmental as well as organismal level *(5)*. Accordingly, it appears that the search for novel secondary metabolites should center on organisms that inhabit unique biotopes. Thus, it behooves the investigator to carefully study and select the biological source before proceeding, rather than to take a totally random approach in selecting the source material. Careful study also indicates that organisms and their biotopes that are subjected to constant metabolic and environmental interactions should produce even more secondary metabolites *(5)*. Endophytes are microbes that inhabit such biotopes, namely higher plants, which is why they are currently considered as a wellspring of novel secondary metabolites offering the potential for exploitation of their medical benefits.

In addition, it also is extremely helpful for the investigator interested in exploiting endophytes to have access to, or have some expertise in, microbial taxonomy, and this includes modern molecular techniques involving sequence analyses of 16S and 18S rDNA. Currently, endophytes are viewed as an outstanding source of bioactive natural products because there are so many of them occupying literally millions of unique biological niches (higher plants) growing in so many unusual environments. Thus, it would appear that a myriad of biotypical factors associated with plants can be important in the selection of a plant for study. It may be the case that these factors may govern which microbes are present in the plant as well as the biological activity of the products associated with these organisms.

Since the discovery of endophytes in Darnel, Germany, in 1904, various investigators have defined endophytes in different ways, which usually depended on the perspective from which the endophytes were being isolated and subsequently examined *(6)*. Bacon et al. give an inclusive and widely accepted definition of endophytes: "Microbes that colonize living, internal tissues of plants without causing any immediate, overt negative effects" *(2)*. While the symptomless nature of endophyte occupation in plant tissue has prompted focus on symbiotic or mutualistic relationships between endophytes and their hosts, the observed biodiversity of endophytes suggests they can also be aggressive saprophytes or opportunistic pathogens. Both fungi and bacteria are the most common microbes existing as endophytes. It would seem that other microbial forms most certainly exist in plants as endophytes, such as mycoplasmas, rickettsia, and archebacteria; however, no evidence for them has yet been presented. The most frequently isolated endophytes are the fungi *(7)*. It turns out that the vast majority of plants have not been studied for their endophytes. Thus, enormous opportunities exist for the recovery of novel fungal forms, including genera, biotypes, as well as species in the myriad of plants yet to be studied. Hawksworth and Rossman estimated there may be as many as 1 million different fungal species, yet only approx 100,000 have been described *(8)*. As more evidence accumulates, estimates keep rising as to the actual number of fungal species. For instance, Dreyfuss and Chapela estimate there may be at least 1 million species of endophytic fungi alone *(9)*. It seems obvious that endophytes are a rich and reliable source of genetic diversity and may represent previously undescribed species. Finally, in our experience, novel microbes (as defined at the morphological and/or molecular levels) often have novel natural products associated with them. This fact alone helps eliminate the problems of dereplication in compound discovery.

4. Rationale for Plant Selection

It is important to understand the methods and rationale used seem to provide the best opportunities to isolate novel endophytic microorganisms at the genus, species, or biotype level. Thus, since the number of plant species in the world is so great, creative and imaginative strategies must be used to quickly narrow the search for endophytes displaying bioactivity *(10)*.

A specific rationale for the collection of each plant for endophyte isolation and natural product discovery is used. Several hypotheses govern this plant selection strategy, and these are as follows:

1. Plants from unique environmental settings, especially those with an unusual biology, and possessing novel strategies for survival, are seriously considered for study.
2. Plants that have an ethnobotanical history (use by indigenous peoples) that are related to the specific uses or applications of interest are selected for study. These plants are chosen either by direct contact with local peoples or via local literature. Ultimately, it may be learned that the healing powers of the botanical source, in fact, may have nothing to do with the natural products of the plant, but of the endophyte inhabiting the plant.
3. Plants that are endemic, having an unusual longevity, or that have occupied a certain ancient land mass, such as Gondwanaland, are also more likely to lodge endophytes with active natural products than other plants.
4. Plants growing in areas of great biodiversity, it follows, also have the prospect of housing endophytes with great biodiversity.

Just as plants from a distinct environmental setting are considered to be a promising source of novel endophytes and their compounds, so too are plants with an unconventional biology. For example, an aquatic plant, *Rhyncholacis penicillata*, was collected from a river system in southwest Venezuela where the harsh aquatic environment subjected the plant to constant beating by virtue of rushing waters, debris, and tumbling rocks and pebbles *(11)*. These environmental insults created many portals through which common phytopathogenic oomycetes could enter the plant. Still, the plant population appeared to be healthy, possibly owing to protection by an endophytic product. This was the environmental biological clue used to pick this plant for a comprehensive study of its endophytes. Eventually, an unusual and potent antifungal strain of *Serratia marcescens*, living both as an epiphyte and an endophyte, was recovered from *R. penicillata*. This bacterium was shown to produce oocydin A, a novel antioomycetous compound having the properties of a chlorinated macrocyclic lactone (**Fig. 1**) *(11)*. It is conceivable that the production of oocydin A by *S. marcescens* is directly related to the endophyte's relationship with its higher-plant host. Currently, oocydin A is being considered for agricultural use to control the ever-threatening presence of oomyceteous

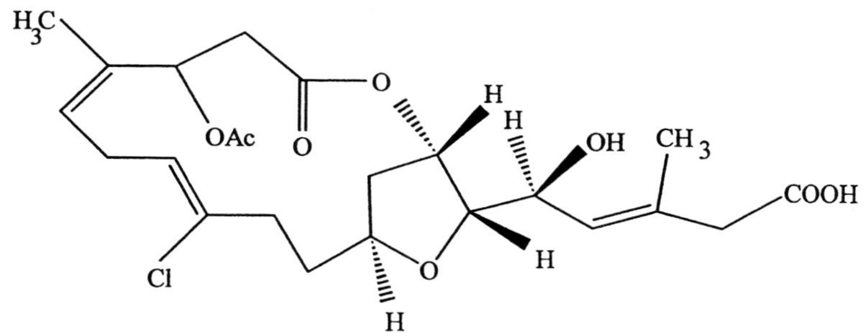

Fig. 1. Oocydin A, a chlorinated macrocyclic lactone isolated and characterized from a strain of *Serratia marcescens*, obtained from *Rhyncholacis penicillata* (stereochemistry is not known).

fungi such as *Pythium* spp. and *Phytophthora* spp. Oocydin A also has activity against a number of rapidly dividing cancer cell lines *(11)*.

Plants with ethnobotanical history, as mentioned above, also are likely candidates for study, since the medical uses for which the plant was selected may relate more to its population of endophytes than to the plant biochemistry itself. For example, a sample of the snakevine, *Kennedia nigriscans*, from the Northern Territory of Australia, was selected for study since its sap has traditionally been used as bush medicine for many millenia. In fact, this area was selected for plant sampling because it has been home to the world's longest standing civilization—the Australian aborigines. The snakevine is harvested, crushed, and heated in an aqueous brew by local aborigines in southwest Arnhemland to treat cuts, wounds, and infections. As it turned out, the plant contained a streptomycete that possessed unique partial 16S rDNA sequences when compared to those in GenBank. The organism was designated *Streptomyces* NRRL 30562, and it produces broad-spectrum novel peptide antibiotics called munumbicins, which are discussed below *(12)*. It seems likely that some of the healing properties in plants, as discovered by indigenous peoples, might be facilitated by compounds produced by one or more specific plant-associated endophytes as well as the plant products themselves.

In addition, it is worthy to note that some plants generating bioactive natural products have associated endophytes that produce the same natural products. Such is the case with taxol, a highly functionalized diterpenoid and famed anticancer agent that is found in each of the world's yew tree species (*Taxus* spp.) *(11,12)*. In 1993, a novel taxol-producing fungus, *Taxomyces andreanae*, from the yew *Taxus brevifolia*, was isolated and characterized *(13)*.

5. Endophytes and Biodiversity

Of the myriad of ecosystems on earth, those having the greatest general biodiversity seem to be the ones also having the greatest number and most diverse endophytes. Tropical and temperate rainforests are the most biologically diverse terrestrial ecosystems on earth. The most threatened of these spots cover only 1.44% of the land's surface, yet they harbor over 60% of the world's terrestrial biodiversity *(10)*. In addition, each of the 20–25 areas identified as supporting the world's greatest biodiversity support unusually high levels of plant endemism *(10)*. As such, one would expect, with high plant endemism, there also should exist specific endophytes that may have evolved with the endemic plant species. Biological diversity implies chemical diversity, because of the constant chemical innovation that is required to survive in ecosystems where the evolutionary race to survive is most active. Tropical rainforests are a remarkable example of this type of environment. Competition is great, resources are limited, and selection pressure is at its peak. This gives rise to a high probability that rainforests are a source of novel molecular structures and biologically active compounds *(14)*.

Bills et al. describe a metabolic distinction between tropical and temperate endophytes through statistical data that compare the number of bioactive natural products isolated from endophytes of tropical regions to the number of those isolated from endophytes of temperate origin *(15)*. Not only did they find that tropical endophytes provide more active natural products than temperate endophytes, but they also noted that a significantly higher number of tropical endophytes produced a larger number of active secondary metabolites than did fungi from other substrata. This observation suggests the importance of the host plant as well as the ecosystem in influencing the general metabolism of endophytic microbes.

6. Endophytes and Phytochemistry

Tan and Zou believe the reason why some endophytes produce certain phytochemicals, originally characteristic of the host, might be related to a genetic recombination of the endophyte with the host that occurred in evolutionary time *(6)*. This is a concept that was originally proposed as a mechanism to explain why *T. andreanae* may be producing taxol *(16)*. Thus, if endophytes can produce the same rare and important bioactive compounds as their host plants, this would not only reduce the need to harvest slow-growing and possibly rare plants, but also help to preserve the world's ever-diminishing biodiversity. Furthermore, it is recognized that a microbial source of a high-value product may be easier and more economical to produce effectively, thereby reducing its market price.

All aspects of the biology and interrelatedness of endophytes with their respective hosts is a vastly under-investigated and exciting field *(17,18)*. Thus, more background information on a given plant species and its microorganismal biology would be exceedingly helpful in directing the search for bioactive products. Presently, no one is quite certain of the role of endophytes in nature and their relationship to various host plant species. Although some endophytic fungi appear to be ubiquitous (e.g., *Fusarium* spp., *Pestalotiopsis* spp., and *Xylaria* spp.), one cannot definitively state that endophytes are truly host-specific or even systemic within plants, any more than one can assume that their associations are chance encounters. Frequently, many endophytes of the same species are isolated from the same plant, and only one or a few biotypes of a given fungus will produce a highly biologically active compound in culture *(19)*. A great deal of uncertainty also exists between what an endophyte produces in culture and what it may produce in nature. It does seem possible that the production of certain bioactive compounds by the endophyte *in situ* may facilitate the domination of its biological niche within the plant or even provide protection to the plant from harmful invading pathogens. Furthermore, little information exists relative to the biochemistry and physiology of the interactions of the endophyte with its host plant. It would seem that many factors changing in the host, related to the season, age, environment, and location, may influence the biology of the endophyte. Indeed, further research at the molecular level must be conducted in the field to study endophyte interactions and ecology. All of these interactions are probably chemically mediated for some purpose in nature. An ecological awareness of the role these organisms play in nature will provide the best clues for targeting particular types of endophytic bioactivity with the greatest potential for bioprospecting.

7. Collection, Isolation, and Preservation of Endophytes

After a plant is selected for study, it is identified, and its location is plotted using a global positioning device. Small stem pieces are cut from the plant and placed in sealed plastic bags after excess moisture is removed. Every attempt is made to store the materials at 4°C until isolation procedures can begin *(20,21)*.

In the laboratory, the surfaces of plant materials are thoroughly treated with 70% ethanol, sometimes flamed, and ultimately they are air dried under a laminar-flow hood. This is done in order to eliminate surface-contaminating microbes *(20)*. Then, with a sterile knife blade, outer tissues are removed from the samples and the inner tissues carefully excised and placed on water agar plates. After several days of incubation, hyphal tips of the fungi are removed and transferred to potato dextrose or other suitable agar. Bacterial forms also emerge from the plant tissues, including, on rare occasions, certain *Streptomyces* spp. The endophytes are encouraged to sporulate on specific plant materials and are eventually identified via standard morphological and molecular biological techniques and methods. Eventually, when an endophyte is acquired in pure culture, it is tested for its ability to be grown in shake or still culture using various media and growth conditions *(21)*. It is also immediately placed in storage under various conditions including 15% glycerol at −70°C. Ultimately, once appropriate growth conditions are found, the microbe is subjected to fermentation, extraction, and the bioactive compounds are isolated and characterized. Virtually all of the common and advanced procedures for product isolation and characterization are utilized in order to acquire the product(s) of interest. Central to the processes of isolation is the establishment of one or more bioassays that will guide the compound purification processes. One cannot put too much emphasis on this point, since the ultimate success of any natural-product isolation activity is directly related to the development or selection of appropriate bioassay procedures. These can involve target organisms, enzymes, tissues, or model chemical systems that relate to the purpose for which the new compound is needed.

8. Natural Products From Endophytic Microbes

The following section shows some examples of natural products obtained from endophytic microbes and their potential in the pharmaceutical and agrochemical arenas. Many of the examples are taken from our work, and thus, this review is by no means inclusive of all natural-product work in endophytes.

8.1. Endophytic Fungal Products As Antibiotics

Fungi are the most commonly isolated endophytic microbes. They usually appear as fine filaments growing from the plant material on the agar surface. Generally, the most commonly isolated fungi are in the group *Fungi imperfecti* or *Deuteromycetes*. Basically, they produce asexual spores in or on various fruiting structures. Also, it is quite common to isolate endophytes that are producing no fruiting structures whatsoever, such as *Mycelia sterilia*. Quite commonly endophytes do produce secondary metabolites when placed in culture. However, the temperature, the composition of the medium, and the degree of aeration will affect the amount and kind of compounds that are produced. Sometimes endophytic fungi produce antibiotics. Natural products from endophytic fungi have been observed to inhibit or kill a wide variety of harmful microorganisms including, but not limited to, phytopathogens, as well as bacteria, fungi, viruses, and protozoans that affect humans and animals. Described below are some examples of bioactive products from endophytic fungi.

Cryptosporiopsis cf. *quercina* is the imperfect stage of *Pezicula cinnamomea*, a fungus commonly associated with hardwood species in Europe. It was isolated as an endophyte from *Tripterigeum wilfordii*, a medicinal plant native to Eurasia *(22)*. On Petri plates, *C. quercina* demonstrated excellent antifungal activity against some important human fungal pathogens, including *Candida albicans* and *Trichophyton* spp. A unique peptide antimycotic, termed "cryptocandin," was isolated and characterized *(22)*. This compound contains a number of peculiar hydoxylated amino acids and a novel amino acid, 3-hydroxy-4-hydroxy methyl proline (**Fig. 2**). The bioactive compound is related to known antimycotics—the echinocandins and the pneumocandins *(23)*. As is generally true, not one but several bioactive and related compounds are produced by an endophytic microbe. Thus, other antifungal agents related to cryptocandin are also produced by *C. quercina*. Cryptocandin is also active against a number of plant pathogenic fungi, including *Sclerotinia sclerotiorum* and *Botrytis cinerea*. Cryptocandin and its related compounds are currently being considered for use against a number of fungi causing diseases of the skin and nails.

Cryptocin, a unique tetramic acid, is also produced by *C. quercina* (discussed previously) (**Fig. 3**)*(24)*. This unusual compound possesses potent activity against *Pyricularia oryzae*, the causal organism of one of the worst plant diseases in the world, as well as a number of other plant pathogenic fungi *(24)*. The compound was generally ineffective against a general array of human pathogenic fungi. Nevertheless, with minimum inhibitory concentrations against *P. oryzae* at 0.39 µg/mL, this compound is being examined as a natural chemical control agent for rice blast and is being used as a platform for the synthesis of other antifungal compounds.

As mentioned earlier, *P. microspora* is a common rainforest endophyte *(17–20)*. It turns out that enormous biochemical diversity does exist in this endophytic fungus, and many secondary metabolites are produced by various strains of this widely dispersed organism. One such secondary metabolite is ambuic acid, an antifungal agent, which has been recently described from several isolates of *P. microspora* found as representative isolates in many of the world's rainforests (**Fig. 4**) *(25)*. This compound as well as another endophyte product, terrein, have been used as models to develop new solid-state nuclear magnetic resonance (NMR) tensor methods to assist in the characterization of the molecular stereochemistry of organic molecules.

A strain of *P. microspora* was also isolated from the endangered tree *Torreya taxifolia* and produced several compounds having antifungal activity, including pestaloside, an aromatic β-glucoside (**Fig. 5**), and two pyrones—pestalopyrone and hydroxypestalopyrone *(26)*. These products also possess phytotoxic properties. Other

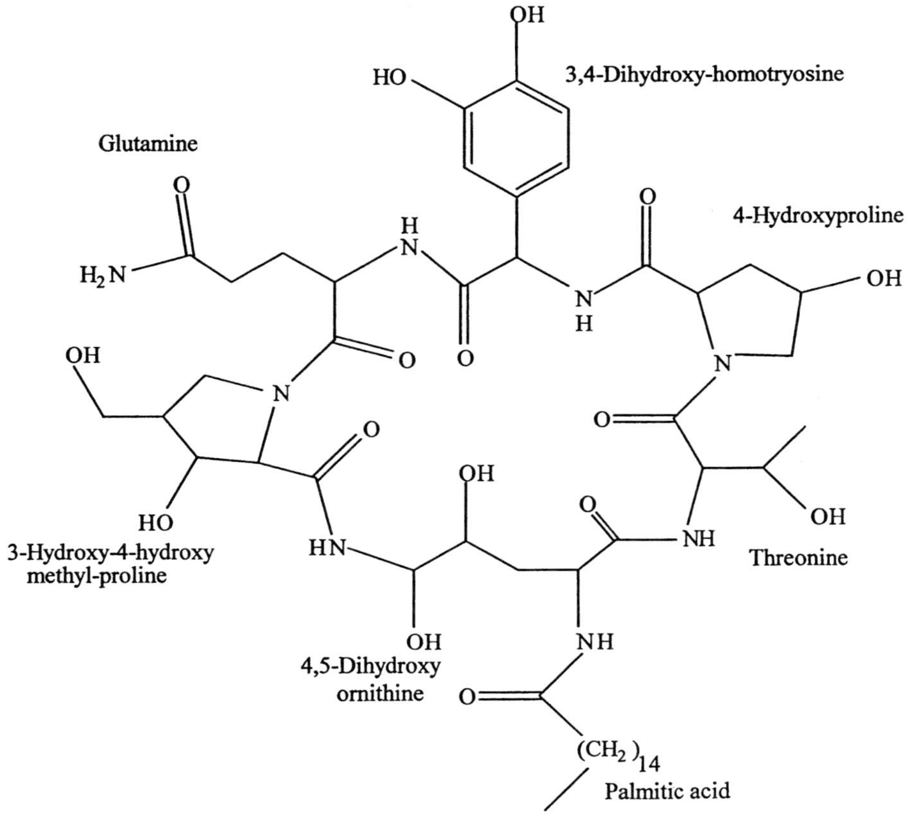

Fig. 2. Cryptocandin A, an antifungal lipopeptide obtained from the endophytic fungus *Cryptosporiopsis* cf. *quercina* (no stereochemistry is intended).

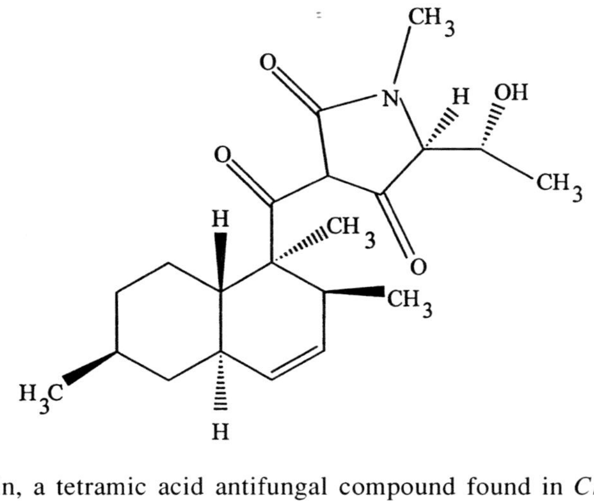

Fig. 3. Cryptocin, a tetramic acid antifungal compound found in *Cryptosporiopsis* cf. *quercina*.

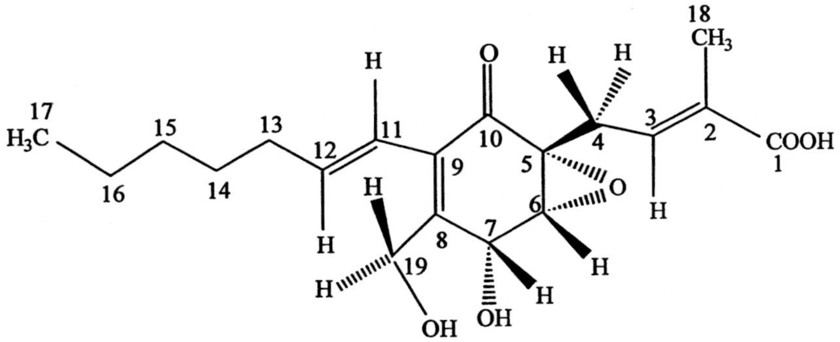

Fig. 4. Ambuic acid, a highly functionalized cyclohexenone produced by a number of isolates of *Pestalotiopsis microspora* found in rainforests around the world. This compound possesses antifungal activity and has been used as a model compound for the development of solid-state nuclear magnetic resonance methods for the structural determination of natural products.

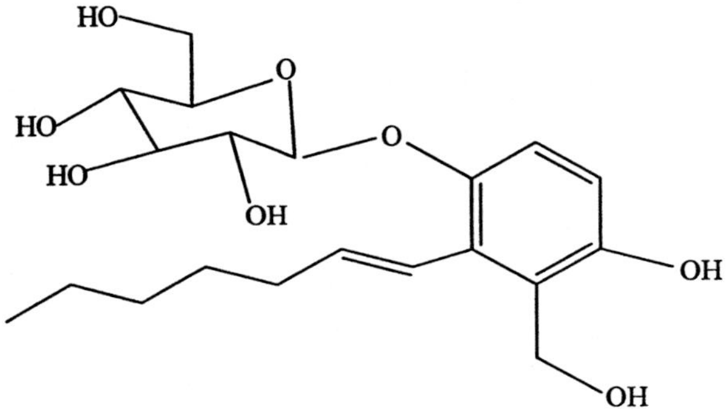

Fig. 5. Pestaloside, a glucosylated aromatic compound with antifungal properties from *Pestalotiopsis microspora*.

newly isolated secondary products obtained from *P. microspora* (endophytic on *Taxus brevifolia*) include two new caryophyllene sesquiterpenes—pestalotiopsins A and B *(27)*. Additional new sesquiterpenes produced by this fungus are 2α-hydroxydimeninol and a highly functionalized humulane *(28,29)*. Variation in the amount and kinds of products found with this fungus depends on both the cultural conditions as well as the original plant source from which it was isolated.

Pestalotiopsis jesteri is a newly described endophytic fungal species from the Sepik River area of Papua New Guinea, and it produces jesterone and hydroxyjesterone, which exhibit antifungal activity against a variety of plant pathogenic fungi *(30)*. These compounds are highly functionalized cyclohexenone epoxides. Jesterone, subsequently, has been prepared by organic synthesis with complete retention of biological activity (**Fig. 6**) *(31)*. Jesterone is one of only a few products from endophytic microbes in which total synthesis of a bioactive product has been successfully accomplished.

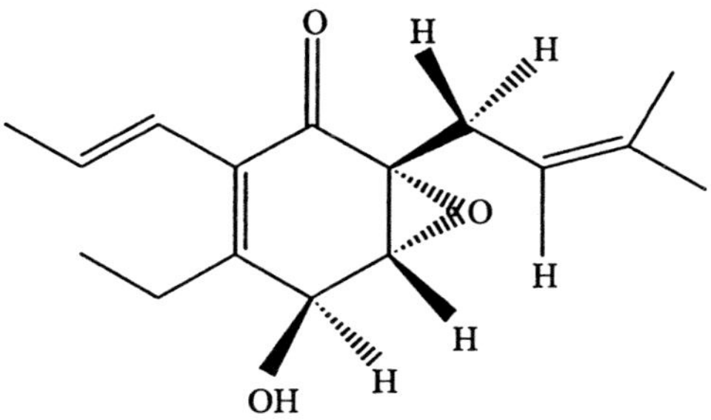

Fig. 6. Jesterone, a cyclohexenone epoxide from *Pestaliotiopsis jesteri*, has antioomycete activity.

Phomopsichalasin, a metabolite from an endophytic *Phomopsis* sp., represents the first cytochalasin-type compound with a three-ring system replacing the cytochalasin macrolide ring. This metabolite exhibits antibacterial activity in disk diffusion assays (at a concentration of 4 µg/disk) against *Bacillus subtilis*, *Salmonella gallinarum*, and *Staphylococcus aureus*. It also displays moderate activity against the yeast *Candida tropicalis* *(32)*.

An endophytic *Fusarium* sp. from the plant *Selaginella pallescens*, collected in the Guanacaste Conservation Area of Costa Rica, was screened for antifungal activity. A new pentaketide antifungal agent, CR377, was isolated from the culture broth of the fungus and showed potent activity against *C. albicans* in agar diffusion assays *(33)*.

Colletotric acid, a metabolite of *Colletotrichum gloeosporioides*, an endophytic fungus isolated from *Artemisia mongolica*, displays activity against bacteria as well as against the fungus *Helminthsporium sativum* *(34)*. Another *Colletotrichum* sp., isolated from *Artemisia annua*, produces antimicrobial metabolites as well. *A. annua* is a traditional Chinese herb that is well recognized for its synthesis of artemisinin (an antimalarial drug) and its ability to inhabit many geographically different areas. Not only did the *Colletotrichum* sp. found in *A. annua* produce metabolites with activity against human pathogenic fungi and bacteria, but also metabolites that were fungistatic to plant pathogenic fungi *(35)*.

8.2. Endophytic Bacterial Products As Antibiotics

There are only a limited number of bacterial species known to be associated with plants, and one of the most common is *Pseudomonas* spp. *Pseudomonas* has representative biotypes and species that are epiphytic, endophytic, and pathogenic. They have been reported from every continent including the Antarctic. Some of these species produce phytotoxic compounds as well as antibiotics. The ecomycins are produced by *Pseudomonas viridiflava* *(36)*. This bacterium is generally associated with the leaves of many grass species and is located on and within the tissues *(36)*. The ecomycins represent a family of novel lipopeptides and have masses of 1153 and 1181. Besides common amino acids such as alanine, serine, threonine, and glycine, some nonprotein amino acids are incorporated into the structure of the ecomycins, including homoserine and β-hydroxyaspartic acid *(36)*. The ecomycins are active against such human pathogenic fungi as *Cryptococcus neoformans* and *C. albicans*.

The pseudomycins produced by a plant-associated pseudomonad are another group of antifungal peptides *(37,38)*. They are active against a variety of plant and human pathogenic fungi, including *Candida albicans*, *Cryptococcus neoformans*, and a variety of plant pathogenic fungi, including *Ceratocystis ulmi* (the Dutch Elm disease pathogen) and *Mycosphaerella fijiensis* (causal agent of Black Sigatoka disease in bananas). The pseudomycins are cyclic depsipeptides formed by acylation of the OH group of the N-terminal serine with the terminal carboxyl group of L-chlorothreonine. Variety in this family of compounds is imparted via *N*-acetylation by one of a series of fatty acids, including 3,4-dihydroxydecanoate, 3-hydroxy-tetradecanoate *(38)*. The pseudomycins contain several nontraditional amino acids, including L-chlorothreonine, L-hydroxyaspartic acid, and both D- and L-diaminobutryic acid. The molecules are candidates for use in human medicine, especially after structural modification by chemical synthesis has successfully eliminated mammalian toxicity *(39)* The pseudomycins are also effective against a number of ascomycetous fungi, and are being considered for agricultural use for the control of the Black Sigatoka disease in bananas (Strobel, unpublished).

8.3. Endophytic Streptomycetes As Antibiotic Producers

Streptomyces spp. are filamentous bacteria, belonging to the order Actinomycetales, that live in widely diverse ecological settings. Generally, this group is Gram positive, has a high G+C content, and does not have an organized nucleus. To date, actinomycetes have been the world's greatest source of natural antibiotics *(40)*. In fact, just one genus, *Streptomyces*, is the source of 80% of these compounds. The majority of the antibiotic producers are from soil sources, and until recently it was not realized that these organisms can exist as endophytes. One of the first endophytic *Streptomyces* spp. isolated was that from *Lolium perenne*, a grass species *(41)*. This isolate produces a diketopiperazine that is a weak antibiotic and has been designated "methylalbonoursin" *(41)*.

Using the ethnobotanical approach to plant selection, the snakevine plant, *K. nigriscans,* was chosen as a possible source of endophytic microbes because of its long-held traditional use by Australian aborigines to treat cuts and open wounds, resulting in reduced infection and rapid healing. This plant was collected near the Aboriginal Community of Manyallaluk in Northern Territory, Australia, and consistently yielded an endophytic actinomycete designated *Streptomyces* NRRL 30562 *(12)*. The organism was not found in several tree species supporting the vine, suggesting a host-selective or -specific association of the endophyte with a specific plant genus. This streptomycete produces a family of extremely potent peptide antibiotics, and these compounds may not only protect the plant from fungal and bacterial infections, but also have unknowingly served the aborigines as a source of bush medicine.

The antibiotics produced by *Streptomyces* NRRL 30562, called "munumbicins," possess widely differing biological activities, depending on the target organism. In general, the munumbicins demonstrate activity against Gram-positive bacteria such as *Bacillus anthracis* and multidrug-resistant *Mycobacterium tuberculosis*, as well as a number of other drug-resistant bacteria. However, the most impressive biological activity of any of the munumbicins is that of munumbicin D against the malarial parasite *Plasmodium falciparum*, having an IC_{50} of 4.5 ± 0.07 ng/mL *(12)*. The munumbicins are highly functionalized peptides, each containing threonine, aspartic acid (or asparagine), and glutamic acid (or glutamine). Since the peptides are yellowish orange, they also contain one or more chromophoric groups, whose structures have not been determined. Their masses range from 1269 to 1326 Da. The isolation of this endophytic streptomycete represents an important finding, providing one of the first examples of plants serving as reservoirs of actinomycetes. More than 40 of these endophytic streptomycetes, now in hand in our laboratory, possess antibiotic activity (Castillo, U., Strobel, G.A., unpublished data). Endophytic actinomycetes are now being tested and considered for use in controlling plant diseases *(42)*.

Another endophytic *Streptomyces* sp. (NRRL 30566), from a fern-leaved grevillea (*Grevillea pteridifolia*) tree growing in the Northern Territory of Australia, produces novel antibiotics called "kakadumycins," which are related to the echinomycins *(43)*. Each of these antibiotics contains alanine, serine, and an unknown amino acid. Kakadumycin A has wide-spectrum antibiotic activity similar to that of munumbicin D, especially against Gram-positive bacteria, and it generally displays better bioactivity than echinomycin. For instance, against *B. anthracis* strains, kakadumycin A has MICs of 0.2–0.3 µg/mL, in contrast to echinomycin at 1.0–1.2 µg/mL. Both echinomycin and kakadumycin A have impressive activity against *P. falciparum*, with LD_{50}s in the range of 7–10 ng/mL *(43)*. Kakadumycin A and echinomycin are related by virtue of their very similar structures (amino acid content and quinoxoline rings), but differ slightly with respect to their elemental compositions, aspects of their spectral qualities, chromatographic retention times, and biological activities *(53)*.

Echinomycin and kakadymycin A were studied as inhibitors of macromolecular synthesis, with control substances such as ciprofloxacin, rifampin, chloramphenicol, and vancomycin used as standards with well-established modes of action. Tests were done for DNA, RNA, protein, and cell-wall synthesis inhibition activities, respectively. Kakadumycin A significantly inhibited RNA synthesis in *B. subtilis (43)*. Kakadumycin A also inhibited protein synthesis and cell-wall synthesis substantially, but had a lower effect on DNA synthesis. Kakadumycin A shares a very similar inhibitory profile with echinomycin in four macromolecular synthesis assays. Kakadumycin A preferentially inhibits RNA synthesis, and may have the same mode of action as echinomycin, which inhibits RNA synthesis by binding to a DNA template *(53)*.

More recently, endophytic streptomycetes have been discovered in an area of the world claimed to be one of the most biologically diverse—the upper Amazon of Peru. The inner tissues of the follow me vine, *Monstera* sp., commonly yielded a verticillated streptomycete with outstanding inhibitory activities against pythiaceous fungi as well as the malarial parasite *Plasmodium falciparum*. The bioactive component is a mixture of lipopeptides named "coronamycins" *(44)*.

8.4. Antiviral Compounds

Another fascinating use of products from endophytic fungi is the inhibition of viruses. Two novel human cytomegalovirus (hCMV) protease inhibitors, cytonic acids A and B, have been isolated from solid-state fermentation of the endophytic fungus *Cytonaema* sp. Their structures were elucidated as *p*-tridepside isomers by MS and NMR methods *(45)*. It is apparent that the potential for the discovery of compounds having antiviral activity from endophytes is in its infancy. The main limitation to com-

8.5. Volatile Antibiotics From Endophytes

Muscodor albus is a newly described endophytic fungus obtained from small limbs of *Cinnamomum zeylanicum* (cinnamon tree) *(46)*. This xylariaceaous (non-spore producing) fungus effectively inhibits and kills certain fungi and bacteria by producing a mixture of volatile compounds *(47)*. The majority of these compounds have been identified by gas chromatography (GC)/MS, synthesized or acquired, and then formulated into an artificial mixture. This mixture not only mimicked the antibiotic effects of the volatile compounds produced by the fungus, but also was used to confirm the identity of the majority of the volatiles emitted by this organism *(47)*. Each of the five classes of volatile compounds produced by the fungus had some microbial effects against the test fungi and bacteria, but none was lethal. However, they acted synergistically to cause death in a broad range of plant and human pathogenic fungi and bacteria. The most effective class of inhibitory compounds was the esters, of which isoamyl acetate was the most biologically active. The composition of the medium on which *M. albus* grows dramatically influences the kind of volatile compounds that are produced *(48)*. The ecological implications and potential practical benefits of the "mycofumigation" effects of *M. albus* are very promising, given the fact that soil fumigation utilizing methyl bromide will soon be illegal in the United States. Methyl bromide is not only a hazard to human health, but it has been implicated in causing destruction of the ozone layer. The potential use of mycofumigation to treat soil, seeds, and plants may soon be a reality. The artificial mixture of volatile compounds may also have usefulness in treating seeds, fruits, and plant parts in storage and while being transported. *Muscodor albus* already has a limited market for the treatment of human wastes. Its gases have both inhibitory and lethal effects on such fecal-inhabiting organisms as *Escherichia coli* and *Vibrio cholera*.

Using *M. albus* as a screening tool, it has now been possible to isolate other endophytic fungi producing volatile antibiotics. The newly described *M. roseus* was obtained twice from tree species growing in the Northern Territory of Australia. This fungus is just as effective in causing inhibition and death of test microbes in the laboratory as *M. albus (49)*. In addition, for the first time, a nonmuscodor species (*Gliocladium* sp.) was discovered as a producer of volatile antibiotics. The volatile components of this organism are totally different from those of either *M. albus* or *M. roseus*. In fact, the most abundant volatile inhibitor is [8]-annulene, formerly used as a rocket fuel and discovered for the first time as a natural product. However, the bioactivity of the volatiles of this *Gliocladium* sp. is not as good or comprehensive as those from *Muscodor* spp. *(21,47)*.

8.6. Endophytic Fungal Products As Anticancer Agents

Taxol and some of its derivatives represent the first major group of anticancer agents that are produced by endophytes (**Fig. 6**). Taxol (**Fig. 7**), a highly functionalized diterpenoid, is found in each of the world's yew (*Taxus*) species, but was originally isolated from *Taxus brevifolia (50)*. The original targets for this compound were ovarian and breast cancers, but now it is used to treat a number of other human tissue

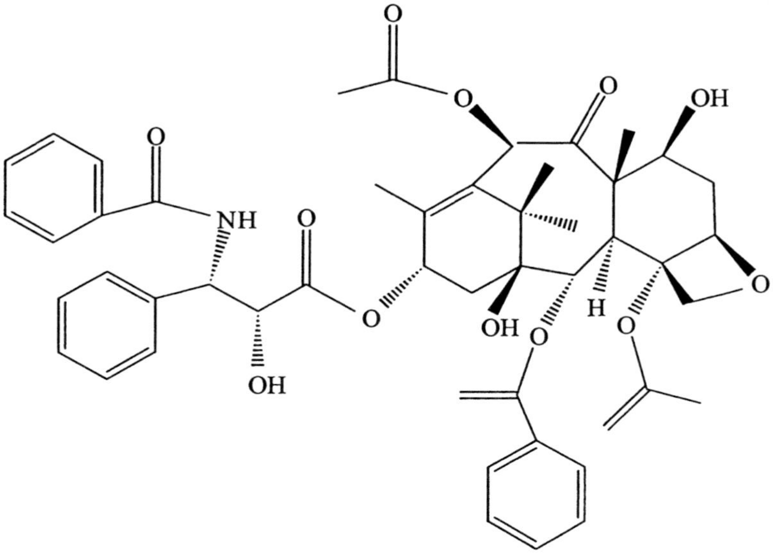

Fig. 7. Taxol, the world's first billion-dollar anticancer drug, is produced by many endophytic fungi. It too, possesses outstanding anti-oomycete activity.

proliferating diseases as well. The presence of taxol in yew species prompted the study of their endophytes. By the early 1990s, however, no endophytic fungi had been isolated from any of the world's representative yew species. After several years of effort, a novel taxol-producing endophytic fungus, *Taxomyces andreanae*, was discovered in *Taxus brevifolia (13)*. The most critical line of evidence for the presence of taxol in the culture fluids of this fungus was the electrospray mass spectrum of the putative taxol isolated from *T. andreanae*. In electrospray mass spectroscopy, taxol usually gives two peaks—one at mass 854, which is M+H$^+$, and the other at 876, which is M+Na$^+$. Fungal taxol had a mass spectrum identical to authentic taxol *(16)*. Then, ^{14}C labeling studies showed the presence of fungal-derived taxol in the culture medium *(26)*. This early work set the stage for a more comprehensive examination of the ability of other *Taxus* species and many other plants to yield endophytes producing taxol.

Some of the most commonly found endophytes of the world's yews and many other plants are *Pestalotiopsis* spp. *(17–20)*. One of the most frequently isolated endophytic species is *Pestalotiopsis microspora (17)*. An examination of the endophytes of *Taxus wallichiana* yielded *P. microspora*, and a preliminary monoclonal antibody test indicated that it might produce taxol *(20)*. After preparative TLC, a compound was isolated and shown by spectroscopic techniques to be taxol. Labeled (^{14}C) taxol was produced by this organism from several ^{14}C precursors *(20)*. Furthermore, several other *P. microspora* isolates that produce taxol were obtained from a bald cypress tree in South Carolina *(19)*. This was the first indication that endophytes, residing in plants other than *Taxus* spp., produce taxol. Therefore, a specific search was conducted for taxol-producing endophytes on continents not known for any indigenous *Taxus* spp., e.g., South America and Australia. From the extremely rare, and previously thought to be extinct, Wollemi pine (*Wollemia nobilis*), *Pestalotiopsis guepini* was isolated, which was shown to produce taxol *(51)*. Also, quite surprisingly, a rubiaceous plant, *Maguireothamnus speciosus*, yielded a novel fungus, *Seimatoantlerium tepuiense*, that produces taxol. This endemic plant grows on the top of the tepuis in the Venzuelan-Guyana border in southwest Venezuela *(52)*. Furthermore, fungal taxol production has also been noted in *Periconia* sp. *(53)* and *Seimatoantlerium nepalense*, another novel endophytic fungal species *(54)*. Simply, it appears that the distribution of taxol-making fungi is worldwide and is not confined to endophytes of yews. The ecological and physiological explanation for fungi making taxol seems to be related to the fact that taxol is a fungicide, and the organisms most sensitive to it are plant pathogens such as *Pythium* spp. and *Phytophthora* spp. *(55)*. These pythiaceous organisms are some of the world's most important plant pathogens and are strong competitors with endophytic fungi for niches within plants. In fact, their sensitivity to taxol is based on their interaction with tubulin, in an identical manner as in rapidly dividing human cancer cells *(55)*. Thus, *bona fide* endophytes may be producing taxol and related taxanes to protect their respective host plant from degradation and disease caused by these pathogens.

Other investigators have also made observations on taxol production by endophytes, including the discovery of taxol production by *Tubercularia* sp. isolated from the Chinese yew (*Taxus mairei*) in the Fujian province of southeastern mainland China *(56)*. At least three endophytes of *Taxus wallichiana* produce taxol, including *Sporormia minima* and *Trichothecium* sp. *(57)*. Using HPLC and ESIMS, taxol has been discovered in *Corylus avellana* cv. Gasaway *(58)*. Several fungal endophytes of this plant (filbert) produce taxol in culture *(58)*. It is important to note, however, that taxol production by all endophytes in culture is in the range of sub-micrograms to micrograms per liter. Also, commonly, the fungi will attenuate taxol production in culture, with some possibility for recovery, if certain activator compounds are added to the medium *(53)*. Efforts are being made to determine the feasibility of making microbial taxol a commercial possibility, e.g., the discovery of endophytes that make large quantities of one or more taxanes that could then be used as intermediates for the organic synthesis of taxol or one of its anticancer relatives.

Torreyanic acid, a selectively cytotoxic quinone dimer and potential anticancer agent, was isolated from a *P. microspora* strain (**Fig. 8**). This strain was originally obtained as an endophyte associated with the endangered tree *Torreya taxifolia* (Florida torreya) *(59)*. Torreyanic acid was tested in several cancer cell lines, and it demonstrated 5 to 10 times more potent cytotoxicity in lines that are sensitive to protein kinase C agonists; it causes cell death by apoptosis. Recently, torreyanic acid has been successfully synthesized by a biomimetic oxidation/dimerization cascade *(60)*.

Alkaloids are also commonly found in endophytic fungi. Fungal genera such as *Xylaria*, *Phoma*, *Hypoxylon*, and *Chalara* are representative producers of a relatively large group of substances known as the cytochalasins, of which more than 20 are now known. Many of these compounds possess antitumor and antibiotic activities, but because of their cellular toxicity they have not been developed into pharmaceuticals. Three novel cytochalasins have recently been reported from *Rhinocladiella* sp., as an endophyte on *Tripterygium wilfordii*. These compounds have antitumor activity and have been identified as 22-oxa-[12]-cytochalasins *(61)*. Thus, it is not uncommon to find one or more cytochalasins in endophytic fungi, and this provides an example of

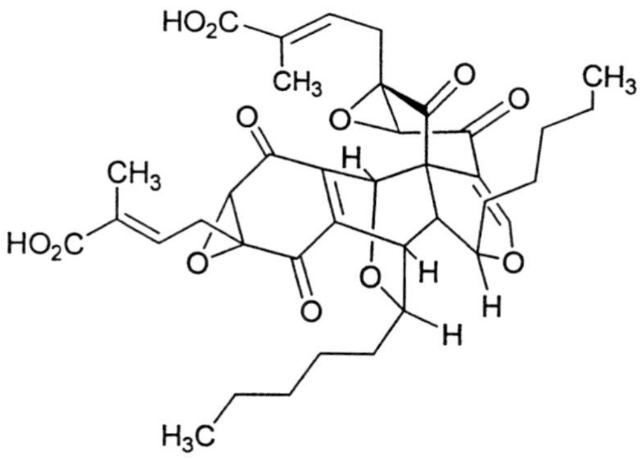

Fig. 8. Torreyanic acid, an anticancer compound, from *Pestalotiopsis microspora*.

the fact that redundancy in discovery does occur, making dereplication an issue even for these under-investigated sources.

8.7. Products From Endophytes As Antioxidants

Two compounds, pestacin and isopestacin, have been obtained from culture fluids of *Pestalotiopsis microspora*, an endophyte isolated from a combretaceaous plant, *Terminalia morobensis*, growing in the Sepik River drainage system of Papua New Guinea *(62,63)*. Both pestacin and isopestacin display antimicrobial as well as antioxidant activity. Isopestacin was attributed with antioxidant activity based on its structural similarity to the flavonoids (**Fig. 9**). Electron spin resonance spectroscopy confirmed this antioxidant activity; the compound is able to scavenge superoxide and hydroxyl free radicals in solution *(62)*. Pestacin was later described from the same culture fluid, occurring naturally as a racemic mixture and also possessing potent antioxidant activity (**Fig. 10**) *(63)*. The proposed antioxidant activity of pestacin arises primarily via cleavage of an unusually reactive C-H bond and, to a lesser extent, through O-H abstraction *(63)*. The antioxidant activity of pestacin is at least one order of magnitude more potent than that of trolox, a vitamin E derivative *(63)*.

8.8. Antidiabetic Agents From Rainforest Fungi

A nonpeptidal fungal metabolite (L-783,281) was isolated from an endophytic fungus (*Pseudomassaria* sp.) collected from an African rainforest near Kinshasa in the Democratic Republic of the Congo *(64)*. This compound acts as an insulin mimetic but, unlike insulin, is not destroyed in the digestive tract and may be given orally. Oral administration of L-783,281 in two mouse models of diabetes resulted in significant lowering of blood glucose levels. These results may lead to new therapies for diabetes.

8.9. Immunosuppressive Compounds From Endophytes

Immunosuppressive drugs are used today to prevent allograft rejection in transplant patients, and in the future they could be used to treat autoimmune diseases such as rheumatoid arthritis and insulin-dependent diabetes. The endophytic fungus *Fusarium subglutinans*, isolated from *T. wilfordii*, produces the immunosuppressive but noncytotoxic diterpene pyrones subglutinols A and B (**Fig. 11**) *(65)*. Subglutinol A and

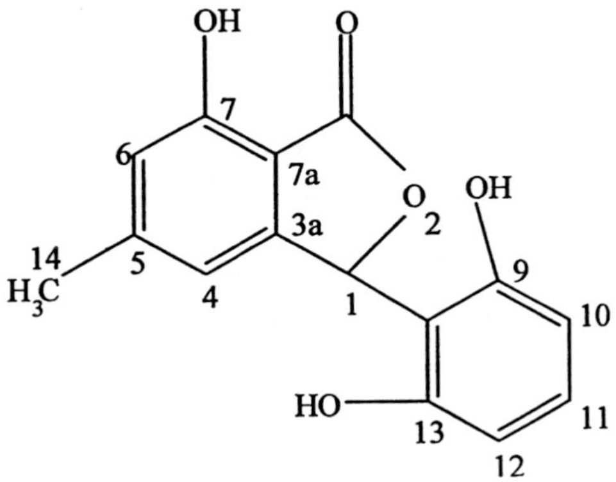

Fig. 9. Isopestacin, an antioxidant produced by an endophytic *Pestalotiopsis microspora* strain, isolated from *Terminalia morobensis* growing on the north coast of Papua New Guinea.

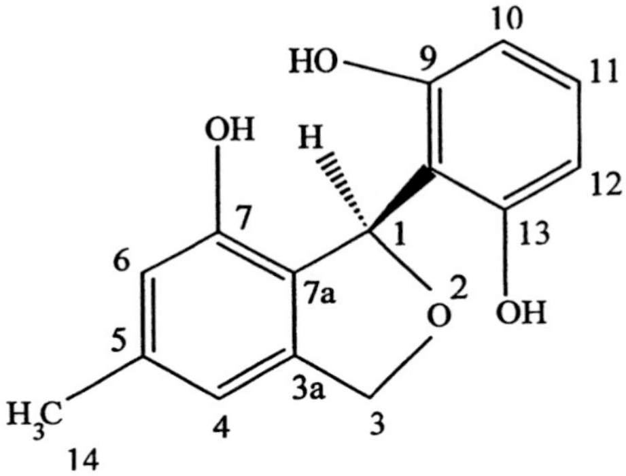

Fig. 10. Pestacin is also produced by *Pestalotiopsis microspora*, and it too is an antioxidant.

B are equipotent in the mixed lymphocyte reaction (MLR) assay and thymocyte proliferation (TP) assay, with an IC_{50} of 0.1 µ*M*. In the same assay systems, the famed immunosuppressant drug cyclosporin A, also a fungal metabolite, was roughly as potent in the MLR assay and 10^4 more potent in the TP assay. Still, the lack of toxicity associated with subglutinols A and B suggests that they should be explored in greater detail as potential immunosuppressants *(65)*.

9. Surprising Results From Molecular Biological Studies on Pestalotiopsis microspora

Of some compelling interest is an explanation as to how the genes for taxol production may have been acquired by *P. microspora (66)*. Although the complete answer to this question is not at hand, relevant genetic studies have been performed on this organism. *P. microspora* Ne 32 is one of the most easily genetically transformable fungi that have been studied to date. In vivo addition of telomeric repeats to foreign DNA generates extrachromosomal DNAs in this fungus *(66)*. Repeats of the telomeric sequence 5'-TTAGGG-3' were appended to nontelomeric transforming DNA termini. The new

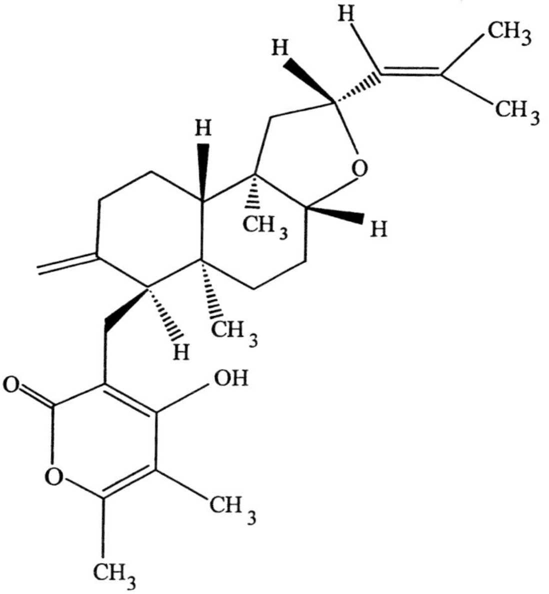

Fig. 11. Subglutinol A, an immunosuppressant, is produced by an endophytic *Fusarium subglutinans* strain.

DNAs, carrying foreign genes and the telomeric repeats, replicated independently of the chromosome and expressed the information carried by the foreign genes. The addition of telomeric repeats to foreign DNA is unusual among fungi. This finding may have important implications in the biology of *P. microspora* Ne 32, because it explains at least one mechanism as to how new DNA can be captured by this organism and eventually expressed and replicated. Such a mechanism may begin to explain how the enormous biochemical variation may have arisen in this fungus *(19)*. Also, this initial work represents a framework to aid in the understanding of how this fungus may adapt itself to the environment of its plant hosts and suggests that the uptake of plant DNA into its own genome may occur. In addition, the telomeric repeats have the same sequence as human telomeres, and this points to the possibility that *P. microspora* may serve as a means to make artificial human chromosomes, a totally unexpected result.

10. Conclusion

Endophytes are a poorly investigated group of microorganisms that represent an abundant and dependable source of bioactive and chemically novel compounds with potential for exploitation in a wide variety of medical applications. The mechanisms through which endophytes exist and respond to their surroundings must be better understood in order to be more predictive about which higher plants to seek, study, and employ in isolating microfloral components. This may facilitate the natural-product discovery process.

Although work on the utilization of this vast resource of poorly understood microorganisms has just begun, it has already become obvious that an enormous potential for organism, product, and utilitarian discovery in this field holds exciting promise. This is evidenced by the discovery of a wide range of products, and microorganisms that present potential. It is important for all involved in this work to realize the importance of acquiring the necessary permits from governmental, local, and other sources to pick and transport plant materials (especially from abroad) from which endophytes are to be eventually isolated. In addition to this aspect of the work is the added activity of producing the necessary agreements and financial sharing arrangements with indigenous peoples or governments in case a product does develop an income stream.

Certainly, one of the major problems facing the future of endophyte biology and natural-product discovery is the rapidly diminishing rainforests, which hold the greatest possible resource for acquiring novel microorganisms and their products. The total land mass of the world that currently supports rainforests is about equal to the area of the United States *(10)*. Each year, an area the size of Vermont or greater is lost to clearing, harvesting, fire, agricultural development, mining, or other human-oriented activities *(10)*. Presently, it is estimated that only a small fraction (10–20%) of what were the original rainforests existing 1000–2000 yr ago, are currently present on the earth *(10)*. The advent of major negative pressures on them from these human-related activities appears to be eliminating entire mega-life forms at an alarming rate. Few have ever expressed information or opinions about what is happening to the potential loss of microbial diversity as entire plant species disappear. It can only be guessed that this loss is also happening, perhaps at the same frequency as the loss of mega-life forms, especially because certain microorganisms may have developed unique specific symbiotic relationships with their plant hosts. Thus, when a plant species disappears, so too does its entire suite of associated endophytes and consequently all of the capabilities that they might possess to make natural products with medicinal potential. Multistep processes are needed now to secure information and life forms before they are lost. Areas of the planet that represent unique places housing biodiversity need immediate preservation. Countries need to establish information bases of their biodiversity and at the same time begin to make national collections of microorganisms that live in these areas. Endophytes are only one example of a life-form source that holds enormous promise to impact many aspects of human existence. The problem of the loss of biodiversity should be one of concern to the entire world.

Acknowledgments

We thank Dr. Gene Ford and Dr. David Ezra for helpful discussions. The authors express appreciation to the NSF, USDA, Novozymes Biotech, NIH, the BARD Foundation of Israel, the R&C Board of the State of Montana, and the Montana Agricultural Experiment Station for providing financial support for some of the work reviewed in this chapter.

References

1. Lane GA, Christensen MJ, Miles, CO. Coevolution of fungal endophytes with grasses: the significance of secondary metabolites. In: Bacon CW, White JF (eds), Microbial Endophytes, Marcel Dekker, New York: 2000.
2. Bacon C, White JF (eds). Microbial Endophytes; Marcel Dekker, New York: 2000.

3. NIAID Global Health Research Plan for HIV/AIDS, Malaria and Tuberculosis. U.S. Department of Health and Human Services, Bethesda, MD, 2001.
4. Demain AL. Industrial microbiology. Science 1981;214:987–994.
5. Schutz, B. in Bioactive Fungal Metabolites—Impact and Exploitation. British Mycological Society, International Symposium Proceedings, Swansea: University of Wales, U.K., 2001, p. 20.
6. Tan R X, Zou WX. Endophytes: a rich source of functional metabolites. Nat Prod Rep 2000;18:448–459.
7. Redlin, SC, Carris LM (eds), Endophytic Fungi in Grasses and Woody Plants. APS, St. Paul: 1996.
8. Hawksworth DC, Rossman AY. Where are the undescribed fungi? Phytopathology 1987;87:888–891.
9. Dreyfuss MM, Chapela IH. Potential of fungi in the discovery of novel, low-molecular weight pharmaceuticals. In: Gullo VP (ed), The Discovery of Natural Products with Therapeutic Potential. Butterworth-Heinemann, Boston: 1994; pp. 49–80.
10. Mittermeier RA, Myers N, Gil PR, Mittermeier CG. Hotspots: Earth's Biologically Richest and Most Endangered Ecoregions. Washington DC. CEMEX Conservation International, 1999.
11. Strobel GA, Li JY, Sugawara F, Koshino H, Harper J, Hess WM. Oocydin A, a chlorinated macrocyclic lactone with potent anti-oomycete activity from *Serratia marcescens*. Microbiololgy 1999;145:3557–3564.
12. Castillo UF, Strobel GA, Ford EJ, et al. Munumbicins, wide-spectrum antibiotics produced by Streptomyces NRRL 30562, endophytic on *Kennedia nigriscans*. Microbiololgy 2002;148:2675–2685.
13. Strobel GA, Stierle A, Stierle D, Hess WM. *Taxomyces andreanae* a proposed new taxon for a bulbilliferous hyphomycete associated with Pacific yew. Mycotaxon 1993;47:71–78.
14. Redell P, Gordon V. Lessons from nature: can ecology provide new leads in the search for novel bioactive chemicals from rainforests? In: Wrigley SK, Hayes MA, Thomas R, Chrystal EJT, Nicholson, N (eds), Biodiversity: New Leads for Pharmaceutical and Agrochemical Industries. The Royal Society of Chemistry: UK, Cambridge, UK: 2000; pp. 205–212.
15. Bills G, Dombrowski A, Pelaez F, Polishook J. Recent and future discoveries of pharmacologically active metabolites from tropical fungi. In: Watling R, Frankland JC, Ainsworth AM, Issac S, Robinson CH, Eda, Z. Tropical Mycology: Micromycetes. New York: CABI Publishing. 2002;2:165–194.
16. Stierle A, Strobel GA, Stierle D. Taxol and taxane production by *Taxomyces andreanae*. Science 1993;260:214–216.
17. Strobel GA. Microbial gifts from rain forests. Can J Plant Path 2002;24:14–20.
18. Strobel GA. Rainforest endophytes and bioactive products. Crit Rev Biotechnol 2002;22:315–333.
19. Li JY, Strobel GA, Sidhu R, Hess WM, Ford E. Endophytic taxol producing fungi from Bald Cypress *Taxodium distichum*. Microbiololgy 1996;142:2223–2226.
20. Strobel G, Yang X, Sears J, Kramer R, Sidhu RS, Hess WM. Taxol from *Pestalotiopsis microspora*, an endophytic fungus of *Taxus wallichiana*. Microbiology 1996;142:435–440.
21. Stinson M, Ezra D, Strobel GA. An endophytic *Gliocladium* sp. of *Eucryphia cordifolia* producing selective volatile antimicrobial compounds. Plant Sci 2003;165:913–922.
22. Strobel GA, Miller RV, Miller C, Condron M, Teplow DB, Hess WM. Cryptocandin, a potent antimycotic from the endophytic fungus *Cryptosporiopsis cf. quercina*. Microbiology 1999;145:1919–1926.
23. Walsh TA. Inhibitors of β-glucan synthesis. In: Sutcliffe JA, Georgopapadakou NH (eds), Emerging Targets in Antibacterial and Antifungal Chemotherapy. Chapman & Hall, London: 1992; pp. 349–373.
24. Li JY, Strobel GA, Harper JK, Lobkovsky E, Clardy J. Cryptocin, a potent tetramic acid antimycotic from the endophytic fungus *Cryptosporiopsis cf. quercina*. Org Lett 2000;2:767–770.

25. Li JY, Harper JK, Grant DM, et al. Ambuic acid, a highly functionalized cyclohexene with antifungal activity from *Pestalotiopsis* spp. and *Monochaetia* sp. Phytochemistry 2001;56:463–468.
26. Lee JC, Yang X, Schwartz M, Strobel GA, Clardy J. The relationship between an endangered North Americn tree and an endophytic fungus. Chem & Biol 1995;2:721–727.
27. Pulici M, Sugawara F, Koshino H, et al. Pestalotiopsin-A and pestalotiopsin-B—new caryophyllenes from an endophytic fungus of *Taxus brevifolia*. J Org Chem 1996;61:2122–2124.
28. Pulici M, Sugawara F, Koshino H, et al. A new isodrimeninol from *Pestalotiopsis* sp. J Nat Prod 1996;59:47–48.
29. Pulici M, Sugawara F, Koshino H, et al. Metabolites of endophytic fungi of *Taxus brevifolia*—the first highly functionalized humulane of fungal origin. J Chem Res 1996;378–379.
30. Li JY, Strobel GA. Jesterone and hydroxy-jesterone antioomycete cyclohexenenone epoxides from the endophytic fungus *Pestalotiopsis jesteri*. Phytochemistry 2001;57:261–265.
31. Hu Y, Chaomin L, Kulkarni B, et al. Exploring chemical diversity of epoxyquinoid natural products: synthesis and biological activity of jesterone and related molecules. J Org Lett 2001;3:1649–1652.
32. Horn WS, Simmonds MSJ, Schwartz RE, Blaney WM, Phomopsichalasin, a novel antimicrobial agent from an endophytic *Phomopsis* sp. Tetrahedron 1995;14:3969–3978.
33. Brady SF, Clardy J. CR377, a new pentaketide antifungal agent isolated from an endophytic fungus. J Nat Prod 2000;63:1447–1448.
34. Zou WX, Meng JC, Lu H, et al. Metabolites of *Colletotrichum gloeosporioides*, an endophytic fungus in *Artemisia mongolica*. J Nat Prod 2000;63:529–1530.
35. Lu H, Zou WX, Meng JC, Hu J, Tan RX. New bioactive metabolites produced by *Colletotrichum* sp., an endophytic fungus in *Artemisia annua*. Plant Sci 2000;151:67–73.
36. Miller RV, Miller CM, Garton-Kinney D, et al. Ecomycins, unique antimycotics from *Pseudomonas viridiflava*. J Appl Microbiol 1998;84:937–944.
37. Harrison L, Teplow D, Rinaldi M, Strobel GA. Pseudomycins, a family of novel peptides from *Pseudomonas syringae*, possessing broad spectrum antifungal activity. J Gen Microbiol 1991;137:2857–2865.
38. Ballio A, Bossa F, DiGiogio P, et al. Structure of the pseudomycins, new lipodepsipeptides produced by *Pseudomonas syringae* MSU 16H. FEBS Lett 1994;355:96–100.
39. Zhang YZ, Sun X, Zechner D, et al. Synthesis and antifungal activities of novel 3-amido bearing pseudomycin analogs. Bioorg & Med Chem 2001;1:903–907.
40. Keiser T, Bibb MJ, Buttner MJ, Charter KF, Hopwood DA, Practical Streptomycetes Genetics. The John Innes Foundation, Norwich: 2000.
41. Guerny KA, Mantle PG. Biosynthesis of 1-N-methylalbonoursin by an endophytic *Streptomyces* sp. J Nat Prod 1993;56:1194–1199.
42. Kunoh HJ. Endophytic actinomycetes: attractive biocontrol agents. Gen Plant Pathol 2002;68:249–252.
43. Castillo U, Harper JK, Strobel GA, et al. Kakadumycins, novel antibiotics from *Streptomyces* sp. NRRL 30566, an endophyte of *Grevillea pteridifolia*. FEMS Lett 2003;224:183–190.
44. Ezra D, Castillo U, Strobel GA, et al. Coronamycins, peptide antibiotics produced by a verticillated *Streptomyces* sp. (MSU-2110) endophytic on *Monstera* sp. Microbiology 2004;150:785–793.
45. Guo B, Dai J, Ng S, et al. Cytonic acids A & B: novel tridepside inhibitors of hCMV protease from the endophytic fungus *Cytonaema* species. J Nat Prod 2000;63:602–604.
46. Worapong J, Strobel GA, Ford EJ, Li JY, Baird G, Hess WM. *Muscodor albus* gen. et sp. nov., an endophyte from *Cinnamomum zeylanicum*. Mycotaxon 2001;79:67–79.

47. Strobel GA, Dirksie E, Sears J, Markworth C. Volatile antimicrobials from a novel endophytic fungus. Microbiology 2001;147:2943–2950.
48. Ezra D, Strobel GA. Effect of substrate on the bioactivity of volatile antimicrobials produced by *Muscodor albus*. Plant Sci 2003;65:1229–1238.
49. Worapong J, Strobel GA, Daisy B, Castillo U, Baird G, Hess WM. *Muscodor roseus* anna. nov. an endophyte from *Grevillea pteridifolia*. Mycotaxon. 2002;81:463–475.
50. Wani, MC, Taylor H L, Wall ME, Goggon P, McPhail AT. Plant antitumor agents,VI. The isolation of taxol, a novel antitumor agent from *Taxus brevifolia*. J Am Chem Soc 1971;93: 2325–2327.
51. Strobel GA, Hess WM, Li JY, et al. *Pestalotiopsis guepinii*, a taxol producing endophyte of the Wollemi Pine, *Wollemia nobilis*. Aust J Bot 1997;45:1073–1082.
52. Strobel GA, Ford E, Li JY, Sears J, Sidhu R, Hess WM. *Seimatoantlerium tepuiense* gen. nov. a unique epiphytic fungus producing taxol from the Venezuelan Guyana. System Appl Microbiol 1999;22:426–433.
53. Li JY, Sidhu RS, Ford E, Hess WM, Strobel GA. The induction of taxol production in the endophytic fungus *Periconia sp.* from *Torreya grandifolia*. J Ind Microbiol 1998;20:259–264.
54. Bashyal B, Li JY, Strobel GA, Hess WM. *Seimatoantlerium nepalense*, an endophytic taxol producing coelomycete from Himalayan yew (*Taxus wallichiana*). Mycotaxon 1999;72:33–42.
55. Young DH, Michelotti EJ, Sivendell CS, Krauss NE. Antifungal properties of taxol and various analogues. Experientia 1992;48:882–885.
56. Wang J, Li G, Lu H, Zheng Z, Huang Y, Su W. Taxol from *Tubercularia* sp. strain TF5, an endophytic fungus of *Taxus mairei*. FEMS Microbiol Lett 2000;193:249–253.
57. Shrestha K, Strobel GA, Prakash S, Gewali M. Evidence for paclitaxel from three new endophytic fungi of Himalayan yew of Nepal. Planta Medica 2001;67:374–376.
58. Hoffman A, Khan W, Worapong J, et al. Bioprospecting for taxol in Angiosperm plant extracts. Spectroscopy 1998;13:22–32.
59. Lee JC, Strobel GA, Lobkovsky E, Clardy JC. Torreyanic acid: a selectively cytotoxic quinone dimer from the endophytic fungus *Pestalotiopsis microspora*. J Org Chem 1996;61:3232–3233.
60. Li C, Johnson RP, Porco JA. Total synthesis of the quinine epoxide dimer (+) torreyanic acid: Application of a biomimetic oxidation/ electrocyclization/Diels-Alder dimerization cascade. J Am Chem Soc 2003;125:5059–5106.
61. Wagenaar M, Corwin J, Strobel GA, Clardy J. Three new chytochalasins produced by an endophytic fungus in the genus *Rhinocladiella*. J Nat Prod 2000;63:1692–1695.
62. Strobel GA, Ford E, Worapong J, et al. Ispoestacin, an isobenzofuranone from *Pestalotiopsis microspora*, possessing antifungal and antioxidant activities. Phytochemistry 2002;60:179–183.
63. Harper JK, Ford EJ, Strobel GA, et al. Pestacin: a 1,3 -dihydro isobenzofuran from *Pestalotiopsis microspora* possessing antioxidant and antimycotic activities. Tetrahedron 2003;59:2471–2476.
64. Zhang B, Salituro G, Szalkowski D, et al. Discovery of small molecule insulin mimetic with antidiabetic activity in mice. Science 1999;284:974–981.
65. Lee J, Lobkovsky E, Pliam NB, Strobel GA, Clardy J. Subglutinols A & B: immunosuppressive compounds from the endophytic fungus *Fusarium subglutinans*. J Org Chem 1995;60:7076–7077.
66. Long DE, Smidmansky ED, Archer AJ, Strobel GA. In vivo addition of telomeric repeats to foreign DNA generates chromosomal DNAs in the taxol-producing fungus Pestalotiopsis microspora. Fungal Genetics Biol 1998;24:335–344.

16

Biological, Economic, Ecological, and Legal Aspects of Harvesting Traditional Medicine in Ecuador

Alexandra Guevara-Aguirre and Ximena Chiriboga

Summary

Harvesting and further developing the traditional herbal remedies in Ecuador (THME) is a very promising area from various perspectives that might be considered and subject to serious analysis. The coincidence of the richest biodiversity in South America confined within a small geographical area presents inherent scientific, pragmatic, and logistical advantages; moreover, Ecuador at this time uses the US dollar as its only currency. This implies that any economic planning can be done by investors using the widest known monetary instrument. Another important development taking place in this country is the effort presently under way to implement proper legal measures to protect foreign investment. At any rate, the endeavor of harvesting and developing THM requires in Ecuador, as in any other fragile location, an integral approach to avoid serious mistakes seen in the past. If such conditions are met, it is very likely that the participating individuals will be conscientious enough to place the interests of collectivity ahead of their own and protect, as their own, the very source that generates any potential earnings: a healthy and rationally used environment.

Key Words: Traditional herbal remedies in Ecuador (THME); drug discovery; plant extracts; Ecuador.

1. Introduction

The knowledge that particular plants have therapeutic value often stems from traditional knowledge and wisdom, which is frequently the true source of information for phytochemical investigation. This ancient knowledge is also the basis for further identification of active components and, in many cases, the foundation for the development of new drugs. In fact, according to experts from the Food and Agricultural Organization (FAO) of the United Nations, 25% of prescriptions written by physicians in the United States and Europe are for compounds derived from plant extracts *(1)*. These are obtained by what appears to be genuine discovery, but should be more accurately seen as clever development derived from observations made on the widespread traditional practices of ancient medicine.

Selection of plants used in traditional practice for the treatment of common illnesses not only suggests a course of treatment to follow for specific pathologies, but also, and more importantly, provides a potential source of mixed active ingredients that may be useful for drug development. As a matter of fact, the use of medicinal plants as raw material for the extraction of pharmacologically active components, as well as of precursors for chemo-pharmaceutical semi-synthesis, is a common procedure utilized in the pharmaceutical industry; moreover, one-third of existing pharmaceutical products are derived from either plants or their components *(2)*.

The above-mentioned facts, taken together with the reality that the vast majority of plant species from the tropical rainforest have not yet been described or classified, highlights the need for further investigation, drug development, and, especially, implementation of regulatory measures. Exploring this potential, with a focus on biological and chemical properties, is a wide-open research field with the promise of more potent, specific, and less toxic drugs.

As an additional and important consideration, several authors have addressed the issue of the economic importance of exploring the qualities of natural products *(3)*. This is supported by the fact that nearly half of the top-selling drugs in the world are either original plant or microbial compounds or their derivatives; moreover, the size of the plant-derived products market is approx $18 billon USD in the United States alone, underscoring the major importance of the worldwide business of plant-derived pharmaceuticals. As a matter of fact, international sales of these products have been estimated to be at least $200 billion USD, with an expanding perspective of growth for the coming years *(4)*.

The present issues about the use and commercialization of medicinal plants must lead to sustainable development strategies that potentially integrate with community participation and involvement, in order to create socially and environmentally sound businesses that protect, rather than harm, nature. This ideal situation can be achieved through international collaboration to design intelligent self-sustainable programs and, at the same time, with commitment to execute them in a responsible manner. Environmental law enforcement, locally and internationally, is the cornerstone to guarantee the proper use and not abuse, of natural resources.

In the last 30 yr, the methodology for separation of active compounds has been dramatically improved through methods of analytical separation, structural elucidation, and quantitative analysis *(5)*. The characterization, in minute detail, of complex chemical compounds can now be accomplished with relative ease, high speed, and reasonable cost. The design and use of potent assays has also achieved modern standards and comprises manually efficient, semi-automatic, and fully automated systems that allow detection of small compounds proven to be active in a wide range of primary biological activities.

Although the focus on production of synthetic drugs by the pharmaceutical industry has until now somehow limited the use of medicinal plants, the industry is again turning to natural products, therefore creating the basis for a new opportunity for plant-derived compounds. In fact, renewed interest in this field has increased substantially in the past few years, emphasizing not only the existing technical feasibility for use and industrial production, but also the important economic potential underlying this matter.

This chapter will discuss the advantages of biodiversity and the possibilities of creating a profitable enterprise while developing sustainable strategies that protect biodiversity.

2. Biological Considerations

2.1 Preserve and Explore the Biodiversity Without Endangering the Environment

Geographical locations, as well as topographical, edaphic, and climatic conditions have resulted in the magnificent ecological mosaic displayed in Ecuadorian biodiversity. These extensive, complex, and inherently rich environmental conditions have also generated an impressive diversity of natural ecosystems in which different species and varieties of plants and animals have successfully evolved, adapted, and flourished *(6)*.

Variations in altitude, humidity, pressure, and precipitation in each zone are reflected in the always-variable vegetation coverage. An entire range of examples of this can be appreciated in a small area; in fact, in Ecuador we observe a dramatic change in ecosystems, from locations with annual grass-dominated deserts, which depend on occasional rain, to very humid tropical forests, including large conglomerates of ancient trees that depend on high precipitation throughout the year, and all of them confined within a restricted geographical location *(7)*. These small examples of very different regions that are generated by a simple gradient in the occidental region within a very small area, emphasize the peculiarities of the country. The Andean foothills also possess these gradients, and they generate natural wonders that range from the typical humid forests to herbaceous-dominated bleak plateaus, and deserts in the high mountains *(8)*.

Ecuador also possesses a great variety of aquatic ecosystems, including marine, coastal, and insular environments, as well as fluvial and continental lagoon areas. These ecosystems shelter a great diversity of organisms, many of which are of economic importance. Although information about diversity at the flora and fauna species level is scarce, dispersed, and heterogeneous, preliminary data confirm the existence of an enormous biological richness *(9)*.

According to Steere *(10)*, Ecuador is the country with the largest number of plant species per unit area in South America. Its flora encompasses approx 20,000 to 25,000 species of vascular plants, many of them unique, and with an estimated endemic value of 20% based on local flora distribution patterns. In the same way, the diversity of nonvascular plants is very high, although no estimate of the total number of species existing in the country is available. These realities highlight the need for investment in science and technology to explore a largely untouched world full of promise. Furthermore, investigations of the genetic diversity of natural plants of Ecuador reveal the presence of an extraordinary diversity in phytogenetic resources that might be important for improving the production of unique natural foods and for diversifying cultivation *(11)*.

Myers *(12)* identified 10 areas that harbor somewhere between 30 and 40% of the world's biological diversity, but some of the locations are the most threatened by human activities. These areas, characterized by an exceptional concentration of flora and fauna and with a large percentage of endemic species, are known in the environmental sciences as "hot zones" or "hot spots," and comprise 1% of the surface of the planet. In spite of its very small surface, Ecuador contains three of these zones which are the very humid tropical forests of the occidental coast, the external flanks of the Andean-range mountain chain, and the Amazon forests of the northeastern area. The protection and preservation of these areas is considered to be one of the top priorities for the conservation of biological diversity at the global level. In this regard, we are witnessing, in the last few years, what seems to be independent events such as variations in weather, alteration in rain patterns, ozone-layer deterioration, polar cap melting, and other environmental tragedies. It is clear now that all of these are related to the nonrational use and abuse of natural resources and destruction of nature *(13)*. Economic interests with no consideration for damage to nature have greatly surpassed the capability of nature to restore its losses, and have led to relentless destruction. Similarly, and among other factors, the destruction of habitats, the overexploitation of resources, the introduction of nonnative species, and environmental contamination are leading to the disappearance of Ecuador's flora and fauna species. This progressive detriment of the habitat is also leading to the irreparable loss of valuable genetic information stored in wild species and biodiversity.

Countless species have already disappeared, and several hundred more are in danger of extinction due to the destruction of occidental forests. An isolated but sad example of this fact is the *Dicliptera dodsoni* species, endangered as a result of the conversion of very humid forests of the coast into banana and African-palm plantations, as well as into land for pasture *(14)*. Some other species, like *Tabebuia chrysantha* (Guayacán), which is the most valued wood in dry tropical forests of the coastal area, have greatly diminished due to indiscriminate exploitation *(14)*. The list of destroyed or permanently damaged plant species is enormous and beyond the scope of this chapter; however, it emphasizes the need for more rational and intelligent development procedures, associated with stronger policies specifically designed to protect extremely fragile resources. In similar considerations, the displacement of native crops as well as the new agricultural practices are causing the accelerated disappearance of genetic resources stored in native cultivated species *(15)*. Indeed, modern agriculture is promoting the replacement of indigenous species and varieties with new species and biotechnologically modified varieties. Even if we consider that under some conditions these newly developed or modified plant species are "more economically efficient" and indeed could raise the production of individual parcels, the unification of crops may result in massive losses of general harvests due to plagues, drought, or frosts induced by severe alterations in the environment caused by modern technologies. This vicious circle can be found in almost every location of Ecuador; moreover, it has greatly contributed to the already massive destruction of large areas of the Ecuadorian rainforest *(16)*.

2.2. Bio-Prospecting for Natural Resources in Developing Countries

Bio-prospecting for natural resources comprises the study and collection of biological specimens from a certain location, aiming to discover biological species with commercial or industrial use *(17)*. One of the oldest and best-known examples is that of quinine, extracted from the "chinchona" or "cascarilla," a common plant of the Andes *(10)*. This life-saving plant was studied and quinine was extracted and used for therapeutic purposes worldwide. Some people from the drug industry, fortunately not all, claim that uneducated people, such as shamans, have never found a pure compound, and that is true. However, it is also true that the scientist would never have found many compounds without ancient knowledge, including the hundreds of years of empirical experimentation with plants.

In Ecuador, recently established companies such as Shaman International demonstrate slight improvements over previous practices *(18)*. This firm specializes in investigating traditional medical knowledge to enhance the possibilities of finding promising active ingredients. Shaman's investigations in Ecuador and Peru yielded two active-ingredient patents from *Croton lechlieri*; the patents belong exclusively to the company. This entity continues funding the production of large quantities of raw material; however, it is expected that this production will generate jobs for people within the area and thereby benefit the local economy. Through their foundation, the Healing Forest Conservancy, Shaman Pharmaceuticals has signed contracts with several Ecuadorian and Peruvian communities in order to cultivate *Croton lechlieri* and buy from them the corresponding raw materials, generating jobs while saving taxes because of the legal advantages that foundations have in Ecuador. These types of new approaches have to be evaluated and regulated by authorities and promoted if proven beneficial to the country.

At any rate, it is important to acknowledge that any new plantations must be developed with great caution, since single-crop farming might cause further damage to an already altered environment. It is also essential to analyze whether these newly developed plantations will eventually replace primary forests or other subsistence plantations. If that is the case, a sound plan for regional development and substitution of primary forests must first be developed.

2.3. An Optimistic Scenario

Aggressive bio-prospecting efforts and leakage of genetic resources out of biodiversity-rich countries, such as Ecuador, are problems that should be addressed through the clear commitment of scientists who have in mind the importance of the development of the therapeutic potential of a large variety of Ecuadorian plant species. Moreover, the rational design of research protocols, aiming to test both safety and efficacy of promising Ecuadorian plant species, will certainly lead to sound proposals for the international community and for the industry. The fact that scientific studies can be done more economically in Ecuador than in other countries is an advantageous feature that might attract potential investors and companies. Nowadays, the existing therapeutic leads, as well as the already proven uses of plant extracts, are provided not only by ancestral knowledge but are also being developed by local scientists. International cooperation and the industry will find a fertile soil for their specific aims, be they conservation, science, education, self-sustainable development, or profit.

2.4. Economical Considerations of Biodiversity

The assignment of an economic value to biodiversity is not an easy task; nevertheless, a few examples can be used to obtain an estimate of the economical potential of this resource. One of the best evaluations was performed by Simpson, Sedjo, and Reid *(17)*, who concluded that the pharmaceutical industry is willing to invest more for conservation of the habitat in Ecuador than in any other part of the world. They showed that the industry will not only greatly benefit in general, but indeed that ultimately it will save money if makes a minimum annual investment of $20 USD per hectare to implement preventive measurements against deforestation. Calculations have been made, e.g., those of Adger et al. *(19)*, that estimate an annual average value of $22.4

USD per hectare for biological conservation of the forests of Mexico. Using parameters such as those used by these authors, but considering that the biodiversity index of the high Amazonian basin (Ecuador) is much higher *(20)*, the corresponding estimate for Ecuador is also much elevated.

Biodiversity poses an enormous economic potential and can produce innumerable benefits by serving not only as the origin of raw material but also as the source of miscellaneous ingredients for the chemical industry, drug production, and other related activities. For this reason, prospecting biodiversity and its commercial application have gained special attention *(21)*. Collaboration between nations and the pharmaceutical industry have met with success; the most well-known example is the agreement between the INBio *(3)* in Costa Rica and Merck Ltd. Unfortunately, Ecuador and most other countries in Latin America have not yet fully developed and benefited from their biological potential. It is clear that during the next few years, cooperative efforts of this kind will flourish more broadly. Therefore, it is necessary to establish means that support and regulate bio-prospecting to make it profitable and allow the use of biodiversity in a sustainable manner. A country such as Ecuador, rich in germoplasm and traditional knowledge, might well be a suitable partner for participation within the borders of the ever-expanding pharmaceutical market. Several studies of medicinal plants have only just begun to describe the natural product potential of Ecuador's biodiversity. In this sense, Vogel *(22)* referred mainly to two studies—that of Ruiz *(23)*, who published a list of the 43 species of the Andean region with 2.7 medicinal properties per species; and that of Mena *(24)*, who documented the useful therapeutic plants in the low north-occidental area, listing 63 species with a median of 2.6 medicinal properties per species. After eliminating duplication of species in the two regions, Vogel calculated a total of 256 medicinal properties.

The multiplication of this number by known current values ($466.00 USD), according to Aylward calculus *(25)*, and by 2500 (the 25% bioactivity calculus in agreement with Balick *(26)*, and by 7.5 (the efficient bonus rate of 15%), results in an expected potential value of this list of traditional plant medicines in the range of $2.24 billion USD *(8)*. Ecuador must increase its efforts to obtain a larger share of this promising natural-products market.

2.5. Alternatives for Development That Preserve the Environment: Sustainable Development

In its classic form, the concept of sustainable development has been described as "The development that meets the needs of the present without compromising the ability of future generations to meet their own needs" *(27)*. However, we are daily witnessing the dramatic consequences of the destruction of the environment, including dramatic changes in weather, rain patterns, and the disappearance of species.

Simple solutions, such as those of cultivation of the desired plant species as an alternative to forest degradation, can be easily implemented. In this respect, an edaphic assessment of the land and rational use of it can be scientifically ascertained and the most appropriate type of cultivation selected; therefore, efficient productivity can be achieved without destruction or major alteration of primary forest areas *(14)*. In accordance with this general philosophy, community participation should be utilized—for example, by combining production with traditional agricultural practices that lead to achievement of an appropriate balance in agro-forestry, therefore providing additional economic income to those communities who wish to collaborate. In this model, local communities can provide raw materials in addition to maintaining their traditionally grown crops. These simple measures will not only create jobs and economic benefits for those communities, but will also serve to dignify, without charity, the lives of these presently impoverished individuals.

3. Medicinal Plants in Drug Development

3.1. Phytotherapeutics As an Alternative System of Primary Medicine

In developing countries, the use of medicinal plants is highly controversial because of lack of knowledge and poor understanding of the contextual circumstances by both the public and the authorities responsible for regulating the use of these types of medicines. In fact, there is an erroneous political attitude supposedly aimed to avoid offenses to the owners of ancestral knowledge, considered to be a discrete minority. In reality, the ministries of health of developing countries display a dual and contradictorily political behavior that never takes a clear and univocal attitude towards proper regulation; moreover, these authorities prefer, most of the time, to simply ignore the widespread nonjudicious use of botanicals that leads to, in many cases, public and specific health dangers and even catastrophes. For instance, and without any objective action being taken from the regulators, it is not uncommon to observe the unaffected practice of some witch-doctors who advise the public against the use of vaccines, and their replacement with herbal teas and ineffective and dangerous measures.

It is true that an ethno-pharmacological tradition exists; however, many recent entrepreneurs of botanicals have created a vast, frequent, and indiscriminate use of plants and plant extracts by millions of people, generating not only health dangers but also permanent destruction of large areas of the precious and irreplaceable primary rainforest. In multiple instances, the real therapeutic potential of these resources is assumed and promoted, but there is little if any systematic scientific proof of both efficacy and safety. The determination of the efficacy of medicinal plants and their therapeutic usefulness is of primary importance for therapy as well as for safety. This is critically important nowadays because of the proliferation of small, unregulated firms that promote and market the use of hundreds of poorly studied natural remedies. In fact, the majority of plants used in folk medicine, and phytomedicines in general, are traditionally sold over the counter, and lack adequate pharmacological, toxicological, and clinical evaluation (28). In this context, self-medication and uncontrolled use of homemade remedies are undeniable facts that sooner or later will generate health deterioration. It must be remembered that, in contrast to wealthy communities, the use of these products in developing regions is not an alternative to other effective medicines, but indeed the only option for primary health care (5). Therefore, implementation of good practices for production, development, and use of these compounds, as well as their scientific evaluation and separation of toxic products, are of primary importance to public health. Recent history has shown that the tools needed to discover, characterize, and fully study a pharmaceutical remedy can be effectively and properly used under cooperative programs between academicians and the pharmaceutical industry; similar measures for proper pharmaceutical development of botanicals need to be implemented. As an example, the previously mentioned discovery of the medical use of quinine by ancient aborigines and further characterization of quinine by the industry clearly epitomizes a strategy that has been successfully used in recent decades *(10)*. We believe that the correct approach is for all parties to share and combine knowledge and technology to obtain new compounds for pharmaceutical use and profitability, both for the pharmaceutical industry as well as for the rightful owners of the initial discovery. In this fair fashion, new remedies obtained from the use of traditional practices and technologically advanced measures, within the framework of an ethically correct approach, might prove to be a helpful strategy to solve the present health crisis in underdeveloped countries. Moreover, these actions will have additional beneficial effects, both by their inherent action and also by indirect mechanisms, such as the generation of jobs. Taken together, these aspects and perspectives suggest that development of national programs to integrate the different partners of this large endeavor would be highly desirable; moreover, governments must also participate in the proper regulation of safety, production, quality control, and marketing of botanicals.

It must be emphasized that a significant percentage of the Latin American population does not have access to proper medical care. It is believed that fully one-half of the people in this part of the continent have limited access, or no access at all, to the most needed medicines and drugs. The direct impact of these statistics falls on children and lactating mothers. In 1980, for example, the worldwide consumption of medicines reached approx $80 billion USD, of which only $6.4 billion (8%) were used in Latin America *(21)*.

In addition, the increased control of the local market by foreign enterprises, the low percentage of local investment, and negligible participation of governmental entities in drug supply have also influenced the shift in preference by the poorest and neediest sectors of the population toward traditional medicine and unvalidated natural products. It is believed, as ascertained by officials of the Institute for International Cooperation and Development, that as much as 70% of the population use plants as their sole source of remedies for miscellaneous illnesses in this part of the world *(29)*. The fact that the vast majority of these products have no controlled proof of efficacy, no studies done on their safety profile, and are used by an extremely fragile part of the population speaks to the dire situation in Latin America in terms of human health and social development.

In response to these circumstances, the World Health Organization (WHO), in the context of its Resolution WHA 31.33, has recognized the importance of medicinal plants for primary healthcare in this part of the planet and has recommended to its Member States the use of a comprehensive approach to medicinal plants *(30)*. Among others, the following recommendations have been made for State Members:

- A periodic update of the inventory and classification of medicinal plants used in different countries.
- The establishment and development of scientific criteria and methods to ensure the quality of medicinal plant preparations as well as the study of their efficacy in the treatment of specific conditions and illnesses.
- The development of international standards and specifications for identification, purity, potency and good manufacturing practices.
- The development of a manual of methodologies for the effective and secure use of phytotherapeutic products by health professionals.

- The distribution of all related information to Member States.
- The allocation of resources by local governments for research and training of human resources for the study of medicinal plants.
- The initiation of global programs for the identification, evaluation, preparation, culture, controlled growth, and conservation of medicinal plants used in traditional medicine.

In addition, and as discussed previously, the market for plant-based medicines in industrialized countries is very important and displays an ever-expanding pattern. Plants are also commonly used by the pharmaceutical industry as raw materials for many of its products. In the last decade, the market for botanicals was estimated in the range of $35 billion USD per year *(3)*, with a considerable increase during the last five years.

In summary, medicinal plants are used in every country as both a raw material and in the form of plant extracts, as semi-purified forms, as an inexpensive source of active compounds for chemical or semi-synthetic procedures, and for other miscellaneous uses. The large increase in the use of medicinal plants for primary healthcare in Latin America, as well as the unusually high interest in naturally derived medications in developed countries, constitute a very promising opportunity not only for scientists and researchers in the pharmaceutical industry but also for entrepreneurs and investors.

3.2. Therapeutic Potential of Plants in Ecuador

In the last few years, the scientific investigation of plants has received wider attention. Innumerable plant species have been selected and studied in the search for therapeutic effects and qualities; as a consequence, interesting and novel medicinal properties have been described. Many active compounds have been further characterized and are used in the therapeutic armamentarium of modern physicians, who are often unaware of the botanical origin of the medications they commonly prescribe.

A vast variety of botanicals are available as nonprescription as well as prescription drugs. In addition to their use in the pharmaceutical industry, plant extracts and their derivatives are also used in the cosmetic and other industries, including the food business. Botanists living in Ecuador have only recently begun to explore its rich resources. In addition to the Amazon Basin, other important plant resources are found in the Ecuadorian Sierra (Andean Region) *(4)*. The following are just a few examples of the many hundreds of interesting plant species that can be found growing wild in the various regions of Ecuador: *Bidems leucanta* Willd, *Bidems pilosa* Linneo, *Ceratocephalus pilosus*, Rich, *Kerneria duvia* Cass, and *Kerneria tetrágona* Moench, commonly named Morisco, Putso, or Shirán. The last-cited plant is described as having hypotensive, hypoglycemic, and antiviral activities. It contains compounds proven to be phototoxic to bacteria and fungi *(31)*. It is common knowledge that among its traditional uses, its infusion produces diuretic effects, and its topical application, as an ointment, helps to relieve pain. Many old people chew the plant leaves to alleviate their rheumatic pain, apparently with good results. Notably, in many instances, it might be assumed that its use is preferred because of its low cost; however, in some locations, people with proper income and access to expensive drugs still prefer to use these leaves instead of sophisticated anti-inflammatory pharmaceutical compounds, due to negligible side effects.

Culcitium reflexum HBK, *Lasiocephalus ovayus* Schlecht, commonly named Arquitecta, has traditional medicinal uses as a diuretic and a "purifier or depurative." It has also been used in the treatment of syphilis *(31)*. *Xantium catharticum* HBK, commonly named Cashamarucha, has been traditionally used in folk medicine as a diuretic and purgative. It is a popular and widespread belief that women should not use this plant, as witch-doctors warn that it also possesses abortive properties. Chuquiragua insignes, commonly named Chuquiragua, have a traditional medicinal use against hepatic problems and also as a diuretic, tonic, and for relief of persistent coughs. An infusion of its flowers is commonly recommended *(32)*.

Bacharis riparia HBK, *Bacharis polyanta, Bacharis latifolia* HBK, commonly named Chilca, Chilca negra, Chiza, or Yana chilca, has a traditional use as an infusion to alleviate diarrhea in children *(31)*. This plant is commonly used throughout the Andean region of Ecuador. Needless to say, this remedy is the only alternative in some impoverished locations of this region. *Senecio vaccinoides* Kunth, commonly named Cubilán, Ayalongo, or Cubillín, is used as a potent analgesic when taken as an infusion *(31)*. The more intense the pain, the more concentrated the infusion must be. *Althernanthera panículata*, commonly named Moradilla, is used as an infusion to alleviate renal and hepatic diseases. *Franseria artemisoides* Willd, commonly named Marco, Altamisa, or Marcu, is commonly used as a topical agent to alleviate pain. The flowers of this plant are used for the preparation. It is also employed as a flea repellent *(32)*. *Aristiguietia glutinoso, Eupatorium glutinosum* HBK, commonly named Matico, Soldier herb, or Chuzalongo, is one of the most popular remedies in the Ecuadorian Sierra (Andean region). Its infusion is traditionally used to cure gastritis and ulcers. A preparation of the steamed or cooked leaves is used to alleviate rheumatic pain, renal disease, and to clean external wounds.

It should be mentioned that some preliminary phytochemical analyses of several of the aforementioned plant species have been carried out. Investigation of *Schistoscarpa aff euphatoroides* and *Grias neubertii* McBride has revealed the presence of numerous terpenoid structures that might indicate an inherent antimycotic activity *(33)*. In addition, the presence of sesquiterpene hydrocarbons among its structural components supports the popular use of this plant as an insecticide. In addition, the antibacterial and antimycotic activity of *Minguartia guianensis* Aubl. and *Clavija procera* have been studied by Villacrés and collaborators *(34)*.

With regard to other Ecuadorian medicinal plants, the anti-inflammatory activities of several species have been studied in some detail by using various in vivo models *(35)*, most frequently of the so-called carragenin-induced edema *(36)*. These models allow the documentation, on a qualitative and semi-quantitative basis, of the relative anti-inflammatory capabilities of plant extracts. The determination of the range of activity and not of the mechanism underlying the anti-inflammatory process by these methods is a limiting technological problem that needs further assessment and refinement to obtain the active principles responsible for the observed effects. At any rate, these models are extremely useful, as they allow an objective evaluation of a discrete activity that can be, at least, a starting point in the elucidation of the active compounds capable of interfering in the genesis of the inflammatory process. In fact, carragenin-induced rat-plantar edema, as described by Wintern *(37)*, is a very sensitive model that allows evaluation of anti-inflammatory substances. Using this methodology, the authors studied Wistar rats distributed in homogeneous groups of 10 animals each *(35)*. Extracts, positive and negative controls, and pattern batches were tested using this model. The

Table 1
Relative Anti-Inflammatory Activity of Aqueous Plant Extracts

Investigated substance	3 h (%)	5 h (%)	7 h (%)
Phenylbutazone	59.32	51.67	62.96
Baccharis trinervis	−11.87	10.83	60.49
Baccharis teindalensis	−25.42	−15.00	56.79
Eupatorium glutinosum	25.42	18.33	43.21
Eupatorium articulatum	37.29	41.67	49.38
Tagetes pusilla	16.95	18.33	20.99
Conyza floribunda	−16.95	−8.33	62.96
Neurolaena lobata	−5.08	20.00	58.02
Urena lobata	8.47	21.67	60.49
Croton draco	−8.47	−8.33	59.26

Negative values indicate larger plantar edema than that seen in the control animals (these received only water *ad libitum*).

results obtained can be seen in **Tables 1–6**. Briefly, **Table 1** shows the anti-inflammatory activity of various aqueous plant extracts studied with the carragenin-induced rat-plantar edema method (reference drug phenylbutazone, a well-known anti-inflammatory agent). Results are expressed as the percent activity of the aqueous plant extracts, at different times (3, 5, and 7 h); phenylbutazone had a percent activity of 59.32, 51.67, and 62.96 at those times. **Table 2** and **Table 3** show the activities generated by the plant extracts obtained by using a dichloro-methanic (**Table 2**) and by an ethanolic (**Table 3**) extraction method, respectively and at the same times.

Tables 4–6 show the relative anti-inflammatory activities (RAA) of different plant extracts (aqueous, dichloro-methanic, and ethanolic, respectively). The common statistical method of the Student's t test was used for all comparisons. In these studies, the RAA of all plant extracts was compared to that of phenylbutazone, to which a value of 1 was putatively ascribed. The statistical analysis included, sequentially, the use of the Kolmogorov test to ascertain whether any two data samples were compatible with being random sampling of the same, unknown distribution, of all the data generated from both the use of the extracts and phenylbutazone as well as from the controls. Thereafter, the Student's *t* test was used to establish the differences between the groups and a statistical *p* value less than 0.05 was set as the observed level of significance. This means that any *p* value similar to or <0.05 or <0.01 found, shown in **Tables 3–6**, corresponding to any of the extracts, signifies that this discrete and particular compound has an anti-inflammatory activity comparable to or better than phenylbutazone. On the other hand, the nonsignificance (abbreviated as ns) of any value generated from a given compound means that such extract does not possess any anti-inflammatory activity, at least, as assessed by this methodology.

In summary, results from **Tables 1–6** lead to the following observations *(35)*:

1. When compared with the pattern drug, phenylbutazone, the anti-inflammatory properties of all aqueous extracts, with the exception of Eupatorium articulatum, showed a relatively low anti-inflammatory activity at 3 and 5 h after administration. At 7 h, these extracts

Table 2
Relative Anti-Inflammatory Activity of Diclorometanic Plant Extracts

Active substance	3 h (%)	5 h (%)	7 h (%)
Phenylbutazone	39.69	53.22	93.75
Baccharis teindalensis	5.00	−3.08	−16.25
Baccharis trinervis	9.37	5.60	−0.94
Eupatorium glutinosum	2.19	31.93	62.50
Tagetes pusilla	46.88	31.93	62.59
Conyza floribunda	42.81	25.21	37.50
Neurolaena lobata	41.56	36.41	33.44
Urena lobata	33.44	44.82	33.44
Croton draco	35.31	28.01	12.50
Eupatorium articulatum	37.50	52.38	41.56

Negative values mean larger plantar edema than that seen in the control animals (these received only water *ad libitum*).

Table 3
Relative Anti-Inflammatory Activity of Ethanolic Plant Extracts

Investigated substance	3 h (%)	5 h (%)	7 h (%)
Phenylbutazone	29.13	50.59	44.69
Baccharis trinervis	12.62	32.94	21.79
Eupatorium glutinosum	26.21	38.82	22.91
Tagetes pusilla	40.78	62.35	64.26
Conyza floribunda	44.57	36.47	29.61
Neurolaena lobata	0.00	14.71	19.55
Urena lobata	6.81	20.59	21.79
Eupatorium articulatum	59.22	42.39	53.63

reach their maximum activity, which was similar to that of phenylbutazone. Six of the nine extracts showed a relative activity of 0.9 to 1 in comparison with 1 of the pattern drug.

2. For the anti-inflammatory activity of the dicloromethanolic extracts, the maximum activity was observed after 3 h of administration, with a decrease at 5 h and a further decrement at 7 h. Interestingly, six of the nine extracts showed their maximum relative activity at 3 h, with a range between 0.84 and 1.18 with respect to 1 of the pattern drug phenylbutazone.

3. The ethanolic plant extracts were the most active: various degrees of important relative activity were documented at 3, 5, and 7 h. Surprisingly, the relative activity of some of these extracts was superior to that of the pattern drug at various times. The plants with the highest relative activities were *Tagetes pusilla* (Asteraceae), with 1.4, 1.23, and 1.44, and *Eupatorium articulatum* (Asteraceae), with 2.03, 0.84, and 1.20, at 3, 5, and 7 h, respectively, and *Conyza floribunda*, with 1.53 at 3 h, as compared with 1 of phenylbutazone.

Table 4
Relative Anti-Inflammatory Activity (aa) of Aqueous Plant Extracts As Compared With Phenylbutazone (aa = 1.0)

Plant species	Relative anti-inflammatory activity			Statistical analysis (student's t test)		
	RAA at 3 h	RAA at 5 h	RAA at 7 h	3 h	5 h	7 h
Croton draco	0.00	0.00	0.94	ns	ns	$p < 0.01$
Tagetes pusilla	0.29	0.36	0.33	$p < 0.05$	ns	$p < 0.01$
E. glutinosum	0.43	0.36	0.69	ns	ns	$p < 0.01$
Baccharis teindalensis	0.00	0.00	0.90	ns	ns	$p < 0.01$
Baccharis trinervis	0.00	0.21	0.96	ns	ns	$p < 0.01$
N. lovata	0.00	0.39	0.92	ns	ns	$p < 0.01$
Conyza floribunda	0.00	0.00	1.00	ns	ns	$p < 0.01$
Urena lobata	0.14	0.42	0.96	ns	ns	$p < 0.01$
Eupatorium articulatum	0.63	0.81	0.68	ns	$p < 0.01$	ns

A p value less than 0.05 ($p < 0.05$) is considered statistically significant.
RAA, relative anti-inflammatory activity; ns, not statistically significant

Table 5
Relative Anti-Inflammatory Activity (aa) of Diclorometanic Plant Extracts As Compared With Phenylbutazone (aa = 1.0)

Plant species	Relative anti-inflammatory activity			Statistical analysis (student's t test)		
	RAA at 3 h	RAA at 5 h	RAA at 7 h	3 h	5 h	7 h
Croton draco	0.89	0.53	0.13	$p < 0.01$	$p < 0.05$	ns
Tagetes pusilla	1.18	0.60	0.67	$p < 0.01$	$p < 0.05$	$p < 0.01$
Eupatorium glutinosum	0.06	0.00	0.03	ns	ns	$p < 0.05$
Baccharis teindalensis	0.13	0.00	0.00	ns	ns	ns
Baccharis trinervis	0.24	0.11	0.00	ns	ns	ns
N. lovata	1.05	0.68	0.36	$p < 0.01$	$p < 0.01$	$p < 0.05$
Conyza floribunda	1.08	0.47	0.40	$p < 0.01$	$p < 0.05$	ns
Urena lobata	0.84	0.84	0.36	$p < 0.01$	$p < 0.01$	$p < 0.01$
Eupatorium articulatum	0.94	0.98	0.44	$p < 0.01$	$p < 0.01$	$p < 0.01$

A p value less than 0.05 ($p < 0.05$) is considered statistically significant.
RAA, relative anti-inflammatory activity; ns, not statistically significant

In a different but related matter, it is important to mention that, among the many Ecuadorian plant-derived products already commercialized, there are some apparently novel products presently being heavily marketed and sold in large quantities. Some of them have been promoted as having protean and intense activities on various bodily systems, including the immune, cardiovascular, gastrointestinal, and dermatological systems. These agents are also not only being sold in Ecuador, but also exported to

Table 6
Relative Anti-Inflammatory Activity (aa) of Ethanolic Plant Extracts As Compared With Phenylbutazone (aa = 1.0)

Plant species	Relative anti-inflammatory activity			Statistical analysis (student's t test)		
	RAA at 3 h	RAA at 5 h	RAA at 7 h	3 h	5 h	7 h
Tagetes pusilla	1.40	1.23	1.44	$p < 0.01$	$p < 0.01$	$p < 0.01$
Eupatorium glutinosum	0.90	0.67	0.51	$p < 0.05$	$p < 0.05$	ns
Conyza floribunda	1.53	0.72	0.66	$p < 0.01$	$p < 0.01$	$p < 0.05$
Baccharis trinervis	0.43	0.65	0.49	ns	$p < 0.05$	ns
N. lovata	0.00	0.29	0.44	ns	ns	ns
Urena lobata	0.23	0.41	0.49	ns	ns	ns
Eupatorium articulatum	2.03	0.84	1.20	$p < 0.01$	$p < 0.05$	$p < 0.01$

A p value less than 0.05 ($p < 0.05$) is considered statistically significant.
RAA, relative anti-inflammatory activity; ns, not statistically significant.

other countries as "magic" entities aimed to cure cancer, all types of rheumatic disorders, as well as other human diseases. In spite of anecdotal accounts of efficacy of these compounds, supposedly by mechanisms known only by their promoters, there is no credible evidence that supports any clear therapeutic activity. For these reasons, we firmly believe that these products need to undergo real scientific testing with rigid regulatory procedures included in the study protocols. This will not only ensure human health and document possible side effects, but will also save money and avoid major fraud to people.

The enormous therapeutic potential derived from plants is a very promising alternative for drug discovery, and could also help to provide a model for genuine sustainable development that preserves, restores, and intelligently uses natural resources. The implementation of such a system will not only help to create jobs, wealth, education, and environmental consciousness in developing nations, but will also create a model of true feasibility of the coexistence of environmentally protective endeavors along with profitable businesses. This is especially important at the present time, when the industrialized and rich nations of the world are mainly responsible for the worst pollution the world has ever seen.

4. Legal Considerations

Besides major traditional legal principles already included or adapted as integral parts of legislation of most countries, and besides declaratory or law-designing meetings, including the very well-known Rio de Janeiro Biodiversity Treaty, the Kyoto Protocol, and other environmental legal milestones, it is pertinent to briefly cite some legal instruments relevant to the Ecuadorian context.

4.1. Ecuadorian Legislation

In 1996, because of the necessity of determining the value of wild species and traditional knowledge in all projects included in the so-called genetic resource area, the Ecuadorian Congress approved the "Law that protects the biodiversity in Ecuador."

Similarly, in 1998, the legislation approved the legal body that addresses intellectual property. It is contained in "The Intellectual Property Law" issued at that time *(13)*.

The law that protects biodiversity was published in Official Register No. 35, dated September 27, 1996. This law defines biodiversity as a national good of public use and ratifies the contents of the Biological Diversity Agreement, issued for those countries with sovereign rights over biological resources. The rights of indigenous communities over all knowledge associated with biodiversity are also recognized in this law. Nevertheless, and because of its brevity and the mainly declarative character of its articles, the law is insufficient to regulate the majority of multifarious and miscellaneous factors related to the conservation and sustainable use of biodiversity.

The Ecuadorian Policy on Forestry, Natural Areas, and Wildlife *(23)* favors the conservation and sustainable use of biodiversity *(38)*. This policy encourages the coordination and participation of institutions involved in this area to design a strategy for the investigation, monitoring, evaluation, and protection of biodiversity. In this sense, this policy is aimed at regulating issues such as the evaluation of markets for products and services derived from biodiversity, the establishment of access or restriction to resources of biodiversity, and the creation of biodiversity investigation programs. It also includes the determination of procedures for economic, scientific, ecological, genetic, and cultural evaluation of the products and services derived from biodiversity, as well as the development of alternatives for the protection and management of biological resources.

4.2. International Legislation: The Andean Pact

In reference to treatment of intellectual property rights in countries belonging to the Andean Pact, two standards were established: Decision 345 and Decision 344 *(38)*. Decision 345 recognizes and guarantees the rights of the developers of new vegetable varieties through an achiever certificate, as follows: "Member States will confer an achiever certificate to those persons who have created vegetable varieties when these are new, homogeneous, distinguishable and stable, and have been assigned a denomination that constitutes its generic designation. It should be understood that the term 'create' refers to the achievement of a new variety through the application of scientific knowledge or the hereditary enhancement of the parts."

Decision 344 establishes patents and states the following: "Member States will confer patents for inventions which are products or procedures in every technological field. Products and procedures should be new, have an inventive level and be susceptible of industrial application. An invention is new when it is not comprehended in the technical statement, meaning that it has not been accessible to the public through an oral or written description, through its use or in any other way before the date of patent solicitation, in the case of recognized priority."

5. Conclusions

Considering that Ecuador possesses some of the greatest biological diversity in the world despite consisting of less than 0.2% of the planet's surface, an investment in developing its natural resources seems appropriate. This small country has the highest biodiversity per hectare in all of South America, and harbors approx 25,000 vegetal species that encompass enormous therapeutic potential.

In this context, the feasibility of implementing research studies locally and at a very reasonable cost is one of the most attractive features for developing partnerships among entities from both the developed and developing countries, including Ecuador. These study protocols, aimed at testing for safety and efficacy of the plant extracts, are the cornerstone of modern and rational drug development. The latest introduction of appropriate review boards that meet international standards, guarantees the enforcement of the most important aspect of medical investigation in the world: the ethical compliance by both local and international partners.

The feasibility of locally implementing sound scientific and well-controlled studies also alleviates the heavy economic burden that companies belonging to the developed world have to bear. As a matter of fact, there are many institutions and young scientists in South America and Ecuador that are waiting their opportunity to collaborate in studies not only of plant extracts, but of pharmaceuticals in general. In this sense, with rigid regulation and extensive monitoring, two low-cost procedures that can be easily implemented at these locations, this strategy will prove to be a cost-saving maneuver, since studies can be done at a fraction of their cost, if compared to the cost of their performance in the developed world; nonetheless, the quality of their results can be preserved and guaranteed.

These simple facts and considerations should lead environmental scientists, economists, researchers, and investors to consider Ecuador as one of the most rewarding places for plant natural-product research in the present and in the future.

References

1. Medicinal plants for conservation and health care. FAO—Food and Agriculture Organization of the United Nations, Corporate document repository, 1995.
2. Said M. Potential of herbal medicines in modern medical therapy. Ancient Sci Life 1984;4(1): 36–47.
3. Burneo D. Mecanismos de financiamiento para la conservación de la biodiversidad. In: Carmen Josse (ed), La biodiversidad del Ecuador Informe 2000. 2000;287–305.
4. Buitrón C. Ecuador: uso y comercio de plantas medicinales, situación actual y aspectos importantes para su conservación. 1999.
5. Bravo E. La problemática mundial de los recursos fitogenéticos. In: Castillo R, et al. (eds), Memorias de la II reunión nacional sobre recursos filogenéticos. 1991.
6. Suárez L. La importancia de la biodiversidad. In: Varea A (ed), Biodiversidad, bioprospección y bioseguridad. 1997;17–32.
7. Estrella J, Tapia C. Investigación y conservación de los recursos filogenéticos: las experiencias del INIAP. In: Mena PA, Suárez L (eds), La investigación para la conservación de la diversidad biológica en Ecuador. 1993.
8. Ministerio del Ambiente (Ministry of the Environment of Ecuador). Estrategia para el desarrollo forestal sustentable del Ecuador. 1999.
9. Proyecto de protección de la Biodiversidad. In: Ministerio del Ambiente del Ecuador (Ministry of the Environment of Ecuador), Estrategia nacional para la protección y el uso sustentable de la vida silvestre en Ecuador. 1999.
10. Steere WC. El descubrimiento y distribución de la *Cinchona pitayensis* en el Ecuador. Flora (Inst. Ecuat.) 1944;4:1–9.
11. Cabarle BJ, et al. An assessment of biological diversity and tropical forests for Ecuador. A World Resources Institute report to USAID/Ecuador 1989.

12. Myers N. Threatened biota: "hot spots" in tropical forests. Environmentalist 1988;8:187–208.
13. Suárez L, Albán MA. La biodiversidad en las políticas y en la legislación. In: Carmen Josse (ed), La biodiversidad del Ecuador Informe 2000. 2000;263–283.
14. El manejo para la protección y el uso sustentable de la vida silvestre en el Ecuador: diagnóstico de la situación actual. Instituto de Ecología Aplicada (ECOLAP) 1998.
15. Thorpe JP, Smartt J. Genetic diversity as a component of biodiversity. In: Heywood VH (ed), Global biodiversity assessment. 1995.
16. Alvarez A, Granizo T. La organización de la información sobre biodiversidad: el Centro de datos para la conservación. In: Mena A, Suárez L (eds), La investigación para la conservación de la diversidad biológica en el Ecuador. 1993.
17. Simpson, D, Sedjo R, Reid J. Valuing biodiversity: an application to genetic prospecting. J Polit Econ 1999;104 (1):163–185.
18. Fernández M. La articulación de la medicina académica con la tradicional como estrategia de salud. In: Naranjo P, Crespo A (eds), Etnomedicina Progresos Italo-Latinoamericanos. 1997;II:193–200.
19. Adger WN, Brown K, Cervigni R, Moran D. Total economic value of forests in Mexico. Ambio 1995;24:286–296.
20. Bush MB. Amazonian conservation in a changing world. Biol Conserv 1996;76(3):219–228.
21. Lapa A, Mesia S, Souccar C. Validation of medicinal plants: a major enterprise. In: Naranjo P, Crespo A (eds), Etnomedicina Progresos Italo-Latinoamericanos. 1997;I:195–206.
22. Vogel J. (1997). "The Successful Use of Economic Instruments to Foster Sustainable Use of Biodiversity: Six Case Studies from Latin America and the Caribbean". Biopolicy Journal, Vol. 2, Paper 5. White Paper, final report, Commissioned by the BiodiversitySupport Program on behalf of the Inter-American Commission on Biodiversity andSustainable Development, for the Summit of the Americas on SustainableDevelopment, Santa Cruz de la Sierra, Bolivia, December 6–8, 1996.
23. Ruiz Pérez M, Sayer J, Cohen S. Extractive reserves, IUCN-Commission of European Communities-CNPT, Gland, Switzerland. 1993.
24. Ellerman D, Jacoby HD, Decaux A. The effects on developing countries of the Kioto Protocol and CO_2 emissions trading. Discussion Article. The Joint Program on the Science and Policy of Global Change 1998.
25. Aylward B. The economic value of pharmaceutical prospecting and its role in biodiversity conservation. 1993.
26. Balick M, Cox PA. Ethnobotanical research and traditional healthcare in developing countries. In: Non-wood forest products 11. FAO-Food and Agriculture Organization of the United Nations. 1995.
27. Our Common Future. Rio Conference on the Environment. Convention text. Rio de Janeiro. 1992.
28. Liebstein AM. Therapeutic effects of various food articles. Amer Med 1927;33–38.
29. Blixt S. The role of genebanks in plant genetic resource conservation under the convention on biological diversity. In: Krattiger AF, McNeely JA, Lesser, Miller KR, St.Hill Y, Senanayake R (eds), Widening Perspectives on Biodiversity. IUCN. 1994.
30. World Health Assembly (WHA 31.33) on Medicinal Plants. World Health Organization (WHO). Geneva. 1978.
31. Abdo S, Játiva C, et al. Importancia etnobotánica de las compuestas en la Sierra Ecuatoriana. In: Biodiversidad, bioprospección y bioseguridad. Varea A. 1997;11–24.
32. Bonifaz C. La diversidad florística del occidente ecuatoriano. In: Suárez L (ed), Ecuador y biodiversidad. 1998.
33. Maldonado ME, Vidari G, et al. Investigación fotoquímica de *Schistoscarpa aff euphatorioides* y *Grias Neubertti McBride*. In: Etnomedicina Progresos Italo-Latinoamericanos. Naranjo P, Crespo A. 1997;I:59–73.

34. Villacrés V, Suárez M, et al. Actividad antimicrobial y prueba de de mortalidad con camarones salinos de la *Minguartia guianensis Aubl.* y *Clavija procera*. In: Naranjo P, Crespo A (ed), Etnomedicina Progresos Italo-Latinoamericanos. 1997;I:87–95.
35. Chiriboga X, Villar del Fresno A, et al. Actividad antinflamatoria de plantas medicinales del Ecuador. In: Naranjo P, Crespo A (eds), Etnomedicina Progresos Italo-Latinoamericanos. 1997;I:49–58.
36. Alacaraz MJ, Moroney, et al. Effects of hypolaetin-8-glucoside and its aglycone in vivo tests and antiinflamatory agents. Planta Médica. 1989;55:107–108.
37. Wintern CA, et al. Carragenin induced edema in hing paw of the rat: an assay for antiinflamatory drugs. Proc Soc Exp Biol Med 111:544–547.
38. Vásquez L. Implicaciones éticas de los derechos de propiedad intelectual. In: Varea A (ed), Biodiversidad, bioprospección y bioseguridad. 1997;143–149.

Index

A

Acarbose, enzyme inhibition, 16
Actinomycetes, *see* Marine actinomycetes
Actinomycin D,
  development, 130
  structure, 135
Acute promyelocytic leukemia (APL),
  all-trans retinoic acid treatment, 252
  arsenic trioxide treatment,
    all-trans retinoic acid combination therapy, 257
    coagulopathy correction, 256
    complete remission rate, 255, 256
    mechanism of action,
      angiogenesis inhibition, 263
      apoptosis induction, 259–261
      differentiation induction, 261, 262
      fusion protein and sensitivity, 262, 263
      in vivo studies, 258, 259
      metabolite effects, 263, 264
    overview, 252, 253
    pharmacokinetics, 254, 255
    postremission treatment, 257
    prospects, 265, 266
    survival, 257
    toxicity, 256, 257
  chromosomal translocation, 252
  clinical features, 252
  tetra-arsenic tetra-sulfide treatment, 257, 258
Adriamycin, *see* Doxorubicin
AE-941, *see* Neovastat
Ambuic acid, discovery and activity, 336, 338
Andean pact, intellectual property rights, 367

Angiogenesis, arsenic trioxide inhibition, 263
Antibiotics,
  anticancer drugs, 12, 13
  classes, 6, 7
  combinatorial biosynthesis, *see* Genetic engineering; Polyketide synthases
  development from natural products, 6–8
  discovery barriers, 6–8
  drug approval decline, 276
  endophyte sources,
    bacterial products, 339–341
    fungal products, 335–339
    volatile antibiotics, 342
  history of development, 33
  market, 7
  resistance,
    design of new drugs, 10, 11
    needs for new drugs, 8, 9
  screening trends, 6–8, 17, 18
  semisynthetic compounds,
    erythromycins, 10
    streptogramins, 9
    tetracyclines, 9
    vancomycin, 9, 10
Antifungal agents,
  classes and mechanisms, 11
  genomics-guided discovery platform,
    Streptomyces aizunensis case study,
      automated analysis of gene clusters and chemical structure prediction, 99, 100
      ECO-02301 discovery, 103, 105
      gene correlation with chemical substructures, 100, 102
      genome scanning, 98, 99

371

genomics-guided purification, 102, 103
immunosuppressive activity, 13, 14
APL, *see* Acute promyelocytic leukemia
Aplidine (dehydrodidemnin B),
cancer clinical trials, 151, 152
structure, 150
Apoptosis, arsenic trioxide induction in leukemia, 259–261
Ara-C, *see* Cytosine arabinoside
Arsenic,
arsenic trioxide for cancer treatment,
acute promyelocytic leukemia,
all-trans retinoic acid combination therapy, 257
coagulopathy correction, 256
complete remission rate, 255, 256
overview, 252, 253
pharmacokinetics, 254, 255
postremission treatment, 257
prospects, 265, 266
survival, 257
toxicity, 256, 257
anticancer activity spectrum, 264, 265
mechanism of action,
angiogenesis inhibition, 263
apoptosis induction, 259–261
differentiation induction, 261, 262
fusion protein and sensitivity, 262, 263
in vivo studies, 258, 259
metabolite effects, 263, 264
medicinal use history, 253, 254
tetra-arsenic tetra-sulfide treatment of acute promyelocytic leukemia, 257, 258
Artemisinin,
applications, 202
chemical synthesis, 211, 212
Aspirin, development, 4

Avarol,
clinical applications, 204
structure, 205
Avarone,
clinical applications, 204
structure, 205
Avermectins,
gene shuffling and dorametin synthesis by Streptomyces avermitilis, 114
market, 114
structures, 115

B

Bacteria,
collection efforts, 6, 276, 315, 316
culture, *see* Culture, microorganisms
discovery, 6, 276
diversity, 276, 278, 296
engineering, *see* Genetic engineering
genera for secondary metabolite production, 275
isolation strategies, 37, 38
phylogenetic analysis, 276–278, 296
sample collection, 35–37
strain engineering, *see* Strain engineering
Bengamides,
cancer clinical trials, 146
structures, 150
Biotechnology, success compared with pharmaceutical industry, 20
Bleomycin,
anticancer activity, 134
structure, 135
Bromosphaerone,
clinical applications, 208
structure, 208
Bryostatin,
cancer clinical trials, 148, 149
structure, 150
Calicheamicin,
anticancer activity, 134
structure, 135

Index

CAM, *see* Complementary and alternative medicine
Camptothecin,
  anticancer activity, 139
  structure, 140
Cancer drug discovery, *see also specific drugs,*
  antibiotics, 12, 13
  arsenic trioxide, *see* Arsenic
  drug source categorization, 129, 130
  endophyte sources, 342–345
  marine sources,
    approved agents, 142–145
    clinical trials of agents, 148, 149, 151–154, 156
    drug source identification, 158, 159
    mechanisms of agents, 141, 142
    preclinical studies, 156, 158
  plants,
    approved agents, 137–139
    clinical trials of agents, 139–141
    mechanisms of agents, 141
  prokaryotes,
    approved agents, 130–134
    clinical trials of agents, 134–137
  prospects, 159, 160
Carvone, applications, 201
Caspases, arsenic trioxide induction, 259
Catharanthus roseus alkaloids, *see* Vinblastine; Vincristine
Cell culture, *see* Culture, microorganisms
Cephalosporin-vinblastine prodrugs, synthesis, 186, 187
China, *see* Traditional Chinese medicine
Clavulanic acid,
  discovery, 15
  mechanism of action, 15
Combinatorial biosynthesis, *see* Genetic engineering; Polyketide synthases

Combrestatins,
  anticancer activity, 141
  structure, 147
Complementary and alternative medicine (CAM), *see also* Traditional Chinese medicine,
  government organizations and oversight,
    Europe, 232
    United States, 231
  phytotherapeutics as alternative system of primary medicine, 359–361
  popularity, 231
Cryptocandin A, discovery and activity, 336, 337
Cryptocin, discovery and activity, 336, 337
Cryptophycins,
  cancer clinical trials, 148
  structures, 150
Culture, microorganisms,
  bias against diversity, 278
  co-cultivation, 42
  economics of fermentation, 108
  endophytes, 335
  marine microorganisms, 279
  metabolically talented strain exploitation, 41, 42
  nonculturable microorganism features, 278, 279
  optimization for secondary metabolite synthesis, 39, 40
  selection of conditions, 40, 41
  strain engineering, *see* Strain engineering
  terpenoid production, 214, 215
Curacin A,
  structure, 157
  tubulin interactions, 156
Cyclosporin A,
  antifungal activity, 13, 14
  immunosuppression, 13
  structure, 5

Cytosine arabinoside (Ara-C),
  anticancer activity, 142
  structure, 147

D

Daunorubicin,
  anticancer activity, 130
  structure, 135
Debromarinone,
  antimicrobial activity, 323
  structure, 324
Dehydrodidemnin B, *see* Aplidine
Dereplication,
  chemical dereplication to prevent
    repeated discovery, 47, 48
  microbial strains, 44, 45
  prescreening of microbial strains and
    extracts, 43, 44
Desferal, indications, 16
Diazonamide A,
  structure, 157
  tubulin interactions, 156
Didemnin B,
  cancer clinical trials, 142, 143
  structure, 147
Digitoxin, history of use, 16
Diisocyanoadociane,
  clinical applications, 208, 209
  structure, 208
Discodermolide,
  cancer clinical trials, 153
  structure, 155
DNA shuffling, strain engineering, 110, 111
Docetaxel, *see* Taxotere
Dolastatins,
  analogs, 146
  cancer clinical trials, 143, 146
  structures, 147
Doramectin, gene shuffling and
  synthesis by *Streptomyces
  avermitilis*, 114
Doxorubicin (adriamycin),
  anticancer activity, 130
  structure, 135
Drug sources, categorization, 129, 130

E

Echinomycin, discovery and activity, 341
ECO-02301, discovery, 103, 105
Ecteinascidin 743
  cancer clinical trials, 149, 151
  structure, 150
Eleutherobin,
  chemical synthesis, 213
  clinical applications, 207
  structure, 207
Ellipticinium acetate,
  anticancer activity, 139
  structure, 140
Endophytes, *see* Rainforest endophytes
Environmental DNA libraries,
  construction,
    DNA extraction, 298
    host strains for expression, 301, 302
    overview, 285, 286
    phylogenetic diversity, 299–301
    transfer and expression of
      recombinant plasmids, 302, 303
    vectors, 298, 299
  prospects, 309, 3310
  rationale, 296, 297
  screening for natural products,
    Escherichia coli colony screens, 304
    expression-based approach, 286, 287
    findings, 305, 306
    high-performance liquid
      chromatography screens, 304, 305
    high throughput screening,
      host strains, 308

Index

prescreening, 307
process, 308, 309
homology screens, 305, 306
hybridization-based approach, 286
overview, 303, 304
sequence-based approach, 286
Epirubicin,
  anticancer activity, 130
  structure, 135
Epothilones,
  analog synthesis with engineered polyketide synthases, 84, 85
  anticancer activity, 13
  antifungal activity, 13
  cancer clinical trials, 135, 136
  structures, 138
Equador,
  biodiversity,
    drug discovery prospects, 367, 368
    economical considerations, 357, 358
    flora, 355, 356, 367
    legal considerations for protection, 366, 367
  bio-prospecting, 356, 357
  phytotherapeutics as alternative system of primary medicine, 359–361
  plants,
    anti-inflammatory activity of extracts,
      aqueous extraction, 363, 365
      dichloromethane extraction, 363–365
      ethanol extraction, 363, 364, 366
    preservation and exploration, 355, 356
    therapeutic potential, 361–366
  sustainable development, 358, 359
Erogorgiaene,
  clinical applications, 208
  structure, 208
Erythromycins,
  analog synthesis with engineered 6-deoxyerythronolide B synthase genes, 82, 83
  semisynthetic compounds, 10

F

Feverfew, indications, 202
FK506 (tacrolimus),
  antifungal activity, 14
  immunosuppression, 13
  structure, 5
FK520, analog synthesis with engineered polyketide synthases, 84
Flavopiridol,
  anticancer activity, 140, 141
  structure, 140
Fungus species, *see also* Antifungal agents,
  automated optimization, *see* High-performance liquid chromatography
  collection efforts, 6
  culture, see Culture, microorganisms
  discovery, 6
  high-throughput cultivation for diversity optimization, 279–283
  infection epidemiology, 11
  isolation,
    automation, 279
    strategies, 37, 38
  sample collection, 35–37

G

Geldanamycin, analog synthesis with engineered polyketide synthases, 85–87
Genetic engineering, *see also* Polyketide synthases; Strain engineering,

combinatorial biosynthesis, 17, 43, 49
heterologous metagenome
    expression in surrogate host,
    42, 43, 283–285
terpenoid synthesis, 213, 214
Genome sequence tag (GST), library
    screening, 43
Genomics, *see also* Environmental
    DNA libraries,
    drug discovery, 19, 20
    genomic libraries,
        construction,
            environmental samples, 285,
                286
            single isolates, 285
        screening for natural products,
            expression-based approach,
                286, 287
            hybridization-based approach,
                286
            sequence-based approach, 286
    microbial strain characterization and
        selection, 44
    natural product discovery,
        genome scanning, 97
        genomics-guided discovery
            platform, 97–99
        Streptomyces aizunensis case
            study,
            automated analysis of gene
                clusters and chemical
                structure prediction, 99, 100
            ECO-02301 discovery, 103,
                105
            gene correlation with chemical
                substructures, 100, 102
            genome scanning, 98, 99
            genomics-guided purification,
                102, 103
Girolline,
    cancer clinical trials, 146
    structure, 150
Glutathione, role in arsenic trioxide
    induction of apoptosis, 260

GST, *see* Genome sequence tag

H

Halichondrin B,
    cancer clinical trials, 152
    structure, 155
Halomin, applications, 202
Halorosellinic acid,
    clinical applications, 210
    structure, 211
Hemiasterlin,
    cancer clinical trials of HTI-286, 154
    structure, 155
High-performance liquid
    chromatography (HPLC),
    automated analysis with HPLC
        Studio,
        applications,
            chemical diversity
                determination, 66, 67
            extraction solvent
                optimization, 64
            fermentation format
                optimization, 64–66
            fermentation optimization for
                single strains, 68
            fermentation optimization for
                strain sets, 68, 69
            relative extracted quantity
                ranking, 68
            secondary metabolite profile
                comparison with geographic
                origin, 71, 73
            secondary metabolite profile
                comparison with taxonomy,
                69, 71
        fermentation condition
            optimization, 60
        overview of steps, 58, 59
        prescreening of small-scale
            fermentation, 60
        software features, 62
    chemical dereplication to prevent
        repeated discovery, 48

Index

environmental DNA library screens, 304, 305
overview of microbial extract analysis, 46
High throughput cultivation, microbial diversity optimization, 279–283
High throughput screening (HTS),
   drug discovery, 18, 19
   environmental DNA libraries,
      host strains, 308
      prescreening, 307
      process, 308, 309
   natural products, 19, 34
Homoarringtonine,
   cancer clinical trials, 139
   structure, 140
HPLC, see High-performance liquid chromatography
HPLC Studio, see High-performance liquid chromatography
HTI-286, see Hemiasterlin
HTS, see High throughput screening

I

Idarubicin,
   anticancer activity, 130
   structure, 135
Iludins,
   clinical applications, 204, 205
   structure, 205
Indolocarbazoles, cancer clinical trials, 136
Intellectual property rights,
   Andean pact, 367
   traditional Chinese medicine, 243
Irinotecan,
   anticancer activity, 139
   structure, 140
Irofulven,
   clinical applications, 204, 205
   structure, 205
Isolation, see Culture, microorganisms
Isopestacin, discovery and activity, 345, 346

J

Jesterone, discovery and activity, 338, 339
JNK, see Jun N-terminal kinase
Jun N-terminal kinase (JNK), arsenic trioxide induction, 261

K

Kahalide F,
   cancer clinical trials, 153
   structure, 155
Kakadymycin A, discovery and activity, 341
Ketoconazole, synergism with F0101604, 48
KRN-7000
   cancer clinical trials, 154
   structure, 157

L

Laulimalide,
   structure, 157
   tubulin interactions, 156
Leukemia, see Acute promyelocytic leukemia
Limonene, applications, 200, 201
Lipstatin, enzyme inhibition, 16
Lomaivitacins,
   antimicrobial activity, 322
   structures, 320
Lovastatin,
   development, 15
   structure, 5
Luffariellin,
   clinical applications, 209
   structure, 210

M

Magicols,
   clinical applications, 210
   structures, 211
Marine actinomycetes,
   drug discovery prospects, 315, 316, 326

"Marinophilus" compounds, 322–325
phylogenetic analysis, 317, 318
"Salinospora" compounds, 317, 318, 320, 321, 322
seawater requirements for growth, 325, 326
Marinone,
antimicrobial activity, 323
structures, 324
Mass spectrometry (MS),
microbial extract characterization following separations, 47
strain engineering screening, 113
Metabolic engineering, see Genetic engineering
Methopterosin, clinical applications, 208
Microtubule,
stabilizing agents in cancer treatment, 156, 169–171
structure, 169
Monoalide,
analog development, 212
clinical applications, 209
structure, 210
MS, see Mass spectrometry
Munubucins, discovery and activity, 340, 341
Mycophenolic acid, history of study, 14
Mycotoxins,
diseases, 16
therapeutic potential, 16, 17

N

Napavin, synthesis, 187, 188
Natural products,
diversity, 4
drug development,
advantages, 34
challenges, 34, 35
overview, 4–6, 77, 78
Neomarinone,
antimicrobial activity, 323
structure, 324
Neovastat (AE-941), cancer clinical trials, 156
NMR, see Nuclear magnetic resonance
Nuclear magnetic resonance (NMR), microbial extract characterization following separations, 47

O

Omuralide,
proteasome inhibition, 321
structure, 320, 321
Oocydin A, discovery and structure, 332, 333

P

Paclitaxel, see Taxol
Peloruside A,
anticancer activity, 156
structure, 159
Perrillyl alcohol, applications, 201
Pestacin, discovery and activity, 345, 346
Pestaloside, discovery and activity, 336, 338
Pharmaceutical industry,
biotechnology success comparison, 20
market size and trends, 229, 230
research and development spending trends, 18
time and costs for drugs to reach market, 18, 230
Phospholipase A2, sestererpene inhibition, 209, 212
Phytol,
clinical applications, 205, 206
structure, 206
Pirirubicin,
anticancer activity, 130
structure, 135
Plants, see also Equador,

Index

cancer drug discovery,
  approved agents, 137–139
  clinical trials of agents, 139–141
  mechanisms of agents, 141
  market for products, 354
  prescription drug source abundance, 353
  rainforest endophyte selection and collection rationale, 332, 333
Podophyllotoxin, anticancer activity, 137, 138
Polyketide synthases,
  acyltransferase specificity, 78, 81
  epothilone analog synthesis, 84, 85
  erythromycin analog synthesis with engineered 6-deoxyerythronolide B synthase genes, 82, 83
  FK520 analog synthesis, 84
  geldanamycin analog synthesis, 85–87
  Kosan Biosciences research and development strategy, 81
  marine organism gene engineering, 87, 88
  modular synthases and modules, 78, 79, 81
  novel gene creation, 89, 90
  recombinant Escherichia coli for large-scale drug production, 88, 89
  type I polyketide synthase mechanism, 78, 80
  type II polyketide synthase mechanism, 78, 79
Pseudopteroxazole,
  clinical applications, 208
  structure, 208
Pyrethrins, applications, 202

Q

Quercetin,
  anticancer activity, 139, 140
  structure, 140

R

Rainforest endophytes,
  antibiotics,
    bacterial products, 339–341
    fungal products, 335–339
    volatile antibiotics, 342
  anticancer agents, 342–345
  antidiabetic agents, 345
  antioxidant products, 345
  antiviral compounds, 341, 342
  collection, isolation and preservation, 335
  diversity, 330–334
  drug discovery prospects, 329, 330, 347, 348
  immunosuppressive compounds, 345, 346
  Pestalotiopsis microspora genetics, 346, 347
  phytochemicals, 334, 335
  plant selection and collection rationale, 332, 333
Rapamycin (sirolimus),
  antifungal activity, 14
  immunosuppression, 13, 14
  structure, 5
Reactive oxygen species (ROS), role in arsenic trioxide induction of apoptosis, 260
Rebaccamycin,
  anticancer activity, 136, 137
  structure and analogs, 138
Research and development,
  time for drugs to reach market, 18
  trends in pharmaceutical company spending, 18
Restriction fragment length polymorphism (RFLP), microbial strains, 45, 46
RFLP, see Restriction fragment length polymorphism

ROS, *see* Reactive oxygen species

S

Salicylic acid, history of use, 4

Salicylihalimides,
    anticancer activity, 158
    structure, 159
Salinosporamide A,
    antimicrobial activity, 318, 320
    structure, 318, 320, 321
Salmahyrtisol A,
    clinical applications, 209, 210
    structure, 211
Sample collection, strategy for natural product drug discovery, 35–37
SAR, *see* Structure–activity relationship
Sarcodictyn,
    anticancer activity, 156
    structure, 159
Scalaradial,
    clinical applications, 209
    structure, 210
Sesquiterpene lactones, applications, 202, 203
Shikimate,
    gene shuffling and synthesis by *Escherichia coli*, 116, 117
    synthetic applications, 116
Sirolimus, *see* Rapamycin
Solenolide A,
    clinical applications, 209
    structure, 208
Sphaerococcenol A,
    clinical applications, 208
    structure, 208
Spisulosine,
    cancer clinical trials, 153
    structure, 155
Spongiadiol,
    clinical applications, 209
    structure, 208

Squalamine,
    cancer clinical trials, 154, 156
    structure, 157
Statins, development from natural products, 14, 15
Strain engineering,
    economics of fermentation, 108
    evolutionary engineering,
        genetic diversification with DNA shuffling, 110, 111
        high-throughput fermentation, 113, 114
        overview, 109, 110
        screening, 111, 113
    gene shuffling and enzyme evolution,
        doramectin synthesis by *Streptomyces avermitilis*, 114
        pathway shuffling and evolution, 117
        shikimate synthesis by *Escherichia coli*, 116, 117
    genome shuffling,
        overview, 117, 118
        tylosin synthesis by *Streptomyces fradiae*, 119, 120
    goals, 190
    prospects, 120–122
    rationale, 107–109
    volumetric productivity, 108, 109
Streptogramins, semisynthetic compounds, 9
Structure–activity relationship (SAR), *see* Taxol; Vinblastine; Vincristine

T

Tacrolimus, *see* FK506
Taxol (paclitaxel),
    anticancer activity, 12, 13, 138, 139, 171, 172
    antifungal activity, 13, 342, 343
    biosynthesis, 206, 207

chemical synthesis, 211
development, 206, 207
mechanism of action, 169, 170
natural source, 138, 169, 171
Pestalotiopsis microspora synthesis, 346, 347
structure, 140, 170, 206, 343
structure–activity relationship studies in analog development,
C-2 studies,
derivatization, 173, 174
structures, 175
tubulin polymerization and cytotoxicity studies, 173, 174
C-4 studies,
derivatization, 174–176
esters and carbonates, anticancer activity, 177–179
tubulin polymerization and cytotoxicity studies, 175, 176
overview, 171–173
oxetane ring modification, 179–181
prospects, 190–192
Taxotere (docetaxel),
chemical synthesis, 211
development, 172
structure, 173
TCM, *see* Traditional Chinese medicine
Terpenoids,
anticancer drugs, 198
biosynthesis, 198, 199
chemical synthesis in drug production and development, 210–213
industrial applications, 197, 198
pharmaceuticals,
diterpenes, 205–209
monoterpenes, 200–202
sesquiterpenes, 202–205
sesterterpenes, 209, 210
prospects for study, 215, 216

sustainable production approaches,
cell culture, 214, 215
metabolic engineering, 213, 214
Tetracyclines, semisynthetic compounds, 9
Thin-layer chromatography (TLC),
chemical dereplication to prevent repeated discovery, 48
microbial extracts, 46, 47
TLC, *see* Thin-layer chromatography
Torreyanic acid, discovery and activity, 344, 345
Traditional Chinese medicine (TCM),
cancer treatment, 252
clinical research status, 239
current good agriculture practice bases in China, 235, 236
demand and Chinese exports, 238, 239, 241, 242
herb preparation industry in China,
granule preparations, 236
limited capability, 244, 245
manufacturers,
sales, 235–240
status, 237–239
quality control, 235, 236
sectors, 235
industry market and growth, 232, 233
intellectual property rights, 243
international validation, 240–242
mechanism studies, 241, 242
modernization in China, 233, 234
prospects, 245
resource status in China, 234, 235
safety issues, 243, 244
techniques, 231
Tuberculosis, terpenoid treatment, 202, 203, 208, 210
Tylosin,
genome shuffling and synthesis by *Streptomyces fradiae*, 119, 120
synthesis, 120

UCN-01
    cancer clinical trials, 136
    structure, 138
Valrubicin,
    anticancer activity, 130
    structure, 135
Vancomycin, semisynthetic
        compounds, 9, 10
Vinamidine, anticancer activity, 181
Vinblastine,
    anticancer activity, 137, 171, 181
    C16-carboxyl amino acid
        derivatives, 186
    cephalosporin-vinblastine prodrugs,
        186, 187
    discovery, 170
    peptidyl prodrugs, 190, 191
    phosphonic acid derivatives, 186, 187
    structure, 138
    structure–activity relationship
            studies in analog development,
        overview, 181
        prospects, 190–192
        velbanamine moiety, 181–183
        vindoline moiety, 183–188, 190
Vincristine,
    anticancer activity, 137, 171, 181
    discovery, 170
    structure, 138
    structure–activity relationship
            studies in analog development,
        overview, 181
        prospects, 190–192
        velbanamine moiety, 181–183
        vindoline moiety, 183–188, 190
Vindesine,
    analogs, 185, 186
    anticancer activity, 185
    synthesis, 184, 185
Vinepidine, synthesis, 183, 184
Vinflunine, synthesis, 182, 183
Vinformide,
    anticancer activity, 184
    synthesis, 184, 185
Vinglycinate,
    anticancer activity, 188, 190
    synthesis, 188, 189
    Vinleurosine, anticancer activity,
        181
    Vinorelbine,
    analogs and anticancer activity, 183
    synthesis, 182
Vinrosidine, anticancer activity, 181
Vinzolidine,
    analog synthesis, 187, 188
    anticancer activity, 188, 189
    discovery, 187, 188
Vitelevuamide,
    structure, 157
    tubulin interactions, 156
Volumetric productivity definition, 108,
        109